Environmental Horticulture

Science and Management of Green Landscapes
2nd Edition

ROSS W.F. CAMERON

This book is dedicated to the memory of Margaret Isobel Cameron. For all those house plants that survived despite the Spartan management regimes imposed!

Environmental Horticulture

Science and Management of Green Landscapes

2nd Edition

Ross W.F. Cameron

Professor of Environmental Horticulture
School of Architecture and Landscape
University of Sheffield
UK

CABI is a trading name of CAB International

CABI
Nosworthy Way
Wallingford
Oxfordshire OX10 8DE
UK

Tel: +44 (0)1491 832111
E-mail: info@cabi.org
Website: www.cabi.org

CABI
200 Portland Street
Boston
MA 02114
USA

Tel: +1 (0)617 682 9015
E-mail: cabi-nao@cabi.org

First published in Great Britain in 2026 by CABI.

Published in association with the Royal Horticultural Society.

A catalogue record for this book is available from the British Library, London, UK.

ISBN-13: 9781836992530 (hardback)
9781800621749 (paperback)
9781800621756 (ePDF)
9781800621763 (ePub)

DOI: 10.1079/9781800621763.0000

Commissioning editor: Rebecca Stubbs
Editorial assistant: Emma McCann
Production editor: Shankari Wilford
Typeset by Straive, Pondicherry, India
Printed in the USA

RHS Publisher Helen Griffin
RHS Consultant Editor Simon Maughan
RHS Head of Editorial Tom Howard

Contents

Foreword

Environmental horticulture plays an increasingly vital role in shaping the liveable, resilient and biodiverse places that people depend upon. As global urbanisation intensifies and climate change accelerates, plants and thoughtfully designed landscapes are no longer simply aesthetic features; they are fundamental green infrastructure that supports health, wellbeing, biodiversity, climate mitigation and the long-term sustainability of our towns and cities.

This second edition of *Environmental Horticulture: Science and Management of Green Landscapes* arrives at a critical moment. For example in the UK, 41 million people – adults and children – are now actively engaged with gardening, as highlighted in the *RHS State of Gardening Report 2025* (rhs.org.uk/stateofgardening, accessed February 2026). Their motivations range from enhancing wellbeing and strengthening connections with nature, to supporting wildlife and improving the quality of their local environments. Importantly, gardens now sustain over 40% of the UK's biodiversity, reinforcing that all biodiversity matters, not only species native to a specific region.

Professor Cameron's comprehensive work brings together an expanding body of scientific evidence, demonstrating how plants and landscapes can cool cities, improve air and water quality, reduce flood risk, support nature-based solutions, enhance biodiversity and promote physical, mental and social wellbeing. It reflects the breadth and depth of modern environmental horticulture – from meadows and urban forests to green roofs, SuDS, interior landscapes and streetscapes – and illustrates the transformative potential of intelligent plant selection, design and management.

Crucially, this book recognises that environmental horticulture is shaped not only by ecological principles, but also by people. It is a discipline grounded in horticultural expertise yet enriched through collaboration with architects, engineers, planners, ecologists, health professionals and communities. It values human agency – our ability to create, curate and care for green spaces – while championing approaches that support resilience, sustainability, regenerative practices and ecological richness.

As Director of Science and Collections at the Royal Horticultural Society, I see daily how plants, horticultural science and human stewardship combine to deliver greener, healthier futures. This book provides essential insight, evidence and inspiration for practitioners, students, policy makers and all who care about the vital role that green landscapes play in society. I warmly commend this important work. It will help guide the creation and care of resilient, biodiverse and beautiful green spaces in which both people and nature can thrive.

Professor Alistair Griffiths
Royal Horticultural Society
Director of Science and Collections

Acknowledgements

I am also grateful to Professor James Hitchmough, who co-authored the first edition of this book, and whose initial content still forms an important 'backbone' to some of the chapters.

I would also like to thank all the following: The Royal Horticultural Society, The Horticultural Trades Association, the Wildlife Trusts, The Royal Society for the Protection Of Birds, Earthwatch, The Agricultural and Horticultural Development Board, UKRI, DEFRA and colleagues at the University of Sheffield for providing information and supporting research that contributed to this book.

1 Introduction to Environmental Horticulture

Abstract

Environmental horticulture is the sustainable management of urban green spaces. Such spaces include parks, gardens, cemeteries, sports facilities, green roofs, green walls, rain gardens, interior landscapes and roadside plantings. Cultivation of plants is the central factor although environmental horticulturists also need to have the skills to deal with a wide range of stakeholders interested in quality green space. Increasingly, urban green spaces are valued for their functionality, over and above that of purely aesthetics. Green spaces are now designed, created and managed to help deal with excess rainwater, pollution, urban heat or to encourage a more active lifestyle and allow people to relax and improve their well-being. Human agency is important in environmental horticulture, but so is care for the environment, and practices should be genuinely sustainable, promote opportunities for biodiversity and support wider ecosystem functions.

1.1 What Is Environmental Horticulture?

Environmental horticulture is the use and management of plants to create visually engaging, functional and environmentally sustainable spaces and locations. It moves the agenda forward compared to similar disciplines of landscape, urban or amenity horticulture (the design and management of green spaces), by aspiring to put the environment first and foremost. Thus, careful management of natural resources, minimising pollution, being aligned with ecosystem functions and caring for the wider biodiversity that utilises these green spaces are paramount considerations. It is not ecological management or wildlife habitat management though, as it recognises that these landscapes are human orientated with space frequently being designed and managed for human use or aesthetics. It does not restrict itself solely to the use of native, ecologically appropriate plants species, but rather acknowledges that in urban environments particularly, non-native plants can enhance the aesthetics and functionality of a given site (Fig.1.1). Indeed, in some situations such as the indoor spaces of buildings located in temperate countries, the interior conditions are actually more conductive to plants from the tropics than those growing naturally outdoors in that region.

The 'green spaces' that embrace environmental horticulture are wide-ranging and cover both public and private space. At one end of the spectrum is elite sports turf that underpins the effective playing of football, rugby, golf, tennis, cricket, horse racing etc. At the other end, environmental horticulturists may be designing and managing 'natural' meadows and woodlands, and increasing support for wildlife through ponds, nectar-rich wildflower beds, coppiced woodlands and adding bird and bat boxes to tree trunks. Management regimes deal with both formal and informal designs (Fig. 1.2). Environmental horticulture often aims to 'green-up' somewhat 'sterile' environments, including the inside of buildings (interior plantscaping), but also the roofs, walls, podiums and entrances of buildings. Almost central to all roles is an understanding of plant biology and the capacity to carry out effective plant husbandry.

As a name 'environmental horticulture' first begins to appear in the literature and educational course terminology in the 1980s, particularly in North America. It is gradually adopted as a term for the subset of horticulture that is concerned with the use and management of plants in public and semi-public environments. In some parts of the world it replaces the term 'urban horticulture', although in many cases these descriptors coexist for long periods of time and in essence cover the same territory. In the UK, the terms environmental and urban never really catch on as educational course descriptors, with the much older term 'amenity horticulture' tending to persist right up to the present day. In North America and

DOI: 10.1079/9781800621763.0001

Fig. 1.1. Designed, aesthetically strong landscapes are a fundamental component within environmental horticulture.

elsewhere, 'landscape horticulture' is often the preferred term to cover the planting and maintenance of landscape plants in public or private space. Over the last 10 years, environmental horticulture, however, has gained greater prominence, with university degrees, academic professorial 'chairs' and scientific research teams housed within the discipline. An Environmental Horticulture Group composed of industrial trade bodies and charities liaises with the UK Government and aims to champion gardens, landscape firms, arboriculture and environmental horticulture and encourage greater governmental support for the sector. Environmental horticulture also encompasses a number of philosophies within wildlife gardening and thus can embrace the landscapes of private (as well as public) gardens.

At one level this evolution in terms is just about marketing and branding and trying to present a modern, culturally responsive face to compete in the business, brand or educational marketplaces. At another level these newer names signify changing ideas within this branch of horticulture. In the UK, the term 'amenity horticulture' first emerges in the 1970s to denote the horticulture that is concerned with public and semi-public landscape spaces, as opposed to various forms of crop production. This particular term develops at the same time as recreation and leisure management and reflects a view of the world that this strand of horticulture exists to provide leisure or amenity benefits to citizens through the cultivation of plants in landscapes (largely urban) visited by the public.

In other parts of the world, for example North America, which have different horticultural traditions, less based on the gardenesque park of 19th-century Britain, 'urban' horticulture develops in the 1970s. Urban horticulture seemed more modern in its perspective. The name explicitly recognised that *urban* places were often more biologically challenging places in which to grow plants than were 'gardens' or 'green sward parks'. This was due to, for example, higher levels of atmospheric pollution, soils destroyed by engineering and construction activities, sealed surfaces, changed urban climatic regimes, and sometimes hostile social contexts such as vandalism and other forms of anti-social behaviour.

Fig. 1.2. Plants are managed within both formal (top) and informal (bottom) designs.

Fig. 1.3. Urban citizens often depend on environmental horticulture and urban green spaces to directly experience the wider natural world.

Urban horticulture also explicitly made connections with the human social, cultural and psychological realms. Horticulture is engaged in and practised because it can make us feel better about our lives; it provides complex stimuli in both time and space that constructively moderates the fundamentally highly artificial experience of living in cities. Horticulture, whether practised by green space professionals or home gardeners, is likely to provide the most immediate experience of 'nature': the patterns and processes associated with interactions between the physical world of our planet and the living organisms that have evolved in response to this (Fig. 1.3).

Perhaps environmental horticulture reflects these transitions and relationships even to a greater degree. It recognises both (i) that the natural world is under pressure and other species are increasingly dependent on human (benign) management processes for their survival and (ii) man-made climate change is altering our environment, and that effective management of green space can provide some, albeit limited, mitigation potential. Thus, environmental horticulture has (or should have) 'values' linked to it that relate to addressing these phenomena.

The exact origins of environmental horticulture as term is uncertain, but it seems to represent an attempt to reposition horticulture to be more open to many of the ideas that developed from the 1990s about biodiversity and ecological processes as well as the human-centred ideas discussed above. In countries where the emergence of ecological consciousness had begun to lead to a wedge between ecological and horticultural thought, environmental horticulture provided a new paradigm whereby the practices involved in cultivation could be applied to systems that might, for example, consist of entirely native plants that would be seen as appropriate by ecologists. There is also an undercurrent that environmental horticulture should and does embrace more 'environmentally benign' actions and adopts procedures that conform to an environmentally sustainable agenda. This includes reduced use of pesticides, use of biological control methods and seeking alternatives to peat as a growing media and other 'non-sustainable' resources.

Green infrastructure

Environmental horticulture is now intrinsically linked to the management of urban green space or green infrastructure, although these terms (see Table 1.1) may encompass woodland and other less intensively managed areas. As such they may not be the exclusive 'domain of the horticulturist'.

Table 1.1. What is meant by open space, green space and green infrastructure?

Open space is any open piece of land that is undeveloped (has no buildings or other built structures) and is generally accepted as being accessible to the public. Typical open spaces are school playgrounds, public squares and plazas, pathways, public seating areas and vacant areas of even brownfield sites (ex-industrial sites).

Green space is a form of open space. This is land that is partly or completely covered with grass, trees, shrubs or other vegetation. Green space includes parks, community gardens and cemeteries. Green space, however, also comprises land that may not be always open to the public, such as private gardens and certain sports facilities, e.g. golf courses or soccer pitches.

Green infrastructure is the term used to combine different forms (typologies) of green space and is often used in conjunction with urban green spaces (i.e. as a juxtaposition to built [grey] infrastructure). It implies a matrix of green spaces and typologies, and increasingly one that is intentionally or strategically planned. It also alludes to the services such spaces should provide to local (human) populations. These include its role in housing and economic growth, the regeneration of urban areas, the protection (or creation) of environmental assets and the underpinning of the sustainability of a town or city.

Natural England, the government body in England, UK, tasked with nature conservation and the management of national nature reserves defines green infrastructure as follows:
'a network of multi-functional green space, both new and existing, both rural and urban, which supports the natural and ecological processes and is integral to the health and quality of life of sustainable communities'.

Included within green infrastructure typology are the following:

- Natural and semi-natural urban green spaces
  - Woodland and shrubs, grassland (e.g. downland and meadow), heath or moor, wetlands, open and running water, wastelands and disturbed ground, and bare rock habitats (e.g. cliffs and quarries)
- Parks and gardens
  - Urban parks, country and regional parks and formal gardens
- Amenity green space
  - Informal recreation spaces, housing green spaces, domestic gardens, village greens, urban commons, pocket parks, other incidental space, green roofs, green walls, green podiums and planted atriums
- Green (and blue) corridors
  - Rivers and canals including their banks, road and rail corridors, cycling routes, pedestrian paths, bridleways and other rights of way
- Other areas
  - Allotments, community gardens, urban farms, cemeteries and churchyards

Urban 'ecosystem services' and 'nature-based solutions'

The beneficial aspects of environmental horticulture also align closely with the *in vogue* terms of 'urban ecosystem services' and 'nature-based solutions'. Ecosystem services are the benefits accrued from the natural world and nature-based solutions are the pragmatic approaches to using the natural world to help humans with set, given problems. An example might include the use of a designed rain garden to filter out pollutants from rainwater running off city streetscapes. Soil and plant roots can help trap and deactivate organic molecules associated with diesel, petrol and oil deposits, as well as chelate toxic heavy metals derived from car catalytic converters. This semi-natural filtration system can reduce the amount of pollutants reaching natural watercourses and damaging aquatic life (Fig. 1.4). An alternative example would be the planting of street trees to provide shade and evapotranspirative cooling to pedestrians along the adjacent pavements.

Interventions such as these mean that environmental horticulturists are invariably working with other disciplines such as city planners, architects, building engineers, etc. as well as developers and commercial entities to help improve the design and functionality of cities. In such circumstances their plant knowledge is essential to ensure the right form of green infrastructure is embedded into the nature-based solution to optimise functionality.

Environmental horticulture is part of a sector with significant economic impact: within the UK for example, landscape horticulture is valued at about £38 billion and supports 722,000 jobs

Fig. 1.4. Urban green spaces help filter out pollutants from watercourses allowing aquatic life to re-establish within cities and towns. As long as there is fresh water, fish and bankside nesting habitat the iconic *Alcedo atthis* (Eurasian kingfisher) can thrive in urban situations. (Used with permission of P. Dixon.)

(Oxford Economics, 2004). Moreover, one in two people (30 million in total) in the UK are keen gardeners. The value of services from urban plants is estimated at £15.7 billion p.a. (Horticultural Trades Association, 2023). UK public gardens and estates are thought to be visited by 30% of overseas tourists to the country. Specialist jobs within the sector include arboriculturists, landscape designers and contractors, golf course managers, nursery and garden centre managers, irrigation designers and biocontrol biologists, to name a few. Environmental horticulture plays a crucial role in creating sustainable, green spaces that benefit both the environment and human well-being. It combines scientific knowledge with practical skills to address challenges driven by climate change and population growth, making it an increasingly important field in today's world.

1.2 Horticulture Involves Human Agency

Whilst environmental horticulture covers a wide range of theoretical and practical territory, these very same practices and understandings of cultivating plants to achieve clearly defined goals lie at its core. Implicit in the idea of cultivation is that human decision making, 'agency', is consciously applied, both as thought and action. This agency may vary from being very occasional through to frequent but potentially significantly impacting on the life of a given plant and or the synthetic plant community that it is part of. This agency process often commences through making decisions on which plants can be used in which environment; indeed in a rational world, the environmental conditions of the planting site should be one of the main factors in plant selection, and this is discussed later in this book. Agency is applied to different degrees in different circumstances, often involving the manipulation of water, nutrients, light and physical removal of plant tissues to control the rate and form of growth, the degree to which plants flower and fruit, and how plants are likely to be perceived by people. The use of these manipulation levers can be either intensive or extensive, sophisticated or crude, i.e. there are inherent gradients across which decision making can or must be made, depending on the needs of the location or context and the resources that are available to decision makers.

At the low-intensity end of the spectrum rather blunt forms of management, such as the non-selective cutting off of plant parts may be used to nudge a plant or a plant community in a chosen direction. This issue of choice and decision making is critical to understanding the horticultural mindset. Within horticulture, agency is seen as an intrinsic part of the process, essentially as a 'good', whilst recognising that no one input or outcome is appropriate in all situations.

Environmental horticulture and relationships with pure (purist) ecology

The involvement of human agency within environmental horticulture is the prime difference to parallel disciplines that are concerned with plants, for example conservation ecology or restoration ecology. Depending on who is practising these activities, and in what context, there is also a gradient in terms of how much human agency can be applied, but in general it is much less in total and often restricted to forms of management which work to kick-start or direct an ecological process, for example, increasing light at ground level in a plant community to benefit species that require this, by removing the canopy in total. These processes inevitably disadvantage some species in order to benefit others. There are winners and losers; at least temporally.

Horticulturists tend to find this a little disturbing; in the horticultural paradigm there is a desire to achieve some form of equivalence of benefit, i.e. to manage to avoid there being obvious losers. This notion of 'care' is much more remote in conservation ecology or restoration ecology than in environmental horticulture, even when the latter practises these same activities. Indeed, because these former disciplines are philosophically deeply rooted in the idea that human agency typically corrupts or damages the organisms and processes that we call nature, they are fundamentally uncomfortable or even hostile to the application of human agency to vegetated systems. They are most strongly developed in large, recently settled (by Europeans) countries with a strong 'wilderness construct' and least developed (but still present) in countries with a long and obvious, often obviously constructive, relationship between the natural vegetation and people through various forms of low-intensity agricultural exploitation and management.

Within ecological science dialogues, 'gardening', i.e. the intense and prolonged application of agency to plants, is often used as a pejorative, rather than a positive concept. These polar views explain why traditional ecological science is sometimes philosophically uncomfortable with horticultural utilisation of plants; it seems at best pointless and at worst almost decadent.

Horticulture, by contrast, is not at all embarrassed by human agency; it recognises that human beings can obtain great pleasure and sometimes much deeper psychological states such as meaning through application of their own agency and that of others (Fig. 1.5). Indeed, creative processes/expression can be an attracting factor for one to engage with environmental horticulture. By adding another trophic level of interaction (human agency) to the ecological food web, it is possible to achieve end points that are impossible through the rather narrow and blinkered processes that underpin conventional ecosystem development. Within the limits of what is possible, horticulture can choose; it does not have to be insular and dogmatic!

Fig. 1.5. Environmental horticulture embraces the concept that green spaces are 'human habitat' as well as habitat for plants and animals.

But choose what? Within ecology and botany, the development of increasingly comparative taxonomy linked to effective field survey and vegetation sampling from the early 18th century onwards allows the construction of reliable lists of the native plants that make up the plant communities of a given region. These floral lists, within a strictly ecological view of the world, circumscribe our choices for us. Deviation from these lists, in terms of species selection, is increasingly seen in the contemporary world of biodiversity as inappropriate, unethical or even plain 'bad'. This process has been in operation, at least to some degree, in the 'wilderness' countries recently settled by Europeans since at least the 1970s, but is increasingly prevalent even in long-settled countries, as a result of biodiversity legislation effected through the planning process.

What genotypes should be planted?

For any given planting site, the range of plants prescribed by the natural distribution of native species as a result of 'natural' ecological process is often relatively small. In the UK for example, if the planting site was for argument's sake the whole of the British Isles, it would correspond to approximately 1100 native species. If a county was chosen, for example Northumberland, plant genotypes native to the area would decrease to *c.*1000, not much of a drop, because in small countries many species are found across the entire territory. If, however, a heavily shaded planting site in woodland in Northumberland was the chosen arena, and hence required heavily shade-tolerant understorey plants, the number of native species would fall to less than 300. If the criterion was then applied that the plants chosen had to be particularly attractive to human beings, this number would fall to less than 30. Continuing the interrogation that any species identified needed to be in flower or otherwise attractive in autumn, the numbers would drop to almost zero.

Hence in the Northumberland context, if one wanted to use species that flower in autumn, it would have to be accepted that this is simply not possible or one would have to look to the floras of other locations, where in response to local evolutionary pressures, plants exist that do what is desirable. In climates where summer rainfall is very high, a niche is created for some woodland plants to flower in autumn because the soil is moist enough to support this and there is less competition for pollinating insects, hence a genotype such as the eastern North American *Aster divaricatus* could be utilised. There will nearly always be a series of locations elsewhere in the world with an analogous climate to the site that needs to be planted, each with their own distinctive flora that are sufficiently fit to grow well on the designated site. For Northumberland, one looks to the montane species of Western China, Japan, and western and eastern North America. This larger pool of species throws up choices such as *Actaea racemosa, Aster divaricatus, Heuchera villosa, Rudbeckia fulgida* (all North America), *Begonia grandis, Impatiens omeiana, Saruma henryi, Nipponanthemum nipponicum, Saxifraga fortunei* (China and Japan) and so on. The quid pro quo in this process of meeting aesthetic aspirations is that these species may not be as well fitted to their planting environment, and hence as robust, as the species native to Northumberland. Hence, a balance must be struck between aesthetic and functional fitness, and where resources are too few to 'care' for the non-native species of lower fitness, then native or other more fitted species that do not meet the original aesthetic specification must be used.

Context therefore becomes very important in making these judgements; urban areas with relatively high expectations of what plantings should look like, and relatively abundant resources for management, are more likely to use non-native species than rural locations. In large countries, for example the USA, an interesting situation has arisen that has fed the nativism concept in plant selection. Because the USA is both very large and biologically at the richer end of the plant diversity spectrum, it is easier to meet aesthetic planting specifications from the politically native flora of that particular very large nation state. This leads to a certain assumption that working purely within the native flora is entirely possible and why don't other countries do the same? This argument plays less well at an intellectual and practical level in small nation states such as the UK, the Netherlands or Monaco and where there are relatively low levels of botanical diversity. The flip side to the USA position on exclusive native plant use is that just because a plant is politically native does not mean it is well fitted and robust to anywhere within this politically defined envelope. *R. fulgida* is native to the USA but is highly unfit for the winter or low rainfall areas of that country.

Whilst horticulture has a strong relationship with the use of plants from other parts of the world for the reasons given above, in counties with large floras, for example South Africa, it is often possible to work purely within the native flora, although again subject to the inevitable issues of poor fitness for distant species, unless altitude and other factors reduce the expected loss of fitness. In this latter sense, the issues facing those who use native species drawn from beyond the local region and those using non-politically native species from further afield are the same. A fascination with the floras of other places whether within the nation state or not, and not simply as part of intellectual curiosity but as something that might be used and useful, is fundamental to the horticultural paradigm.

This extends to the manufacture of new genotypes through collection in the wild, breeding or selection; horticulture has an endless appetite for the new, whether the plants are originally native or non-native. This is in stark contrast to ecological thought, which is often horrified by the manufacture of the new and would like to believe that species represented in lists (floras and the like) are intrinsically right to that location and always have been. When one considers these issues in the context of geological time, it is clear that floras only represent ephemeral moments in time, not immortal certainties.

Can one apply moral or ethical notions to these contrasting paradigms? Debates in the media appear to assume that one can do this. Over the past 30 years there has been a tendency to see species drawn from regional lists as somehow right and appropriate, perhaps even good, and species from outside of these regions as inappropriate, perhaps even unethical. Such positions are entirely human constructs; one cannot find evidence for the righteousness of this within ecological science per se. One can measure the negative consequences of these choices in terms of species that are insufficiently or too well fitted and have naturalised amongst extant native species, although even here it is often difficult to measure harm per se? (Thompson, 2014). The basis of measuring harm is that a native plant is better for the native animals that depend on that vegetation. Whilst in highly specialised ancient floras, such as in the case of South African, this may well be the case, it is looking increasingly difficult, however, to argue this as a generalisation in northern Europe. A paper by Hanley *et al.* (2014) on the 'goodness' of native plants for generalist and highly specialised pollinators (bee species) found that the idea of localness at the scale of the nation state is often ecologically meaningless because many plants and pollinators have very wide overlap over their ecological history; a UK native bee might be just as comfortable with the plants of distant portions of the Palearctic (the temperate band of vegetation running from Western Europe to Japan north of 33°) as those found in the UK.

The fundamental right of 'nativeness' (i.e. one should solely plant out species native to that region) is also increasingly challenged by a changing world, and specifically by a changing climate within that world. Plant species that have been well fitted to their location for the last 12,000 years (i.e. the end of the last Ice Age) may not be fit for it in as little as 50 years' time, due to a warming climate (e.g. Warszawski *et al.*, 2013). 'Hanging on' to a philosophy that 'native is always best' seems somewhat naïve in the face of such dramatic, fundamental climatic (and subsequently ecological) change.

1.3 Future Directions

So what does the future hold for environmental horticulture? In some ways the discipline is caught in a dilemma. The coronavirus 2 (SARS-CoV-2) 'COVID-19' pandemic from 2019–2023 with its resultant restrictions on human movement brought home sharply an urban citizen's need and desire for green space (Berdejo-Espinola *et al.*, 2021). Parks and other local green spaces were never more popular as people were desperate to leave the confines of their home and experience 'fresh air' and nature. Those lucky enough to have access to a garden, appreciated what a precious resource it was for their well-being and mental health – either by simply sitting in it or engaging with garden activities to pass the time. Many, worried about the impact of the pandemic itself, sought gardening as a distraction and remedy to the constant flow of bad news emanating from news outlets and social media. Urban green space has probably never been more appreciated. In contrast, 'economic downturns' in many parts of the world over the last 10–15 years have meant less public money is available to support public-run parks, gardens and other forms of green infrastructure. Key civic landscapes such as city botanic gardens are dependent on volunteer labour to keep the shrub borders clear of weeds and the entrance beds filled with colourful bedding plants each season. Professional environmental horticulturists may move into management positions, but they frequently have limited numbers of professional staff to manage. Many of the core attitudes and values of landscape horticulture were fashioned in a time when resources were much more abundant than they now are. This includes energy, water and the affordability of human labour. Diminishing resource availability has not made the end points traditionally valued by horticulture irrelevant; however, they have made it much more difficult to deliver on these at the required level. Traditional management tasks are reducing in scale and landscape design is evolving to promote lower-cost processes, with plantings being less demanding in terms of maintenance. Approaches to delivery vary from various forms of rationing through to more radical approaches in which completely new types of vegetation are employed. Rationing is the simplest and most obvious means to reduce resource input: the area of the most intensive to establish and manage vegetation is reduced to meet the budget available. Traditional 'staples' such as annual bedding plants are shrunk down to small roundels in traffic islands sponsored by local businesses. This process tends ultimately to result in landscapes composed primarily of trees and mown grass, since everything else horticultural has higher establishment and management costs than these. This not only reduces the potential for seasonal change, drama and engagement for people but also reduces the amount of pollen and nectar, foraging space and volume available to pollination and herbivorous invertebrates that ultimately builds urban food chains. Rationing, no matter how thoughtfully undertaken, ultimately only gets one so far and ultimately leads to impoverishment over much of the urban estate.

The next step in a rationing process is to look critically at how to manage existing horticultural vegetation to achieve similar outputs with fewer inputs. Sometimes this employs technical understanding, for example, to allow a reduction in watering, or fertilising frequency, by looking at this problem through a social or cultural lens to decide what is the aesthetic threshold for vegetation to be seen as sufficiently appealing to people. To some extent, as 'necessity is the mother of invention', this can lead to technological improvements – for example, in more efficient irrigation equipment – or the adoption of new vegetation types – for example, drought-adapted species that are still colourful.

In addition to the resources issue, there is also the question of horticulture having to deal with and respond to new policy or drivers, for example, biodiversity and sustainability legislation, health and safety legislation in relation to staff or the public, and the ongoing reduction in various biocides used, when staffing levels are very low to manage weed competition manually.

Functional horticulture

A counterbalance to these points, however, is an increased awareness around biodiversity loss, climate change and the causes of poor health in humans and the positive impact environmental horticultural activities can have on agendas dealing with one or more of these phenomena. Recent discussions have cited how urban green spaces could be used more to conserve rare plant species. The environmental benefits are moving from academic dialogues to practical measures, e.g. the use of green barriers to improve air quality around school playgrounds. How urban green spaces can store and slow the flow of runoff water and reduce flooding; this aspect perhaps justifying their retention within urban conurbations more so than just 'costly' recreational green space (Fig. 1.6). Stronger evidential links between green space and human well-being are resulting in traditional 'clinical-orientated' medical practices adopting green prescribing as an antidote to depression and other disorders (Robinson *et al.*, 2020).

Historically, horticulture has mainly seen vegetation in terms of cosmetic beauty; does it look attractive? This is a critical factor to maintain and indeed obtain public support, but particularly when vegetation is to be used on a large scale there is also a need to gain function. Hence, urban designed woodlands have to fix carbon, reduce local heat island effects, support as much animal biodiversity as possible and be harvestable for wood to meet local needs. Woodlands based on riparian species that tolerate cycles of flooding can be used to clothe and provide these functions at the same time as acting as infiltration swales for urban runoff. The same parallels can be drawn with herbaceous vegetation, so, for example, the species used for sustainable drainage swales, which are temporally very wet then gradually dry out before the next storm-water event, must irrespective of where the species come from look attractive for as long as possible, and in particular must not wilt and

Fig. 1.6. Planting increasingly justifies its inclusion in urban spaces due to its functionality. Here, a series of detention basins in Sheffield, UK are used to capture rainwater and slow its runoff into the nearby river. Overflowing water enters the basin from the aperture at the top and only leaves once the basin is filled and the water reaches the overflow at the bottom. Plants help the system dry out more quickly (thus improving its potential to hold more water from the next storm) as well as filter out and deactivate pollutants.

collapse when subject to a lack of storm-water in the height of summer. This mixture of ecological and functional characteristics forces a different approach to plant selection and plant use.

Horticultural skills

Because of climate change and associated sustainability and biodiversity agendas, vegetation design, establishment and management skills are becoming increasingly important in urban areas offering the potential for a horticultural urban renaissance. This will, however, only be the case where horticultural skills are honed to match the current political and policy agenda. Environmental horticulture will maximise the employability of its graduates by looking to forge more relationships with other disciplines which are also key actors in the development and management of urban landscapes and have shared interests. These are likely to be landscape architecture, architecture, engineering and hydrological science but also increasingly, psychologists, social workers and medical experts. The sector will need to address rapidly moving technologies, such as artificial intelligence – already large swathes of lawn turf are managed by robotic grass cutters. Technical specifications may become more demanding. New green wall systems must address fire safety to the building they are housed on, be accessible for easy maintenance, and perhaps provide (grey) water cleaning capacity as well as irrigate plants effectively. They may need to be redesigned to act as better hosts for invertebrates, small mammals and birds, or to optimise the insulation capacity to the host building. Green walls have been largely designed for aesthetics and conveying corporate power (or [pseudo-] 'green credentials') in the company buildings they front. But how about green wall systems that provide ecological functionality and community structures akin to prairies, calcareous chalk grasslands or coastal saltmarsh communities?

Communication and interaction with less conventional partners will also become paramount. Environmental horticulturists need to be able to liaise with architects and engineers, for example, to further break down the barriers between grey and green infrastructure, so in future it may be possible literally to entirely cover a building with a green mantel of vegetation (not just provide a rather tokenistic panel or two on the front façade). Similarly, better dialogue with sociologists and health experts is required for horticultural activities to be more effectively implemented to address issues around mental health, well-being, physical activity, crime avoidance and enhanced social integration. To make these collaborations work, however, horticulture will need to be able to speak at least some of the language of these other disciplines, and in particular, it will need to demonstrate that in addition to experience of plants in practice, it also has a research-based understanding of vegetation and establishment in the city. Overall, environmental horticulture has a key role in making sure our cities remain 'liveable' over the 21st century.

Conclusions

- The term 'environmental horticulture' was coined in the 1980s and broadly correlates with the management of landscape plants in public and semi-public arenas. Over the subsequent years it has also become linked with a philosophy for more environmentally sustainable practices within urban horticulture.
- Environmental horticulture has a number of similarities with other forms of green space or environmental management but differs in that it actively promotes the cultivation of plants and does not necessarily restrict itself to the use of native genotypes. It tends to bring humans and their perceptions and activities more actively into the framework of green space design and management.
- Environmental horticulture recognises that human beings can obtain great pleasure and sometimes much deeper psychological states, such as a sense of 'meaning' or 'worth', through application of their own interventions in the green space, and indeed from that of others.
- There is an acknowledgement within the discipline that the urban environment is unique in terms of its conditions and pressures, and that plant choice in this environment needs to reflect this, in terms of fitness, but also in the wider rationale of public appreciation or functionality.
- Functionality of green space (e.g. its use as a nature-based solution) is increasingly important within environmental horticulture. Plants (both native and non-native) are now used to provide resources for wildlife, trap and hold back rainwater, reduce urban heat and pollution, mitigate noise, insulate buildings and provide recreational and relaxing environments that improve mood and encourage people to relax (and by doing so, improve longer-term health outcomes).
- Those wishing to develop a career in environmental horticulture need to be adept at plant husbandry skills but also are increasingly required to embrace other disciplines and widen skills. Horticulturists of the 21st century need to work in partnership with landscape architects, ecologists, land or civil engineers, health and social work professionals and indeed many other disciplines.

References

Berdejo-Espinola, V., Suárez-Castro, A.F., Amano, T., Fielding, K.S., Oh, R.R.Y. and Fuller, R.A. (2021) Urban green space use during a time of stress: A case study during the COVID-19 pandemic in Brisbane, Australia. *People and Nature* 3(3), 597–609.

Hanley, M.E., Awbi, A.J. and Franco, M. (2014) Going native? Flower use by bumblebees in English urban gardens. *Annals of Botany* 113(5), 799–806.

Horticultural Trades Association (2023) From Nursery to Nature – The Value of Plants. Available at: https://hta.org.uk/value-of-plants (accessed 11 September 2024).

Oxford Economics (2004) *The Economic Impact of Environmental Horticulture and Landscaping in the UK*. Report for the Environmental Horticultural Group. November 2024.

Robinson, J.M., Jorgensen, A., Cameron, R. and Brindley, P. (2020) Let nature be thy medicine: A socio-oecological exploration of green prescribing in the UK. *International Journal of Environmental Research and Public Health* 17(10), 3460.

Thompson, K. (2014) *Where Do Camels Belong?: Why Invasive Species Aren't All Bad*. Greystone Books, Vancouver, Canada.

Warszawski, L., Friend, A., Ostberg, S., Frieler, K., Lucht, W. *et al.* (2013) A multi-model analysis of risk of ecosystem shifts under climate change. *Environmental Research Letters* 8(4), 044018.

2 Succeeding with Landscape Plants – Context and Conditions

Abstract

Landscape plants are composed of true species (i.e. genotypes found occurring naturally) or plants selected/bred by humans (cultivars). True species have evolved in different biomes and are adapted to certain environmental and biotic conditions. When used in horticulture, such species may be growing outside their normal ecological and climatic niches. Similarly, the bred cultivated forms may have lost those traits that helped their parents survive in the wild. This chapter outlines where landscape plants are found in human-dominated landscapes and the challenges they face in these locations. Most challenges are associated with human-built (grey) infrastructure. For plants to survive and thrive in our cities, they need to cope with impoverished, damaged soils, inadequate/infrequent access to water, insufficient light, and altered temperature regimes as well as 'by-products' of human society such as atmospheric or soil pollutants. The chapter illustrates these problems and how plant choice and management can help overcome these aspects.

2.1 Ornamental Plants in the Landscape – Rural and Urban Locations

Horticulture takes place across the globe and is embedded in both urban and rural environments. In rural environments it is a component of commercial or subsistence food growing. Landscape or ornamental horticulture specifically spans a spectrum of rural landscapes and gardens, city parks, city gardens, and streetscapes, as well as in and on buildings. Attractive ornamental landscapes frequently complement large- and small-dwelling properties in the country or peri-urban locations. Indeed, the majority of historically important heritage gardens within Europe and North America are probably found in rural locations. These are locations where, traditionally, people of affluence had space to 'play' with the landscape, due to the scope of the land available and its relatively low cost. It is also where, simply, many inherited estates were located, e.g. relating back to feudal times in the UK and France, where wealth and status were partially linked to agricultural production. As castles gave way to palaces and manor houses in response to the countryside became a less violent place to live, so the landscapes around these dwellings took on a more ornamental form and become places of leisure – at least for the landowners (Olivadese and Dindo, 2022)! Many garden styles originated and evolved in the countryside – with the views of the surrounding landscape often being a key feature (Goulty, 2003). In Europe this has included the Italian renaissance gardens, French formal gardens and the English landscape garden movement. Even at the small scale – village cottagers (cottage garden style) were instrumental in cultivating many wild plants and over time selecting naturally occurring 'improved' forms. Many rural garden soils are naturally rich and fertile – being a legacy of well-structured, brown woodland soils and manured ex-agricultural fields. Sadly, the same is not necessarily true of urban soils, where soil structure here can be damaged, with knock-on effects for drainage and root development. Space can also be at a premium in towns and cities, with the temptation to plant material that ultimately grows too large for the space provided. However, changing planning priorities, such as the pedestrianisation of roadways does provide opportunities for larger and more imaginative green infrastructure (Fig. 2.1). Both rural and urban landscapes have their challenges for plant development (Table 2.1), but those in our towns and cities can have perhaps the greater impact on determining plant viability at all and thus are the focus of this chapter.

Challenges of the urban environment

Urban plants can have a 'hard time of it'! There are a number of locations in the urban environment where plants are 'asked to grow', yet the conditions

DOI: 10.1079/9781800621763.0002

Fig. 2.1. New developments within cities and towns, including the prioritisation of pedestrian flows over vehicle traffic, can create opportunities for attractive and functional plantings.

are not very conducive for growth. Compacted, low-organic soils may be a result of urbanisation processes, where the prevalence of heavy machinery has compacted the top strata of the soil, resulting in poor aeration, and restricted the capacity of water to filter through. Heavily compacted clay soils will have a high bulk density, challenging roots to physically penetrate the soil structure. Rubble, sand and residual concrete, tarmac, brickwork, etc. are often the residue of the construction process: materials that provide little opportunity for holding the water or nutrients that plant roots seek. Such soils can seem almost lifeless with little earthworm activity (important for improving drainage [via macropores] and incorporating organic matter into the lower layers of the topsoil). Microbial life is different from woodland brown earth soils and tree roots, for example, may have limited ability to join up with their symbiotic mycorrhizae species (Buil *et al.*, 2021). A lack of organic matter and limited biological activity, impede or interrupt nutrient cycling processes and impermeable crusts can at times form on the soil surface (Mullaney *et al.*, 2015; O'Riordan *et al.*, 2021). Trees and other woody plant types tend to be the ones that suffer most if soil conditions are not appropriate (Fig. 2.2).

2.2 What a Landscape Plant Needs to Thrive

A landscape plant needs very much the same things as any other plant. This is primarily space to grow, light (irradiance), water, carbon dioxide, nutrients absorbed through its roots (macro elements – nitrogen, phosphorus and potassium) and a variety of micronutrients – primarily magnesium, calcium, manganese, iron, boron, zinc, copper, molybdenum, chlorine

Table 2.1. Comparison of the challenges of growing plants in rural and urban settings

Challenges in the rural environment	Challenges in the urban environment
Herbivory pressure from wildlife. Deer, rabbits, hares, voles, etc. are more prevalent and browse garden plants when accessible. Birds may eat flower buds and fruit of orchard trees.	Vandalism and damage from people creating 'desire lines' and cutting across planting beds. Tree trunks can be damaged by vehicle collisions into them. Litter can accumulate on foliage and around the base of plants.
More investment in fencing and netting required.	More investment in staking and protection of park and street trees from vandalism.
Gardens are on average larger, but that means more investment in local infrastructure, e.g. irrigation systems, pathways and machinery, and greater energy use. Bulky garden materials may need to be moved further.	Land has limited availability, and so gardens may be smaller. People in apartments and flats may need to rely on balconies and alleyways to grow their plants. Public green space can be constrained by buildings and other infrastructure.
Rural gardens may be in more exposed sites (windy, exposed hillsides, or prone to damaging salt spray by the coast) or in river valley bottoms, with more frost episodes (cold air gets trapped in valley bottoms, causing 'frost pockets').	Urban sites are often warmer than equivalent rural locations. This means a longer growing season, and a wider range of plants can be grown (in temperate regions). Excessive heat though can be a problem for plants in summer (e.g. street trees). Downdraughts and canyon-ways make city centre streetscapes windy locations.
Incidental pesticide sprays from nearby agricultural land. This may cause foliage damage to boundary plants.	Urban areas my experience high pollution levels – both in air and soil. Urban soils frequently contain contaminants such as heavy metals, pesticides and industrial chemicals. Plants may also suffer salt damage from road de-icing salts.
Weed pressure may be greater – both with weed seed being more prevalent but also larger areas of land to be weeded/cultivated.	As well as contaminants in the soil, soil structures are often poor, with machinery compacting the structures, increasing their bulk density and impairing air and water movement into and through the soil (see below).
Potentially further away from resources – such as locations to purchase garden machinery or plants.	Access to clean water: public plantings, especially street trees, may not be watered regularly and have limited space for rainwater to access the roots. Urban gardeners may struggle with limited access to safe water sources. Rainwater can be polluted. Efficient irrigation systems are essential but can be costly to install.

and nickel. Plants acquire most of these nutrients from the soil – a medium that also allows them to anchor and stabilise themselves. Landscape plants also regulate their own growth through their phytohormones. These organic molecules that flush through their cells and phloem system control root development and leaf and stem growth, as well as key processes such as flowering and dormancy induction. They also have a role in helping regulate a plant's defence against pathogens and herbivores and organise repair to damaged tissues.

Water

Rainfall is a key factor influencing the location of the world's different biomes, and forest/woodland systems usually require at least 750–1000 mm of annual rainfall to sustain themselves. Although irrigation can be applied to those cities located in drier areas, often tree cover is lower in cities found in hot, arid regions (Endreny *et al.*, 2020) (Fig. 2.3).

Water plays a number of roles in landscape plant viability and performance. Water is essential for maintaining cell turgor pressure, which keeps plants upright and rigid. Water regulates cell expansion and thus stem, leaf and shoot growth. Water is moved through the plant via long tubes in the stem – xylem vessels that transport water up from the roots and is then transpired through the leaves back out into the atmosphere as water vapour. Water also moves through the phloem vessels – living cells that use water to transport nutrients and phytohormones from one part of the

Fig. 2.2. Urban environments can be stressful places for trees and other landscape plants. Providing adequate space for root systems that allow for both good water supply and aeration are vital components. A lack of sufficient space and protection against soil compaction has caused the death of the tree shown on the left: for comparison, better, if not optimal, protection is shown on the right.

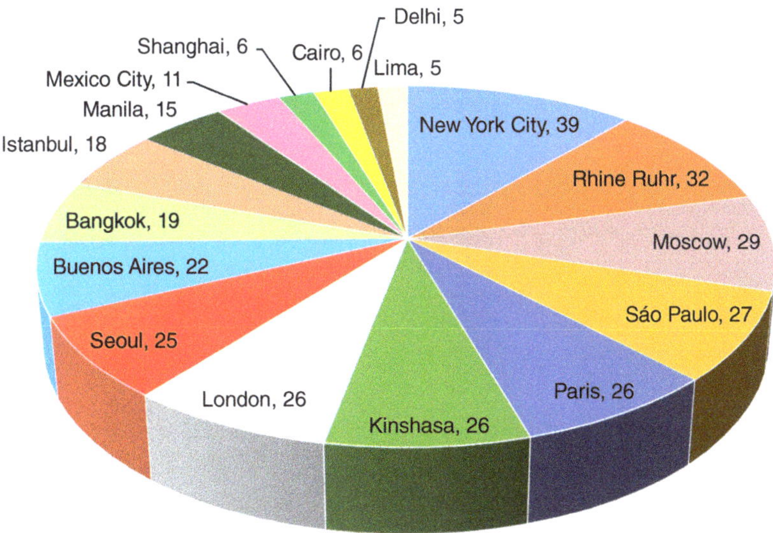

Fig. 2.3. Percentage tree canopy cover in a cross-section of global cities. Those cities found in drier climates tend to have less tree cover. Numbers relate to % cover. (Modified with permission from Endreny *et al*., 2020.) CC BY 4.0.

plant to another (movement is bi-directional in the phloem). All plant cells need to be bathed in water (aqueous solution) as this is the medium that allows biochemical reactions to take place. Plant cells have both a cell wall (giving structure and rigidity) and a cell membrane that regulates ion and molecule movement in and out of the cell. Many plant cells of course are green, and this is because they house chlorophyll within the chloroplasts; chloroplasts being special organelles (plastids) that move around the cell structure. It is interesting to note that chloroplasts evolved from an ancient cyanobacterium that was engulfed by an early eukaryotic cell (i.e. a cell with a discrete nucleus). Because of this symbiotic origin, chloroplasts contain their own DNA separate from the cell nucleus. Chloroplasts cannot be made anew by the plant cell and must be inherited by each daughter cell during cell division.

Chlorophyll molecules are arranged within thylakoids (membrane, sack-like structures) that often stack together in grana within the chloroplast. A photon of light hitting a chlorophyll molecule confers energy to it, which is then used to split a molecule of water – creating hydrogen ions and oxygen molecules (the latter 'by-product' we, along with many other animals use to breathe). The hydrogen ion is a form of energy itself and is used to drive a biochemical chain (the Calvin cycle) that ultimately results in the formation of simple sugars such as glucose. Chloroplasts are essentially 'sugar-producing factories', although they usually store their carbohydrates in less soluble, starch granules. The whole process is known as photosynthesis.

Chloroplasts carry other pigments – many of which also harvest light. Landscapes plants with high numbers of these other pigments may have red, purple or yellow leaves (and thus court favour for being unusual). Sometimes these pigments simply disguise the presence of chlorophyll, but in others there may actually be less chlorophyll present. Due to the fact that some landscape plants have gold or pale, white bars (variegated) leaves, these cultivars are less vigorous in growth, largely due to more inefficient carbohydrate production caused by a lack of chlorophyll.

As well as being a medium to transport molecules and ions, water helps keep a plant cool. Plants use water for evaporative cooling (energy lost as latent heat) through their transpiration stream. As water evaporates from leaf surfaces, it helps regulate the plant's temperature, preventing overheating in hot conditions. The rate of moisture loss from the leaves is determined by the pores, called stomata, on (usually the underside) of leaves.

Matching water supply to the requirements of urban plants can be challenging. The urban heat island effect drives up the demand for water, and restricted root runs can limit the capacity for a plant to consistently access soil water. Small-volume tree pits and limited surface areas that can harvest rainfall or catch runoff water mean supplies of water to street trees can be limited at critical times. Such trees often possess a poor root-to-shoot ratio and as such are more prone to drought stress. When studying city trees of different ages, Rosenberger *et al.* (2024) indicated that older, larger trees were more prone to severe soil moisture deficits, but that larger tree pits with greater surface permeability helped trees, especially the younger specimens, to access water (and survive drought) (Fig. 2.4). Ensuring good root space and avoiding excessive sealing of the soil surface helps tree survival.

For landscape plants grown in soil directly and where irrigation is applied, the convention is to water thoroughly but less frequently to encourage deeper root growth. Roots growing deeper down the soil profile are thus able to access water more easily during dry weather or drought events. Young, newly planted plants are the most prone to drought stress as often the root system has not made sufficient contact with the main soil matrix to start absorbing water and pulling it towards the developing roots via a water potential gradient. Thus, it is imperative that all new plantings are watered thoroughly on planting and usually for a number of weeks afterwards if planting takes place in the late spring or summer. Plants grown from seed or transplanted as very young specimens will need regular irrigation until the roots starts to develop and penetrate lower soil strata. Once an area has been planted, then to help avoid moisture loss from the soil per se, mulches can be applied over the soil and around the base of the plants.

Soil aeration and avoidance of waterlogged conditions

Roots respire, like any other part of a plant. As such, they require access to oxygen. Soils with good structure facilitate the infusion of oxygen from the atmosphere and allow it to diffuse through the soil matrix. Macro- and micropores within the soil aid

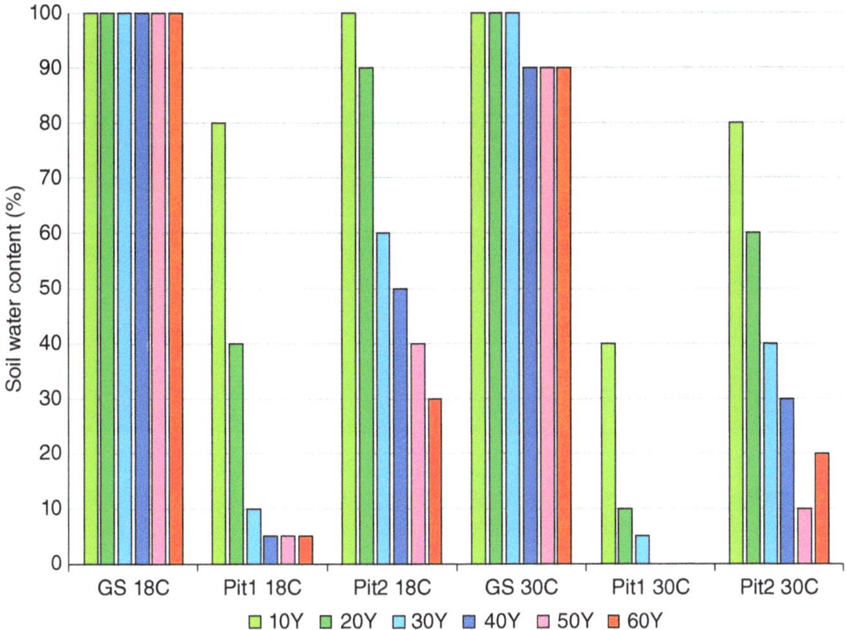

Fig. 2.4. Soil moisture content (% of available pore space) in tree root pits with trees of different ages (e.g. 10Y = 10 years old; 60Y = 60 years old). Data from scenarios with maximum daily temperature of approximately 18°C (18C) or approximately 30°C (30C). Key: GS = tree pits in green space where 1.5 × the area of the crown projection area is not sealed; Pit1 = tree in a two-thirds' partially sealed area according to the German standard (FLL guidelines, 2010) i.e. 8 m^2 and a maximum root volume of 12 m^3; Pit2 = optimized tree pit of 24 m^2 and a maximum root volume of 36 m^3. (Modified with permission from Rosenberger *et al.*, 2024.) CC BY.

this transport, as does the highly textured structure of soil crumbs (mineral and organic material bound together, creating a porous 3D structure). Well-aerated soils tend to have an open structure with plenty of organic matter (Fig. 2.5). 'Heavy' clay soils, on the other hand, are composed of smaller mineral particles and are more prone to being compressed, thus limiting the pores present and creating a poor oxygen-deficient medium, especially when wet. As mentioned previously, machinery, road traffic and even pedestrian trampling are all likely to compress clay soils. This can lead to anaerobic soil conditions and a high soil bulk density – both factors not conducive to healthy root development. Poor street tree growth is often linked to compacted, compressed soils undermining root proliferation and extension. Under wet conditions too, clay soils will puddle – due to the lack of soil pores, leading to further conditions of anaerobism for root systems.

Waterlogging also changes the biochemistry within the rhizosphere. Microbial activity alters to favour those microbes that work best under anaerobic conditions. During anaerobic respiration, organic matter is oxidized, and the components of the soil are chemically reduced, with waterlogged soils being associated with denitrification, the reduction of nitrate and nitrite ions to ammonium, the reduction of iron hydroxide and the evolution of methane. Concentrations of ethylene, ethanol, lactic acid and ammonia can build up and be toxic to root cells. As respiration is faster at warmer temperatures, more severe anaerobic soil conditions are often linked to flooding during summer. The lack of biologically activity in winter means that paradoxically winter flooding is less damaging to trees and other plants.

The paradox with waterlogging is that the foliar parts of many plants actually experience 'physiological drought', as root cells fail to maintain the transpiration stream. This results in water no longer effectively being transported into and up the xylem tissues. There are a number of stress factors that impair plant development during waterlogging,

Fig. 2.5. Soils with a high organic matter content resist compression and retain aeration via drainage pores and other spaces. Roots exploit these spaces in their search for nutrients and water. The proliferation of root hairs along the expanding roots is an indication of healthy root development.

but a lack of oxygen in the root zone alters respiration with less energy available for cells to regulate water uptake. Oxygen diffuses through air much more effectively than it does water, so when soils become saturated diffusion rates decrease dramatically, leading to oxygen starvation and anaerobic conditions at the root–soil interface. Flooding (or more appropriately in physiological terms 'anaerobism') affects a number of processes in the transpiration stream, not only water absorption into cells but also root hydraulic conductance and stomatal opening. Not all plants respond in the same way, however, leading to some debate as to what precisely the stress factors are and how plants respond to them. This is accentuated by the fact that not all species demonstrate the symptoms of water deficit during flooding. One theory is this may be due to subtle effects relating to soil oxygen (O_2) and carbon dioxide (CO_2) partial pressures as well as differences due to plant species. High CO_2 levels that accumulate during flooding due to respiration from both soil microflora and roots may be transformed to carbonic acid (H_2CO_3). This in turn is transported to root cells and acidifies the cytoplasm. This is thought to inhibit aquaporin activity; these are the pores by which water moves between one cell and the next. Tournaire-Roux *et al.* (2003) claim it is this transformation of CO_2 to H_2CO_3 and the resultant acidification of the cell cytoplasm that inhibits root hydraulic conductivity under anaerobic conditions. Unfortunately, many studies on waterlogging document O_2 depletion, but do not monitor for CO_2 accumulation and activity. As such it is still difficult to say whether loss of water uptake is directly due to a lack of O_2 or alternatively a build-up of CO_2.

Others argue that stomatal closure (and hence transpiration) is mediated by root-derived chemical signals (Jackson and Ricard, 2003), and although the role of the stress hormone abscisic acid (ABA) is implicated, the exact nature of the possible chemical signals is still unknown. A further mechanism may be that phytotoxic compounds generated in the roots during anaerobic conditions are responsible for interfering with water absorption and movement within the roots, for example, ethanol, acetaldehyde (ACC) or lactic acid (Kamaluddin and Zwiazek, 2001). Irrespective of the mechanisms involved, the consequences are usually the same – significant and drastic reductions in growth in non-adapted species.

Some landscape plants are more prone to flooding injury than others, Mediterranean species have a reputation of not tolerating waterlogged heavy clay soils, and such genotypes are best grown on free-draining slopes, in raised beds or even in containers with free-draining growing media. The art of ornamental plant cultivation is to understand where species derive from in the wild and understand how similar conditions can be provided in the urban landscape (Figs. 2.6 and 2.7). Many other landscape species of course do well in damp or even waterlogged conditions – because they have ecological adaptations to deal with these scenarios in nature.

Fig. 2.6. Understanding where plants come from in nature and then matching the plant community to the different conditions that are encountered in the urban environment helps to ensure a successful and enduring plant composition. Top: Low-growing alpine plants can be used on free-draining raised beds, podiums, green roofs and walls; Bottom: Plants adapted to warm Mediterranean conditions, often with hairy, grey or narrow leaves, can be used in warm locations with good drainage.

Fig. 2.7. Top: Woodland plants like this *Rhododendron* are useful in locations with partial shade. Bottom: Bog plants and wetland riverine species can be accommodated along artificial or designed watercourses.

Other soil conditions and problems

As well as compaction and waterlogging, plants growing in urban conditions can face toxicity from roadway de-icing salts, roadway water runoff or a legacy of contaminants in the soil left over from previous industrial activity. These can damage all types of landscape plants (Czaja *et al.*, 2020; Fan *et al.*, 2020; Kisvarga *et al.*, 2023), but plants that are grown for food also pose health risks for humans consuming any subsequent harvested produce (López *et al.*, 2019). Contaminants include heavy metals ions (lead, mercury, arsenic, cadmium, chromium, copper, zinc, boron, and nickel), oil-based residues, other hydrocarbons, pesticides, paints and creosote, as well as some chemicals that can radically alter the soil pH, such as calcium hydroxide (Dickinson *et al.*, 2000). It is difficult to define specific toxicity levels as physico-chemical and biological variables control bioavailability and potential toxicity. The toxicity of metals to landscape plants and the bioavailability of metals for uptake is controlled by their total concentration and changing chemical forms (referred to as speciation) between solid and solution phases. The balance between these two phases and hence exposure of plants to specific ions is affected by absorption in the soil, complexation, precipitation, leaching, organic matter content, clay mineral content, iron, manganese and aluminium oxides, pH and soil redox potential (Mench *et al.*, 1998). As such the amounts of different metals taken up by plants from the exchangeable and readily mobile soil metal pools soils vary markedly.

Evidence of heavy metal injury at the whole plant level includes leaf chlorosis, necrotic lesions or spots and leaf abscission. At the plant cellular level, toxic metal ions block the action of important molecules (e.g. enzymes and polynucleotides), interfere with ion transport, displace essential ions within larger molecules (such as magnesium from chlorophyll), denature or inactivate proteins, and disrupt cell membranes or organelles (Azevedo and Rodriguez, 2012). Salt damage from de-icing salts (sodium chloride) is often diagnosed by browning of the foliage at the base of shrubs and other roadside vegetation, with a gradient of damage leading away from the road (i.e. as the salt concentration decreases in line with distance from the road).

A variety of different salts are used for de-icing roads across the globe, although sodium chloride tends to be the most common in Europe. Different salts have different plant toxicities; for example, Ke *et al.* (2019) suggested that sodium chloride was less damaging to grass than ethylene glycol, urea and sodium formate but more injurious than calcium magnesium acetate, magnesium chloride and calcium chloride. Concentration and length of exposure is important of course, and misuse can cause significant plant damage (Czerniawska-Kusza *et al.*, 2004). In the spring of 2005, the use of de-icing salt in the urban area of Beijing was linked to the ultimate death of 11,000 street trees, 1.5 million shrubs and 200,000 m^2 of lawn, resulting in the economic losses of more than 30 million Yuan (Ke *et al.*, 2019). Prolonged or frequent applications can cause the salt concentration of groundwater to increase which induces physiological drought in the plants (the roots cannot extract water from the saline solution), and the prevalence of sodium and chloride ions disrupts the absorption of key plant nutrients, such as potassium, as well as directly damaging cell membranes when absorbed in high concentrations.

Road and building construction and the introduction of sewage, gas and communication pipes and cabling physically damage tree roots as trenches are cut through the soil to lay infrastructure or prepare road/building foundations. Tree root systems are generally shallow and widespread (Day *et al.*, 2010), and trenching decreases tree health and reduces longevity compared to trees grown on undisturbed natural soils. Construction/repair of infrastructure often severs the main arterial 'primary' tree roots, resulting in loss of water and nutrient uptake and inducing localised die-back in the tree canopy. Even when trenching is not employed just the physical presence of a building can restrict root development or alter behaviour (Watson *et al.*, 2014). This not only has a detrimental effect on the tree but also causes the roots to damage the building foundations, or associated infrastructure, such as utility pipework.

Light (irradiance)

Urban plants need irradiance to capture energy and grow (photosynthesis) and to control developmental processes (photomorphogenesis). Intensity of radiation (based on the number of photons of light present or more simply the brightness of light), quality of light (radiation of set wavelengths) and duration or frequency of irradiance (for example, the number of sunlight hours in a given day) all affect plant development. Photosynthesis is associated with the absorption of light between 400 and 700 nanometres (nm) (roughly the equivalent of the visible light spectrum humans perceive). This is known as photosynthetically active radiation or PAR. In reality, for most plants photon energy capture is optimised by their chlorophyll molecules at 430–450 nm (blue light) and 640–660 nm (red light). Accessory pigments like carotenes and xanthophylls can absorb other wavelengths, including some green light. The amount of energy absorbed by plants is referred to as the PAR irradiance: this measures the radiative energy received by unit surface area in unit time, expressed as watts per square metre ($W\ m^{-2}$). PAR availability for urban plants is determined by factors such as time of day, season, latitude, cloud cover, the presence of other light sources, shade from buildings/other plants and air pollution. An allied term is photosynthetic photon flux density (PPFD) and this is a measure of the number of photons in the PAR range that reach a specific surface area per unit of time, expressed in micromoles per square metre per second ($\mu mol\ m^{-2}\ s^{-1}$). The rate of photosynthesis is directly influenced by light intensity, with higher intensities generally leading to increased photosynthetic activity, although very high irradiance intensities (and other stress factors) can impair photosynthesis through a process known as photoinhibition. Some species of plants are adapted to high solar radiation conditions, such as desert perennials, whereas others are shade adapted, for example, those adapted to growing in the understorey of a woodland or rainforest. Environmental horticulturists exploit these traits, so desert or steppe perennials may be used on open, exposed 'high-light' green roofs, e.g. *Perovskia atriplicifolia, Salvia nemorosa, Echinops* spp., and *Festuca glauca*, yet woodland plants such as *Pachysandra terminalis*, *Brunnera*, *Hosta* and *Corydalis* spp. might be used on a north-facing green wall (i.e. predominately shaded in the northern hemisphere).

Irradiance also affects urban plants by regulating development. Plants grown under low light intensities may stretch upwards towards the light, resulting in long internodes and relatively small leaves – this 'etiolation' being a consequence of the low red to far-red light ratio the plants' photoreceptors are receiving. Tall, straggly plants are not an optimal growth form – but are a requirement if the plant is to reach enough light to survive at all. Irradiance quality controls other plant development responses such as seed germination. Daylength (photoperiod) is important in flower bud initiation and development, bud dormancy induction, cold tolerance acclimation and leaf abscission. There is some evidence that street lamps in cities affect aspects of photoperiodic response, such as time of autumnal leaf colouration, leaf fall and dormancy induction (Piccolo *et al*., 2023) in certain tree species. Czaja and Kołton (2022) showed that light pollution within the urban environment affected sugar metabolism in spring, with woody plants breaking bud earlier and coming into full leaf before other plants kept in normal, dark-night conditions (Fig. 2.8). For some species, e.g. *Tilia* and *Kerria*, this could be as much as 2–3 weeks earlier than normal. Early leaf development can make the plant more susceptible to inclement weather, such as frost, but also puts it out of sync with other natural rhythms, such as the timing of aphid proliferation: an important food source for insectivorous birds. Exposure to artificial night lights has been linked to delays in the time by which leaves turn colour in the autumn (by 6.0 ± 11.9 days according to Meng *et al*., 2022) (Fig. 2.9). These authors also noted that the magnitude of phenological changes was significantly correlated with the intensity of night light pollution.

Temperature

Plants have optimum growing temperatures based on their natural ecological niche, so plants from the

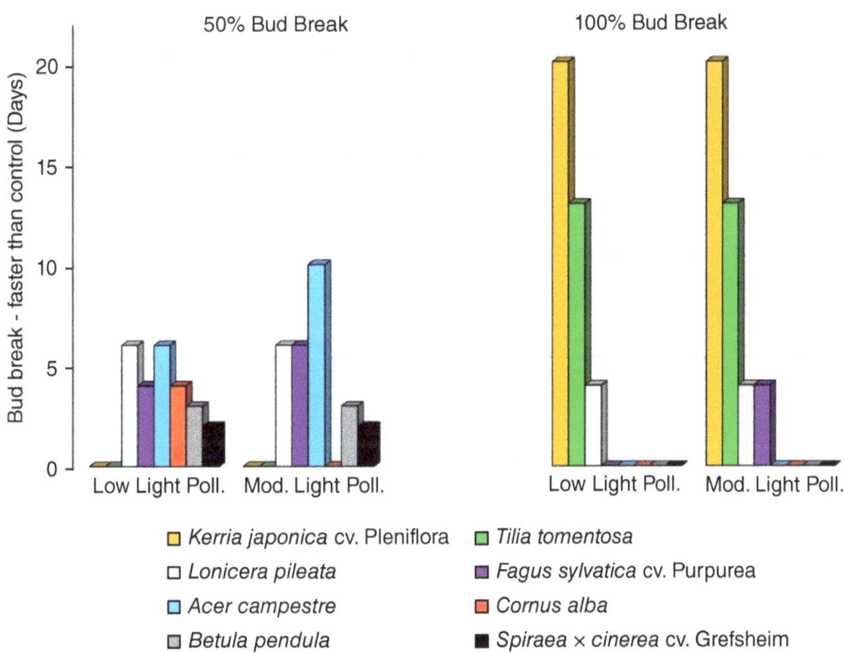

Fig. 2.8. The effects of artificial night illumination from street lamps (Mod., low or moderate; Poll., light pollution) on bud break in a range of woody plant taxa. Bars represent the numbers of days that bud break and leaf development is advanced compared to control plants kept completely dark at night. Bars on the left represent days faster for 50% of the total buds to break and on the right for 100% of buds to break. (Modified with permission from Czaja and Kołton, 2022.) CC BY.

Fig. 2.9. Artificial irradiance at night can interfere with timing of autumn leaf colouration and abscission, with leaves hanging on longer than normal.

tropics tend to have higher optimum temperatures than those from cooler regions. Thus, within cities in temperate or cold regions, tropical plants would be grown indoors (see Chapter 11) or within heated glasshouses. Most plants found outdoors in temperate climates will have an optimal temperate growing range of approximately 21–27°C day and 15–20°C night temperatures. However, some plants, for example a number of 'cool-season' grasses, may start growing at temperatures as low as 2–3°C (Bartholomew and Williams, 2005). Age and growth stage of the plant affect response to temperatures, with seeds of many deciduous woody species requiring chilling periods before germination can take place (to remove endogenous seed dormancy). Taxa vary in their requirements as each species has enzymes optimized for specific temperature ranges (based on their original habitats), affecting photosynthesis and other crucial processes. Many temperate species actually alter their optimal developmental temperatures between seasons, this being conducted in tandem with their cold acclimation (hardening) processes between summer and winter. Thus a plant that survives a –15°C frost in December (mid-winter), would be killed outright if it should be so unlucky as to experience a frost in June (mid-summer). Its 'programming' for summer metabolisms would not be able to cope with lower temperatures and ice formation.

Urban city centres can be on average 1–3°C warmer than surrounding rural areas, with differences at night being as much as 12°C warmer, due to the thermodynamics of the city (e.g. heat being released slowly in the evening from buildings and other hard infrastructure). This is the urban heat island effect. The predominance of concrete, tarmac, bricks, etc. that absorb solar radiation and re-emit it as heat also means that plants in cities can be exposed to strong surface heat sources during the day.

Although plants can become stressed by high temperatures impairing transpiration and causing localised water shortages within tissues, direct heat stress (heat shock) tends to become most apparent at air temperatures >40°C. Membrane systems are particularly vulnerable, and damage to photosystem II in the chloroplast is often an early indication of heat stress damage. In such scenarios, injury is greater with longer duration to the excessive heat, although pre-adaption (acclimation to heat) of plants – by growing them at warmer ambient temperature before a heat stress event – aids tolerance against the stress. Data from Way and Sage (2008) indicated that warmer growing temperatures (30°C during the day) conferred some heat tolerance in *Picea mariana* (black spruce) compared to plants kept at 22°C (Fig. 2.10).

Urban heat islands are accentuated by larger conurbations with greater population densities, a high proportion of built grey infrastructure compared to green/blue infrastructure (plants and water), the presence of industrialised zones that also emit heat from machinery/chemical activity, heavier and more congested traffic flows, and a large proportion of buildings with air conditioning units, as well as the location of the city (e.g. the extent it is exposed to prevalent cooling winds).

2.3 A Changing Climate

As with many other aspects of modern life, climate change poses a significant challenge to urban plantings. It has been cited by the naturalist, David Attenborough, as 'the biggest threat to security that modern humans have ever faced' (United Nations,

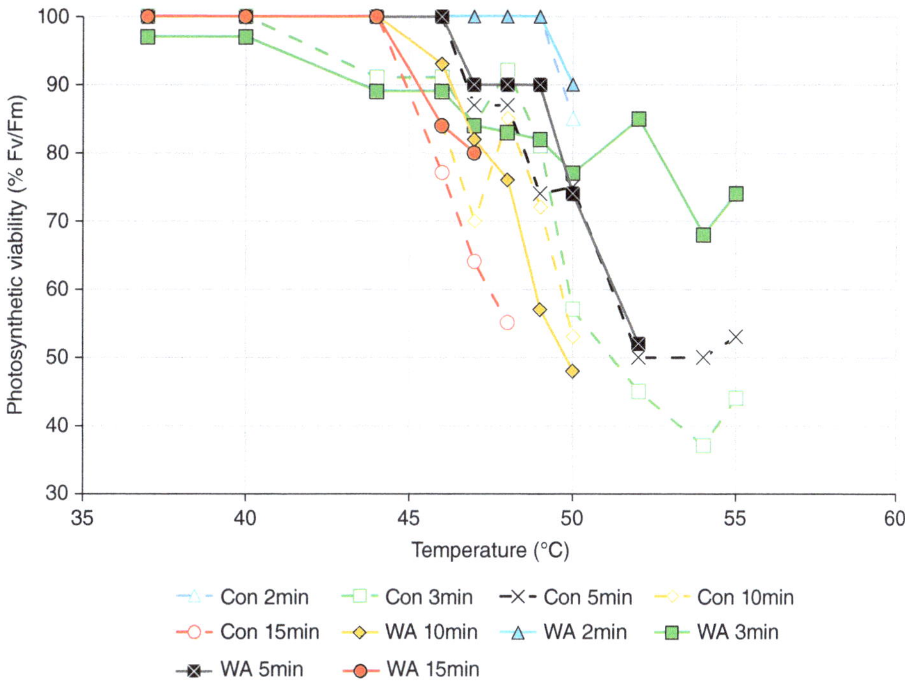

Fig. 2.10. Loss of photochemical efficiency of photosystem II (Fv/Fm) after an acute heat stress on the needles of *Picea mariana* (black spruce). Min = the time needles were exposed to different temperatures (in minutes). Needles were derived from Con = control plants grown at 22°C/16°C day/night regime or WA = warm-acclimated plants grown at 30°C/24°C day/night regime normal temperatures. NB: Needles were not exposed to all possible temperature/duration combinations. (Modified with permission from Way and Sage, 2008.)

2021). The threats to urban plants are threefold: (i) the altered climate may make it difficult for native and traditionally planted species to survive (and thus theoretically in some locations *any* plants to survive) through climate impacts such as heat stress, drought and flooding, (ii) biological interactions, such as new pests and pathogens, may undermine plant species that have little natural tolerance to them, and (iii) pressure on human survival (lack of food, political instability, insecure water resources, etc.) may mean non-edible plants are given little priority or space in the urban environment (Kemp, 2023).

Environmental horticulturists face a dilemma – just what sorts of scenario are landscape plants being asked to address and cope with in the future? This largely depends on the rate of warming, future socio-political action/inaction on carbon emissions and how far ahead we are trying to plan for. There are also as yet unknown uncertainties. A stable climate seems unlikely with most models predicting more extreme weather events, including increasing oscillation between wet and dry regimes, i.e. paradoxically, both more frequent flooding and more frequent droughts for many locations. In the short to medium term, planting plant taxa that have enhanced tolerance to abiotic stress or come from warmer/drier climatic zones would seem prudent. Ultimately, though, some locations may themselves become almost unviable for certain types of plant to grow. It is hard to imagine a tree species that tolerates 40°C for prolonged periods, 2 months of drought, followed by a further month of waterlogging, whilst also experiencing 'autumnal' storms with 200 km h^{-1} winds. It is the intensity, regularity and unpredictability of severe weather events that proves most challenging for a living organism that is fundamentally 'anchored' to the one location. In reality, the earth has experienced hotter, drier, stormier conditions with higher levels of atmospheric CO_2 in the past, but it is difficult to ascertain if the planet's current flora can cope with these conditions. Given time, it is likely plants would adapt, but the current anthropogenically driven

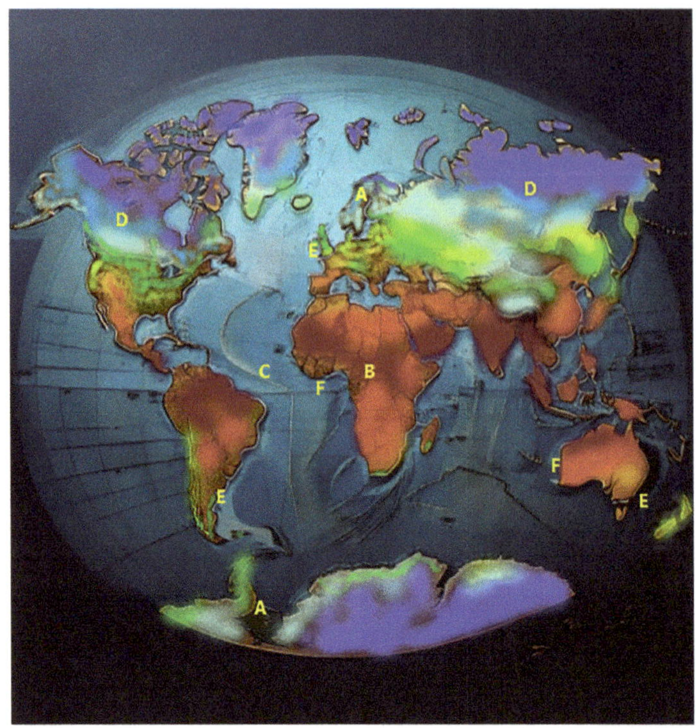

Fig. 2.11. Future scenarios based on climate change depend on future greenhouse gas emissions and the political will to curb these. The scenario above is based on the Intergovernmental Panel on Climate Change (IPCC) data for RCP 8.5 in the year 2100 showing mean surface air temperatures. Predicted average mean temperature increases are superimposed on current mean temperatures. Mean temperatures 2100: red and brown = 30 to 40°C, orange and yellow = 20 to 30°C, greens 10 to 20°C, pale blue = 0 to 10°C and dark blue <0°C. These are mean temperatures, and in the hottest regions maximum temperatures start to exceed those for human thermal comfort and typical plant tolerance ranges. No one can predict the future, but a number of plausible scenarios are being envisaged. These include the following: A – New, densely populated cities have developed around the Arctic and Antarctic regions. These have been facilitated by the loss of ice and natural plant colonisation in these zones. Temperate trees and other plants are used in these cities. B – Large parts of the equatorial region (red) are uninhabitable due to air temperatures exceeding 40°C regularly. These areas are now dominated by solar and wind farms to generate energy that is conducted to the cities in the far north and south. C – Some models suggest algal blooms develop in the tropical oceans, depleting the water of oxygen, when they decay – essentially depleting much of the existing marine life in this zone. D – Human food crops have 'migrated' north – with the 'Wheat Belt' now located in central Canada and Siberia. E – What were once temperate cities like London and Paris are now at the edge of deserts – urban landscapes use drought-adapted xeric plant taxa. F – Some models suggest tropical heat-adapted forests develop in West Africa and Western Australia – supported by relatively cooler ocean currents around the coasts. It is not clear how high temperatures and high humidity affect animal life in these though. (Author's own using Adobe Firefly.)

climate change is likely to alter conditions so quickly that adaptive evolutionary process may not have time to raise the required tolerance levels sufficiently.

Based on current CO_2 and other greenhouse gas emissions (the Representative Concentration Pathway [RCP] 8.5 scenario), large parts of the globe would be considered too warm to support human life by 2100 (Sherwood and Huber, 2010; Mahlstein *et al.*, 2013) (Fig. 2.11). So trees and much other natural and designed vegetation is likely to be lost from such regions. People will migrate towards the polar regions, which by this time will have become temperate climates (although 'temperate' may be somewhat of a misnomer, as these locations may still experience fierce storm events). New cities will develop along the coastal seaboards of what once were the colder polar regions of the planet. These new cities will require green infrastructure to be fully functional. Cities which are now located in temperate zones – e.g. London, Paris, Montreal, Seattle, Melbourne, Wellington and Buenos Aires – will be cities at the edge of desert regions and will rely on much more xeric-based flora.

In the intervening years, climate change will impose pressures and stresses on the existing urban plantings. They will need to cope with rapid transitions in weather events and more extreme conditions. They also may experience additional biotic stress due to invasive pest and pathogen species, many of which will gain a foothold because the flora is stressed by abiotic factors. Researchers are beginning to identify and search for plant species that will cope with warmer and drier conditions, e.g. for north-west Europe this means examining the flora of south-east Europe and the Mediterranean. Cameron, however, argues that plants that possess more generic resilient traits, i.e. those that are essentially quite 'weedy in character', need to be examined too for potential future use (Lewis *et al.*, 2019). These 'weedy' traits include: the capacity to store carbohydrate reserves effectively, spread rapidly via seed and colonise new areas quickly, spend less resources on large flowers or fruit, have widespread natural distributions and/or tolerate a wide range of stress factors. Overall, increasing the diversity of urban plant taxa utilised within our cities will help enhance overall resilience to climate change impacts.

Green landscapes themselves are likely to change to deal with the vagaries of more variable climate. For those regions predicted to get wetter, drainage becomes more important, with slopes and raised beds being relied on to a greater degree for plant communities. Swales and rain gardens are intertwined between buildings to hold and slow excess rainwater. Trees and climbers are used to provide localised cooling to buildings and recreational areas. Garden design is already acknowledging that water may be scarcer in summer, with more reliance on catching and holding water from the wetter winter months (Fig. 2.12).

The UK is one of the countries that may sit on the boundary line between a wetter maritime climate in its north-west region and a drier climate to the south-east, with landscapes exposed to more droughts. Cameron and Clayden (see Webster *et al.*, 2017) depicted how garden landscapes might respond to these climatic alterations in the UK. The scenarios relate to a future date of 2100 and assume certain trajectories (the different Representative Concentration Pathways – RCPs). These scenarios are not designed to predict the future accurately, but rather attempt to demonstrate the trends in garden style that may occur as the climate progressively changes towards the end of the 21st century.

UK West Country garden – wetter and windier

Key points: by 2100, mean temperature 3°C warmer than current conditions; heavy winter precipitation and moderate summer precipitation – storms more frequent and prevalent westerly winds; a frost-free climate with mild, high-humidity winters; periodic, but variable periods of warm dry weather in summer; growing season longer. Modifications may include the following (Fig. 2.13):

Garden features:

- Traditional lawns have been converted into woodland or shrub borders, due to the demands of mowing all year round, and where the frequently wet turf is not conducive to cutting.
- Ditches, rills, swales, pond and rain gardens are introduced – this provides greater accommodation of wetland plant species (1).
- Large trees on wet soils are more prone to wind-toppling so there is a transition to smaller species, or larger species are coppiced/pollarded to reduce crown weight (2).
- There is some loss of patios and other sunspots, especially in winter.
- Slippery paths are more of a problem and wooden features have reduced longevity due to higher humidity.
- Green roofs are widely adopted to minimise run-off rates from built structures. Roofs accommodate a range of wet montane plant species, not just drought-adapted ones (3).
- Gravel raised beds are used above heavy soils to allow traditional favourite garden plants to grow in the wetter climate, e.g. *Allium*, *Dianthus* (pinks), *Geranium* (cranesbill), *Pelargonium*, *Symphyotrichum* (aster) and *Tulipa* (tulips) (4).
- Gardens on slopes are frequently terraced to stop soil erosion (5).
- Fashion for 'natural' water features, such as ponds and streams increases, encouraged by legislation to reduce the rate of runoff water from land after heavy precipitation (6).

Fig. 2.12. How a domestic garden might deal with the impacts of climate change. Top: A – Rainwater is caught in a canvas canopy keeping the patio area dry. B – There is a small aperture at the lowest point of the canopy and water runs down a chain and enters water storage tanks placed below the patio. C – Here water is stored and can be used for garden irrigation in summer. D – Overflow water from the tanks feeds an ornamental pond. E – When this itself overflows it feeds another pond and then is diverted across the garden path. Bottom: F – Water then enters a series of planted rain gardens. G – Berms are used to ensure each section is filled with water before the overflow starts filling the next lower level. H – Free-draining gravel supports the movement and ingress of water, whilst also providing a medium for the plants. Images are of the Royal Horticultural Society, Climate Change Garden, Chatsworth Flower Show 2017 designed by Clayden and Cameron.

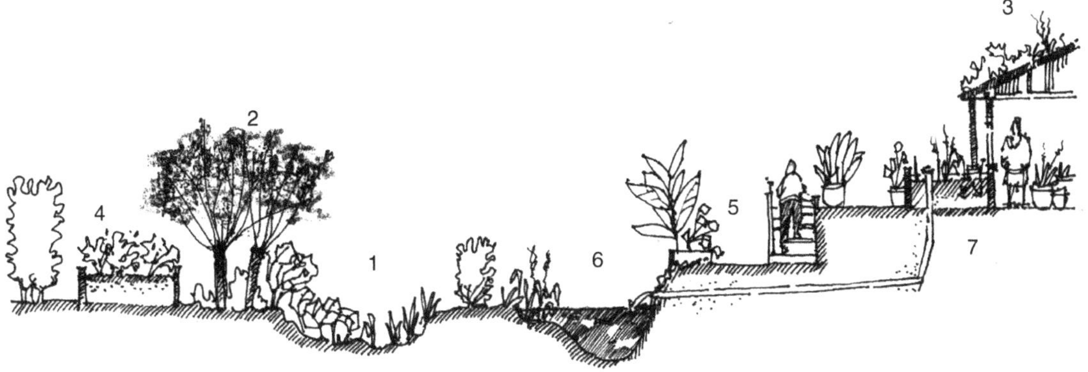

Fig. 2.13. Scenarios for UK West Country garden. Numbers relate to text on p.25 and below. (Used with permission from Clayden as depicted in Webster et al., 2017.)

- Gutters and downpipes feed into planted storm water retainers. Non-native 'tropical' fish species such as *Tanichthys albonubes* (white cloud minnow) and *Poecilia reticulata* (guppy) are used to control mosquito larvae in standing water features (7).

Plantings:

- There is greater pathogen pressure. Many existing tree and shrub species are prone to more fungal and bacteria pathogens, e.g. *Phytopthora* and *Pythium* spp.
- Lichens and moss are a more common and distinctive feature on garden plants.
- *Camellia*, *Escallonia*, *Fuchsia* and other 'temperate' species become dominant hedge and boundary plants.
- Groves of *Acacia* (wattle), *Embothrium* (Chilean firebush), *Eucalyptus* and similarly adapted species have become commonplace where gardens and other landscapes provide sufficient space and shelter from the wind.
- Shelter belts are used to protect large-leaved, semi-tropical species from wind tear. This includes *Brugmansia* (angel's trumpet) *Canna, Colocasia esculenta*, *Hedychium* (gingers), *Musa basjoo* (Japanese banana), but also *Ensete ventricosum* 'Maurelii' (Abyssinian banana).
- Pot-grown *Monstera delicosa* (Swiss cheese plant) are moved indoors in winter to protect them from storm damage.
- *Crocosmia* spp. (montbretia) are now considered a landscape weed.
- Apart from the most disease-resistant rose cultivars, many cultivars of rose have disappeared due to black spot disease (*Diplocarpon rosae*).
- It is too wet for dessert apple cultivars and these have given way to juicing and cider varieties.
- Ornamental cherries (*Prunus*) and crab apples (*Malus*) have lost some of their popularity due to wind damage to blossom and bacterial/fungal pathogens.

UK south-eastern garden – warmer and drier

Key points: by 2100, mean temperature 4°C warmer than current conditions; moderate winter precipitation with limited summer precipitation – soil moisture deficits common in summer; frost uncommon and temperatures rarely below 4°C; growing season longer by 6–8 weeks. Modifications may include the following (Fig. 2.14):

Garden features:

- Some lawns have been converted to dry steppe meadows, with bulb species being used to extend the flowering period. Dry grasses and other stems and seeds heads are used to provide form during mid-late summer after the peak flowering period. Spot plants include *Agave americana*, *Eremurus* (foxtail lily) and *Echium* spp. (1).
- In other gardens the lawn has also disappeared, to be replaced by gravel beds and hardy 'cornfield' annuals; the latter providing peak flowering in May – June.
- Downpipes from the house roof are connected to an underground tank to store rainwater runoff that can be used in the garden during summer (2).
- Trees are planted with a perforated watering pipe penetrating into the rootball to provide deep watering during hot periods. Although this

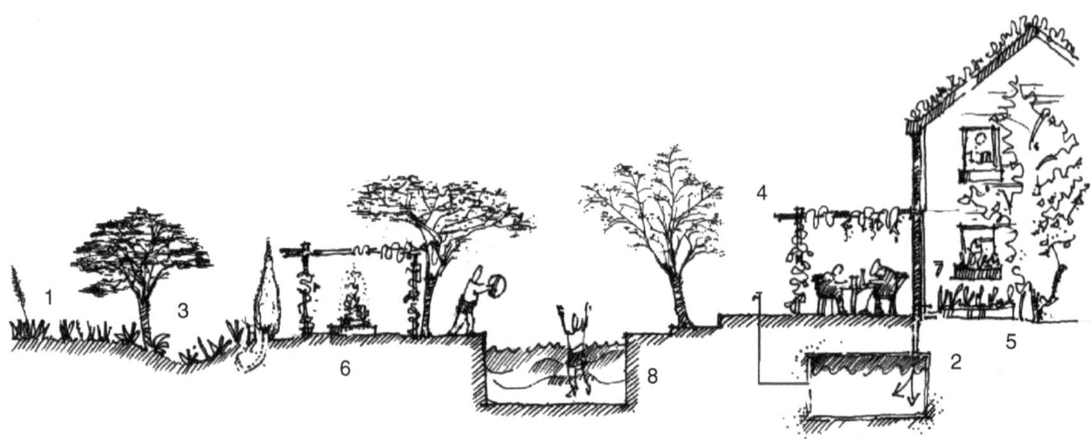

Fig. 2.14. Scenarios for UK south-eastern garden. Numbers relate to text on p.27 and below. (Used with permission from Clayden as depicted in Webster *et al*., 2017.)

may encourage localised rooting around the pipe, the benefits outweigh the disadvantages.

- Plants bought from nurseries will be preconditioned to drought stress during the production phase (but brought back up to full water status before shipping to the sales bench). Drying down the nursery plants in a controlled way preconditions plants to subsequent drought phases and aids survival after planting in the garden, the establishment phases being the most vulnerable time for many ornamental plants.
- Gardens are screened to the south and west to provide mid-day shade. Species such as *Citrus*, *Cercis siliquastrum* (Judas tree), *Prunus dulcis* (almond), *P. persica* (peach) and *Olea europea* (olive) are commonly used. In some gardens these are located in shallow scrapes of which recycled grey water from the local authority communal tank is added when required (3).
- Patios are used more frequently and have become important hubs for the social activities of the household. Shade is provided by species such as *Catalpa* (Indian bean tree), *Koelreuteria* and *Paulownia* (foxglove tree) where moisture availability allows (4).
- Wall climbers are planted around the air conditioning units of houses to improve their cooling efficiency further. A mixed collection of *Bougainvillea, Ipomea learii, Jasminum* (jasmine) and *Wisteria* provide both shade and evapotranspirational cooling to the unit. These are again linked to irrigation systems using recycled water to ensure plants remain adequately watered during heat waves and continue to provide a cooling service to the house (5).
- Shallow swales and depressions are used to recreate 'dry' riverbed landscapes, with prominent use of gravel, stones and 'driftwood'. Planting is infrequent but used to good effect to promote form (e.g. grasses such as *Miscanthus sinensis*) and brief interludes of colour via flowers (e.g. *Eschscholzia californica,* [Californian poppy] and *Osteospermum* spp. [Cape daisy]).
- Although water will be a challenging feature to manage due to high evaporation rates, rills, small pools, bubble fountains, etc. become increasing important to provide relief from the heat and dryness experienced in summer. Windbreaks and sheltering walls reduce the amount of water lost to wind drift and evaporation from the water feature (6).
- Colourful patio plantings and window boxes are located close to the house to help ease of watering. Plants currently semi-tender become mainstream. *Bidens*, for example, flowers all year around (7).
- Borders and beds are more frequently mulched to reduce soil temperatures and inhibit evaporation of moisture from the soil. Mulch will include organic materials, e.g. pine bark, but also inorganic materials such as pebbles, crushed glass, etc.
- Shade becomes a more important element in the garden to help mitigate effects due to dry soil (although there will be less cloud cover – solar irradiance itself will not increase). Nevertheless, drought-sensitive species will be given some

respite by providing shade/semi-shade especially during the middle of the day.

- Where water supplies allow, a number of gardens accommodate a natural swimming pool to capitalise on the warmer summers now prevalent (8).

Plantings:

- Higher temperatures increase the range of plants that can be grown, although this is somewhat confounded by greater incidence of soil moisture deficit in summer. Evergreen sclerophyllous species (e.g. *Arbutus* spp. [strawberry tree], *Ceanothus* [Californian lilac], *Eucalyptus* spp., *Laurus nobilis* and *Quercus ilex*) and smaller or pinnate leaved-species (*Acacia dealbata* [mimosa], *Albizzia julibrissin* [Persian silk tree], *Cytisus battandieri*, *Leptospermum* spp. and *Tamarix* spp.) as well as drought-tolerant pines (e.g. *Pinus aleppo*, *P. haleppensis*, *P. pinaster* and *P. pinea*) and palms (e.g. *Phoenix canariensis* and *Trachycarpus fortunei*) become more common.
- Summer soil moisture deficits mean that conventional garden shrubs such as *Cotinus* (smoke bush), *Cotoneaster*, *Photinia* and *Syringa* (lilac) have shorter internodes and tend to be smaller and more compact plants than is currently the case. Even climbers such as *Clematis*, *Wisteria* and *Vitis* (grape) are less vigorous and have reduced shoot extension. This compaction, however, results in a more intense flower display, as the flowering nodes are not spaced so far apart. Heat though reduces the time each bloom lasts.
- The drier climate not only reduces growth of plants but increases the longevity of species adapted to such climates, such as Mediterranean shrubs/sub-shrubs (*Ceanothus*, *Cistus*, *Fremontodendron*, *Lavandula* and *Salvia* spp.) as growth is more in line with quiescent/dormancy phases, and there is less pathogen pressure brought on by damp weather.
- On light soils, mixed borders have lost their hybrid tea roses in favour of species roses and drought-tolerant climbers such as *Rosa* 'Veilchenblau', 'Alberic Barbier' and 'American Pillar'. Border *Dahlia* have been replaced by *Osteospermum*. Grey and evergreen foliage becomes more prominent in mixed borders as greater reliance falls on Mediterranean shrubs and sub-shrubs.
- Herbaceous perennial borders still occur, but the range of species tends to omit those that require good levels of moisture such as *Aconitum* (monkshood), *Astilbe*, *Dicentra* (bleeding heart), *Dodecatheon*, *Filipendula* (meadowsweet), *Galega* (goat's rue), *Heuchera*, *Hosta*, *Primula*, *Pulmonaria* and *Trollius* (globeflower).
- Species such as *Ceratostigma plumbaginoides*, *Cotoneaster horizontalis*, *Hypericum olympicum*, *Salvia rosmarinus* 'Corsican Blue' (rosemary), *Stachys byzantina* (lamb's ear) and *Vinca major* (periwinkle) remain reliable ground coverers, but others such as *Ajuga* (bugle), *Bergenia* (elephant's ear), *Polygonatum* spp. and dwarf bamboos become less common.
- Rock gardens have lost many of their European and Asian alpine species, but now host xerophytes from Australia, North America and South Africa.

UK north-eastern garden – more variable weather

Key points: by 2100, temperatures 2°C warmer than current, winter mean rainfall increase, summer mean rainfall similar to current; more extreme weather events, with moderate risk of significant frost due to a weakened Atlantic meridional overturning circulation (AMOC), greater frequency of heavy precipitation events and storm surges and more frequent occurrence of summer soil moisture deficits and heat waves; oscillations and rapid transitions between different weather events become more common. Modifications may include the following (Fig. 2.15):

Garden features:

- Garden features in general are similar to current. Gardeners still desire open areas of lawn, tree and shrub cover and interest provided by flower beds and borders.
- Gardeners are more challenged in the choice of species and cultivars that will cope with the potential extremes in temperature (–15°C in winter to +40°C in summer), wetness and dryness as well as more rapid oscillations and cycling of different weather events (e.g. drought followed by flood).
- Some garden features are designed to deal with extremes of climate, for example raised beds with free-draining soil are now provided with a network of seep irrigation hose to provide water during warm, dry periods (1).
- Due to the potential of heavy precipitation, gardens are designed to cope with rapid runoff, with drains and catchment areas used to divert water away from built structures and vegetation

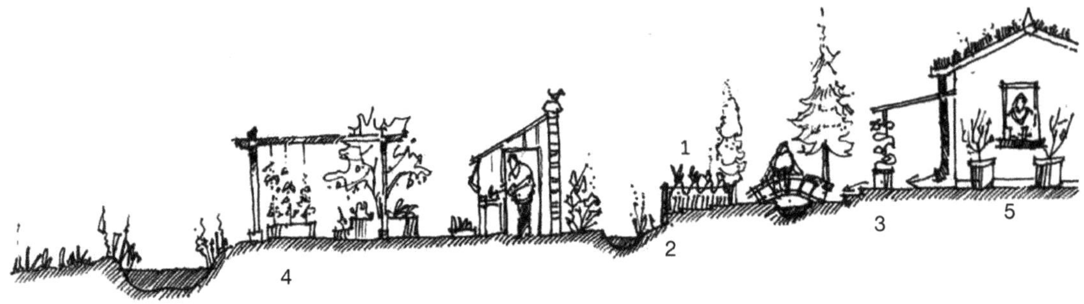

Fig. 2.15. Scenarios for UK north-eastern Scottish garden. Numbers relate to text on p.29 and below. (Used with permission from Clayden as depicted in Webster *et al.*, 2017.)

used to slow the rate of water released into the drainage system (2).

- Flood and storm water surges will leave deposits of silt (and potentially salt) on the soil and on plant foliage. Paradoxically, once flood waters ebb away, the garden may need to be flushed with fresh water to remove or dilute these deposits.
- Gardening under protection or semi-protection becomes more in vogue, although this is tempered by greater risk of structural damage to features such as glasshouses, polythene tunnels and conservatories. Less robust species may be located next to walls and shelter belts, under pergolas or by the house where the physical structures give a degree of protection against frost and strong winds (3).
- Fruit and vegetables are grown by walls, raised beds and along sheltered paths, alleyways and courtyards (4).
- More plants are grown in containers to allow movement between protected and less protected areas. For example, some species are moved into glasshouses in winter, patio in spring for peak flowering display, and then under a shady arbour in mid-summer (5).

Plantings:

- Plants are used that possess a high degree of resilience to stress in general and have a natural range that covers a wide geographical or habitat range. This includes current common taxa such as *Amelanchier, Berberis* (barberry), *Buddleia* (butterfly bush), *Cotinus* (smoke bush), *Cotoneaster, Forsythia, Hypericum, Lavatera* (mallow), *Magnolia stellata, Philadelphus* (mock orange), *Potentilla, Ribes* (flowering currant), *Sambucus* (elder), *Spirea* spp., *Viburnum tinus* and *V. opulus*, as well as climbers such as *Clematis montana, Hedera* (ivy), *Lonicera* (honeysuckle), *Parthenocissus* spp. and *Solanum crispum.*
- Garden trees and those used in the wider landscape tend to be robust species, with good levels of cold and heat tolerance (e.g. *Abies, Acer pseudoplatanus* [sycamore], *A. platanoides* [Norway maple], *Cedrus atlantica* [cedar], *Laburnum, Picea, Pinus sylvestris* [Scots pine], *P. nigra, P. mugo* and cvs., *Platanus* × *acerifolia, Robinia* spp. (where protected from the wind) and *Tilia* (lime).
- Resilient herbaceous taxa are employed including *Acanthus, Aquilegia, Aruncus, Eupatorium, Euphorbia, Filipendula, Geranium, Helleborus, Hemerocallis, Hesperis, Leucanthemum, Lychnis, Papaver, Persicaria, Salvia, Solidago and Stachys* spp.
- Due to the increased variability in climate, there may be a trend towards short-term plantings, so annual plants and short-lived patio plants remain popular, despite their relatively intensive management requirements (annual planting, or sowing, irrigation and nutrition). Therefore, *Antirrhinum* (snapdragon) *Bellis* (daisy), *Calendula* (marigold), *Polyanthus* and *Viola* (pansy and viola) retain their popularity.

Conclusions

- Landscape plants require the same elements to survive as most other plants, This incudes access to light (irradiance), a regular supply of water, appropriate temperature regimes (for growth, but also potentially for other physiological processes, e.g. dormancy induction and release) and nutrients from the soil.

- Most challenges for plant growth and development come from urban environments. These are heterogeneous in nature but can be difficult for plant growth, due to compacted, poorly-structured soils; urban pollution; vandalism; lack of root volumes and little ability to capture or hold rainwater; a lack of symbiotic organisms such as mycorrhizae; injury to shoots and roots due to competition for space with human infrastructures; excessively high temperatures and wind speeds (where buildings funnel the wind); and simply being removed to make way for buildings and roads.
- For urban trees, a lack of available space for shoots, but particularly for roots, constrains growth and can reduce the longevity of the tree. Trees are better managed if they are provided with designated, large, and ideally connected tree pits. A large permeable aperture around the trunk of the tree aids the ingress of rainwater, helping to avoid drought stress. Effective tree support infrastructures such as these help with storm water management too, reducing the risk from flash flooding in highly built-up neighbourhoods. Irrigation management is a key factor underscoring the success or failure of many urban landscape plants.
- Enhanced temperatures in urban environments (the urban heat island effect) corresponds to a longer growing season (in temperate and cold regions), with many plant taxa flowering earlier, and dropping their leaves later compared to contemporary specimens outside the city. Some city centres are effectively frost-free zones, with a different spectrum of landscape plants available to grow compared to nearby rural areas. For temperate cities, the value of plants and urban green spaces to keep the city cool and partially mitigate some of the effects of climate change is becoming more widely recognised.
- Climate change is continuing to drive global temperatures up and test plant resilience to atypical and extreme weather patterns. In the interim time period, many cities will see their plant palette change, and ultimately the conditions in some cities may become so extreme that conventional landscape plants fail altogether. Even xeric (drought adapted) species with tolerance to high temperature may struggle to survive when temperatures are consistently over 45–50°C. Over the next few decades, environmental horticulturists will need to be innovative and adaptable if they are to keep cities 'green' especially those located within 40° N–40° S latitudes.

References

Azevedo, R. and Rodriguez, E. (2012) Phytotoxicity of mercury in plants: A review. *Journal of Botany* 2012(1), 848614.

Bartholomew, P.W. and Williams, R.D. (2005) Cool-season grass development response to accumulated temperature under a range of temperature regimes. *Crop Science* 45, 529–534.

Buil, P.A., Renison, D. and Becerra, A.G. (2021) Soil infectivity and *Arbuscular mycorrhizal* fungi communities in four urban green sites in central Argentina. *Urban Forestry & Urban Greening* 64, 127285.

Czaja, M. and Kołton, A. (2022) How light pollution can affect spring development of urban trees and shrubs. *Urban Forestry & Urban Greening* 77, 127753.

Czaja, M., Kołton, A. and Muras, P. (2020) The complex issue of urban trees—Stress factor accumulation and ecological service possibilities. *Forests* 11, 932.

Czerniawska-Kusza, I., Kusza, G. and Duzynski, M. (2004) Effect of deicing salts on urban soils and health status of roadside trees in the Opole region. *Environmental Toxicology* 19, 296–301.

Day, S.D., Wiseman, P.E., Dickinson, S.B. and Harris, J.R. (2010) Tree root ecology in the urban environment and implications for a sustainable rhizosphere. *Arboriculture & Urban Forestry* 36, 193–205.

Dickinson, N.M., Mackay, J.M., Goodman, A. and Putwain, P. (2000) Planting trees on contaminated soils: Issues and guidelines. *Land Contamination & Reclamation* 8, 87–102.

Endreny, T., Sica, F. and Nowak, D. (2020) Tree cover is unevenly distributed across cities globally, with lowest levels near highway pollution sources. *Frontiers in Sustainable Cities* 2, 16.

Fan, J., Zhang, W., Amombo, E., Hu, L., Kjorven, J.O. and Chen, L. (2020) Mechanisms of environmental stress tolerance in turfgrass. *Agronomy* 10, 522.

FLL - Forschungsgesellschaft Landschaftsentwicklung, Landschaftsbau (2010): Empfehlungen für Baumpflanzungen. Teil 2: Standortvorbereitungen für Neupflanzungen, Pflanzgruben und Wurzelraume rweiterung, Bauweisen und Substrate. 2. Ausg., Ausg. 2010. Edited by FLL. Bonn.

Goulty, S.M. (2003) *Heritage Gardens: Care, Conservation, Management.* Routledge, Abingdon, UK.

Jackson, M.B. and Ricard, B. (2003) Physiology, biochemistry and molecular biology of plant root systems subjected to flooding of the soil. In: *Root Ecology*. Springer, Berlin, Heidelberg, pp. 193–213.

Kamaluddin, M. and Zwiazek, J.J. (2001) Metabolic inhibition of root water flow in red-osier dogwood (*Cornus stolonifera*) seedlings. *Journal of Experimental Botany* 52, 739–745.

Ke, G., Zhang, J. and Tian, B. (2019) Evaluation and selection of de-icing salt based on multi-factor. *Materials* 12, 912.

Kemp, L. (2023) Uninhabitable Futures on a Habitable Earth. PERN Cyberseminar. chrome-extension://hbgjioklmpbdmemlmbkfckopochbgjpl/https://www.populationenvironmentresearch.org/pern_files/statements/CyberseminarExpertPaper_Kemp_Uninhabitable_Futures.pdf (accessed 17 August 2024).

Kisvarga, S., Horotán, K., Wani, M.A. and Orlóci, L. (2023) Plant responses to global climate change and urbanization: Implications for sustainable urban landscapes. *Horticulturae* 9, 1051.

Lewis, E., Phoenix, G.K., Alexander, P., David, J. and Cameron, R.W. (2019) Rewilding in the Garden: Are garden hybrid plants (cultivars) less resilient to the effects of hydrological extremes than their parent species? A case study with Primula. *Urban Ecosystems* 22, 841–854.

López, R., Hallat, J., Castro, A., Miras, A. and Burgos, P. (2019) Heavy metal pollution in soils and urban-grown organic vegetables in the province of Sevilla, Spain. *Biological Agriculture & Horticulture* 35, 219–237.

Mahlstein, I., Daniel, J.S. and Solomon, S. (2013) Pace of shifts in climate regions increases with global temperature. *Nature Climate Change* 3, 739–743.

Mench, M., Vangronsveld, J.C.H.M., Lepp, N.W. and Edwards, R. (1998) Physico-chemical aspects and efficiency of trace element immobilization by soil amendments. In: Vangronsveld J. and Cunningham S.D. (eds) *Metal-contaminated Soils: In situ Inactivation and Phytorestoration.* Springer-Verlag and R.G. Landes Company, Georgetown, TX, pp. 151–182.

Meng, L., Zhou, Y., Román, M.O., Stokes, E.C., Wang, Z. *et al.* (2022) Artificial light at night: An underappreciated effect on phenology of deciduous woody plants. *PNAS Nexus* 1, pgac046.

Mullaney, J., Lucke, T. and Trueman, S.J. (2015) A review of benefits and challenges in growing street trees in paved urban environments. *Landscape and Urban Planning* 134, 157–166.

Olivadese, M. and Dindo, M.L. (2022) Historic and contemporary gardens: A humanistic approach to evaluate their role in enhancing cultural, natural and social heritage. *Land* 11, 2214.

O'Riordan, R., Davies, J., Stevens, C., Quinton, J.N. and Boyko, C. (2021) The ecosystem services of urban soils: A review. *Geoderma* 395, 115076.

Piccolo, E.L., Lauria, G., Guidi, L., Remorini, D., Massai, R. and Landi, M. (2023) Shedding light on the effects of LED streetlamps on trees in urban areas: Friends or foes? *Science of the Total Environment* 865, 161200.

Rosenberger, L., Leandro, J., Wood, R., Rötzer, T. and Helmreich, B. (2024) Influence of age, soil volume, and climate change on water availability at urban tree sites. *Sustainable Cities and Society* 113, 105680.

Sherwood, S.C. and Huber, M. (2010) An adaptability limit to climate change due to heat stress. *Proceedings of the National Academy of Sciences* 107, 9552–9555.

Tournaire-Roux, C., Sutka, M., Javot, H., Gout, E., Gerbeau, P., Luu, D.T. and Maurel, C. (2003) Cytosolic pH regulates root water transport during anoxic stress through gating of aquaporins. *Nature* 425, 393–397.

United Nations Press Release (2021) Climate Change 'Biggest Threat Modern Humans Have Ever Faced', World-Renowned Naturalist Tells Security Council, Calls for Greater Global Cooperation. Available at: https://press.un.org/en/2021/sc14445.doc.htm (accessed 5 November 2024).

Watson, G.W., Hewitt, A.M., Custic, M. and Lo, M. (2014) The management of tree root systems in urban and suburban settings II: A review of strategies to mitigate human impacts. *Arboriculture & Urban Forestry* 40, 249–271.

Way, D.A. and Sage, R.F. (2008) Elevated growth temperatures reduce the carbon gain of black spruce (*Picea mariana* [mill.] B.S.P.). *Global Change Biology* 14, 624–636.

Webster, E., Cameron, R. and Culham, A. (2017) *Gardening in a Changing Climate*. Royal Horticultural Society, London.

3 Environmental Horticulture: Benefits and Impacts

Abstract

Environmental horticulture provides many benefits to humans (ecosystem services), particularly, but not exclusively, to those living in towns and cities. Landscape plants have the capacity to cool our cities and buildings in summer and insulate them against cold and wind in winter. They capture the rain, slow its runoff and reduce the risk of flash flooding. Some plants improve local air quality, reduce the amount of chemical contaminants entering water courses, reduce the noise in certain locations, and even make some places seem quieter than they actually are. Plant choice and careful landscape design are critical in optimising these services. Urban plantings come with drawbacks too though (disservices), such as the use of pesticides, excessive use of potable water or adding to pollution issues through synthetic fertilisers. Environmental horticulturists have a duty of care to mitigate these factors where possible.

3.1 Ecosystem Services and Disservices

Environmental horticulture has a key role to play in effective ecosystem service provision, especially within an urban context. The term 'ecosystem service' has been developed to help understand and quantify the benefits derived from natural resources and systems (in essence what benefits nature provides for us). Another allied term has come into the vocabulary in recent years, that of 'nature-based solutions'. These are situations where a specific societal problem is solved largely, but not necessarily exclusively, by the action of ecosystem services. The Millennium Ecosystem Assessment divides ecosystem services into the following:

- Provisioning services: the products obtained from ecosystems including food, fibre, fuel, genetic resources, biochemicals, natural medicines, pharmaceuticals, ornamental resources and fresh water.
- Supporting services: the services that are necessary for the production of all other ecosystem services including water and nutrient cycling, photosynthesis and soil formation.
- Regulating services: the benefits obtained from the regulation of ecosystem processes, including the regulation of air quality, climate, water, erosion, pests and disease and natural hazards.
- Cultural services: the non-material benefits people obtain from ecosystems through spiritual enrichment, cognitive development, reflection, recreation and aesthetic experiences – thereby taking account of landscape values.

An example of horticultural ecosystem service would be that woody plants can trap aerial pollutants such as particulate matter (PM) derived from car exhausts, and an associated nature-based solution might be to surround a children's playground with woody plants to reduce the amount of polluted air that enters from a nearby roadway.

It should be noted too that horticultural activities, however, can cause or contribute to problems (i.e. ecosystem disservices, for example fruit from street trees may rot on streets/pavements and require cleaning up), or their positives roles can be over-emphasised or exaggerated ('green-wash'). This chapter aims to highlight the genuine current and potential future benefits, as well as some of the drawbacks associated with environmental horticulture.

The principal benefits of vegetation in urban areas include the following (Cameron 2023; Cameron and Blanuša, 2016):

- reduction in urban air temperatures helping to mitigate urban heat island effects, and lower the surface temperatures of buildings thereby reducing the reliance on mechanised air conditioning;

 Environmental Horticulture: Science and Management of Green Landscapes (R.W.F. Cameron)
DOI: 10.1079/9781800621763.0003

- screening out aerial particulate matter and improving air quality;
- de-activating chemical compounds and recycling nutrients;
- intercepting precipitation, reducing runoff rates, mitigating flash flooding and removing pollutants from watercourses;
- regulating water flow and providing potential storage of water in wetlands and watersheds;
- acting as a physical barrier to wind;
- screening off visual intrusions;
- reducing noise and seeming to make locations quieter;
- providing habitat and food sources for urban biodiversity (see Chapter 5);
- enhancing health and well-being (see Chapter 4);
- providing food, fibre and medicinal resources for humans;
- aesthetic enhancement and recreation; and
- educational and cultural opportunities.

A number of environmental services and disservices are covered in this chapter.

3.2 Microclimate Modification

The composition of plant communities and rate of plant development is dictated strongly by climate, yet plants themselves can affect climate. In urban areas the provision of vegetation influences the local microclimate, by altering temperature, humidity and wind speed or direction. The precise role of green infrastructure to influence city cooling, reduce energy loads on buildings and improve human thermal comfort has warranted attention over the last two decades, largely driven by concerns over climate change, urban expansion and density and the desire for more environmentally sustainable buildings.

Urban heat islands

Urban areas are typically 2°C warmer than neighbouring rural areas (the urban heat island [UHI] effect), although differences of as much as 7°C are not uncommon under certain weather conditions (Wilby, 2007; Martin-Vide and Moreno-Garcia, 2020). Urban areas are warmer due to buildings and other hard surfaces trapping solar energy (and dissipating it more slowly) as well as heat derived from human-related activities (building heating, air conditioning units, vehicle engines, heat from computers and industrial machinery). Construction materials can be important with light-coloured ones having a higher albedo effect (reflecting more solar energy back to space without local heating). In contrast, dark materials such as tarmac roads and dark bitumen roofs absorb high levels of short-wave radiation but conduct long-wave radiation back (i.e. release more heat). Globally, the urban heat island effect is increasing as urbanisation is causing buildings to increase in number within each city, to become larger and to be more densely packed together (Kim and Brown, 2021). Microclimates across the urban locale vary widely; for example, railway lines are amongst the hottest areas, contributing to local heat island effects over the day, but cooling rapidly in the evening and remaining relatively cool overnight. Conversely, high-rise buildings trap short-wave solar radiation, which is absorbed and radiated out as warm, long-wave radiation thus creating heat pockets during the day, which subsequent develop into nocturnal heat islands. Heat islands form around consistent sources of heat such as major roads with heavy volumes of traffic or where there is a predominance of low-albedo building materials. As buildings themselves act as wind shields, urban areas have typically lower wind speeds than more open countryside; the lack of air movement causing 'canyon effects' where warm air gets trapped at street level. Inner city shops, malls and offices are particularly dependent on mechanical air conditioning to provide interior cooling, but in doing so produce surplus heat externally. When these factors combine, they can cause specific 'hot spots' within the city centre (Chun and Guldmann, 2014). Surface temperatures are more variable (due to the properties of individual materials) and these variations are seen over small scales (Fig. 3.1).

Overall, the compound effect of these factors plays a major role in urban temperature flux and results in significant increases in interior temperatures and human thermal discomfort. Urban heat stress is becoming more prominent with climate change, with a number of the world's mega-cities experiencing >40°C air temperatures for prolonged periods (e.g. Delhi, India in May 2022) or shorter but higher peaks elsewhere (e.g. Syracuse in Italy holds the European record with 48.8°C in August 2021). These effects have direct implications for human health as mortality rates rise during heatwaves, especially in urban areas where there is little respite

Fig. 3.1. Typical variation of surface temperatures as affected by solar radiation and materials (°C at midday).

from the heat at night. Elderly and very young citizens or those with respiratory problems or heart disease are most vulnerable to excessive heat. Effects can be dramatic in countries which are unaccustomed to heatwaves, e.g. Scandinavian countries, which although traditionally have cool climates, are experiencing rapidly warming summers, current ones being >5°C warmer than those for the period 1981–2010 (Parsons, 2019). Indeed, Europe with a largely urban and elderly citizenship accounted for more than a third of global heat-related deaths in the elderly in 2018. The UHI effect compounds other problems too, via increasing energy consumption through mechanized cooling, lowering air quality and increasing urban water use.

Increasing the component of green infrastructure within cities mitigates against heat island effects and is advocated to help moderate the more extreme temperature profiles predicted by climate change and increased urbanisation. Plants provide cooling to their immediate environment but the mechanisms to do so vary with the type of vegetation. Cooling can be provided by direct shading of the ground and other surfaces, evaporation of leaf surface moisture and transpiration of internal water, including that being transferred through the plant's water columns, i.e. ultimately from the soil. Plant groupings also modify air flow and can promote insulating layers of still ('dead') air within a building envelope; their foliage alters the albedo of the land surface (i.e. the amount of irradiance that is reflected back into space). Vegetation that is transpiring is also photosynthesising, i.e. converting light energy into sugars and hence creating biomass. Photosynthetic inefficiencies are such, however, that a proportion of irradiance captured by the leaf can still be lost as heat, e.g. 40–60% depending on plant species and prevailing environmental conditions.

The cooling influence of different forms of green infrastructure has been studied, including urban forests, street trees, parks, lawns and meadows, green roofs, domestic and public gardens and green walls. The addition of trees to urban areas is linked to reducing mean air temperatures by between 0.2 and 2.3°C, with the degree of cooling dependent on number, size and species of tree (Balany *et al.*, 2020; Meili *et al.*, 2021). Human thermal comfort is also enhanced by increasing tree quantity. The physiological equivalent temperature (PET) is a tool to measure how the body experiences outdoor thermal conditions. (It uses the heat budget of the human body under ambient indoor conditions as a reference point.) Plants can reduce PET in high-density urban areas by 2.2–3.4°C, with street trees being reported to reduce PET by 10–14°C in tropical countries. Trees (and other vegetation) reduce wind speed and increase local humidity though, and this can have a negative effect on human comfort (Meili *et al.*, 2021).

Within any given form of green infrastructure, the predominant plant type and interactions with other factors such as soil moisture content are likely to strongly influence the cooling potential. Even relative contributions of these cooling mechanisms will depend on plant form, species, canopy cover, moisture availability, seasonality and plant vigour. Vegetation such as trees also alters the urban climate via different mechanisms. It blocks/absorbs/reflects solar radiation, which both shades and cools urban surfaces. Evapotranspiration (ET) provides localised cooling through latent heat transformations (incoming energy is used to convert liquid water to vapour in or on the surface of the leaf, and subsequently this energy does not raise air temperatures). Evapotranspirational cooling though might be affected by environmental factors, such as water availability to the root or high humidity around a leaf. Of the two key factors, shading is often credited with the greater cooling effects (80% of cooling potential), for trees at least (Tan *et al.*, 2018). All the interactions are not necessarily positive. At larger scales, trees block air flow and alter the aerodynamic roughness of the urban fabric which tends to promote local warming, not cooling.

City cooling may be one of the key justifications for keeping large-stature trees within urban conurbations, despite increasing pressures for space and perceived conflicts with infrastructure. Cooling influences are not restricted to urban trees alone and other forms of vegetation can aid localized cooling, e.g. through planting shrubs, climbers, turf, etc. next to, or on, a building.

Compared to concrete surfaces used as a reference point, grass swards reduced ground surface temperatures by up to 24°C, whereas shade from trees only cooled surfaces by up to 19°C (Armson *et al.*, 2012). Trees, however, were shown to have a more pronounced effect (5–7°C) on air temperatures recorded above the surface of the ground. In those regions where it is difficult to grow full-canopy trees, due to arid soils and high temperatures, Yilmaz *et al.* (2008) argue that the presence of (irrigated) grass is beneficial as an alternative form of green infrastructure. In these more extreme environments, their work demonstrated that grass swards were able to both reduce ground surface temperatures (11.8°C cooler compared to asphalt/concrete) and also air temperatures 2 m above the surface of the grass (7.5°C cooler). So urban plants have significant capacity to cool, but the type chosen and their design in the landscape is important to optimise the effect.

Building insulation, cooling and energy efficiency

Trees and other vegetation significantly alter the microclimate in their locality. Including trees in the urban matrix and considering their location carefully with respect to adjacent buildings can provide a real bonus for the energy management for the aforementioned buildings. Trees through both their shade and evapotranspirative cooling can offset the requirement for artificial air conditioning during warm weather. In warm climates, savings of 40% (USA: Huang et al., 1987); 49% (Italy: Calcerano and Martinellie, 2016); 50% (USA: Meerow and Black, 1993) and 54% (Greece: Tsoka *et al.*, 2021) are cited via empirical and modelling studies. A review by Ko (2018) illustrates energy savings between 2% and 90% depending on climate, method of assessment and the assumptions made about both buildings and trees but overall confirms that the concept is solid. Scaling up data based on a 30–40% saving, Akbari *et al.* (2001) estimates a 20% overall reduction in cooling energy demand within the USA, amounting to over US$10B savings p.a. Similarly, the provision of four carefully positioned ‘shade’ trees per house would result in annual savings of carbon emissions of up to 41,000 tonnes per city (Akbari, 2002). Larger-stature trees and greater numbers of trees that help shade entire buildings promote the energy-saving components (Fig. 3.2). Choosing species that possess denser canopies or transpire high volumes of water aids the cooling effect (Tsoka *et al.*, 2021). Moss *et al.* (2019) found that for London, UK, *Castanea sativa* (sweet chestnut),

Fig. 3.2. The North American culture is one that appreciates the cooling and shading effects of trees on residential properties.

Prunus avium (cherry), *Quercus petraea* (sessile oak), *Platanus* × *hispanica* (London plane) and *Fagus sylvatica* (beech) provided more evapotranspirative cooling than other commonly occurring tree species found in the urban forest.

Although the value of trees in providing localised cooling is now recognized, there are conflicts with enhanced tree planting in urban areas. Urbanisation is encouraging more compact housing (high densification) with less room for large tree species. Compact developments often feature medium- to high-density building footprints, smaller open spaces and buildings located closer to roads and pathways. Where trees are still encouraged, smaller-growing species (e.g. *Malus*, *Prunus*, *Sorbus*, and *Cercis*) are often advocated over the taller, larger-canopy species, resulting in trees growing in parallel to the building rather than providing shade to the roof.

To avoid urban sprawl, planners are encouraging the use of more compact developments, with mixed-use functions, e.g. housing alongside retail or business structures. These trends are likely to continue; urban areas in the conterminous USA having doubled in size between 1969 and 1994 and are set to increase by a further 50% by 2050. More pressure on urban space reduces opportunities for new tree plantings and increased density of build infrastructure constrains both tree roots and canopies, providing insufficient room for healthy canopy development and root growth and creating a greater likelihood of infrastructure conflicts. The close proximity of roots to cables, pipes, building foundations and pathways can lead to structural damage. Similarly, the encouragement of large canopies does not sit comfortably with the desire to maximise irradiance levels for energy generation through roof-mounted solar panels. Allowing the development of solar technology within residential areas, whilst also retaining the cooling effect that trees provide for buildings, will require better designed green infrastructure (i.e. taking greater account of aspect, wind direction and diurnal shading patterns). As solar panels themselves can cause localised microclimate warming, other forms of green infrastructure that do not shade the panels directly but provide nearby cooling are worth investigating (e.g. green roofs/green façades).

Vegetation, and trees in particular, are vulnerable themselves to the effects of climate change, and large trees will not serve a city's cooling needs if there is no water available to ensure their survival. Further research is required as to what species may be suitable in the future as local climates change and also how water reserves can be 're-managed' to ensure some are allocated to support green infrastructure.

Green roofs and green wall microclimates

A green mantle over the soil or a building surface (green walls and green roofs) reduces surface temperatures. In a tropical climate such as Singapore, vegetation provides comparable cooling to painting surface materials either white or with reflective paints (Tan *et al.*, 2021). Green roofs (see Chapter 10) are often advocated due to their capacity to affect the thermal insulation of a building and the surrounding air temperatures. Much of the thermal buffering comes from the substrate – thicker substrates giving greater insulation to the building. Permpituck and Namprakai (2012) demonstrated that a 10 cm deep substrate gave a 59% reduction and a 20 cm deep substrate a 96% reduction in heat transfer compared to a conventional roof. The vegetation has a role to play too though, especially if the plants have access to water to allow for high rates of ET. Modelling approaches suggest 58% of heat loss can be attributed to ET, 31% to long-wave re-radiation and 11% heats the building through the roof (Castleton *et al.*, 2010). Plant shading provides additional cooling. Raji *et al.* (2015) reported that when surface temperatures of an uninsulated roof were 57°C they were 42°C for a roof with bare soil, and only 26.5°C for a green roof, where the plants had a high leaf area index (i.e. a strong shade effect due to multiple leaf layers in the canopy). The type of green roof is important; those with a pale-coloured substrate (e.g. limestone gravel) may reflect more thermal energy away from the building, whereas those with deep substrates and dense plant canopies (e.g. an intensive green roof with larger plants) may provide most of their cooling through ET and shading (Jamei *et al.*, 2021).

Although surface temperatures and the subsequent heat flux into the building are largely affected by solar intensity, air humidity, green roof type, substrate depth substrate moisture content, canopy cover/density and plant choice also dictate thermal properties. Vaz Monteiro *et al.* (2017) considered pale-leaved Mediterranean species such as *Salvia* and *Stachys* to be superior at cooling compared to the succulent species that are often used in extensive green roofs. Across a range of shrub taxa used for roof cooling in the UK, the evergreen *Viburnum tinus* was considered optimal (Love V., 2021, personal communication).

Green walls (Chapter 10) are often cited as providing effective cooling to buildings, although the type of green wall may influence the extent of the cooling (and the amount of resources required to do so). Green walls tend to be divided into two main categories: (i) 'green façades' which are composed of climbing plants (vines) grown in soil at ground level and (ii) 'living wall systems' where the plants are rooted into cellular modules or placed in hydroponic systems (see Chapter 10). Green walls shield buildings from high temperatures by reducing short-wave heat gain and, depending on the system, can provide a layer of 'dead' air as insulation between the wall and the vegetation. Living wall systems in China have been shown to reduce exterior wall temperatures by a maximum of 20.8°C, and the corresponding interior of the wall by 7.7°C. Air temperatures between the wall and vegetation were on average 3.1°C cooler than ambient air. Most studies on green walls focus on wall surface temperatures, with maximum differences between vegetated and non-vegetated cited as 2.7°C (the Netherlands), 8.3°C (Greece), 11.6°C (Singapore), 15.2°C (Spain), 18°C (Japan), 20°C (Italy) and 26°C (UK) (Kunasingam *et al.*, 2024). As with the influence of nearby trees, effective wall cover on buildings is likely to reduce the reliance on mechanised air-conditioning. For example, computer models using *Hedera helix* (English ivy) have suggested solar gain is reduced on a south-facing wall by up to 37%.

The extent vegetation covers a wall dictates the amount of energy saved. Covering a single south-facing wall only saved 6% of cooling energy, whereas covering three walls and letting the foliage increase in depth resulted in a 36% reduction in energy use (Coma *et al.*, 2014; Pérez *et al.*, 2017) – a point reinforced with studies on *Parthenocissus tricuspidata* where maximum summer energy savings of 34% were achieved in Spain (Perini *et al.*, 2017). Using green façades and walls more widely may accentuate the benefits. Modelling city-wide green wall applications for Abu Dhabi (United Arab Emirates) suggested an 8% reduction in air-conditioning energy consumption (Afshari, 2017) and retrofitting Hong Kong's high-rise residential buildings with green walls was calculated to save 2651×10^6 kWh p.a. in cooling (equivalent to 2200×10^6 kg of carbon dioxide, CO_2) (Wong and Baldwin, 2016).

Further studies show maximum temperature differences between vegetated wall and non-vegetated walls varied due to the plant taxa being utilized (Fig. 3.3), although some of the differences were explained by variation in the percentage of canopy cover over the wall, rather than any other plant trait (Kunasingam *et al.*, 2024). Cameron *et al.* (2014) demonstrated that plants could cool walls through both shading and ET, but the proportional contribution of these two factors altered between taxa. Shading provided the greatest cooling influence with taxa such as *Hedera*, *Jasminum* and *Lonicera*. In contrast, *Fuchsia* was strongly reliant on ET, with this accounting for 3°C cooling compared to only 1.5°C associated with shading.

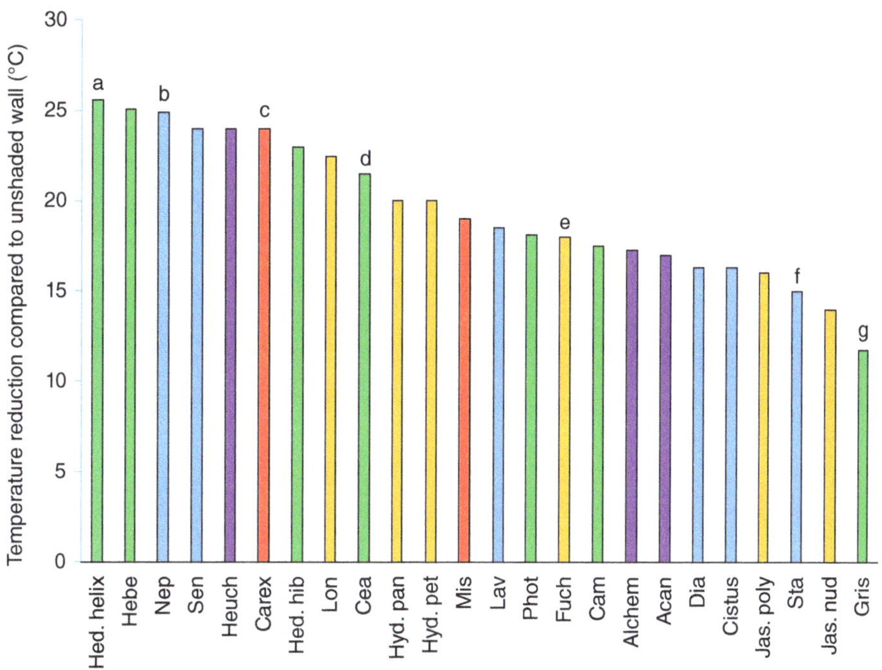

Fig. 3.3. The effects of different plant taxa in reducing wall temperature under warm conditions in the UK (the unshaded control wall was 55°C). Note the lack of pattern based on plant type, e.g. the evergreen *Hedera helix* cooled the wall from 55°C to 30°C, whereas another evergreen with similar dimensions – *Griselinia littoralis* – only cooled it from 55°C to 43°C. Evergreen shrubs/climbers (green): Hed. helix = *Hedera helix,* Hebe = *Hebe* cv. Mrs Winder, Hed. hib = *Hedera hibernica,* Cea = *Ceanothus* cv. Concha, Phot = *Photinia fraseri* cv. Little Red Robin, Cam = *Camellia japonica* cv. Doctor King, Gris = *Griselinia littoralis*. Mediterranean herbs – grey-leaved plants (blue): Nep = *Nepeta racemosa,* Sen = *Senecio candicans* cv. Angel wings, Lav = *Lavandula angustifolia,* Dia = *Dianthus plumarius,* Cistus = *Cistus purpureus,* Sta = *Stachys byzantine*. Herbaceous perennials (purple): Heuch = *Heuchera micrantha* cv. Palace Purple, Alchem = *Alchemilla mollis,* Acan = *Acanthus mollis* cv. Morning Candle. Grass family (red): Carex = *Carex buchananii* cv. Red Rooster, Mis = *Miscanthus sinensis* cv. Flamingo. Deciduous shrubs/climbers (yellow): Lon = *Lonicera japonica* var. *repens,* Hyd. pan = *Hydrangea paniculata* cv. Tardivar, Hyd. pet = *Hydrangea petiolaris,* Fuch = *Fuchsia* cv. Midnight, Jas. poly = *Jasminum polyanthum,* Jas. nud = *Jasminum nudiflorum.* Letters a-g denote significant differences in cooling – representative sample only. (Used with permission from Kunasingam *et al.*, 2024.)

Cooling effects with *Stachys* and *Prunus* were equally attributable to shading and ET. With the silver, pubescent-leaved *Stachys*, greater reflection of irradiance may also have attributed to the cooling influence (i.e. the leaves providing a greater albedo effect than conventional green leaves). Cameron has defined plant taxa that possess certain greater capacities for cooling (or indeed any other ecosystem service) as functional ornamental plants (FOPS). This concept helping the sector identify optimal taxa for each service.

The precise location of plants with respect to an individual building, or the directional aspect of a wall that a green façade is placed upon, strongly determines the extent of any benefit. Planting can restrict the flow of cool air to a wall, particularly in the early and later parts of the day but reducing air flow and trapping shaded (hence cooler) air at later times during the day is desirable (Yoshimi and Altan, 2011). In the northern hemisphere, façades that are south-facing benefit from being planted with deciduous climbers where leaf cover blocks excessive solar irradiance in summer, yet the absence of foliage in winter enables irradiance to penetrate through to the building, thus contributing to heating the building during the coldest months. Conversely, evergreen foliage may be used on a north aspect to provide insulation all year round.

Building insulation in winter

Indeed, green infrastructure has the advantage of influencing a building's energy dynamics in winter as well as summer. Again, appropriate choice of vegetation and careful positioning can reduce energy loss in cold weather conditions. This is accomplished through a number of mechanisms. Vegetation acts as a wind break helping to maintain a positive microclimate surrounding the building through the trapping of warm air. It reduces the speed of air masses moving over a building and thus the rate to which heat is lost. Finally, vegetation slows the velocity of air directly and reduces the effect of cold draughts entering the building and warm air leaking out. Dense planting close to a building such as occurs with climbing perennials and wall shrubs can create dead-air space, reducing air movement in the immediate vicinity of walls (Hutchison and Taylor, 1983). The physical presence of the vegetation and the associated dead-air space provides a barrier which reduces cold air infiltration through cracks and apertures by up to 40% (DeWalle and Heisler, 1980).

Using an *in vivo* model system of brick cuboids Cameron *et al.* (2015) demonstrated a green façade system using *Hedera helix* (Fig. 3.4) could reduce heating energy use by as much as 40% and enhance wall surface temperatures by 3°C. Temperature differences were affected by weather parameters, aspect, diurnal time and canopy density. Largest savings in energy due to vegetation were associated with more extreme weather, such as cold temperatures, strong wind or rain (Fig. 3.5).

Trees around buildings as bespoke shelter belts are also useful (Fig. 3.6). Approximately 8% of

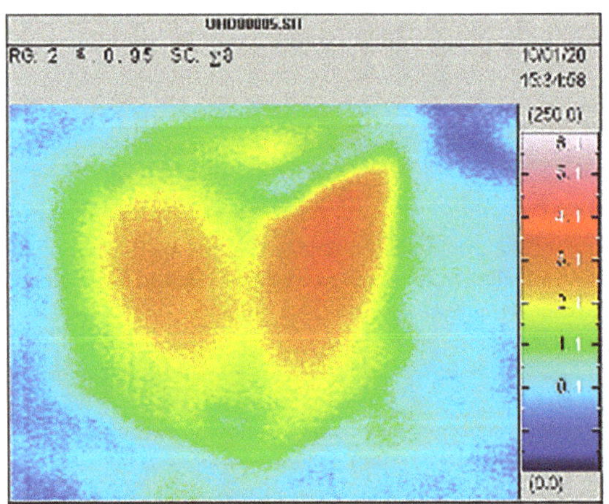

Fig. 3.4. Thermal image of a brick cuboid (left) showing heat energy escaping through the walls and population of brick cuboids (right), with those covered by a green façade showing greater snow retention around the base, due to better insulation.

convective heat loss is preventable with shelter belts composed of trees and hedges located in a manner to directly intercept any particularly strong or cold winds. Strategically placed vegetation, which takes account of prevailing wind directions and solar irradiance, can reduce overall winter energy costs by 17% to 45% (Jo and McPherson, 2001; Liu and Harris, 2008; Rouhollahi *et al.*, 2022). The benefits of shelter belts and other woodland features are most apparent in the winter. Even woodland shelter belts located some distance from the building can have a marked effect, for example, Wang (2006) showed that trees 50 m away reduced energy costs by 4.5% when they were in line to intercept the prevailing winds. In terms of reducing wind (and thus heat energy loss by convection), Li *et al.* (2022) consider a plantation of six trees/shrubs depth as optimal in their wind tunnel tests. For a single shelter belt, a mix of trees and shrubs was best, but if a double shelter system was employed (i.e. a gap employed between two plantations) then more uniform plant types (i.e. all trees or all shrubs) was seen as preferential.

Wind amelioration

In urban open spaces trees and hedges can increase the microclimatic temperature on the leeward side of the wind by 1°C, with potential for further warming on southern aspects where solar irradiance heats pockets of retained air. Hedges and other forms of screening minimise evaporative losses from landscape plants and allied soils/substrates and reduce abrasion to foliage and flowers. Similarly, they help reduce water loss from ponds, fountains and other water features and, hence, the requirement to continually top these up with supplementary water.

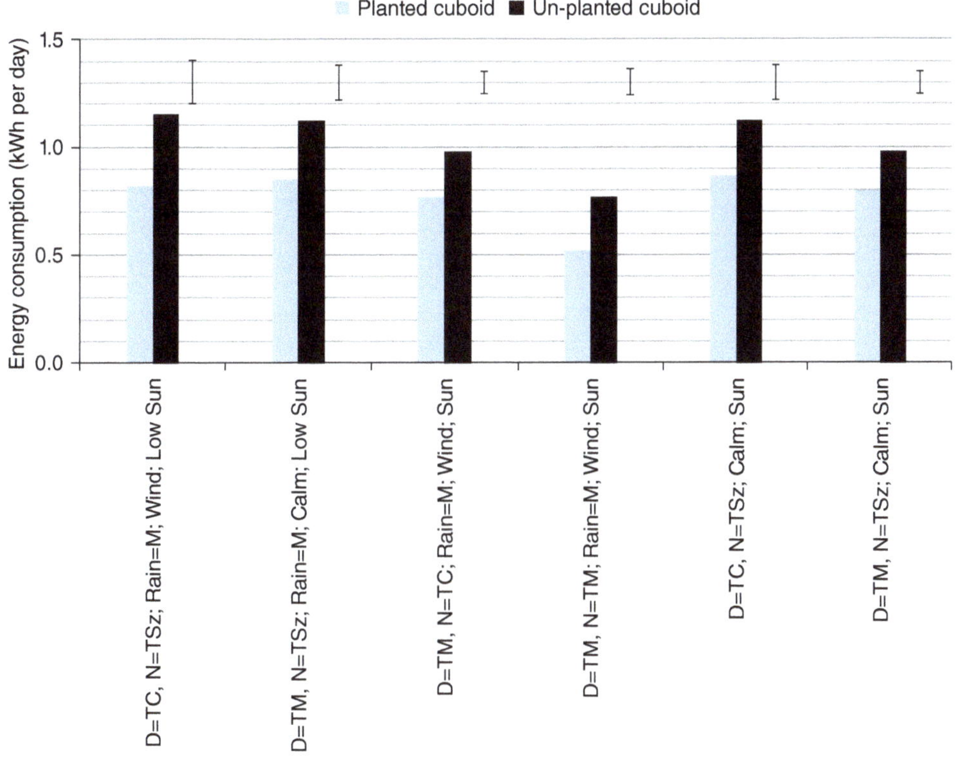

Fig. 3.5. Comparison of diurnal energy consumption per cuboid for selected 24 h periods with associated weather (D = day, N = night, T = temperature, Sz = sub-zero, C = cold, M = moderate). (Used with permission from Cameron *et al.*, 2015.)

Fig. 3.6. Trees can be used to protect buildings from prevailing winds, reducing the effect of rapid air movement dissipating heat from the building 'envelope'. If well designed, such shelter belts can reduce noise and road traffic pollution from entering the building too.

Trees, of course, can be damaged by the wind and hence pose a potential risk to human health and property. During severe storm events falling trees and broken boughs are linked to human deaths and serious injury. For example, winter storms that crossed central Europe in 1999 were estimated to have killed 137 people, a proportion of which were killed by either dropped branches or entire trees falling onto houses and roads (Lopes *et al.*, 2009). Fortunately, serious accidents triggered by windstorms are relatively rare. In the UK, the mortality rate is 1 in 10 million inhabitants p.a. and in Australia 0.6 per 10 million (urban areas) and 1.9 per 10 million (rural areas) (Portoghesi *et al.*, 2023). Urban trees often have a greater disposition to wind damage compared to those in forest situations due to a lack of protection from neighbouring trees and the fact that strong winds can be made even stronger by channelling along roads and urban canyons. The integral strength of street trees may also be compromised by a range of urban stress factors including reduced or restricted root space ('root-run'), drought stress, urban heat, compacted soil, air pollution, road salt damage and root damage through trenching activities for cables and pipes (see Chapters 2 and 7). The greatest storm damage in higher latitudes is often associated with large deciduous trees, which have retained a full canopy of leaves at the onset of autumnal storms. The increased resistance to the wind, due to the presence of a full canopy, results in the whole tree being thrown over. It has been estimated that the majority of tree falls are linked to wind speeds in excess of 7 ms^{-1} for a duration of 6 h or more (Lopes *et al.*, 2009). Other aspects such as heavy precipitation that loosens the soil around tree roots or ice storms that add weight to the tree branches also increase the likelihood of trees being thrown over.

3.3 Noise Amelioration

The effects of excessive noise on human health are often underestimated. Noise can be a killer. Persistent or high levels of environmental noise, such as that from road traffic or aircraft, negatively affect sleep quality and cardiovascular health. In the European Union (EU), the number of 'years of quality life' lost due to noise are 903,000 (via sleep disturbance) and 61,000 (via heart disease). Urban woodland and other types of vegetation can reduce the level of noise people are exposed too (Fig. 3.7). In Birmingham, UK, it is estimated that >177,000 people benefit from noise mitigation provided by nearby woodland, worth £3.83 million p.a. (Fletcher *et al.*, 2022).

Fang and Ling (2003) suggested that dense plantings of shrubs and trees, where the plants were a few metres higher than the receiver of the noise, were optimal for reducing noise. They studied tree belts composed of 35 different plant species, and those genotypes characterised by dense foliage and branches correlated with greatest noise reduction Species choice was important as *Bambusa dolichociada* reduced noise by 9.3 dBA over 20 m compared to only 5.0 dBA with *Nerium indicum* or 1.0 dBA with *Ficus microcarpa*. Woody plants that possess a low-forking habit seem to have the best noise attenuation characteristics. If this branching and stem density is higher up in the canopy, then noise waves can still travel along just above ground level, thus transferring more sound waves. Overall, greater density, height, length and width of shelter belts help diffuse noise more effectively, whereas increasing the leaf size and branching characteristics aid absorption of sound waves. In practice a belt of trees underplanted with a dense shrub layer may be considered the best design to intercept sounds waves from the source of the noise (Fang and Ling, 2003).

Even relatively small-scale green infrastructure interventions aid noise control. Using an urban courtyard as a scenario for modelling noise, Van Renterghem *et al.* (2013) studied the effect of surrounding green roofs, green façade walls and low, vegetated screens at the edges of flat roofs to mitigate the effects of adjacent road noise. The study suggested green roofs have the potential for attenuating noise and on certain roof designs could reduce noise by 7.5 decibels (dB). As is the case with thermal insulation of green roofs, a number of advantages associated with noise attenuation were linked with the substrate rather than the plants. A combination of modelling and field studies with green walls indicate they reduce single-point sources of noise by 2–5 dB and continuous road traffic by 1.6–10 dB depending on the geometry of a site and the depth of vegetation present on the wall (Paull *et al.*, 2020). However good the plants and substrate are at intercepting and attenuating noise, thought also needs to be given to the building construction and materials used. For example, vegetated screens on roof edges were only effective when the supporting structures were made from absorbent materials too as opposed to rigid materials. Adjacent rigid materials can actually increase noise directed into nearby open spaces or along the wall of a building. In overall design terms, soft roof edge

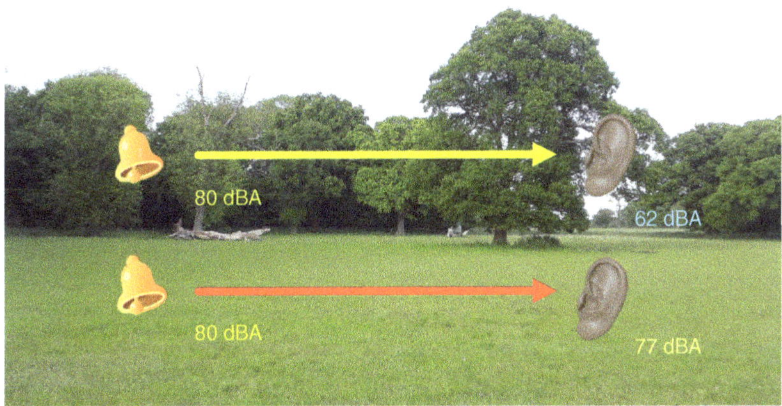

Fig. 3.7. An example of how noise level changes over in dense woodland (top line) compared to open air (bottom line). A noise of 80 dBA (weighted decibel – loudness the human ear perceives) would be reduced to 62 dBA in woodland, but only to 77 dBA over 50 m of open ground under the same weather conditions (Fletcher *et al.*, 2022).

screens combined with either green roofs or walls were the most effective means to reduce noise.

Perceptions of noise

Noise is both a physical and psychological factor; with perceptions of noise being a source of stress as well as actual volume. As such, the presence of natural elements can have a moderating influence on people's noise responses. High-quality neighbourhoods with a high proportion of attractive parks and other green spaces tend to lower dissatisfaction with traffic noise to a significant degree (Kastka and Noack, 1987). 'Better' availability of green space has been shown to reduce noise-related stress symptoms in residents compared to areas with less available green areas (Gidlöf-Gunnarsson and Öhrström, 2007). These authors concluded that local accessible green space is a protective factor, partially moderating for the adverse impact of road traffic noise; adding that vegetation has a positive effect on people's perceptions of 'noise-contaminated' environments.

3.4 Carbon Sequestration and Release

Urban green space is often quoted as being beneficial in terms of sequestering atmospheric carbon. Plants use CO_2 absorption as part of photosynthesis and will 'fix' some of this carbon in their tissues as they grow. In urban areas, 17% of terrestrial carbon is found in trees and shrubs and only 0.5% in grass and herbaceous plants. In woody plants the carbon in dry biomass is fairly consistent across species at 49% to 50% (Whitford *et al.*, 2001), although bulbs, for example, can be lower depending on their growth stage (e.g. 35%). Carbon is also embedded in the soil through the action of plants, most notably adding to soil carbon stocks through leaf drop, dead wood and mucilage release from roots. Herbivory from animals and interactions via mycorrhizal fungi may also be routes that plant-derived carbon ends up in the soil. Approximately 83% of urban terrestrial carbon is found in the top 600 mm of soil. It is important to remember that the soil is the main sink for carbon in terrestrial systems; for example, typical percentages in a woodland might be plant above ground biomass 17%, roots 6%, surface leaf litter and dead wood 5% and soil 72%.

So, while there is no argument that plants fix atmospheric carbon and store this both within their own tissues and within the organic matter of the soil, questions arise around to what extent this helps offset emissions from human activities. The capacity to sequester carbon also depends on factors such as the scale of green infrastructure, growth rate of plants (especially trees, e.g. young fast-growing trees will fix more carbon than mature trees that are no longer significantly increasing their biomass), the capacity of a given soil to store more carbon, respiration rates of soil macro- and microbiota (which can be temperature and moisture dependent), what happens to the wood at the end of a tree's life, and indeed what happens to urban soils – for example, cultivation of soils can lead to carbon being released back into the atmosphere again. Finally, urban green space that is intensely managed using hydrocarbon-powered machinery is also going to release significant amounts of carbon.

Urban green space has been estimated to store 13.5 t ha^{-1} C (5.47 kg m^{-2} C, Italy) to 25.1 t ha^{-1} C (USA), with the average carbon storage per area of tree cover being 5.47 Italy vs 7.69 kg m^{-2} C for US cities (Speak *et al.*, 2020). In the Italian scenario, the amount of carbon sequestered only equated to 0.6% of vehicle emissions (or 0.17% of total emissions). Other studies even suggest the urban green space may not act as a sink for carbon, but rather as a source. When Velasco *et al.* (2016) compared the CO_2 flux data from two residential neighbourhoods in different geographical and contextual locations, they found the biogenic component (vegetation and soil) was found to be a sink of 1 Mg km^{-2} day^{-1} CO_2 in Mexico City, but an emission source of 0.8 Mg km^{-2} day^{-1} CO_2 in Singapore. Results were thought to vary based on the number and type of tree species present and how the soils were managed. Again, such studies imply plant cover, soil condition and management are key (Ariluoma *et al.*, 2021; Thölix *et al.*, 2024). Woody plants provide significant carbon sequestration improvement over transient planting schemes due to soil carbon being reduced significantly by soil disturbance and tillage during replanting (Jandl *et al.*, 2007). Sequestration therefore would seem to be optimised by long-term, stress-tolerant species, which require minimum soil disturbance and management requirements – typical of perennial plantings that reflect permanent woodland or long-term meadow systems. Trees that grow fast (and degrade slowly) have a high wood density, maximise overall biomass and show most resilience to abiotic and biotic stress factors may be the best to choose for longer-term C fixation. Such traits are commonly found in taxa such as *Liquidamber styraciflua* (American sweetgum), *Quercus coccinea* (scarlet oak), *Q. robur* (English oak), *Q. rubra* (red oak), *Q. virginiana* (Virginia live oak), *Platanus x acerifolia*, *Juglans nigra* (black walnut), *Pseudotsuga menziesii* (Douglas fir), *Taxodium distichum* (bald cypress), *Pinus ponderosa* (Western yellow pine), *P. resinosa* (red pine) and *P. strobus* (white pine).

How timber is disposed of at the end of life is considered important too. Burning with energy recovery for electricity was the most efficient way to deal with carbon, with a carbon emissions/input ratio of 0.5. In contrast, putting timber to landfill was the least efficient with a ratio of 121.9, largely due to the wood being converted to the more potent greenhouse gases of methane (CH_4) and nitrous oxide (N_2O), which have a warming potential of 25× and 235× greater than that of CO_2, respectively (Speak *et al.*, 2020). Converting wood to biochar (a form of locked carbon) and adding this to soil has been cited as making the carbon storage and sequestration of urban green space more effective, with biochar accounting for 65% of the storage capacity for the carbon (Ariluoma *et al.*, 2021).

Carbon and vegetation management

Overall carbon balance (life cycle analysis) within green space is also strongly influenced by vegetation management, e.g. fossil fuel use with respect to mowers, hedge-strimmers and chainsaws and also the use of vehicles to visit sites and remove debris, dispose of biomass, manage diseases and pests, and local geographical/sociological factors (Nowak *et al.*, 2002; Liu *et al.*, 2023). Pesticides, fertilisers and potable water used in the landscape provide their own carbon footprint. The production/use of artificial fertilisers contribute significantly to greenhouse gas emissions (Howarth *et al.*, 2002). The Haber–Bosch process is particularly high. This converts atmospheric nitrogen into ammonia which is the basis for inorganic nitrogen fertilisers. In contrast, manufacture of phosphate and potassium fertilisers is 10- to 20-fold less energy intensive. The use of composted organic matter in the garden offers a lower carbon cost alternative for supplementing nitrogen than using an energy-intensive inorganic nitrogen fertiliser such as ammonium nitrate (NH_4NO_3). The incorporation of green and other appropriate organic wastes into the soil can also reduce energy consumption embedded in moving such materials off-site; they frequently have the additional advantage of improving soil structure in the process. Composting green waste before applying it back to the soil is a traditional agricultural/gardening technique and has its uses today for capturing and recycling carbon. But the composting process requires good aeration (oxygen availability) to minimize the generation of the more potent greenhouse gases CH_4 and N_2O.

Further studies are warranted on the role of urban green space and carbon balance – not least because some research suggests that sensitively managed green space can achieve carbon contents commensurate with certain woodland systems, i.e. ranging between 15.2 and 72.1 kg m^{-2} C. Caution is required in that planting up and managing more urban green space carefully should not be seen as an alternative to reducing carbon emissions in the first place. Within an urban context, horticultural activities are now scrutinised more closely to help promote sustainable approaches to design and management, and the advent of a greater component of urban green space is generally seen as a virtue. The advantages in terms of sequestering atmospheric carbon, however, are strongly determined by the specifics of the activities involved. Indeed, the advantages and disadvantages of certain activities are still difficult to assess as they often depend on incomplete calculations or certain assumptions being made within life cycle analyses (LCAs). The point too that the global climate is changing also infers that activities now that are considered positive, need not necessarily remain so. For example, the desire to increase soil carbon pools which are currently being encouraged through management activities (adding organic manures, not removing dead wood or other vegetation from sites and reducing soil tillage) may only enhance soil carbon levels for it to be vulnerable to oxidation later (i.e. released back into the atmosphere). This may be a scenario if soil temperatures increase or microorganism dynamics alter with climate change.

Outwith vegetation management per se, other high carbon costs are associated with construction materials. The constituents of paving, footpaths, walls, etc. may have high embodied energy associated with their transport (many being bulky, heavy materials) or, as in the case of cement, energy used in their manufacture.

Carbon and lawns

The use of fertilisers and pesticides, combined with petrol-driven mowers has been estimated to negate the carbon sequestration potential of lawn turf (Wang *et al.*, 2022; Gillman *et al.*, 2023). Domestic lawns in the USA are thought to hold approx. 4.96×10^{14} g C, but fertiliser application and mowing may release between 2.5 and 7.6×10^{12} g C each year. In effect the carbon footprint of lawn maintenance would mean that any benefits of carbon storage within the lawn would be lost after 66–199 years of maintenance (shorter if the lawns are irrigated with potable water too). Lawn fertiliser use alone has a significant effect on greenhouse gas emissions (Livesley *et al.*, 2010) with lawns emitting up to 10× more N_2O than neighbouring agricultural grassland. The higher emissions are thought to be due to more frequent irrigation and higher soil temperatures found in urban lawns (Bijoor *et al.*, 2008). Naturalized grasslands (meadows) have a low environmental footprint, in contrast to heavily maintained lawns (Trémeau *et al.*, 2024). The former requires a single annual cut, compared to the high-frequency mowing associated with formal lawns (e.g. weekly during optimum growth), although accurate estimations of carbon use through mowing are difficult to assess, due to differences in type and energy supply to the mower, the frequency of cutting, climatic factors and what is the ultimate fate of the grass clippings (Reid *et al.*, 2010). It has been thought that petrol-powered mowing may release 50% more carbon than the lawn itself can sequester. Mowers with two-stroke engines particularly are considered detrimental with a sevenfold higher by-product emission compared to four-stroke engines (Volckens *et al.*, 2007).

Despite intensively managed lawns being associated with high fertiliser and herbicide inputs (and in warmer, affluent countries, high water use through artificial irrigation), where urbanisation results in housing development on land previously degraded or heavily cultivated, the creation of a lawn can help replace soil organic carbon originally lost through oxidation (Zirkle *et al.*, 2011). Pouyat *et al.* (2009) claim that the relatively high organic carbon component within residential soils in the USA is due to lawn management. They relate this to management activities that typically include supplements of water and nutrients which maximises the productivity of the grass. Indeed, they state that soils of residential lawns appear to have the highest density of C in urban landscapes – higher than many forest soils in the USA. Soil carbon pools though vary widely with location and climate. For example, urbanisation (including the promotion of gardens and other green space), in the north-eastern USA is thought to reduce overall soil carbon pools. This is largely due to the parent soil being either brown podzols (woodland) or dark mollisols (prairie grasslands), as such soils have a relatively large natural organic carbon content. In contrast on more arid soils in the southern states (where carbon storage is naturally lower), urbanisation can increase soil carbon content through more prevalent lawn culture. In such situations the age of the lawn often determines the level of stored carbon.

3.5 Water Management

Water management and sustainable drainage systems (SuDS)

Towns and cities are increasingly prone to flooding due to climate change and the widespread occurrence of impermeable surfaces (Fig. 3.8). These surfaces do not allow rainwater to soak into the soil but rather runoff the surface (surface flow). In China, the number of cities suffering from urban floods has increased by approximately 30% since the 1980s, with frequency of urban flooding enhanced from an average of once in every 2 years to more than three times each year in some of the worst affected cities (Yan *et al.*, 2020). Much of the impact is climate related but city densification, deforestation in upper river catchments and reductions in urban green space, including the paving over of gardens to increase 'off-road' car parking space have played their part too. In the UK, one in ten front gardens has been paved and in congested cities, such as London, it is one in five. As gardens have got smaller and more people have desired a patio seating area, then the proportion of area devoted to vegetation has also decreased, with many gardens now having as much as 50% of the garden paved over. In the USA, there are 110,000 km^{-2} of impermeable surface which equates to approximately the same surface area of the state of Ohio, USA (Frazer, 2005).

Traditional means of dealing with rainwater in urban areas has been to catch it into hard-surfaced drainage systems and move it off-site as quickly as

Fig. 3.8. Urbanisation, especially the wide use of impermeable materials, combined with more frequent and severe rainfall events are increasing the risk of flooding. In this particular case, the domestic livestock – ducks and geese – were able to cope!

possible. As rainfall events have increased in intensity and more zones have been hard-surfaced, this has really just resulted in moving the problem of flooding somewhere else downstream. 'Hard'-engineered approaches have simply involved increasing the number of drains present, widening the diameter of drainpipes, diverting river flows, deepening channels and increasing the complexity of the storm channels and culvert networks. Such interventions, however, tend to be expensive and do little for urban ecology and in reality may not deal with the increasing precipitation effectively. Increasing such grey infrastructure can exacerbate habitat loss, increase erosion in other parts of the river system (by increasing volumes of and rates of flow) and concentrate pollutants in the waterways. Canalisation removes natural features such as gravel beds and permeable soil banks, thus reducing opportunities for vegetation establishment and the development of feeding and resting areas for animal life. Deeper, straight water channels alter water temperature and flow rate, again interfering with natural colonisation of the river. Culverting eliminates light from the waterway, excluding any plant growth and inhibiting animals from migrating through the 'barren zone'. Canalisation also provides little suitable habitat for native stream flora and fauna. Both terrestrial animal behaviour and aquatic life forms can be negatively affected, as many species use watercourses as highways and networks within their territories.

Urban storm water runoff is a significant component of non-point-source water pollution, being the third most important factor contributing to overall lake and river pollution. Rapid runoff erodes riverbanks leading to increased sedimentation of the watercourse, which in turn reduces irradiance levels in the water column and alters water chemistry, with adverse effects on aquatic plants and invertebrates. Canalising and waterways may help divert water but rarely slow the rate of passage, potentially exacerbating problems for sites downstream.

Problems with water flow tend to occur during (i) intense storms with high but short-term precipitation rates or (ii) prolonged periods of moderate rainfall which saturate the soil profile and natural aquifers but then cause rivers and other watercourses to overflow. Either scenario can induce surface flows that exceed the drainage capacity of the urban area. Green spaces provide storm attenuation 'services' to the urban matrix in a variety of ways. Vegetation intercepts intense precipitation and holds water temporarily within the foliar canopy and along the branch and trunk network, thus reducing peak flow and easing demand on the city drains (Xiao and McPherson, 2002). In addition, vegetation mitigates flood risk by increasing infiltration into the soil and diminishing surface flow (Schuch *et al.*, 2017). Roots provide passageways for rainwater to drain through the soil and reach deeper profiles. Vegetation and leaf litter reduce the physical impact of precipitation on soil particles, helping to retain the integrity of the soil structure. They lessen surface flow too and reduce the dissolved pollution load in the runoff, by aiding the capture of soil particulates and any embedded contaminants: the larger the single area of green cover, the greater the reduction of pollutants. The ability of vegetation to reduce flooding is influenced by factors such as rainfall intensity and duration, previous soil moisture content (saturated soils allowing little capacity for further storage, whilst excessively dry soil inhibits infiltration and encourages surface runoff), canopy morphology, plant type and species and the presence or absence of leaves on the branches.

The increasing use of vegetation as part of sustainable drainage systems (SuDS) demonstrates the significant role vegetation plays in providing solutions to runoff-related problems (Charlesworth and Booth, 2012). Increasingly more sustainable and integrated approaches are being adopted to reduce surface flow rates, temporarily retain flood water and increase infiltration rates to soil. This not only includes using vegetation to trap and slow rainwater runoff but in addition encourages the evaporation and transpiration of water back into the atmosphere directly (Nur Hannah Ismail *et al.*, 2023). A range of measures can be employed: green roofs to capture and slow water coming off roof systems, tree pits to trap water from roadways and pavements, storm water planters attached to downpipes of houses, swales and rain gardens which help to divert water away from the main drainage systems. Many of these features facilitate a period of temporary flooding of the green space, then allow water to soak away slowly after that.

Rates of runoff in housing with larger gardens have been found to be one-third less than that from dense urban development (Pauleit and Duhme, 2000). Using 9 m^{-2} plots, Armson *et al.* (2013) showed that turf almost completely eliminated runoff, whereas a tree planted within asphalt reduced it by 62% compared to control plots composed entirely of asphalt. Infiltration of water into the tree pit itself was thought to improve the retention rates associated with the tree plots, as the reduction in runoff was more than interception alone could have produced. Reductions relative to the canopy area were also greater than estimated by many previous studies, suggesting the planting pit was playing a pivotal role. Trees are often cited as one of the best forms of vegetation to reduce runoff, but various factors influence runoff between different plant species. In addition to variations in the precipitation itself (raindrop size, frequency, duration wind direction, etc.), the morphology of the tree, including branch structure and number of branch/leaf nodes, foliar characteristics including leaf size, shape and texture, leaf distribution and density, the overall size/spread of crown and trunk surface area (rough and smooth bark types) all affect moisture retention (Fig. 3.9). Precipitation falling on a tree is characterised in three ways:

1. Through-fall which includes water passing through the canopy or is caught then drips off.
2. Stem-flow which is water caught in the canopy that flows down the branches and trunk.
3. Interception which is water that is held in the canopy and subsequently lost via evaporation or absorbed to be later used in cell hydrology and/or enters the plant's transpiration stream.

Livesley *et al.* (2014) compared two forms of *Eucalyptus* with rainfall interception over a continuous 5-month period. *E. nicholii* with a dense canopy, rough bark and greater plant area index intercepted 44% of rainfall in small rainfall events, whereas *E. saligna* with a more open canopy and smooth bark only caught 29%. Trees also dictate other important indirect factors such as the composition of the herb layer, depth of leaf litter and soil structure. Leaf litter, for example, helps act as a 'sponge' for moisture on the ground. Soil and plant water status determine the proportion of water that runs off the soil surface and how much is absorbed by soil and roots. The presence of a leaf canopy in winter is important; Xiao and McPherson, (2011) found that broadleaved evergreen trees intercept 27% of annual gross rainfall compared to a deciduous species (15%), largely due to differences in winter precipitation capture. As trees age, their interception capacity increases (Fig. 3.10). The type of rainfall can be important too. Work on tropical

Fig. 3.9. Variation in leaf size, branch arrangement and canopy density between tree species affects the amount of precipitation that is intercepted or falls through.

tree species by Nytch *et al.* (2019) demonstrated how trees were quite effective at rainwater capture during relatively small and less intense storms, but interception capacity was much more limited during longer or more intense storms.

Trees are particularly important in urban areas, as otherwise, the soil's capacity to absorb water is frequently compromised by both compaction and sealing. Changes in soil infiltration capacity give large impacts on the total water balance and avoid ponding. Moreover, soil moisture losses through ET recharges the capacity of the parent soil to hold more water after a subsequent storm. Modelling work in Berlin, Germany (Ouédraogo *et al.*, 2022) showed that transpiration was the dominant factor in determining water dissipation, accounting for more than 50% of ET loss in the modelled area. Water fluxes were in general greater from trees compared with grass. Interception evaporation, i.e. rainwater held on the leaves then evaporated, was also larger for trees compared with grass. ET from tree-covered areas comprised 80% of the total ET yet the tree canopies only made up less than 30% of the total surface cover.

In an attempt to help water capture in street situations, the use of 'structured soil' in tree planting pits has become prevalent. These structured soils mix gravel aggregates with soil to improve the drainage and storage characteristics of the medium within the pit. They tend to improve porosity and moisture holding capacity over often poorly structured urban soils (see Chapter 2).

In many developed countries, increased urbanisation through extensive, but low-density development has correlated with greater land cover put down to lawns. The total area of turf grass in the USA, including residential, commercial, institutional lawns, golf courses and parks occupies 163,800 km^2 (± 35,850 km^2) (Milesi *et al.*, 2005). This area of turf grass alleviates some of the problems associated with the comparable increases in impermeable soil coverage (asphalt, concrete, stone paving, etc.) brought on by urbanisation. As with other vegetation, grass foliage intercepts and

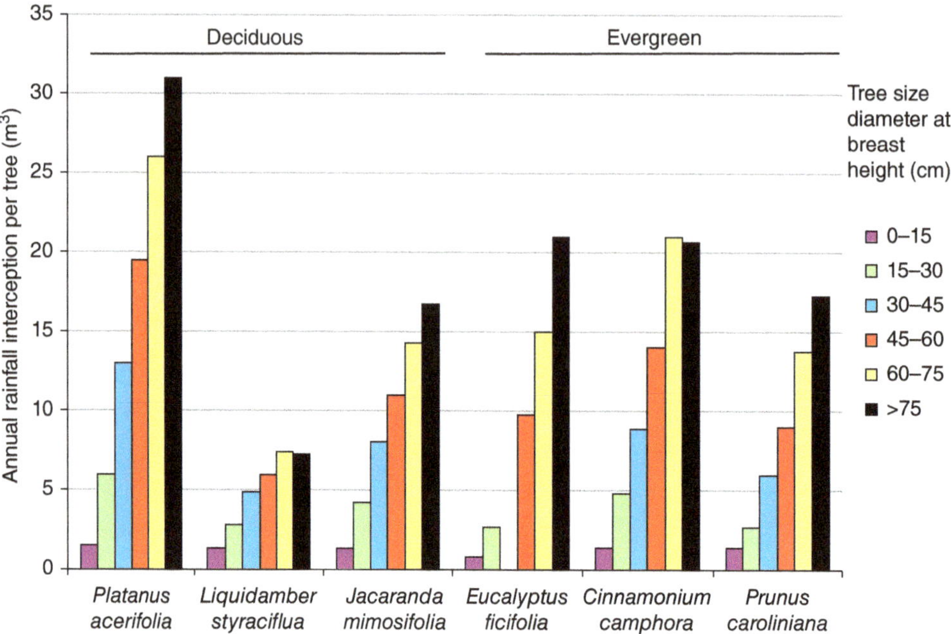

Fig. 3.10. Tree size (diameter at breast height) and character (leaf/branch habit and duration of leaf retention) affect rainfall interception. Larger trees capture more rainfall but also note the differences between deciduous species. (Modified with permission from Xiao and McPherson, 2002.)

absorbs raindrop impact and provides resistance to water runoff (Krenitsky *et al.*, 1998). Large areas of rough grass transpire more water than highly maintained, recreational turf. Grass also lets more water evaporate directly from the soil than other forms of green infrastructure (due probably to less shading and large total surface areas). There are instances where grass is cited as being superior to trees. One study on sub-tropical China (Zou *et al.*, 2021) suggested that in practice, grass swards lost more water than trees. In this study, the daily average ET for a lawn was 2.50 mm day^{-1} and for trees it was 1.40 mm day^{-1}. The results indicated that the daily ET for lawn was determined more by solar energy and diffusion conditions (vapour pressure deficit, VPD), while the daily ET of trees was more restricted by VPD but also the amount of relative extractable water (REW) available in the urban area. The trees appeared to be the more sensitive to soil moisture deficits and reduced ET through stomatal regulation, once they sensed drier soils.

Rain gardens are also effective at removing water. Studies on rain gardens showed estimated ETs account for 61–90% of the collected rainfall. Typical ET rates were 2.4 mm day^{-1} to 3.78 mm day^{-1}, depending on the maintenance regime for the plants, with proportionally high values (3–3.78 mm day^{-1}) when water was stored within the rain garden's lower levels compared to when the water drained away (2.4–2.5 mm day^{-1}) (Ouédraogo *et al.*, 2022). Plant composition plays a factor. In studies of free-draining rain gardens in the Midwest of the USA, Nocco *et al.* (2016) compared the hydrological performance of grasslands, shrubs, managed turf and bare soil over one summer (July–September). They found that the effect of vegetation was significant but could vary across the season as evaporative demand changed. Shrubs were most consistent with constant ET rates of around 6 mm day^{-1} for the months of July and August. Grasslands showed higher ET than the others and this increased from 7 mm day^{-1} in July to 9 mm day^{-1} in August. Losses were less in bare soil where ET rates were 3 mm day^{-1} (July) and 4 mm day^{-1} (August).

Soil type is important, as are the interactions between plants roots and hydraulic processes, e.g. plants with deeper roots allowing better drainage and being able to extract more water. Le Coustumer *et al.* (2012) observed that a taxon with thick roots significantly maintained the permeability of the soil over time. Ouédraogo *et al.* (2022) argue that the capacity of plants to remove water through ET activities (rather than just capture and drainage) needs to be valued more and built into urban hydraulic equations more effectively. A point reinforced by Nur Hannah Ismail *et al.* (2023) for ground-cover plants. This study also showed that plant leaf type and plant ecophysiology (the type of environment the landscape plant evolved in, e.g. woodland understorey vs open habitat) could influence hydrological processes (Figs 3.11 and 3.12).

Green roofs are frequently advocated due to their ability to detain and retain storm water flows. Some studies show retention rates of up to 70% during light to medium precipitation events (i.e. when the soil medium is unsaturated) (Hutchinson *et al.*, 2003) and the average is 62% (Zheng *et al.*, 2021) The ability to detain/retain water depends again on factors such as rainfall intensity and the amount of rainfall in the preceding period, season, size of the green roof, water-holding capacity of the substrate (affected by substrate type, structure and depth) and plant cover/type. Deeper substrates appear to provide an advantage in water-holding capacity (although there are implications for the weight load on the roof with deeper substrates) (VanWoert *et al.*, 2005; Liu *et al.*, 2019). Comparisons of green roof modules (the trays in which moisture reservoirs, substrate and plants are normally held within) indicated that depth of substrate was important. Studies in the absence of vegetation demonstrated that the amount of precipitation intercepted and accumulated increased with deeper substrate (200 mm depth captured 83% of precipitation, whereas 120 mm module only stored 63% [Nardini *et al.*, 2012]). Developing green roof substrates that are lightweight yet provide good water-holding and water supply (to the plants) characteristics remains a key objective for this sector.

The contribution of vegetation per se on green roofs remains unclear. VanWoert *et al.* (2005) reported that vegetation contributed relatively little

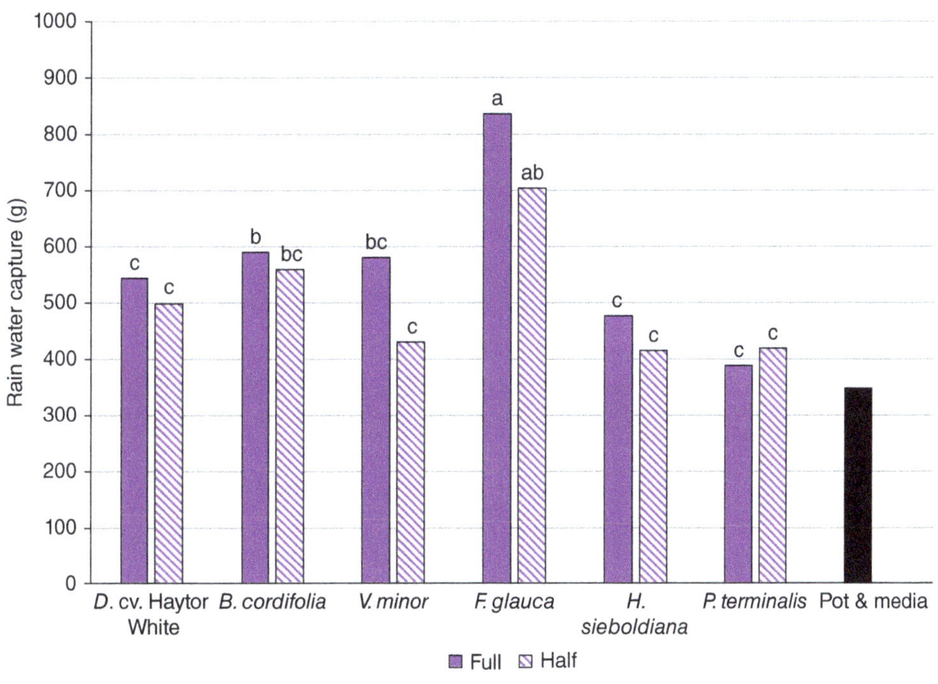

Fig. 3.11. Rainwater interception by different ground cover taxa (*Dianthus* cv. Haytor White, *Bergenia cordifolia*, *Vinca minor*, *Festuca glauca*, *Hosta sieboldiana* and *Pachysandra terminalis*) across 24 h and total rainfall depth of 11mm. Pot-grown plants either have natural canopies (full) or trimmed canopies (half) to give similar surface areas across taxa. Different letters denote significant differences ($P \leq 0.05$) when comparing mean values. (Used with permission from Nur Hannah Ismail *et al.*, 2023.)

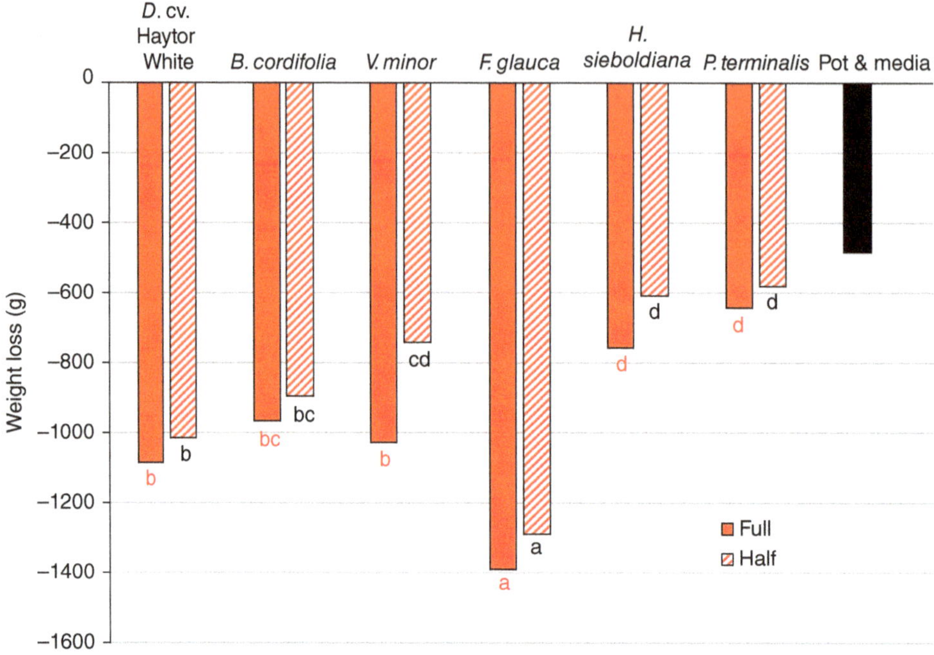

Fig. 3.12. Water loss in ground-cover taxa (*Dianthus* cv. Haytor White, *Bergenia cordifolia*, *Vinca minor*, *Festuca glauca*, *Hosta sieboldiana* and *Pachysandra terminalis*) through ET over a 7-day period. Pot-grown plants with either natural canopies (full) or trimmed canopies (half) to give similar surface areas across taxa. Different letters denote significant differences (P≤0.05) when comparing mean values. (Used with permission from Nur Hannah Ismail *et al.*, 2023).

to the amount of runoff reduction by green roofs. In contrast, Schroll *et al.* (2011) reported that vegetation has a significant effect on runoff when compared with substrate-only roofs. Dunnett *et al.* (2008) indicated that short or prostrate plants appeared to shed the greatest amount of water, whereas broadleaved grasses intercepted and retained the greatest amount. Comparisons of green roofs composed of native North American plant species, which were kept alive through irrigation, and a non-irrigated extensive green roof composed of *Sedum* spp. showed the native species roof to retain more rainwater (64% p.a.) compared to the *Sedum* one (54%) (Shetty *et al.*, 2022). The ET capacity (crop co-efficient) of the native species was considered about twice that of the *Sedum* (1.13 vs 0.57). Studies comparing modules planted with shrubs and modules planted with herbaceous perennials indicated that both types intercepted and stored more than 90% of rainfall during intense precipitation events, with no significant difference between the two vegetation types despite different substrate depths (Nardini *et al.*, 2012). Larger plant canopies may hold more water on their surfaces, but active root systems allow for effective uptake into the transpiration stream. Indeed, rapid depletion of substrate water content following precipitation through transpiration subsequently increases the substrate's capacity to cope with any follow-on rain events (Dunnett *et al.*, 2008; Schroll *et al.*, 2011). Defining plant taxa that have a large elasticity in stomatal control and can rapidly alter the volume of moisture transpired depending on the water available in the substrate would be useful in this respect for future roof designs.

Runoff water quality

Plant communities can aid the quality of the water running off the land. Grass swards protect the soil structure, reduce chemical movement and filter out pollutants from runoff (Beard and Green, 1994). Ground cover does not need to rely on turf – any extensive herbaceous ground cover that covers most of the soil will aid water capture, ET dissipation and the filtration of potential pollutants (Fig. 3.13). Associated microbial activity in the rhizosphere degrades organic chemicals relatively rapidly, thus

Fig. 3.13. Good ground cover and a heterogeneity of plant forms help stabilise soil structure, improve rainwater capture, filter out pollutants and aid the ET of water back to the atmosphere.

improving the runoff water quality and reducing groundwater contamination. Lawn management, however, can impact on the benefits, for example many lawns are over-fertilised with resultant potential for leaching of nitrate and phosphate ions, especially if fertiliser application precedes heavy irrigation or precipitation (see Chapter 9).

Due to such factors, nitrogen pollution of water courses can be higher in residential areas than equivalent rural areas of forests or even agricultural land (Kaushal *et al.*, 2008). Urban development results in the stripping off of topsoil, leaving a compacted sub-soil characterised by a low proportion of organic matter present and a poor pore structure: factors that reduce water retention and infiltration. Rainfall simulations have shown that runoff and pollution levels are influenced markedly based on whether a lawn is established on topsoil or sub-soil, and how lawns are fertilised. Cheng *et al.* (2014) reported in their research that surface runoff was initiated after 13 min of overhead irrigation in lawns established above sub-soil, whereas it was only apparent after 25 min irrigation in lawns established over topsoil. The total runoff volume and sediment loss were significantly larger in sub-soil lawns too (11,900 ml and 3700 mg, respectively) than in topsoil lawns (2200 ml and 900 mg, respectively). Although nutrient losses were relatively low in this study, leachate values were still significantly higher from sub-soil than topsoil lawns, and from lawns treated with inorganic fertiliser compared to those applied with organic forms. Clearly, amelioration of poorly structured urban soils before the establishment of vegetation pays dividends in terms of offsetting problems with nutrient and water applications.

Urban water use

Green spaces, and domestic gardens in particular, are frequently associated with high volumes of water use, especially during periods of dry weather. Domestic water use can be broken down into indoor consumption (e.g. food preparation and washing) and outdoor consumption (garden irrigation, car washing and garden pools). Outdoor consumption is a concern because it can represent up to 70% of residential water use and is expanding in demand, despite climate change resulting in supplies being less reliable in some regions (Hof and Wolf, 2014). The proportion of potable water used for the domestic garden escalates with increased aridity, with water consumption being greatest in warm regions such as southern USA, Australia and southern Europe, where 50% of all water use relates to outdoor activities (Padullés Cubino *et al.*, 2017). In Western Australia, 56% of domestic water use is associated with domestic gardens (Syme *et al.*, 2004). In Spain, 30% of total household water use is consumed within the garden, and this rises to 50% during the summer (Domene and Saurí, 2006). Estimates for the USA are even higher; landscape irrigation can vary from 40% to 70% of household use of water (St. Hilaire *et al.*, 2008). In Mediterranean regions (Reyes-Paecke *et al.*, 2019), more water is consumed in private gardens than in public parks and other amenity areas. Rates of water use tend to be 0.53–0.77 m^3 m^{-2} for small parks and 0.29–0.53 m^3 m^{-2} for larger parks. This is compared to lawn areas in private gardens where consumption is 25.0–31.4 m^3 m^{-2} per year.

Recent plantings are particularly dependent on irrigation water to help new roots develop and allow the plant to establish effectively. Transplanted trees can suffer up to 95% root loss after being removed from nursery soil and the remaining roots need to access large volumes of water to meet the demands of the canopy, once the plant is *in situ*. With container-grown specimens, the open, porous growing media frequently used will only provide moisture for a short period of time post-planting. Some of these growing media too, such as peat, become hydrophobic if they dry out, exacerbating the problems of absorbing water from the surrounding soil. Shrinkage of the pot medium within the soil may mean that roots need to cross an air gap to access the parent soil, a process that is unlikely unless the gap is filled with water or air of a very high humidity. Where ET rates are high, permanent irrigation infrastructure (pipes, drip nozzles, pumps, weather or evaporation-based controllers) may be required to ensure the new landscape plantings have longevity.

Climate change will increase demand for garden water in temperate regions. There will also be enhanced competition for this resource from industrial processes, food production and other domestic requirements, under future scenarios. On a more positive note, it is likely that many landscapes are currently overwatered compared to the need to maintain ecosystem services such as carbon regulation, climate control, and preservation of aesthetic appearance (St. Hilaire *et al.*, 2008). Garden landscapes are overwatered primarily due to the fear of losing plants to drought, but unintentionally this practice may predispose specimens to greater susceptibility to stress long term, due to limited opportunities for the plants to express adaptive traits, such as deep rooting, effective stomatal control and cellular osmotic adjustment. Changes in management practice can save water. In lawns, organic vs intensive style of turf management can reduce water consumption up to tenfold (Morris and Bagby, 2008). Mulching flower and shrub borders too minimises moisture evaporation and has the added benefit of limiting gaseous emissions (Livesley *et al.*, 2010).

Water consumption for use in UK domestic gardens is estimated to be in the region of 12–15 l per person per day. In many parts of the world the pressure on water resource will increase with climate change, as rainfall patterns become more unpredictable. For example, the UK Climate Impacts Programme forecasts that summer rainfall in southern UK will fall by between 30% and 40% by 2090 while winter rainfall may rise by 30% (Hulme *et al.*, 2002). Consequently, as droughts become more prevalent water storage will be necessary to maintain the benefits of green infrastructure (Gill *et al.*, 2007). Water availability will become a key issue in heat island and climate change mitigation (Lallana *et al.*, 1999) and is an important consideration in the efficacy of urban planting to keep the locale cool. Stored supplies could aid urban cooling through evaporation from ponds, rills, and/or SuDS (Charlesworth, 2010). Use of stored runoff water in preference to mains supplies provides energy conservation benefits too in that there is no energy-intensive process required to clean the water to the standard required for drinking water. Increasingly, water storage may be incorporated into new homes or at a local community level, but for most existing homes storage is limited currently to water butts (Syme *et al.*, 2004).

Grey water

There are opportunities to irrigate urban green spaces with grey water. Grey water is waste derived from domestic water use but excludes sewage water. Normally it is collected from hand-basins, kitchen sinks, baths, showers, washing machines and dishwashers. Grey water constitutes 65% of total household wastewater (Vuppaladadiyam *et al.*, 2019). Water properties vary with the source from within the house and grey water can be divided into two categories: that with a high pollutant load (e.g. from laundry and/or kitchen) and that with a lower pollutant load (e.g. from bathrooms and/or washrooms). Kitchen greywater is the most problematic – with potentially oils and fats included – and tends to be produced in rather low volumes. Often it has a high chemical oxygen demand (COD) and direct reuse is not highly recommended. Family lifestyle also strongly influences water volume and composition, providing inter-house variation. Grey water tends to show good biodegradability in terms of the chemical oxygen demand to biological oxygen demand (COD:BOD) ratios but can have low nitrogen and phosphorous proportions (e.g. in the latter case if phosphate-free detergents are used). Reclaimed grey water should fulfil four criteria (hygienic safety, aesthetics, environmental tolerance and economic feasibility) before it can be justified for reuse. A lack of appropriate water quality research, standards or guidelines, however, has hampered appropriate grey water reuse (Radingoana *et al.*, 2020). Despite omitting the waste stream from toilets, grey water frequently still contains faecal coliform microorganisms. Pathogenic organisms of concern associated with grey water reuse include enterotoxigenic *Escheria coli*, *Salmonella*, *Shigella*, *Legionella* and enteric viruses (Ottoson and Stenström, 2003). *Cryptosporidium*, *Giardia* and *Clostridium perfringens* (a protozoa) are comparatively resistant to disinfection and can survive for longer periods of time within a system. *Legionella* is problematic as it spreads via aerosols and thus limits the use of, for example, sprinkler irrigation. Faecal *Streptococci* and faecal coliforms are frequently assessed as indicator organisms to signal the presence of the pathogens (Ottoson and Stenström, 2003). Indeed, a number of studies have found that grey water is only mildly less contaminated (Jefferson *et al.*, 2004), and in some cases more contaminated, than waste that includes sewage. As it can contain hazardous agents, including pathogens and heavy metals, grey water should be treated before application to land where there is risk of human contact.

The level of treatment and techniques employed to improve the quality of grey water vary with its potential re-use. For landscape irrigation, systems such as sand/gravel filtration or settlement and flotation are operated and take out much of the solids within the water. To remove bacteria, biological systems are used (Sijimol and Joseph, 2021). These promote degradation of pathogens by the action of other microorganisms and are useful in removing residual organic matter and other pollutants too. Biological processes are often preceded by a physical pretreatment step such as filtering/screening or the use of sedimentation or septic tanks. Biological action is encouraged by allowing grey water to flow over surfaces covered with beneficial microbial populations (biofilms), such as a membrane bioreactor or through pond/sludge systems. Perhaps the most environmentally friendly, holistic and cost- effective techniques are those that mimic natural processes, for example constructed wetlands used in conjunction with sewage plants. These are suitable for grey water treatment (Table 3.1), although even here coliform bacteria are reduced but not eliminated. Wetland plant species that aid in the remediating of grey water include *Phragmites australis* (common reed), *Juncus* spp. (rush), *Typha* spp, (bulrush, reedmace and cattail), *Scirpus* spp. (bulrush and club-rush), *Carex* spp. (sedge), *Salix* spp. (willow), *Glyceria maxima* (reed grass), *Iris pseudacorus* (yellow flag) and *Acorus calamus* (sweet flag).

How grey water is applied to the landscape will influence the risks involved. Overhead sprinklers should be avoided and even surface application may pose problems through the formation of

Table 3.1. The effect of a wetland treatment in reducing pollutant loads in grey water (Gross *et al.*, 2007)

Pollutant	Grey water	Effluent
Total suspended solids	158 mg l^{-1}	3 mg l^{-1}
Biological oxygen demand (BOD)	466 mg l^{-1}	0.7 mg l^{-1}
Chemical oxygen demand (COD)	839 mg l^{-1}	157 mg l^{-1}
Total nitrogen	34.3 mg l^{-1}	10.8 mg l^{-1}
Total phosphate	22.8 mg l^{-1}	6.6 mg l^{-1}
Anionic surfactants	7.9 mg l^{-1}	0.6 mg l^{-1}
Boron	1.6 mg l^{-1}	0.6 mg l^{-1}
Faecal coliforms	5×10^7 / 100 ml	2×10^5 / 100 ml

aerosols, whereas risks can be reduced by application via sub-irrigation pipes or seep hose. Investigations on turf grass irrigated with water containing bacteriophages (Enriquez *et al.*, 2003) showed that there were lower transmission rates in sub-surface compared to surface irrigation. Even so, pathogens can accumulate in soil, for example *E. coli* is thought to be able to survive 230 days in warm soil, although these environmental conditions may not be as conducive for all potential pathogens.

The effects of grey water on plants and soil biology will depend on the constitution of the water, but common concerns relate to changes in soil pH (usually increasing pH), build-up of ionic levels (e.g. aluminium, sodium and zinc) and alterations in relative ratios of sodium to calcium and magnesium (sodium adsorption ratio, SAR). Organic components may also disrupt/alter soil microbial populations. The phytotoxicity of grey water is mainly due to the anionic surfactant content that alters the microbial communities associated with the rhizosphere (Pinto *et al.*, 2010). Phytotoxic effects can vary between species. No negative effects of grey water have been noted for studies in plants including *Lycopersicum esculentum* (tomato), *Lolium perenne* (ryegrass), *Beta vulgaris* var. *cicla* (Swiss chard) or *Daucus carota* (carrot), but kitchen and laundry waste waters have been shown to be toxic to both algae and *Salix* trees, while bathroom water was toxic to algae only (Eriksson *et al.*, 2006). Increasing soil pH to ≥9 resulted in a substantial reduction in transpiration rate in *Salix*. Non-replicated field observations of garden plants supplied with grey water have suggested species such as *Chrysanthemum*, *Euonymus*, *Hibiscus* and *Juniperus* (juniper) have high tolerance levels, whereas *Prunus* spp., *Valeriana californica*, *Iris germanica* (bearded iris) and *Pinus mugo* are intermediate in tolerance, with *Pinus sylvestris* (Scots pine), *Persea americana* (avocado) and *Citrus limonium* (lemon) showing higher sensitivity (Sharvelle *et al.*, 2012). Longer-term studies, e.g. over 5 years, suggest build-up of detrimental elements such as chloride and sodium ions and increases in the SAR and soil electrical conductivity (EC) can occur with regular irrigation with grey water. Differences can be detected in trees (foliar analyses) growing in these areas, although physical leaf damage was not observed. Nevertheless, Zalacáin *et al.* (2019) advise that soils should be flushed from time to time with fresh water to avoid toxic ion accumulation. In shorter-lifespan plants, e.g. *Lycopersicum* (tomato) irrigated with grey water, these showed no negative symptoms but did increase in uptake of key ions such as phosphate, sodium and iron compared to tap water controls. Indeed, the application of grey water has been linked with a moderate fertiliser effect (Gholami *et al.*, 2023).

Interest has increased recently in using grey water to irrigate green roofs and green walls, thus saving potable water. Pucher *et al.* (2022) indicated that not only did grey water provide a source of irrigation but the water itself was partially cleaned by running it through the rhizosphere of the wall plants. The optimum treatment reduced the COD in the grey water by 80%, total organic carbon (TOC) by 74%, total bonded nitrogen (TNb) by 70%, ammonium nitrogen (NH_4-N) by 81% and turbidity by 79%. Thus, the water reached the threshold values for water reuse in several EU countries. However, a decrease in performance over time was noted. Similarly, Kotsia *et al.* (2020) found phosphorus removal gradually decreased from 100% during the first months of operation to 15% during the second year of operation, and further long-term studies are needed to understand the longer-term dynamics of chemical removal. More positively, a number of studies have shown no additional damage to plants irrigated with grey water compared to tap water. Studies in Melbourne, Australia (Prodanovic *et al.*, 2019) showed that green-wall plants could successfully adapt to greywater-only irrigation and regulated nitrogen and phosphorus uptake from greywater. Higher-performing plants, *Nephrolepis obliterate, Carex appressa, Dianella tasmanica, Myoporum, Parvifolium, Agapanthus praecox, Phormium tenax* and *Liriope muscari,* had on average 7–10% higher total nitrogen removal than un-vegetated portions of the wall (over 88% removal overall), which is attributed to consistently high oxidised nitrogen uptake during a 2-year study. *Pittosporum tobira* and *Hedera helix* (ivy) grew well under greywater irrigation, whereas partial defoliation of *Polygala myrtifolia* was noted during winter (Kotsia *et al.*, 2020).

3.6 Air Pollution

Globally, poor air quality is a significant environmental problem. The World Health Organisation states nine out of ten people breathe polluted air and air pollution kills 7 million people every year. It is thought to contribute to one-third of all deaths due to stroke, lung cancer and heart disease.

Industrial processes, vehicle emissions and wood-fired cooking stoves are the main contributors. Key urban air pollutants include carbon monoxide (CO), volatile organic compounds (VOCs), nitrous oxides (NO_x), sulphur dioxide (SO_2), ozone (O_3), benzene (C_6H_6), various heavy metals including lead (Pb) and particulate matter (PM). Particulate matter is a complex mixture of extremely small particles and liquid droplets and can constitute a number of components, including acids (derived from nitrates and sulphates), organic chemicals, metals and soil or dust particles. Particles smaller than 10 µm (PM_{10}) are considered particularly damaging as they generally pass through the throat and nose and enter the lungs. Many of the pollutants are generated by vehicle traffic as well as industrial processes.

Green interventions

At a local level, green infrastructure can mitigate the effects of poor-quality air, but the extent to which it does this depends on the following:

- type of pollutant;
- wind speed and direction (vegetation needs to block or disperse the pollutants, not trap and concentrate them in locations that citizens frequent);
- temperature;
- building and road 'geometry';
- type of green (vegetation) intervention – trees, hedges, green walls or roofs;
- plant species used in this phytoremediation;
- whether leaves are present or not (e.g. seasonal effects); and
- existing pollutant loads on the plants.

Vehicles are a major source of pollution with diesel engines particularly associated with smuts and PM pollution. Many governments are discouraging the use of diesel vehicles and fossil fuel cars in general. The car market is changing due to government action to encourage the purchase of electric vehicles (EVs) to reduce carbon emissions and improve air quality. But, the relatively high weight of EVs can result in PM emissions that are equal or even higher than those from combustion engine vehicles, with pollutants still being released from tyres and brake dust (Woo *et al.*, 2022).

The effectiveness of plants to mitigate air pollution does depend on the pollutant in question, its concentration and the species of plant being used. Some pollutants are actually toxic to plants. Many plant species are sensitive to O_3 or SO_2, for example, when exposed to higher levels of these compounds. Conversely, plants can filter and absorb pollutants such as nitrogen dioxide (NO_2), SO_2 (lower concentrations), certain heavy metals and vehicle particulates. The capacity to do so can be genotype specific, but in general species that have waxy cuticles, hairs present on their leaves or lots of fine leaves (e.g. conifers) tend to be best at dealing with pollutants (Fig. 3.14).

Plants remove gaseous pollutants by absorption through leaf stomata or leaf epidermal surfaces. Studying downward pollutant flux as the combination of deposition velocity and pollutant concentration, Nowak *et al.* (2014) found pollution removal for NO_2, SO_2, CO, O_3 and PM_{10} varied across cities due to background pollution levels, tree canopy

Fig. 3.14. Fine leaves such as conifer needles as effective at capturing pollutants, in this case *Pinus sylvestris* var. Aurea.

cover, the duration plants were in leaf, the amount of precipitation and other meteorological variables that affect tree transpiration and deposition velocities. Moreover, deposition levels can be imprecise as PM deposited on leaves is retained, but only temporarily, and then can be resuspended to the atmosphere by high wind speed. A lot of the pollutants reach the soil by being washed off leaves by rain or transferred with the abscised leaves in autumn (Nowak *et al.*, 2014).

Using electron microscopy, del Carmen Redondo-Bermúdez *et al.* (2021) showed that different plant species were effective at capturing different sizes of PM (Fig. 3.15) and by using a community of species could widen the spectrum of particles caught. The width, depth and frequency of the grooves typically found on leaves of different plant species were important to explain PM capture across the species. Overall, this study found more PM adhered to *Thuja occidentalis cv.* Smaragd followed by *Hedera helix* cv. Woerner, whereas *Phyllostachys nigra* had the lowest overall particulate numbers.

Ensuring green space helps improve air quality is one situation where landscape designers can really 'earn their spurs'. Good design will improve local air quality, but poor design can make things worse. Key is understanding where a pollutant is being sourced from, and how that relates to prevailing winds and the effects of urban geometry on where pollutants may accumulate or indeed disperse. In narrow urban roadways, where the roads are hemmed in by multi-storey buildings (so-called urban 'canyons'), then the use of broad-canopy trees inhibits air movement and does not let pollutants escape effectively from the canyon (Fig. 3.16). In effect, the trees inhibit the turbulence above the road that can aid dispersal. In contrast, low-level green barriers (hedges) and perhaps green walls can improve air quality in such canyon environments. In more open locations, wide, tall hedges of low porosity can protect pedestrians from vehicle emissions (Abhijith *et al.*, 2017; Kończak *et al.*, 2021) (Fig 3.17). Filtration effects tend to be improved by widening the hedges (barriers 6–7 m wide would be ideal but not always practical in a dense street canyon) and increasing density to 90–95% foliar density (i.e. 5–10 air porosity) (Abhijith *et al.*, 2017). An understanding of the specific site conditions is important though, because a hedge or green barrier can also affect windspeed or alter the eddying

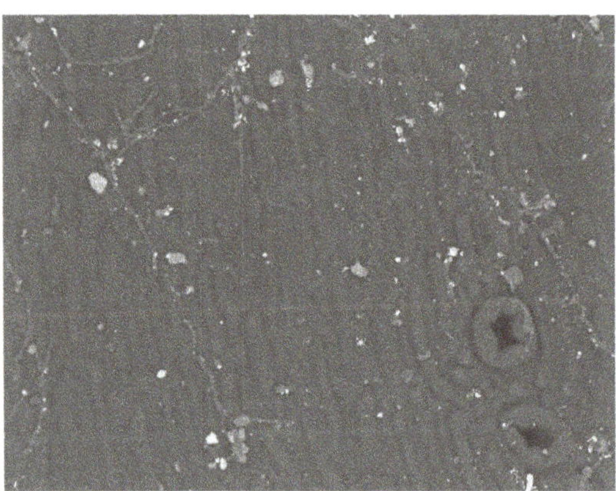

Fig. 3.15. Particulate matter (pale grey) adhering to the leaf surface of *Thuja occidentalis* ×600. (Used with permission from M. Carmen Redondo-Bermúdez.)

Fig. 3.16. In narrow city canyons, broad-canopy trees tend to concentrate air pollutants from traffic and do not allow wind to disperse them effectively (top). Use of narrow, fastigiated trees is preferential here (bottom).

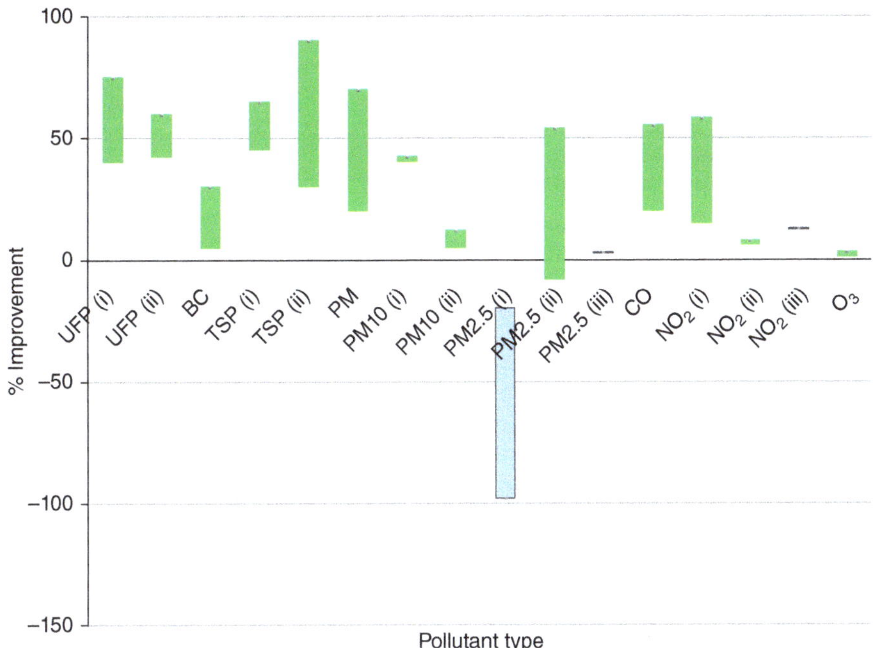

Fig. 3.17. Improvements in local air quality by providing hedges adjacent to major roadways, as evidenced by different studies. Change is based on similar locations without the hedge. Ranges are depicted as weather and other factors can affect the functionality of the hedges. Note one study reported the introduction of hedge increased exposure to $PM_{2.5}$. Key: UFP = ultra fine particles, BC = black carbon, TSP = total solid particles, PM = particulate matter, CO = carbon monoxide, NO_2 = nitrogen dioxide, O_3 = ozone. (Modified with permission from Abhijith *et al.*, 2017.)

effects of the wind, meaning that local pockets of poor-quality air arise. Understanding the source of pollutant, local wind direction and air flow and where people frequently reside is key.

Green roofs and green walls may help urban air quality (Joshi and Ghosh, 2014) or in the combinations of green infrastructure with other passive pollution control methods. Again, there are examples of green infrastructure helping but also hindering air quality control measures. Litschike and Kuttler (2008) recommended green walls as they reduce aerial particulates through deposition but unlike trees do not alter the air exchange between the street canyon and the air above it. Jayasooriya *et al.* (2017) working at a city scale suggested significant improvement in air quality with green walls, but pollutant reductions were not as great as that due to trees. In a street canyon environment, green walls improved air quality in different street canyon aspect ratios (height to width ratios of 1.25 and 2), with reductions of up to 35% for NO_2 concentration and 50% in PM_{10} concentration (Pugh *et al.*, 2012). Climbing vegetation on walls has been found to trap particulate matter, reducing pollutants in the interior living space with the associated implications for human health (Ottelé *et al.*, 2010). As with other green infrastructure interventions, plants with waxy cuticles or fine leaves seem appropriate, e.g. climbers such as *Hedera* or *Jasminum* (Chen *et al.*, 2016).

Green roofs are less likely to be in the direct 'firing line' of poor-quality air but may still contribute to local mitigation. Lower impact from (extensive) green roofs relates to both a greater distance away from the pollutant source and also because they can have a low surface roughness (Speak *et al.*, 2012). Intensive green roofs (i.e. with larger and taller plants) increase pollutant removal (Currie and Bass, 2008; Yang *et al.*, 2008). Baik *et al.* (2012) identified the cooling effect of green roofs may partially explain a greater capacity to reduce pollution. They found a 32% reduction in pollutant concentrations with 2°C cooling intensity at breathing level, due to enhanced canyon vortices and higher vertical dispersion arising from the downward movement of cool air from the green roof. In contrast, other studies have shown only minimal benefits (Pugh *et al.*, 2012). Species differences on green roofs arose from differences in macro- and micro-morphology and that the grasses *Agrostis stolonifera* and *Festuca rubra* were more effective than broadleaved *Plantago lanceolata* and *Sedum album* at PM_{10} capture. Quantification of the annual PM_{10} removal potential was calculated under a maximum *Sedum* green roof installation scenario for an area of the city centre, which totalled 325 ha. Remediation of 2.3% (±0.1%) of 9.18-tonne PM_{10} inputs for this area could be achieved under this scenario. Yang *et al.* (2008) found that 1675 kg of air pollutants, such as NO_2, SO_2 and PM_{10}, were removed by 19.8 ha of green roofs in 1 year. A study in Toronto found that 58,000 kg of air pollutants could be removed if all the roofs in the city were converted to green roofs, with intensive green roofs having a higher impact than extensive green roofs (Currie and Bass, 2008). In contrast, Rafael *et al.* (2021) considered the impact of green roofs on air quality was negligible. While these studies offer promising results, they are based on modelling alone and, to date, relatively little empirical data on green-roof removal of air pollution has been published.

Although it is clear plants trap/absorb aerial pollutants, the key question becomes one of volume and how much green space is required to solicit a large enough impact to help reduce human health risks. Tallis *et al.* (2011), modelling the role of urban forests in London, UK, indicate that the tree canopy removes between 0.7% and 1.4% of PM_{10} (852 to 2121 tonnes) annually from the urban boundary layer. Increasing the urban tree component from 20% to 30% of total land area would only increase removal to 1.1–2.6% of the total by 2050 (1109–2379 tonnes). Sicard *et al.* (2018) considered that the annual percent air quality improvement in cities due to trees and shrubs is less than 2%. Somewhat more positively, studies in Leicester, UK showed that the aerodynamic dispersive effect of trees on $PM_{2.5}$ concentrations result in a 9.0% reduction. There was also a reduction in $PM_{2.5}$ by 2.8% due to direct deposition on trees (11.8 t $year^{-1}$) and 0.6% owing to deposition on grass (2.5 t $year^{-1}$). Data suggest, therefore, that tree planting should focus on the more polluted areas – e.g. beside roads – and use species that maximise particulate deposition all year round (e.g. coniferous spp.). Some research studies suggest that relatively wide belts of woodlands are required to elicit measurable benefits (Pataki *et al.*, 2011), at least for some of the pollutants encountered. Work on a forested, peri-urban national park near Mexico City, Mexico showed that total annual air quality improvement due to park vegetation was detectable but only accounted for approximately 0.02% of carbon monoxide, 1% of O_3 and 2% of PM_{10} of annual concentrations for these pollutants (Baumgardner *et al.*, 2012). Setälä *et al.* (2013) claim the 'empirical evidence on the potential of urban trees to mitigate air pollution is meagre, particularly in northern climates with a short growing season'. Their study monitored levels of NO_2, anthropogenic VOCs and particle deposition using passive sampling systems in two Finnish cities. Concentrations of each pollutant in August (summer, leaf period) and March (winter, leaf-free period) were slightly but not significantly lower under tree canopies than in adjacent open areas. Furthermore, vegetation-related environmental variables (canopy closure, number and size of trees, and density of understorey vegetation) did not explain variation in pollution concentrations. They concluded that the ability of urban park/forest vegetation to remove air pollutants was insignificant, at least in northern climates.

Vegetation as a source of pollution

Certain plant species may contribute to atmospheric pollution through being strong emitters of biogenic volatile organic compounds (BVOCs), such as isoprene and monoterpenes. For example, isoprene is emitted from trees such as *Populus, Quercus, Platanus* (Fig. 3.18) or *Eucalyptus* (Wagner and Kuttler, 2014; Fitzky *et al.*, 2019) and this compound reacts rapidly in the atmosphere, contributing to O_3 and secondary organic aerosol (SOA) formation. They have been implicated in contributing to the formation of photochemical smog and increasing levels of O_3 in the urban environment (Peñuelas and Staudt, 2010). There is a complex relationship between BVOC emissions from trees and O_3 formation and then O_3 absorption by the trees themselves. For cities with a large component of forests, isoprene and other BVOCs can contribute to peak summertime O_3 levels (Utembe *et al.*, 2018) and early-morning O_3 spikes, although these can be mitigated by high atmospheric hydroxyl radical (·OH) activity. Trees that emit high BVOC concentrations may pose a risk of

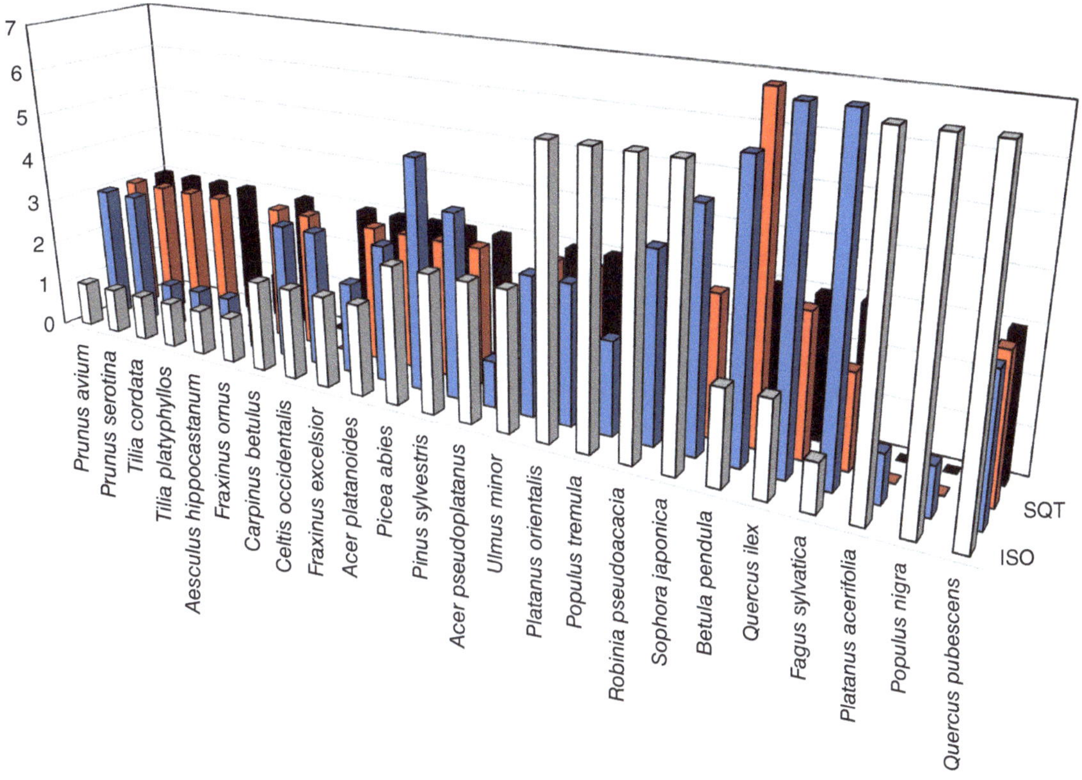

Fig. 3.18. The relative emission levels of BVOCs from temperate trees species. Comparisons between isoprene (ISO, white), monoterpene (MT, blue), sesquiterpenes (SQT, red), and oxygenated VOCs (OVOC, black). (Modified with permission from Fitzky *et al.*, 2019.)

reducing air quality and choice of tree taxa to be planted should reflect this.

Trees and grass, of course, are emitters of one of the world's most common allergens – pollen. Pollen is the cause of hay fever (allergic rhinitis) which not only causes discomfort to millions of sufferers around the globe but can also be a precursor for more serious respiratory diseases such as asthma and allergic bronchitis. In some areas, one in four people are thought to suffer hay fever. In these terms, trees and meadows must be seen to provide a considerable ecosystem disservice. Some trees are worse than others and within temperate regions members of the birch family Betulaceae (*Betula, Alnus, Corylus* and *Carpinus* spp.) and the Fagaceae (*Quercus, Castanea* and *Fagus*) are most commonly implicated in allergic rhinitis, with the most notorious being the birch allergen Bet v 1 associated with *Betula verrucosa* (Biedermann *et al.*, 2019). Thus, green space typology and species composition become important to ensure plants are improving urban air quality, not contributing to its poor quality.

3.7 Pesticides and Other Interactions with Chemicals

In many counties, maintenance of green space in the public realm is less reliant on chemical inputs than in the past. Increasingly, integrated control measures and more targeted applications of fertilisers and pesticides aim to improve environmental performance as well as reduce costs. This is perhaps in contrast to applications in the domestic context where pesticide sales are valued at US$8,580M and are growing (Anon., 2023). Garden chemicals have widespread use, e.g. 50% of UK and 74% of USA homeowners (Robbins *et al.*, 2001; Grey *et al.*, 2006). Their application, particularly on lawns, is apparently linked to home property values, with more affluent residents using them more frequently (Robbins *et al.*, 2001). In the USA, local by-laws encouraging tidy, well-maintained gardens tend to increase the use of garden chemicals and promote intensive management (e.g. lawns regularly mown and weed-free: Robbins *et al.*, 2001; Clayton, 2007). Urbanisation has also increased the area of land devoted to lawns in the USA. It is estimated that nearly a quarter of urban land cover in the USA is dedicated to lawn turf (Fuentes, 2021), which amounts to 164,000 km^{-2}. Pesticides can be applied unintentionally at higher concentrations in small urban areas such as lawns, gardens and impermeable surfaces (Diaz *et al.*, 2020). Concerns over the use of pesticides in public green space has resulted in their prohibition in France, even for compounds such as glyphosate with links to cancer still debated (Cavalier *et al.*, 2022).

Garden and landscape pesticides are still significant contributors to non-point-source water pollution in the USA (Robbins *et al.*, 2001). Nutrient and pesticide runoff contribute to eutrophication of rivers, lakes and coastal zones. Different formulations vary widely in levels of toxicity, health effects and environmental impact. Risks involved are also dependent on correct or inappropriate use by homeowners. Common garden pesticides include 2,4-D, dicamba, glyphosate and chlorpyrifos with potential for mammalian toxicity. Pesticides that may only be mildly toxic to human health may be detrimental to the integrity of water resources, including the biological health of streams, fish and macro-invertebrates. Direct application of pesticides may also be problematic. Neonicotinoids – a range of systemic insecticides (e.g. imidacloprid, clothianidin and thiomethoxam) – have been linked to dramatic falls in bee populations. Use in domestic settings may be higher (perhaps 120 times higher) than that used in commercial agriculture and horticulture. Direct contact with foliar neonicotinoid sprays is hazardous to pollinating insects, and residues can remain on leaf tissues for a number of days after spraying (Porseryd *et al.*, 2024). Bumble bees in urban UK gardens had detectable levels of neonicotonoid insecticides and fungicides, although lower levels than for bees in agricultural areas. Neonicotinoid pesticide use in urban lawns can negatively affect bumble bee colony growth and new queen production when these chemicals get onto flowering plants. Similarly, there are negative correlations between bumble bee and butterfly abundance and use of insecticides and herbicides in French gardens (Baldock, 2020). As neonicotinoids are systemic in the plant, they redistribute themselves around the tissues, and runoff chemicals are thought to persist in the soil for a number of months. The UK aims to ban the use of neonicotinoids in the near future.

Most public green space has low fertiliser and pesticide use but not all. The most intensively managed grass swards, such as bowling greens, golf courses (particularly golf greens and tees) and premier soccer pitches have higher than average chemical inputs. Kearns and Prior (2013) carried out a questionnaire on pesticide use on golf courses in Northern Ireland, UK. They showed that almost 48% of the total area of golf courses assessed received a pesticide treatment in the year preceding the survey. Overall, treated areas of the course received an average of 2.2 kg ha^{-1} p.a., with most intensive application being on the greens with a mean of 7.5 kg ha^{-1}. Herbicides tend to comprise the largest component of pesticides (Fig. 3.19). Kearns and Prior (2013) commented that the rates of pesticide use were relatively high, for example, higher than used in pasture management in agricultural systems. Work in the USA indicated that pesticide use may be even more intensive (Bekken *et al.*, 2021). Golf greens could receive 43 pesticide applications a year, compared to tees at 19 applications, fairways at 18 applications and even the rough grass at the edge of the fairway received an application. Concerns over high levels of pesticide use on golf courses has resulted in some European countries banning (e.g. Spain and the Wallonia region of Belgium) or restricting (e.g. Italy and Norway) pesticides available for use on golf courses (Bekken *et al.*, 2021). Kearns and Prior (2013) highlighted three areas of risk/concern with pesticide application on golf courses:

1. Although the pesticides used are considered low risk to humans when instructions are followed, golfers can access the sites quite shortly after application of the pesticides (21% of courses had removed 'sprayed areas' signs within 4 h of application) with potential for direct contact with the pesticide as golfers placed/lifted their ball from the turf.
2. Many courses have water bodies present with the potential for applied chemicals to enter aquatic systems.
3. Almost 33% of all rough areas on the courses also received a pesticide treatment (mean 1.6 kg ha^{-1}); as these rough areas can be important as wildlife habitats in their own right, this was highlighted as a point of concern.

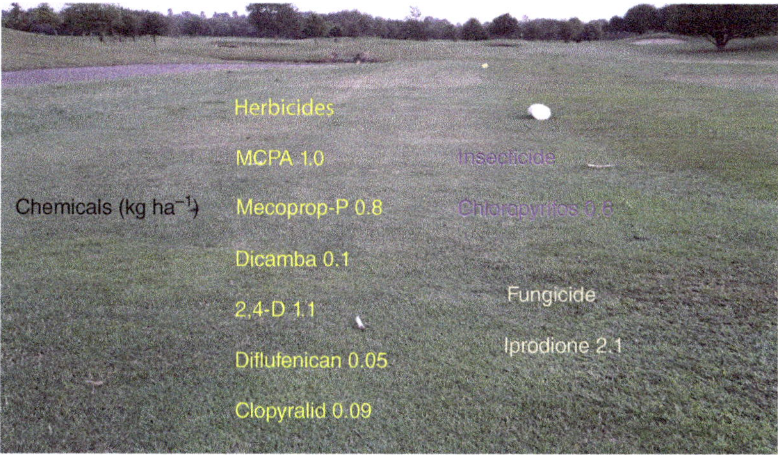

Fig. 3.19. Typical pesticide active ingredients and volumes used on golf courses per year in UK and Northern Ireland.

Despite these data sets, many golf courses have tried to reduce their environmental impact over the last two decades, due to the original criticisms associated with high chemical use and non-sustainable use of irrigation water.

Surface runoff of pesticides, fertilisers and particulate matter from urban areas can contribute to the pollution of watercourses. Parts of the gardening community are, therefore, encouraging lower chemical use and organic approaches, largely due to concerns over environmental issues. For example, the use of organic composts (green waste and spent mushroom) to support pest (e.g. aphid) control by encouraging predatory species (e.g. spiders and beetles) has shown some success. Furthermore, green spaces themselves can reduce pollution through capture of particulates and dissolved pollutants.

3.8 Non-native 'Alien' Species

One of ornamental horticulture's more notorious legacies has been that some introduced plant ('alien') species have become invasive in climatic zones outwith their natural distribution. With reduced natural completion and less pressure from herbivores and pathogens, such plants can exploit new ecological niches, outcompeting native flora. Many species stay restrained within the garden environment and rely on careful cultivation to maintain a presence in their new country; others, however, can 'cross the garden' fence and become invasive alien weeds. Notable examples include *Acacia* spp. in South Africa, *Clematis vitalba* in Australia and New Zealand, *Solidago gigantea* in Japan, *Ailanthus altissima, Eichhornia crassipes, Lantana camara, Lonicera japonica* in the USA and *Rhododendron superponticum, Fallopia japonica* and *Impatiens glandulifera* in the UK. Even though, only a small proportion of introduced plants may become invasive (<10%), the ecological and cost implications are vast (Bigirimana *et al.*, 2012); with <1% contributing to the main costs (Haubrock *et al.*, 2021). In the UK, plant invasions contribute about 16%% of total costs, i.e. £622M p.a. (Eschen et al., 2023). Some counties have been disproportionally affected by non-native species, e.g. weed control in Australia alone costing US$3,306M in 2008.

Increasing international trade in plants also introduces non-native pests and pathogens. The ornamental trade contributes 90% of the human-assisted introductions of plant pests in the UK (Smith *et al.*, 2007), although horticultural associations are trying to address this through new policies and practices. Most new plant pathogens tend to be found on ornamental plants (53%) with lower incidences on native species (15%) (Jones and Baker, 2007). The relatively high plant diversity found in urban environments, most notably gardens, may also provide opportunities for pest and pathogen species to move across to new host plant genotypes. Cultivated forms of wild plants commonly grown in gardens may have more invertebrate herbivores associated with them than has been recorded in the less-widespread, native wild form. With *Potentilla fruticosa*, for example, nine species of moth have been recording feeding off the cultivated ornamental genotypes but none on the native species itself (Owen, 2010). Increasing movement of people and goods including plants and derived products across the globe are contributing to greater incidences of pests and pathogens. This has corresponded to some countries tightening regulations about the movements of plants, including the requirement for sanitary certificates for imported plant materials, more inspections of incoming material by plant health officials and greater use of quarantine facilities to hold plants until they can be shown to be free of key pests and pathogens. In the UK alone, significant landscape tree species such as *Fraxinus excelsior* (ash), *Larix kaempferi* (Japanese larch) and *Pinus sylvestris* (Scots pine) are under threat from pathogens such as *Chalara fraxinea* (ash die-back), *Phytopthora ramorum* (sudden oak death) and *Dothistroma septosporum* (Dothistroma needle blight), respectively – the likely sources of initial infection coming in from overseas. Similar problems exist elsewhere. Some pathogens have more than one host and can be devastating to more than one crop type. *Xylella fastidiosa* is the causal agent of olive quick decline syndrome – a major concern for olive growers (with potential losses to the European olive industry estimated at between €1.9B and €5.2B over the next 50 years), but the sub-strains of the pathogen are thought to already affect 595 plant species globally, covering 275 genera and 85 distinct families (EFSA, 2020).

The other driver for enhanced pest and pathogens incidences is climate change. Alterations in climate are providing new niches for pathogenic organisms, or increasing the likelihood of pest species being able to complete their life cycle or colonise new territory. In parallel, such climatic shifts may induce greater susceptibility in the host plants, due to stress from abiotic factors (Cregg and Dix, 2001). Greater occurrence of heat island effects and allied abiotic stresses have been linked to a higher incidence of pest species, such as clearwing moth *Podesia syringae* on *Fraxinus pennsylvatica* in the UK. Warmer winter temperatures and a lack of frost in urban centres means that pest species that were once restricted to protected environments such as glasshouses and polytunnels can now survive all year round outdoors. This includes species such as *Trialeurodes vaporariorum* (glasshouse whitefly), *Icerya purchasi* (cottony cushion scale) and *Tetranychus urticae* (red spider mite). When consideration is given to controlling and eradicating all invasive species (plants, animals and pathogens), the UK estimates to spend £4,014M (Eschen *et al.*, 2023) and the global cost is thought to be US$29,198M (Diagne *et al.*, 2021).

Conclusions

- Urban plants provide a wide range of benefits (called 'ecosystem services') to the urban environment but also in some instances certain drawbacks (called 'ecosystem disservices').
- Some taxa are better at providing specific ecosystem services than other taxa. Those with cited benefits are categorised as functional ornamental plants (FOPs).
- Urban plants help keep the city and adjacent buildings cool during hot weather and warm during cold or windy periods. In doing so, they can reduce the energy consumed within buildings; for example, a well-designed planted landscape may reduce a house's winter heating bills by 40%.
- Vegetation can interfere with sound waves and make a location physically quieter. Green spaces though also seem to have a psychological effect on people visiting them, and these places can seem quieter, even perhaps when they are not.
- Urban plants sequester carbon out of the atmosphere and fix it within their tissues and within the soil. However, the extent they do this may be relatively limited at a global scale, especially when the amount of carbon emitted through landscape maintenance activities is considered.
- Plants are increasingly being used within the urban environment (and in river catchments in general) to help hydrological flows, particularly the slowing down of rainwater runoff. Plants hold rainwater in their canopies, direct water that runs off their leaves and stems to the soil and allow better infiltration of water to lower soil levels through pores and channels created by their roots. The action of roots and vapour pressure deficits around the plant leaf canopy draw soil water back up to the atmosphere (transpiration), which in turn dries out the soil. Thus drier soils, in turn, can hold more rainwater again, should a subsequent storm event occur, thus helping reduce surface flooding.
- Urban plants act as filters, with potential to remove, deactivate or lock up both aerial and waterborne pollutants. Polluted surface runoff

water is filtered through the action of roots and soil biology. Green barriers, green walls and tree/shrub shelter belts can be used with careful design to partially mitigate point sources of aerial pollutants – for example, screening pedestrians from vehicle fumes originating from a road junction. Absorbance and deactivation rates vary with plant species and the landscape design employed.

- Functional plants require water, and high water use can be a consequence of urban plantings, especially in hot/arid climates. Water use in the landscape can be both wasteful and expensive and distract water resources away from other important uses. Future urban design will need to link the management of green infrastructure more effectively to grey water supplies, ensuring these are filtered and treated effectively so as not to cause risk to the public or the plants.
- Although many green spaces have no or limited reliance on pesticides or fertilisers, their use can still be surprisingly high in urban contexts. This includes chemical use in home gardens. To be truly environmental, urban horticulture needs to identify and reduce the misuse of chemicals and look for alternative, more benign and environmentally sensitive ways of managing green space. For example, developing more biodiverse spaces, are likely to help keep pest populations down naturally.
- Environmental horticulture also needs to take care that 'a balance of nature' is maintained when introducing new plant genotypes to the environment. New genotypes may well be a requirement as the planet warms, but it is important that any new plants are not invasive or accelerate the decline of native plant communities and ecosystems.

References

Abhijith, K.V., Kumar, P., Gallagher, J., McNabola, A., Baldauf, R. *et al.* (2017) Air pollution abatement performances of green infrastructure in open road and built-up street canyon environments – A review. *Atmospheric Environment* 162, 71–86.

Afshari, A. (2017) A new model of urban cooling demand and heat island – Application to vertical greenery systems (VGS). *Energy and Buildings* 157, 204–217.

Akbari, H. (2002) Shade trees reduce building energy use and CO_2 emissions from power plants. *Environmental Pollution* 116, S119–S126.

Akbari, H., Pomerantz, M. and Taha, H. (2001) Cool surfaces and shade trees to reduce energy use and improve air quality in urban areas. *Solar Energy* 70, 295–310.

Anon. (2023) *Home and Garden Pesticides Market Size, Share & Trends Analysis Report: Forecasts, 2024–2030*. Grand View Research. Available at: https://www.grandviewresearch.com/industry-analysis/home-garden-pesticides-market.

Ariluoma, M., Ottelin, J., Hautamäki, R., Tuhkanen, E.M. and Mänttäri, M. (2021) Carbon sequestration and storage potential of urban green in residential yards: A case study from Helsinki. *Urban Forestry & Urban Greening* 57, 126939.

Armson, D., Stringer, P. and Ennos, A.R. (2012) The effect of tree shade and grass on surface and globe temperatures in an urban area. *Urban Forestry & Urban Greening* 11, 245–255.

Armson, D., Stringer, P. and Ennos, A.R. (2013) The effect of street trees and amenity grass on urban surface water runoff in Manchester, UK. *Urban Forestry & Urban Greening* 12, 282–286.

Baik, J.J., Kwak, K.H., Park, S.B. and Ryu, Y.H. (2012) Effects of building roof greening on air quality in street canyons. *Atmospheric Environment* 61, 48–55.

Balany, F., Ng, A.W., Muttil, N., Muthukumaran, S. and Wong, M.S. (2020) Green infrastructure as an urban heat island mitigation strategy – A review. *Water* 12, 3577.

Baldock, K.C. (2020) Opportunities and threats for pollinator conservation in global towns and cities. *Current Opinion in Insect Science* 38, 63–71.

Baumgardner, D., Varela, S., Escobedo, F.J., Chacalo, A. and Ochoa, C. (2012) The role of a peri-urban forest on air quality improvement in the Mexico City megalopolis. *Environmental Pollution* 163, 174–183.

Beard, J.B. and Green, R.L. (1994) The roles of turfgrasses in environmental protection and their benefits to humans. *Journal of Environmental Quality* 23, 452–460.

Bekken, M.A., Schimenti, C.S., Soldat, D.J. and Rossi, F.S. (2021) A novel framework for estimating and analyzing pesticide risk on golf courses. *Science of The Total Environment* 783, 146840.

Biedermann, T., Winther, L., Till, S.J., Panzner, P., Knulst, A. and Valovirta, E. (2019) Birch pollen allergy in Europe. *Allergy* 74, 1237–1248.

Bigirimana, J., Bogaert, J., De Cannière, C., Bigendako, M.J. and Parmentier, I. (2012) Domestic garden plant diversity in Bujumbura, Burundi: Role of the socio-economical status of the neighborhood and alien species invasion risk. *Landscape and Urban Planning* 107, 118–126.

Bijoor, N.S., Czimczik, C.I., Pataki, D.E. and Billings, S.A. (2008) Effects of temperature and fertilization on nitrogen cycling and community composition of an urban lawn. *Global Change Biology* 14, 2119–2131.

Calcerano, F. and Martinelli, L. (2016) Numerical optimisation through dynamic simulation of the position of trees around a stand-alone building to reduce cooling energy consumption. *Energy and Buildings* 112, 234–243.

Cameron, R. (2023) "Do we need to see gardens in a new light?" Recommendations for policy and practice to improve the ecosystem services derived from domestic gardens. *Urban Forestry & Urban Greening* 80, 127820.

Cameron, R.W. and Blanuša, T. (2016) Green infrastructure and ecosystem services – Is the devil in the detail? *Annals of Botany* 118, 377–391.

Cameron, R.W., Taylor, J.E. and Emmett, M.R. (2014) What's 'cool' in the world of green façades? How plant choice influences the cooling properties of green walls. *Building and Environment* 73, 198–207.

Cameron, R.W., Taylor, J. and Emmett, M. (2015) A *Hedera* green façade – Energy performance and saving under different maritime-temperate, winter weather conditions. *Building and Environment* 92, 111–121.

Castleton, H.F., Stovin, V., Beck, S.B.M. and Davison J.B. (2010) Green roofs; building energy savings and the potential for retrofit. *Energy and Building* 42, 1582–1591.

Cavalier, H., Trasande, L. and Porta, M. (2022) Exposures to pesticides and risk of cancer: Evaluation of recent epidemiological evidence in humans and paths forward. *International Journal of Cancer* 152, 879–912.

Charlesworth, S.M. (2010) A review of the adaptation and mitigation of global climate change using sustainable drainage in cities. *Journal of Water and Climate Change* 1, 165–180.

Charlesworth, S.M. and Booth, C.A. (2012) The benefits of green infrastructure in towns and cities. In: *Solutions to Climate Change Challenges in the Built Environment*. Eds Ames and Iowa, Wiley-Blackwell, pp. 163–180.

Chen, L., Liu, C., Zou, R., Yang, M. and Zhang, Z. (2016) Experimental examination of effectiveness of vegetation as bio-filter of particulate matters in the urban environment. *Environmental Pollution* 208, 198–208.

Cheng, Z., McCoy, E.L. and Grewal, P.S. (2014) Water, sediment, and nutrient runoff from urban lawns established on disturbed subsoil or topsoil and managed with inorganic or organic fertilizers. *Urban Ecosystems* 17, 277–289.

Chun, B. and Guldmann J.M. (2014) Spatial statistical analysis and simulation of the urban heat island in high-density central cities. *Landscape and Urban Planning* 125, 76–88.

Clayton, S. (2007) Domesticated nature: Motivations for gardening and perceptions of environmental impact. *Journal of Environmental Psychology* 27, 215–224.

Coma, J., Solé, C., Castell, A. and Cabeza, L.F. (2014) New green facades as passive systems for energy savings on buildings. *Energy Procedia* 57, 1851–1859.

Cregg, B.M. and Dix, M.E. (2001) Tree moisture stress and insect damage in urban areas in relation to heat island effects. *Journal of Arboriculture* 27, 8–17.

Currie, B.A. and Bass, B. (2008) Estimates of air pollution mitigation with green plants and green roofs using the UFORE model. *Urban Ecosystems* 11, 409–422.

del Carmen Redondo-Bermúdez, M., Gulenc, I.T., Cameron, R.W. and Inkson, B.J. (2021) 'Green barriers' for air pollutant capture: Leaf micromorphology as a mechanism to explain plants capacity to capture particulate matter. *Environmental Pollution* 288, 117809.

DeWalle, D.R. and Heisler, G.M. (1980) *Water vapor mass balance method for determining air infiltration rates in houses.* Report 301. Northeastern Forest Experiment Station, Forest Service, US Department of Agriculture.

Diagne, C., Leroy, B., Vaissière, AC. *et al.* (2021) High and rising economic costs of biological invasions worldwide. *Nature* 592, 571–576.

Diaz, J.M., Warner, L.A., Oi, F. and Gusto, C. (2020) What do they know and what do they do? A national evaluation of landscape integrated pest management knowledge and use in the United States. *Journal of Integrated Pest Management* 11, 19.

Domene, E. and Saurí, D. (2006) Urbanisation and water consumption: Influencing factors in the metropolitan region of Barcelona. *Urban Studies* 43, 1605–1623.

Dunnett, N., Nagase, A., Booth, R. and Grime, P. (2008) Influence of vegetation composition on runoff in two simulated green roof experiments. *Urban Ecosystems* 11, 385–398.

European Food Safety Authority (EFSA) (2020) Update of the *Xylella* spp. host plant database – Systematic literature search up to 30 June 2019. *EFSA Journal* 18, e06114.

Enriquez, C., Alum, A., Suarez-Rey, E.M., Choi, C.Y., Oron, G. and Gerba, C.P. (2003) Bacteriophages MS2 and PRD1 in turfgrass by subsurface drip irrigation. *Journal of Environmental Engineering* 129, 852–857.

Eriksson, E., Baun, A., Henze, M. and Ledin, A. (2006) Phytotoxicity of grey wastewater evaluated by toxicity tests. *Urban Water Journal* 3, 13–20.

Eschen, R., Kadzamira, M., Stutz, S., Ogunmodede, A., Djeddour, D., Shaw, R., Pratt, C., Varia, S., Constantine, K. and Williams, F. (2023) An updated assessment of the direct costs of invasive non-native species to the United Kingdom. *Biological Invasions* 25, 3265–3276.

Fang, C.F. and Ling, D.L. (2003) Investigation of the noise reduction provided by tree belts. *Landscape and Urban Planning* 63, 187–195.

Fitzky, A.C., Sandén, H., Karl, T., Fares, S., Calfapietra, C. *et al.* (2019) The interplay between ozone and urban vegetation – BVOC emissions, ozone deposition, and tree ecophysiology. *Frontiers in Forests and Global Change* 2, 50.

Fletcher, D.H., Garrett, J.K., Thomas, A., Fitch, A., Cryle, P., Shilton, S. and Jones, L. (2022) Location, location, location: Modelling of noise mitigation by urban woodland shows the benefit of targeted tree planting in cities. *Sustainability* 14, 7079.

Frazer, L. (2005) Paving paradise: The peril of impervious surfaces. *Environmental Health Perspectives* 113, A456–A462.

Fuentes, T.L. (2021) Homeowner preferences drive lawn care practices and species diversity patterns in new lawn floras, *Journal of Urban Ecology* 7, juab015.

Gholami, M., O'Sullivan, A.D. and Mackey, H.R. (2023) Nutrient treatment of greywater in green wall systems: A critical review of removal mechanisms, performance efficiencies and system design parameters. *Journal of Environmental Management* 345, 118917.

Gidlöf-Gunnarsson, A. and Öhrström, E. (2007) Noise and well-being in urban residential environments: The potential role of perceived availability to nearby green areas. *Landscape and Urban Planning* 83, 115–126.

Gill, S.E., Handley, J.F., Ennos, A.R. and Pauleit, S. (2007) Adapting cities for climate change: The role of the green infrastructure. *Built Environment* 33, 115–133.

Gillman, L.N., Bollard, B. and Leuzinger, S. (2023) Calling time on the imperial lawn and the imperative for greenhouse gas mitigation. *Global Sustainability* 6, e3.

Grey, C.N.B., Nieuwenhuijsen, M.J., Golding, J. and Team, A. (2006) Use and storage of domestic pesticides in the UK. *Science of the Total Environment* 368, 465–470.

Gross, A., Shmueli, O., Ronen, Z. and Raveh, E. (2007) Recycled vertical flow constructed wetland (RVFCW) – A novel method of recycling greywater for irrigation in small communities and households. *Chemosphere* 66, 916–923.

Haubrock, P., Cuthbert, R., Sundermann, A., Diagne, C., Golivets, M. and Courchamp, F. (2021) Economic costs of invasive species in Germany. *NeoBiota* 67, 225–246.

Hof, A. and Wolf, N. (2014) Estimating potential outdoor water consumption in private urban landscapes by coupling high-resolution image analysis, irrigation water needs and evaporation estimation in Spain. *Landscape and Urban Planning* 123, 61–72.

Howarth, R.W., Boyer, E.W., Pabich, W.J. and Galloway, J.N. (2002) Nitrogen use in the United States from 1961–2000 and potential future trends. *AMBIO: A Journal of the Human Environment* 31, 88–96.

Huang, Y.J., Akbari, H., Taha, H. and Rosenfeld, A.H. (1987) The potential of vegetation in reducing summer cooling loads in residential buildings. *Journal of Climate and Applied Meteorology* 26, 1103–1116.

Hulme, M., Jenkins, G.J., Lu, X., Turnpenny, J.R., Mitchell, T.D. *et al.* (2002) *Climate Change Scenarios for the United Kingdom: The UKCIP02 Scientific Report* 120. Tyndall Centre for Climate Change Research, School of Environmental Sciences, University of East Anglia, Norwich.

Hutchison, B.A. and Taylor, F.G. (1983) Energy conservation mechanisms and potentials of landscape design to ameliorate building microclimates. *Landscape Journal* 2, 19–39.

Hutchinson, D., Abrams, P., Retzlaff, R. and Liptan, T. (2003) Stormwater monitoring two ecoroofs in Portland, Oregon, USA. *City of Portland Bureau of Environmental Services*.

Jamei, E., Chau H.W., Seyedmahmoudian, M. and Stojcevski, A. (2021) Review on the cooling potential of green roofs in different climates. *Science of The Total Environment* 791, 148407.

Jandl, R., Lindner, M., Vesterdal, L., Bauwens, B., Baritz, R., Hagedorn, F. and Byrne, K.A. (2007) How strongly can forest management influence soil carbon sequestration? *Geoderma* 137, 253–268.

Jayasooriya, V.M., Ng, A.W.M., Muthukumaran, S. and Perera, B.J.C. (2017) Green infrastructure practices for improvement of urban air quality. *Urban Forestry & Urban Greening* 21, 34–47.

Jefferson, B., Palmer, A., Jeffrey, P., Stuetz, R. and Judd, S. (2004) Grey water characterisation and its impact on the selection and operation of technologies for urban reuse. *Water Science and Technology* 50, 157–164.

Jo, H.K. and McPherson, E.G. (2001) Indirect carbon reduction by residential vegetation and planting strategies in Chicago, USA. *Journal of Environmental Management* 61,165–177.

Jones, D.R. and Baker, R.H.A. (2007) Introductions of non-native plant pathogens into Great Britain, 1970–2004. *Plant Pathology* 56, 891–910.

Joshi, S.V. and Ghosh, S. (2014) On the air cleansing efficiency of an extended green wall: A CFD analysis of mechanistic details of transport processes. *Journal of Theoretical Biology* 361, 101–110.

Kastka, J. and Noack, R. (1987) On the interaction of sensory experience, causal attributive cognitions and visual context parameters in noise annoyance. *Developments in Toxicology and Environmental Science* 15, 345–362.

Kaushal, S.S., Groffman, P.M., Band, L.E., Shields, C.A., Morgan, R.P., Palmer, M.A. and Fisher, G.T. (2008) Interaction between urbanization and climate variability amplifies watershed nitrate export in Maryland. *Environmental Science and Technology* 42, 5872–5878.

Kearns, C.A. and Prior, L. (2013) Toxic greens: A preliminary study on pesticide usage on golf courses in Northern Ireland and potential risks to golfers and the environment. *Safety and Security Engineering* V 134, 173.

Kim, S.W. and Brown, R.D. (2021) Urban heat island (UHI) intensity and magnitude estimations: A systematic literature review. *Science of The Total Environment* 779, 146389.

Ko, Y. (2018) Trees and vegetation for residential energy conservation: A critical review for evidence-based urban greening in North America. *Urban Forestry & Urban Greening* 34, 318–335.

Kończak, B., Cempa, M. and Deska, M. (2021) Assessment of the ability of roadside vegetation to remove particulate matter from the urban air. *Environmental Pollution* 268, 115465.

Kotsia, D., Deligianni, A., Fyllas, N.M., Stasinakis, A.S. and Fountoulakis, M.S. (2020) Converting treatment wetlands into "treatment gardens": Use of ornamental plants for greywater treatment. *Science of The Total Environment* 744, 140889.

Krenitsky, E.C, Carroll, M.J., Hill, R.L., Krouse, J.M. (1998) Runoff and sediment losses from natural and man-made erosion control materials. *Crop Science* 38, 1042–1046.

Kunasingam, P., Clayden, A. and Cameron, R. (2024) How does plant taxonomic choice affect building wall panel cooling? *Building and Environment* 256, 111493.

Lallana, C., Estrela, T., Nixon, S., Zabel, T., Laffon, L., Rees, G. and Cole, G. (1999) *Sustainable Water Use in Europe.* In: Krinner, W. (ed.) Office for Official Publ. of the Europ. Communities.

Le Coustumer, S., Fletcher, T.D., Deletic, A., Barraud, S. and Poelsma, P. (2012) The influence of design parameters on clogging of stormwater biofilters: A large-scale column study. *Water Research* 46, 6743–6752.

Li, H., Wang, Y., Li, S., Askar, A. and Wang, H. (2022) Shelter efficiency of various shelterbelt configurations: A wind tunnel study. *Atmosphere* 13, 1022.

Litschike, T. and Kuttler, W. (2008) On the reduction of urban particle concentration by vegetation – A review. *Meteorologische Zeitschrift* 17, 229–240.

Liu, Y. and Harris, D.J. (2008) Effects of shelterbelt trees on reducing heating-energy consumption of office buildings in Scotland. *Applied Energy* 85, 115–127.

Liu, W., Feng, Q., Chen, W., Wei, W. and Deo, R.C. (2019) The influence of structural factors on stormwater runoff retention of extensive green roofs: New evidence from scale-based models and real experiments. *Journal of Hydrology* 569, 230–238.

Liu, Y., Lei, Y., Johnston, D.M., Jiang, M., Dong, N. *et al.* (2023) Quantitative study on GHG emissions and the GWP influence of cemetery green space maintenance based on LCA. *Journal of Environmental Engineering and Landscape Management* 31, 67–81.

Livesley, S.J., Dougherty, B.J., Smith, A.J., Navaud, D., Wylie, L.J. and Arndt, S.K. (2010) Soil-atmosphere exchange of carbon dioxide, methane and nitrous oxide in urban garden systems: impact of irrigation, fertiliser and mulch. *Urban Ecosystems* 13, 273–293.

Livesley, S.J., Baudinette, B. and Glover, D. (2014) Rainfall interception and stem flow by eucalypt street trees – The impacts of canopy density and bark type. *Urban Forestry & Urban Greening* 13, 192–197.

Lopes, A., Oliveira, S., Fragoso, M., Andrade, J.A. and Pedro, P. (2009) Wind risk assessment in urban environments: The case of falling trees during windstorm events in Lisbon. In: *Bioclimatology and Natural Hazards*. Springer, Netherlands, pp. 55–74.

Martin-Vide, J. and Moreno-Garcia, M.C. (2020) Probability values for the intensity of Barcelona's urban heat island (Spain). *Atmospheric Research* 240, 104877.

Meili, N., Manoli, G., Burlando, P., Carmeliet, J., Chow, W.T. *et al.* (2021) Tree effects on urban microclimate: Diurnal, seasonal, and climatic temperature differences explained by separating radiation, evapotranspiration, and roughness effects. *Urban Forestry & Urban Greening* 58, 126970.

Meerow, A.W. and Black, R.J. (1993) *Enviroscaping to Conserve Energy: Guide to Microclimate Modification.* University of Florida Cooperative Extension Service, Institute of Food and Agriculture Sciences, EDIS.

Milesi, C., Running, S.W., Elvidge, C.D., Dietz, J.B., Tuttle, B.T. and Nemani, R.R. (2005) Mapping and modeling the biogeochemical cycling of turf grasses in the United States. *Environmental Management* 36, 426–438.

Morris, J. and Bagby, J. (2008) Measuring environmental value for natural lawn and garden care practices. *The International Journal of Life Cycle Assessment* 13, 226–234.

Moss, J.L., Doick, K.J., Smith, S. and Shahrestani, M. (2019) Influence of evaporative cooling by urban forests on cooling demand in cities. *Urban Forestry & Urban Greening* 37, 65–73.

Nardini, A., Andri, S. and Crasso, M. (2012) Influence of substrate depth and vegetation type on temperature and water runoff mitigation by extensive green roofs: Shrubs versus herbaceous plants. *Urban Ecosystems* 15, 697–708.

Nocco, M.A., Rouse, S.E. and Balster, N.J. (2016) Vegetation type alters water and nitrogen budgets in a controlled, replicated experiment on residential-sized rain gardens planted with prairie, shrub, and turfgrass. *Urban Ecosystems* 19, 1665–1691.

Nowak, D.J., Stevens, J.C., Sisinni, S.M. and Luley, C.J. (2002) Effects of urban tree management and species selection on atmospheric carbon dioxide. *Arboriculture & Urban Forestry* 28, 113–122.

Nowak, D.J., Hirabayashi, S., Bodine, A. and Greenfield, E. (2014) Tree and forest effects on air quality and human health in the United States. *Environmental Pollution* 193, 119–129.

Nur Hannah Ismail, S., Stovin, V. and Cameron, R.W. (2023) Functional urban ground-cover plants: Identifying traits that promote rainwater retention and dissipation. *Urban Ecosystems* 26, 1709–1724.

Nytch, C.J., Meléndez-Ackerman, E.J., Pérez, M.E. and Ortiz-Zayas, J.R. (2019) Rainfall interception by six urban trees in San Juan, Puerto Rico. *Urban Ecosystems* 22, 103–115.

Ottelé, M., van Bohemen, H.D. and Fraaij, A.L. (2010) Quantifying the deposition of particulate matter on climber vegetation on living walls. *Ecological Engineering* 36, 154–162.

Ottoson, J. and Stenström, T.A. (2003) Faecal contamination of greywater and associated microbial risks. *Water Research* 37, 645–655.

Ouédraogo, A.A., Berthier, E., Durand, B. and Gromaire, M.C. (2022) Determinants of evapotranspiration in urban rain gardens: A case study with lysimeters under temperate climate. *Hydrology* 9, 42.

Owen J. (2010) *Wildlife of a Garden: A Thirty-Year Study*. Royal Horticultural Society, London.

Padullés Cubino, J., Kirkpatrick, J.B. and Vila Subirós, J. (2017) Do water requirements of Mediterranean gardens relate to socio-economic and demographic factors? *Urban Water Journal* 14, 401–408.

Pataki, D.E., Carreiro, M.M., Cherrier, J., Grulke, N.E., Jennings, V. *et al.* (2011) Coupling biogeochemical cycles in urban environments: Ecosystem services, green solutions, and misconceptions. *Frontiers in Ecology and the Environment* 9, 27–36.

Parsons, K. (2019) *Human Heat Stress*. CRC Press, Boca Raton, Florida, USA.

Pauleit, S. and Duhme, F. (2000) Assessing the environmental performance of land cover types for urban planning. *Landscape and Urban Planning* 52, 1–20.

Paull, N., Krix, D., Torpy, F. and Irga, P. (2020) Can green walls reduce outdoor ambient particulate matter, noise pollution and temperature? *International Journal of Environmental Research and Public Health* 17, 5084.

Peñuelas, J. and Staudt, M. (2010) BVOCs and global change. *Trends in Plant Science* 15, 133–144.

Pérez, G., Coma, J., Sol, S. and Cabeza, L.F. (2017) Green facade for energy savings in buildings: The influence of leaf area index and facade orientation on the shadow effect. *Applied Energy* 187, 424–437.

Perini, K. Bazzocchi, F., Croci, L. Magliocco, A. and Cattaneo, E. (2017) The use of vertical greening systems to reduce the energy demand for air conditioning. Field monitoring in Mediterranean climate. *Energy and Building* 143, 35–42.

Permpituck, S. and Namprakai, P. (2012) The energy consumption performance of roof lawn gardens in Thailand. *Renewable Energy* 40, 98–103.

Pinto, U., Maheshwari, B.L. and Grewal, H.S. (2010) Effects of greywater irrigation on plant growth, water use and soil properties. *Resources, Conservation and Recycling* 54, 429–435.

Porseryd, T., Hellström, K.V. and Dinnétz, P. (2024) Pesticide residues in ornamental plants marketed as bee friendly: Levels in flowers, leaves, roots and soil. *Environmental Pollution* 345, 123466.

Portoghesi, L., Masini, E., Tomao, A. and Agrimi, M. (2023) Could climate change and urban growth make Europeans regard urban trees as an additional source of danger? *Frontiers in Forests and Global Change* 6, 1155016.

Pouyat, R.V., Yesilonis, I.D. and Golubiewski, N.E. (2009) A comparison of soil organic carbon stocks between residential turf grass and native soil. *Urban Ecosystems* 12, 45–62.

Prodanovic, V., McCarthy, D., Hatt, B. and Deletic, A. (2019) Designing green walls for greywater treatment: The role of plants and operational factors on nutrient removal. *Ecological Engineering* 130, 184–195.

Pucher, B., Zluwa, I., Spörl, P., Pitha, U. and Langergraber, G. (2022) Evaluation of the multifunctionality of a vertical greening system using different irrigation strategies on cooling, plant development and greywater use. *Science of the Total Environment* 849, 157842.

Pugh, T.A., MacKenzie, A.R., Whyatt, J.D. and Hewitt, C.N. (2012) Effectiveness of green infrastructure for improvement of air quality in urban street canyons. *Environmental Science and Technology* 46, 7692–7699.

Radingoana, M.P., Dube, T. and Mazvimavi, D. (2020) Progress in greywater reuse for home gardening: Opportunities, perceptions and challenges. *Physics and Chemistry of the Earth* 116, 102853.

Rafael, S., Correia, L.P., Ascenso, A., Augusto, B., Lopes, D. and Miranda, A.I. (2021) Are green roofs the path to clean air and low carbon cities? *Science of the Total Environment* 798, 149313.

Raji, B., Tenpierik, M.J. and Van Den Dobbelsteen, A. (2015) The impact of greening systems on building energy performance: A literature review. *Renewable and Sustainable Energy Reviews* 45, 610–623.

Reid, S.B., Pollard, E.K., Sullivan, D.C. and Shaw, S.L. (2010) Improvements to lawn and garden equipment emissions estimates for Baltimore, Maryland. *Journal of the Air & Waste Management Association* 60, 1452–1462.

Reyes-Paecke, S., Gironas, J., Melo, O., Vicuna, S. and Herrera, J. (2019) Irrigation of green spaces and residential gardens in a Mediterranean metropolis: Gaps and opportunities for climate change adaptation. *Landscape and Urban Planning* 182, 34–43.

Robbins, P., Polderman, A. and Birkenholtz, T. (2001) Lawns and toxins: An ecology of the city. *Cities* 18, 369–380.

Rouhollahi, M., Whaley, D., Byrne, J. and Boland, J. (2022) Potential residential tree arrangement to optimise dwelling energy efficiency. *Energy and Buildings* 261, 111962.

Schroll, E., Lambrinos, J., Righetti, T. and Sandrock, D. (2011) The role of vegetation in regulating stormwater runoff from green roofs in a winter rainfall climate. *Ecological Engineering* 37, 595–600.

Schuch, G., Serrao-Neumann, S., Morgan, E. and Choy, D.L. (2017) Water in the city: Green open spaces, land use planning and flood management – An Australian case study. *Land Use Policy* 63, 539–550.

Setälä, H., Viippola, V., Rantalainen, A.L., Pennanen, A. and Yli-Pelkonen, V. (2013) Does urban vegetation mitigate air pollution in northern conditions? *Environmental Pollution* 183, 104–112.

Sharvelle, S., Roesner, L.A., Qian, Y., Stromberger, M. and Azar, M.N. (2012) *Long-Term Study on Landscape Irrigation Using Household Graywater-Experimental Study.* The Urban Water Center, Colorado State University.

Shetty, N.H., Elliott, R.M., Wang, M., Palmer, M.I. and Culligan, P.J. (2022) Comparing the hydrological performance of an irrigated native vegetation green roof with a conventional *Sedum* spp. green roof in New York City. *Plos One* 17, e0266593.

Sicard, P., Agathokleous, E., Araminiene, V., Carrari, E., Hoshika, Y., De Marco, A. and Paoletti, E. (2018) Should we see urban trees as effective solutions to reduce increasing ozone levels in cities? *Environmental Pollution* 243, 163–176.

Sijimol, M.R. and Joseph, S. (2021) Constructed wetland systems for greywater treatment and reuse: A review. *International Journal of Energy and Water Resources* 5, 357–369.

Smith, R.M., Baker, R.H., Malumphy, C.P., Hockland, S., Hammon, R.P., Ostojá-Starzewski, J.C. and Collins, D.W. (2007) Recent non-native invertebrate plant pest establishments in Great Britain: Origins, pathways, and trends. *Agricultural and Forest Entomology* 9, 307–326.

Speak, A.F., Rothwell, J.J., Lindley, S.J. and Smith, C.L. (2012) Urban particulate pollution reduction by four species of green roof vegetation in a UK city. *Atmospheric Environment* 61, 283–293.

Speak, A., Escobedo, F.J., Russo, A. and Zerbe, S. (2020) Total urban tree carbon storage and waste management emissions estimated using a combination of LiDAR, field measurements and an end-of-life wood approach. *Journal of Cleaner Production* 256,120420.

St Hilaire, R., Arnold, M.A., Wilkerson, D.C., Devitt, D.A., Hurd, B.H., Lesikar, B.J. and Zoldoske, D.F. (2008) Efficient water use in residential urban landscapes. *HortScience* 43, 2081–2092.

Syme, G. J., Shao, Q., Po, M. and Campbell, E. (2004) Predicting and understanding home garden water use. *Landscape and Urban Planning* 68, 121–128.

Tallis, M., Taylor, G., Sinnett, D. and Freer-Smith, P. (2011) Estimating the removal of atmospheric particulate pollution by the urban tree canopy of London, under current and future environments. *Landscape and Urban Planning* 103, 129–138.

Tan, P.Y., Wong, N.H., Tan, C.L., Jusuf, S.K., Chang, M.F. and Chiam Q.Z. (2018) A method to partition the relative effects of evaporative cooling and shading on air temperature within vegetation canopy. *Journal of Urban Ecology* 4, 1–11.

Tan, J.K., Belcher, R.N., Tan, H.T., Menz, S. and Schroepfer, T. (2021) The urban heat island mitigation potential of vegetation depends on local surface type and shade. *Urban Forestry & Urban Greening* 62, 127128.

Thölix, L., Backman, L., Havu, M., Karvinen, E., Soininen, J. *et al.* (2024) Carbon sequestration in different urban vegetation types in Southern Finland. *EGUsphere* 2024, 1–38.

Trémeau, J., Olascoaga, B., Backman, L., Karvinen, E., Vekuri, H. and Kulmala, L. (2024) Lawns and meadows in urban green space – A comparison from perspectives of greenhouse gases, drought resilience and plant functional types. *Biogeosciences* 21, 949–972.

Tsoka, S., Leduc, T. and Rodler, A. (2021) Assessing the effects of urban street trees on building cooling energy needs: The role of foliage density and planting pattern. *Sustainable Cities and Society* 65, 102633.

Utembe, S.R., Rayner, P.J., Silver, J.D., Guerette, E.A., Fisher, J.A. *et al.* (2018) Hot Summers: Effect of extreme temperatures on ozone in Sydney, Australia. *Atmosphere* 9, 25.

VanWoert, N.D., Rowe, D.B., Andresen, J.A., Rugh, C.L., Fernandez, R.T. and Xiao, L. (2005) Green roof stormwater retention. *Journal of Environmental Quality* 34, 1036–1044.

Van Renterghem, T., Hornikx, M., Forssen, J. and Botteldooren, D. (2013) The potential of building envelope greening to achieve quietness. *Building and Environment* 61, 34–44.

Vaz Monteiro, M., Blanuša, T., Verhoef, A., Richardson, M., Hadley, P. and Cameron, R.W.F. (2017) Functional green roofs: Importance of plant choice in maximising summertime environmental cooling and substrate insulation potential. *Energy and Buildings* 141, 56–68.

Velasco, E., Roth, M., Norford, L. and Molina, L.T. (2016) Does urban vegetation enhance carbon sequestration? *Landscape and Urban Planning* 148, 99–107.

Volckens, J., Braddock, J., Snow, R. and Crews, W. (2007) Emissions profile from new and in-use handheld, two-stroke engines. *Atmospheric Environment* 41, 640–649.

Vuppaladadiyam, A.K., Merayo, N., Prinsen, P., Luque, R., Blanco, A. and Zhao, M. (2019) A review on greywater reuse: Quality, risks, barriers and global scenarios. *Reviews in Environmental Science and Bio/ Technology* 18, 77–99.

Wagner, P. and Kuttler, W. (2014) Biogenic and anthropogenic isoprene in the near-surface urban atmosphere – A case study in Essen, Germany. *Science of the Total Environment* 475, 104–115.

Wang, F. (2006) Modelling sheltering effects of trees on reducing space heating in office buildings in a windy city. *Energy and Buildings* 38, 1443–1454.

Wang, R., Mattox, C.M., Phillips, C.L. and Kowalewski, A.R. (2022) Carbon sequestration in turfgrass–soil systems. *Plants* 11, 2478.

Whitford, V., Ennos, A.R. and Handley, J.F. (2001) "City form and natural process" – Indicators for the ecological performance of urban areas and their application to Merseyside, UK. *Landscape and Urban Planning* 57, 91–103.

Wilby, R.L. (2007) A review of climate change impacts on the built environment. *Built Environment* 33, 31–45.

Wong, I. and Baldwin, A.N. (2016) Investigating the potential of applying vertical green walls to high-rise residential buildings for energy-saving in sub-tropical region. *Building and Environment* 97, 34–39.

Woo, S.H., Jang, H., Lee, S.B. and Lee, S. (2022) Comparison of total PM emissions emitted from electric and internal combustion engine vehicles: An experimental analysis. *Science of The Total Environment* 842, 156961.

Xiao, Q. and McPherson, E.G. (2002) Rainfall interception by Santa Monica's municipal urban forest. *Urban Ecosystems* 6, 291–302.

Xiao, Q. and McPherson, E.G. (2011) Performance of engineered soil and trees in a parking lot bioswale. *Urban Water Journal* 8, 241–253.

Yan, D., Liu, J., Shao, W. and Mei, C. (2020) Evolution of urban flooding in China. *Proceedings of the International Association of Hydrological Sciences* 383,193–199.

Yang, J., Yu, Q. and Gong, P. (2008) Quantifying air pollution removal by green roofs in Chicago. *Atmospheric Environment* 42, 7266–7273.

Yilmaz, H., Toy, S., Irmak, M. A., Yilmaz, S. and Bulut, Y. (2008) Determination of temperature differences between asphalt concrete, soil and grass surfaces of the City of Erzurum, Turkey. *Atmósfera* 21, 135–146.

Yoshimi, J. and Altan, H. (2011) Thermal simulations on the effects of vegetated walls on indoor building environments. In: *Proceedings of Building Simulation 2011: 12th Conference of International Building Performance Simulation Association, Sydney*.

Zalacáin, D., Martínez-Pérez, S., Bienes, R., García-Díaz, A. and Sastre-Merlín, A. (2019) Salt accumulation in soils and plants under reclaimed water irrigation in urban parks of Madrid (Spain). *Agricultural Water Management* 213, 468–476.

Zheng, X., Zou, Y., Lounsbury, A.W., Wang, C. and Wang, R. (2021) Green roofs for stormwater runoff retention: A global quantitative synthesis of the performance. *Resources, Conservation and Recycling* 170, 105577.

Zirkle, G., Lal, R. and Augustin, B. (2011) Modeling carbon sequestration in home lawns. *HortScience* 46, 808–814.

Zou, Z., Yan, C., Yu, L., Jiang, X., Ding, J., Ding, J. and Qiu, G. (2021) Different responses of evapotranspiration rates of urban lawn and tree to meteorological factors and soil water in hot summer in a subtropical megacity. *Forests* 12, 1463.

4 The Impact of Green Space on Human Health and Well-being

Abstract

Humans are part of the natural world, so it should be no surprise that we are heavily dependent on it for our health and well-being. Plants provide us with oxygen, food and fibre, but increasingly too, we value green space as an antidote to modern urban life. Green spaces relax us, provide emotional uplift, act as a healthy distraction, encourage creativity, promote physical exercise and provide biogenic chemicals/microbial organisms that regulate and support our physiological processes. Viewing, and activity within, green spaces is linked to better prevention against key diseases (heart disease, stroke and diabetes) and promoting longevity of life. There are some risks with green space too though – poisonous plants and exposure to allergens (e.g. hay fever). A number of theories underpin our relationship with the natural world – and these to some extent determine how we design and manage urban green spaces to optimise the human health benefits.

4.1 Introduction

Environmental horticulture has a central role in improving the environment for humans and promoting human health and well-being. This is especially so for the 57% of the global population who now live in towns and cities and have no regular access to 'countryside'. Environmental horticulture and similar disciplines (landscape architecture, urban forestry, urban ecology, etc.) have a responsibility to expand and protect the elements of nature that exist within the city, whether this relates to cultivated green spaces or those left to natural processes. This is part of a nature conservation agenda, but it also increasingly about making people aware of the value of nature and how it impacts on human health and well-being. Humans rely on natural processes and systems to provide water, food, clothing, timber products, oxygen to breathe and sunlight to synthesise vitamin D. We also actively engage with nature as it provides recreational opportunities or acts as a medium for artistic or cultural expression. More of a controversial point, however, is the extent to which we *need* nature for our psychological and spiritual well-being. There is increasing evidence that nature can help provide balance and mental well-being to many, if not all, citizens. This point was brought into sharp focus during the COVID-19 pandemic 'lockdowns' when people were restricted to visiting outdoor spaces for limited time periods. Those citizens lucky enough to have a garden or nearby green space reported better coping mechanisms (less stress, less anxiety and greater opportunities for mental distraction from the impacts of the virus) to those that did not have such resources. This has raised the profile or urban green space from a well-being perspective, with policy makers beginning to understand better the multi-beneficial roles green space provides. This chapter explores the relationships between humans and green space and how the form of green space, and the activities that occur within it, can influence human well-being, including improvements to health (physical and psychological), social factors and self-fulfilment.

A number of health and social benefits have been cited with respect to engagement with green space and nature (e.g. Kuo and Sullivan, 2001a; 2001b; Donovan *et al.*, 2013; Amoly *et al.*, 2014; Hystad *et al.*, 2014; Dadvand *et al.*, 2015; Romagosa *et al.*, 2015; Nguyen *et al.*, 2021; Daiz *et al.*, 2022; Robinson *et al.*, 2020a; Zhang *et al.*, 2023; Farris *et al.*, 2024b; Vilcins *et al.*, 2024) and include the following:

- reduced incidence or severity of depression;
- reduced anxiety and stress;
- promoting feelings of joy and happiness (positive affect);
- delays in the onset of dementia;
- less incidence of heart disease, stroke and certain cancers;
- improvements in physical fitness;

DOI: 10.1079/9781800621763.0004

- better diet;
- longevity of life;
- better birthweight;
- cognitive skills and children's development;
- enhanced self-esteem;
- better communication skills;
- greater opportunities for social interaction;
- better educational performance;
- longer attention span and productivity in the workplace; and
- less crime.

4.2 Theoretical Context

There are a number of theories that help explain the relationship between green space and human health and well-being. These can be summarised as

- Biophilia – an innate relationship with nature, where evolutionary processes determine how we perceive and respond to natural features.
- Attention restoration theory – the natural world allows us to unwind from stress, by providing a capacity to escape from the source of our stress (being away), being distracted by other features (soft fascination): features that allow you to engage with/be immersed by them (extent), and that you find interesting (compatibility).
- Stress reduction theory – viewing green features affects our physiology and induces a relaxation response.
- Positive affect – essentially, nature provides us with short bursts of happiness that aids mental health.
- Physical activity – keeps us physically fit through an interest in active outdoor pursuits.
- Promoting healthy eating – by eating plant products, we ingest compounds that promote a healthy balanced diet and regulate normal cell function.
- Physically reducing environmental stress factors by reducing noise or improving air quality or thermal comfort (see Chapter 3).
- Phytoncides (forest bathing) – plants release chemicals with anti-cancer and other beneficial properties.
- Microbial interactions with the human microbiome – naturally occurring microbial communities that influence our gut activity, and thus other aspects of our health.

Biophilia

The biophila hypothesis suggests that exposure to nature is an intrinsic requirement for effective human development, and conversely, disengagement with nature can lead to psychological problems. Wilson (1984) stated that biophilia is determined by biological needs and is an emotional or spiritual relationship between humans and nature. It has been cited as not only helping human development including aspects such as cognitive skills but also facilitates a stronger appreciation of the natural world and an empathy for other living organisms, factors important to the long-term sustainability of natural resources, and hence, in the longer term, human survival. Biophilia is strongly linked to human evolution, with human development accelerating once our ancestors moved out of forests and dwelt in the 'parkland-like' African savannahs. As much of our fundamental development as *Homo sapiens* relates to the savannah habitat, it is argued we still have a close affinity with such landscapes or components of them. Studies have shown that humans have preferences, even today, for open park-like landscapes and other iconic features typical of the savannah. For example, these landscape features include flat-top trees (reminiscent of the thorn acacias, *Acacia tortilis*, that epitomise the savannahs of East Africa), vantage points with vistas (mimicking the rocky outcrops or 'kopjes' – where game animals and or potential threats could be viewed) and the element of water as typified by ponds and rivers (evolutionary important as a source of drinking water and effective locations to trap/ambush game animals) (Fig. 4.1).

Some of these relationships may be explained by fractals. Fractals are patterns that repeat themselves, often at increasingly fine sizes, and result in shapes of rich visual complexity. Fractal objectives are prevalent in nature and include elements such as mountains, clouds and trees. Such fractals stimulate our visual experiences and may even influence the human visual system – from how the eye moves when acquiring the visual data of fractal patterns through to how the brain responds when processing their characteristics (Taylor, 2021). The biophilia theory suggest that such natural shapes make sense to the human brain and that the brain processes these features efficiently, in contrast to non-natural shapes and stimuli which may over-excite the brain and increase physiological stress. Biophilic

Fig. 4.1. How much does biophilia (our emotional ties with the landscape) determine garden design? Due to our evolutionary needs, psychologists argue that we still look for a place of safety (with an enclosed space at our backs), with a view or advantage point over the landscape, and place we can source water or trap game – elements that can still be seen in conventional 21st-century gardens (top) or even rewilded theme gardens (bottom).

design is a term now used when natural elements such as water, plants, images of natural objects or natural processes are used as design elements in the built environment in an attempt to enhance feelings of nature and induce restorative responses.

Attention restoration theory

Attention restoration theory (ART; Kaplan and Kaplan, 1989; Kaplan, 1995) describes how natural elements can be mentally restorative when the brain suffers from fatigue. Long periods of mental fatigue or frequent occurrences can lead to physiological stress, and over-prolonged periods to more severe mental health problems. The brain suffers fatigue when it is required to focus on complex, challenging or repetitive tasks for a long period; this focus often being referred to as 'directed attention' (i.e. the need to concentrate fully on a particular task). Attention restoration refreshes or restores the brain by providing distracting elements ('soft fascination') that relate to involuntary or non-directed attention. Green space and the natural world is full of these non-stressful distractions. Originally it was thought such distractions worked through visual cues, but there is evidence now that certain natural sounds, smells and even touch-based sensory experiences may also help the mind restore its full attention capacity. ART emphasizes the power of plants, (benign) animals and other natural features to attract and hold a person's attention without effort, thereby enabling the neurocognitive mechanisms to rest and recuperate. A useful example of this theory working in practice is how an encounter with green space helps subsequent directed attention, for example completing a standardised cognitive test more effectively or quickly after a walk along an attractive riverbank. ART states that nature or green space provides a restorative opportunity by acting at four levels; allowing patients to 'be away' (removed from stress-inducing factors either physically or psychologically), promotes 'fascination' (effortless, interest-driven engagement with natural objects), provides 'extent' (scope for exploration and curiosity-driven discovery) and is 'compatible' (how closely the environment matches an individual's needs or inclinations at that time) (Table 4.1). Grahn and Stigsdotter (2010) argue that how people collate and store information is important too. The human brain relies on three different processes:

1. Subsymbolic processes (sensory, motor and somatic modes, as perceived by muscles, inner organs and skin).
2. Symbolic-imagery (visual pictures in a person's mind).
3. Symbolic-verbal (concepts discussed verbally and their interpretations).

Natural areas and green spaces have the advantage in that they promote simple relationships that do not overload these sensory processes, particularly the symbolic-imagery and symbolic-verbal. In contrast, complex interactions (for example, such as people may experience during an intense conversation or argument) overtax these processes. In effect, those environments (natural landscapes, wilderness,

Table 4.1. The key components to attention restoration theory (Kaplan, 1995)

Component	
Being away	At first impression this seem to intimate the need to remove oneself to destinations that represent wilderness or the expansive qualities of the natural world – wildscapes including mountains, coastlines, forests, meadows, lakes, streams, etc. In reality, it is more closely aligned with the ability 'to escape' mentally through engagement with nature at a range of different levels and scales. In practical terms, it can mean involvement with green spaces that are more familiar or easily accessible (e.g. town park, garden or even a picture or photograph of a natural feature that becomes absorbing). In essence, any visual cues that offer an opportunity for respite from single-minded concentration, i.e. 'directed attention'.
Fascination	The ability of nature to provide absorbing processes and objects that help distract the mind from problems and anxious thoughts. Many of the fascinations afforded by the natural setting qualify as 'soft' fascinations: processions of caterpillars moving along the ground, shoals of fish, cloud patterns, leaves blowing in the breeze, snowflake patterns, birdsong, sunsets, etc. Attending to these patterns is effortless, and they do not overly tax the brain.
Extent	The impression of the vastness of nature – the ability to extend onwards. Again, in a similar manner to 'being away' although there are advantages with larger scales and extended wilderness, this aspect is about how the mind can extend its thinking. Even a relatively small area can provide a sense of extent. Trails and paths can be designed so that small areas seem much larger. Diversity of form or miniaturization are devices that can help create illusions of different worlds or timescales. Japanese gardens, for example, can mimic landscapes at a much grander scale, or remind viewers of cultural, symbolic or historical components. Landscapes include historic artefacts or geographical features that promote a sense of being connected to past eras and other environments and thus to a larger world.
Compatibility	For many individuals the natural environment is almost the 'default' environment. Functioning in the natural setting seems to require less effort than functioning in more 'civilized' urban settings, even though they have much greater familiarity with the latter. Kaplan and Kaplan (1989) relate this to the many patterns humans use when relating to natural environments. These include a range of roles or activities, e.g. 'predatory' (hunting/fishing), 'locomotion' (boating/hiking/cycling), 'domestication/dominance of the wild' (gardening/caring for pets), 'observing' (birdwatching and visiting zoos), 'survival' (fire building and constructing shelter), social interaction at meals (picnics and barbeques). People often approach natural areas with the purposes that these patterns readily fulfil already in mind, thus increasing compatibility.

parks and gardens) that optimise the simple processes are deemed to have greater restoration potential than those that require more complex sensory procedures (shopping malls, busy roadways, railway stations, airports, business and work environments) (Hartig *et al.*, 2003).

Stragà *et al.* (2023) used montages of photographs to determine how components of ART and other restorative factors compared when viewing different landscape forms, with natural places that look hospitable scoring highest overall (Fig. 4.2).

Stress reduction theory

Stress reduction theory (SRT) was developed by Roger Ulrich in the 1980s (Ulrich, 1984; Ulrich *et al.*, 1991) and focuses on the capacity of green space to deal with psychophysiological stress. Building on evolutionary behavioural responses relevant for survival (i.e. biophilic elements), SRT proposes that viewing vegetation and other elements of nature can elicit positive emotions that block negative thoughts and emotions, thereby ameliorating or shutting down a stress response. SRT is thus linked with the capacity of green space to alter certain physiological activities (hormonal, cardiovascular and musculoskeletal parameters) that results in reducing physiological stress (or nullifies pain). Ulrich cited his theory after researching hospital patients' responses after gall bladder surgery. He noted those that had a view of parkland and trees from their hospital bed recovered more quickly from surgery and required less painkiller compared to those who had a view of a brick courtyard.

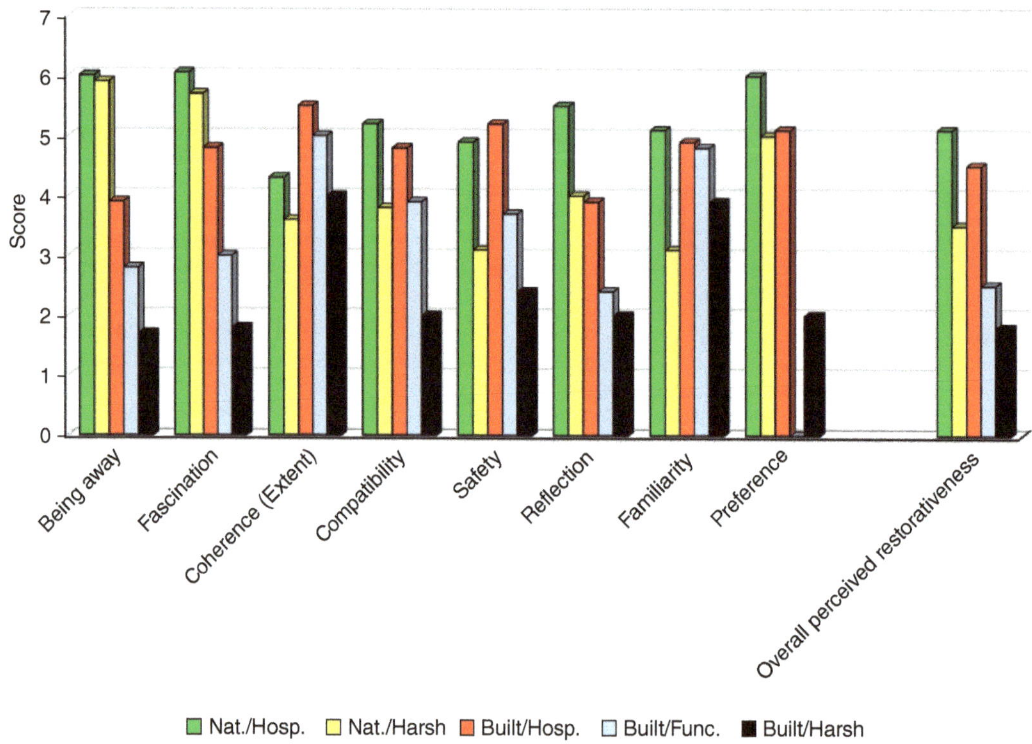

Fig. 4.2. How images of different types of landscape were scored by participants against the different components of attention restoration theory and other subjective feelings about the landscapes. The landscapes were grouped into Natural & Hospitable (Nat./Hosp.), Natural & Harsh (Nat./Harsh), Built Environment & Hospitable (Built/Hosp.), Built Environment & Functional (Built/Func.) and Built Environment & Harsh (Built/Harsh). (Modified with permission from Stragà *et al.*, 2023.)

Positive affect

Other researchers cite evidence that you do not need to be stressed to get a health boost from green space. Positive feelings (short bursts of happiness) or so-called 'positive affect' can help build up resilience against poor mental health. Elements of nature that provide positive emotions or a sense of wonder are useful in providing balance to an individual's life and act as a countermeasure to negative feelings and emotions. These small positive emotional uplifts can be induced by seeing some rare or unusual organism, but also for many people, it is just everyday occurrences that can give lift – such as watching the wind in the trees, noticing a colourful flower or enjoying the close presence of an iconic garden bird (e.g. *Erithacus rubecula* – European robin).

If elements in the natural world provide joy, then it makes sense that objects you already have an emotional relationship with or preference for may provide greater opportunity for this joy. Studies on people's response to flower colours (Zhang *et al.*, 2023) suggested that certain colours were relaxing (blues and whites – Fig. 4.3) and others uplifting (orange, red and yellow – Fig. 4.4), but if you had a particular favourite colour, that colour could also provide mentally positive responses. For example, people who liked red could find red flowers relaxing. Preference may explain why some elements in green space do not universally elicit a salutogenic response.

Physical activity

Green space provides opportunities for physical activity, whether it is park runs, dog walking, gardening or tree climbing or more enticements to run,

Fig. 4.3. Blue, white and green are considered cool and relaxing colours in the landscape. Environmental horticulturists often complement the soothing effect of green foliage with a cool flower colour to maximise the effect.

Fig. 4.4. Red flowers are considered uplifting and stimulating to the general population. Zhang *et al.* (2023), however, showed that if you favoured a particular colour, i.e. if red was your favourite colour, you could potentially also find this hue relaxing.

cycle or walk for longer due to the beautiful landscape one is traversing. Green and natural environments encourage physical activity and those activities can seem effortless compared to comparable activities indoor or in an urban setting. Physical movement is vitally important to us. Globally, 33% of adults do not do enough physical exercise. Modern lifestyles mean sedentary behaviour (desk work, watching TV, video games, etc.) accounts for about 8 h of the typical daytime activity. Such physical inactivity is the fourth leading cause of premature death and is the cause of numerous preventable physical and mental disorders. Inactivity is linked to increased instances of heart disease, diabetes, key cancers (bowel, breast and prostate), the onset of dementia and a plethora of mental health problems. This is because a lack of physical activity affects physiological processes within the body, for example it reduces lipoprotein lipase activity, protein transport and carbohydrate metabolism, and impairs lipid metabolism and sex hormone circulation. Surveys in the UK suggest, that a 10% increase in the population's physical activity would increase lifespan for 6000 people and save the health service £500M.

Activities around home or community gardening or maintaining an allotment can be seen as light physical activity. Hand-pushing a lawn mower, raking up leaves, hoeing weeds, digging the soil or similar physical work can use up to 400 calories h^{-1} and improve cardiovascular performance. Gardening encourages regular activity, and frequent, short bouts of activity are effective at maintaining fitness. Studies where patients were recovering from a stroke or a heart attack have found that exercise in a garden was more effective, enjoyable and sustainable than therapy in traditional formal recuperation settings. As gardening aids muscle strength, dexterity and strengthens bones in old age, it is an activity that can be promoted to middle-aged people and the elderly. There are links between physical health and psychological health. Gardening is considered more effective than walking, self-education activities or moderate alcohol consumption in protecting against dementia.

Promoting better diet

Partaking in home or community food gardens can encourage a healthier diet, through access to fresh produce but also through greater awareness of what is integral to a 'healthy' diet. Vegetables, fruit, seeds and nuts provide essential vitamins, fibre, minerals and phytochemicals. The provision of better, balanced diets, with less reliance on fat or starch-based foods has considerable health-enhancing attributes. Inadequate consumption of fruit and vegetables is estimated to contribute to 2.6 million premature deaths globally a year. Encouraging people to consume up to 600 g per day of fresh fruit and vegetables has been estimated to reduce the instance of

- cardiovascular heart disease by 31%;
- ischaemic stroke by 19%;
- stomach cancer by 19%;
- oesophageal cancer by 20%;
- lung cancer by 12%; and
- colorectal cancer by 2%.

Certain fruit/vegetables species are noted for the presence of specific beneficial compounds, such as antioxidants, e.g. broccoli (*Brassica oleracea*) – sulphoraphane, blackberries (*Rubus fruticosus*) – anthocyanins and phenolics, ginger (*Zingiber officinale*) – gingerol, and tomatoes (*Lycopersicon esculentum*) – lycopene. Other provide fibre, e.g. peas (*Pisum sativum*), beans (*Phaseolus vulgaris*), garlic (*Allium sativum*) and sweet potato (*Ipomoea batatas*).

Growing one's own food is linked to the adoption of healthier eating habits. Due to concerns of obesity and an increasingly sedentary lifestyle in children, this group has been specifically targeted for intervention schemes. School gardening provides positive benefits in terms of improved nutritional intake for children, more awareness about diet and encourages children to be more adventurous when tasting new fruit and vegetables (Schreinemachers *et al.*, 2017). In a review by Davis *et al.* (2015), 54% of studies found school gardening resulted in increased vegetable consumption, and 88% indicated pupils showed a more positive attitude/increased preference for vegetables. Gardening programmes result in better identification of different crop types and greater willingness for pupils to prepare/cook fruit and vegetables themselves. Schreinemachers *et al.* (2017) found gardening programmes in Bhutan increased the probability of vegetables being consumed by 12% but did not alter the number of different fruits or vegetables that were consumed. Whilst school gardening and other diet awareness programmes have, in general, increased pupils' awareness about food and healthy eating, not all have seen a significant uptake of fruit and vegetable consumption (Christian *et al.*, 2014; Khan and Bell, 2019). Ohly *et al.* (2016) were concerned that the methodologies used in school gardening/nutrition studies were not always appropriate (e.g. an inability to back up reported quantitative changes with effective qualitative methods) and direct evidence for changes in fruit and vegetable intake was limited, and too often based on self-reporting. As the interactions around school gardening projects and outcomes can be complex (age of pupil, level and type of intervention, gardening activities aligned with supportive cooking programmes, parental support, etc.), further research is required to optimize how these programmes improve diet directly. The data to date, however, suggests that success can be improved when

- Initiatives last ≥12 months.
- The whole school community is engaged, and teachers and pupils have ownership of the concept and direct the development of the gardens.
- Teachers/catering staff are provided with additional training.
- Gardening is effectively embedded in the curriculum.
- Pupils are formally educated on the growing, cooking and nutritional aspects of food.
- The initiatives are backed up by activities in the home environment.

The beneficial trends seem less ambiguous when dealing with adult engagement with garden or urban food production. Here, increases in fruit and vegetable consumption are linked to urban food-growing initiatives. Community gardens and allotments allow people to access fresh produce and improve their diet, as well as yield opportunities to improve physical fitness and create social networks. In less affluent locations, urban gardeners and the wider community use these growing opportunities to guarantee food security whilst adopting healthier lifestyles. As with children, if community food growing is allied with cooking skills, the take-up and benefits are enhanced (Cruz-Piedrahita *et al.*, 2020). It needs to be recognised that motivations for urban food growing vary. For some it is a recreational pastime and for others it is an essential process to meet their nutritional needs. Countries such as Cuba have 'gardened' out of political necessity, after external trade collapsed with the end of the Soviet Union. There are about 800 million people involved in urban food production globally, with many using it to meet their own nutritional requirements, while others see it as a business opportunity through selling excess produce. In less affluent developing countries, 25–66% of the entire population may be involved in urban/peri-urban food production (Cruz-Piedrahita *et al.*, 2020). For more affluent 'gardeners' however, motivations relate to perceptions of better-quality food or political ideology around sustainable use of resources or perhaps out of the simple enjoyment of growing their own food and watching it come to fruition. Home food growing rose during the 2020–2021 Covid pandemic, as people had time on their hands and looked for opportunities to be outdoors, yet remained socially distanced from one another.

Phytoncides

Plants produce phytoncides (a group of biogenic volatile organic compounds [BVOCs] or aromatic compounds commonly referred to as 'essential oils'). These help plants protect themselves from herbivorous insects or microbial pathogens. Such chemicals have physiological and psychological effects on the human body too, but positive ones. Exposure to natural environments, especially forests, have through the action of phytoncides such as α-pinene and β-pinene been linked with an increase in natural killer (NK) cell activity within human tissues (Li *et al.*, 2009; Lee *et al.*, 2011). Killer cells (lymphocytes) focus on and terminate the early development of cancer cells. Exposure to forest environments also elicits changes to adrenaline, noradrenaline and cortisol, physiological activated compounds associated with stress management. Being 'immersed' in forests ('forest bathing' or 'shinrin-yoku' in Japan where the phenomena was first observed) is linked to lower blood pressure and more regular heart rate (Table 4.2). Studies compared these physiological processes in those participants who walked in forest settings to those who walked around built environments, with significantly more positive responses in the former. Subsequent studies where people were exposed to phytoncide vapours (in a hotel room overnight) compared to controls proved that the phytoncides were the key factor. Phytoncides though provide a good example of the complex links between physiological and psychological processes. Antonelli *et al.* (2020) claim that inhaling phytoncides like limonene and pinene results in enhanced antioxidant and anti-inflammatory effects within the respiratory airways, and moreover, the pharmacological activity of some terpenes may promote brain function by decreasing mental fatigue, inducing relaxation, and improving cognitive performance and mood.

The tree composition influences the concentration of specific phytoncides in the forest air and concentrations vary across the time course of a day. Based on phytocide profiles within China, Zhu *et al.* (2021) claimed that subtropical, evergreen broadleaved forests and *Cunninghamia lanceolata* forests were superior to forests composed primarily of *Phyllostachys edulis* or *Liquidambar formosana* or even coniferous and broadleaved mixed forests. Importantly, all forests though were better than non-forested areas. Phytoncides were generally highest in summer followed by spring then autumn and winter. Herbs and other aromatic Mediterranean species are also linked with phytoncides (aromatic essential oils) such as the *Citrus* family (Fig. 4.5).

Table 4.2. Summary of health benefits of forest-based activity reported in the literature

Body system/illness	Reported health benefits	Putative mechanisms
Cardiovascular system	Decreased systolic and diastolic blood pressure. Decreased heart rate.	Reduced sympathetic activity. Increased parasympathetic activity.
Glycaemic control	Decreased blood glucose. Decreased glycosylated haemoglobin levels.	Autonomic effects. Improved insulin sensitivity.
Immune system	Increased natural killer cell activity. Decreased cytokine levels. Decreased level of C-reactive protein.	Possible antioxidant and anti-inflammatory effects of volatile phytoncides emitted by trees and plants.
Neuropsychiatric	Reduction in stress biomarkers. Reduced levels of anxiety and depression. Decreased negative emotions.	Unclear mechanisms but may be secondary to diminished levels of cortisol and reduced acoustic overload.
Pain management	Decreased self-reported pain using a visual analogue scale.	Increased pain threshold. Improved affective dimension of pain.

(From Le Gear *et al.*, 2023)

Fig. 4.5. The aromatic oils from *Citrus* are thought to have health beneficial properties.

Microbial interactions with the human microbiome

Human beings are essentially symbionts – we are a 'walking city' composed of human cells but supported by microbiota including bacteria, archaea, fungi, protists and viruses. Our health, and particularly our immunity, is dependent on many of the symbiotic relationships that occur within our bodies. We evolved in tandem with much of this microbiota and are dependent on them for our 'full health'. Microbiota are involved in training and regulating our immune systems, ensure good gut health and affect our emotions and feelings by the way they regulate hormonal actions within the body – the so-called atmosphere–gut–brain axis. Through this hormonal regulation our microbiome influences our longer-term mental health.

Many microbiota come from our natural environment and colonise us, particularly during the early years of development, and a lack of exposure to these organisms due to poor diet and an over-sanitised home environment (the hygiene hypothesis) has implications for non-communicable diseases in later life. The global burden of 'urban-associated diseases', such as certain autoimmune and inflammatory diseases (Flies *et al.* 2019), have been linked, at least in part, to a loss of human microbial diversity. Von Hertzen *et al.* (2011) make the connection with human evolution. 'Given that most of human evolution has occurred in biodiverse and wild environments, it is not surprising that non-communicable disease rates are disproportionately rising in industrialized urban populations more than in rural populations.' For example, good respiratory health positively correlates with biodiversity exposure in rural residents (Stein *et al.*, 2016; Liddicoat *et al.*, 2018). This loss in health is evident in the decrease of diversity in urban versus rural populations (Von Hertzen *et al.*, 2011; Lehtimäki *et al.*, 2017, 2023).

The rise of allergen-based problems (through an overactive immune response) may be partly due to poor immunological training in early years (Sbihi *et al.* 2019). Studies in Finland back these theories up. The addition of forest soil to (normally 'clean') kindergarten play areas altered the skin and gut microbiomes of the children who attended (Roslund *et al.*, 2020). Corresponding positive changes in the children's immunity were observed within a month. These included alterations to plasma cytokines (proteins involved in cell-to-cell signalling; often linked to regulating immunity and inflammation responses), an increase in bacterial diversity (Gammaproteobacteria) and more regulatory T cells within the blood, suggesting that 'playing in dirt' had stimulated immunoregulatory pathways. (Regulatory T cells control the immune response to self-generated and foreign particles and help prevent autoimmune disease.) But it is not just babies and children that benefit from playing in nature, being in nature also allows adults to 'top-up' their 'health-giving' skin and gut microbiomes from time to time too.

In adulthood we seem to maintain good gut health by regular exposure to green natural environments. We inoculate our gut microbiome by ingesting or breathing in certain microbial groups; many of which are associated with natural green spaces (Robinson *et al.*, 2020a, 2021a). Thus, to maximise exposure to the beneficial groups, it seems we need to be in regular contact with diverse plant and soil-organism communities. Robinson *et al.*, (2021b) suggested that exposure to beneficial microbiota increased with closer proximity to the soil or to tree canopies (Fig. 4.6). Other studies indicate that more biologically diverse green areas support a greater variety of the beneficial microbiota. Comparison of different urban green space typologies in Australia showed that soil microbiotas in remnants of native woodlands and re-vegetated urban green spaces were greater than parkland, lawns and vacant plots of land (Mills *et al.*, 2020). The re-vegetated areas were restored woodlands which used locally native species and had been planted at least 15 years before the experiments took place. The authors argue that those areas with more biologically rich and more geometrically complex plant communities (i.e. the remnant and

Fig. 4.6. A conceptual image of how microbiological communities around tree canopies and soil affect the aerobiome around them. It is thought being in close proximity to plants and soils increases exposure to microorganisms that help regulate the human microbiome. The various colours highlight that beneficial microbes are derived from a range of different taxonomic groups.

restored woodlands) also had more microbiota. Further research is required on what types of urban green space might optimise the microbiome-based phenomena, under different climates and vegetation communities.

Irrespective of the mechanisms by which (and allied theories of why) green space/nature confers health benefits, it is an increasingly important political driver in holistic healthcare strategies, and indeed, for the justification for urban green space within city planning policies.

4.3 Green Space Typology and Human Health

As outlined above the mechanisms by which green space can confer health and well-being benefits may be numerous and varied and may also depend on wider contextual factors. Factors such as climate, culture, gender, location and type of upbringing, etc. may all affect how we perceive green space and even the types of green space we may prefer. Such factors make it difficult for the landscape professional to prescribe a specific landscape type or activity that optimises the well-being benefits for everyone. Nevertheless, there are some broad themes that are worth considering when designing and managing therapeutic green spaces. Different types of green space may promote more or less of these themes (assets) based on their primary design and functionality.

Wild spaces – blue and green

A number of theories highlight the importance of 'getting away' to 'real nature' to help instil or restore well-being. Wild nature or wilderness do represent the 'least benign' aspects of the natural world and in many cases may induce a sense of awe. Engagement with wild places might include ocean swimming (e.g. with dolphins or whales), mountain climbing, kayaking, walking safaris or simply trekking through landscapes largely devoid of humans and their influences. Part of the attraction of such places relates to risk-taking, being away from regulations or divorced from one's normal comfort zone. Such wild places provide a challenge and involve activities/experiences that promote adrenalin bursts, and where of course there may be a genuine risk to life. Even if not involved in dangerous activities, these locations can provide a sense of wonder, positive emotions, appreciation of beauty, moments of self-reflection and insights into one's relationship with nature. Wilderness instils complex experiences, with opportunities for 'silence, comfort, and contemplation on one hand, and challenging, even terrifying, surprising and overwhelming situations on the other' (Løvoll *et al.*, 2020).

The challenge provided by such environments/activities may partially relate to the well-being benefits ascribed to wild nature. Wild swimming, for example, has been linked to mental health benefits such as mindfulness promotion, resilience-building and increasing one's ability to listen to one's body, as well as developing an enhanced physical condition (McDougall *et al.*, 2022). Wilderness expeditions for adolescents help develop self-esteem as well as better connectedness to nature. A study by Barton *et al.* (2016) showed that all adolescents benefited from partaking in a wilderness expedition, but females particularly showed the largest increases in self-esteem. Similarly, wilderness activities which included hiking, backpacking, kayaking, rock climbing, bushcraft and mindfulness programmes for adolescents who had survived cancer showed well-being benefits (Jong *et al.*, 2022). The aim of this work with cancer patients was to increase physical activity, self-confidence, personal growth, joy, safety within nature, meaningful relationships and self-efficacy. Results showed strong group bonding within the participants, as well as resilience that helped individuals develop their own personal outdoor practices that remained after the end of the programme. Similar approaches by Warber *et al.* (2015) were linked to reductions in both perceived stress and negative emotions, with improvements in positive emotions and relaxation, as well as a sense of wholeness and transcendence. Connections to nature were enhanced, as knowledge and skills relating to the natural environment improved; this was also linked to attributes such as willingness to lead, perceived safety and a better sense of place.

Some elements associated with wild landscapes may be particularly beneficial – for example, the presence of water (blue therapy) or predominance of natural sounds and aromas. The presence of water has been linked to providing greater restoration potential than simply the presence of vegetation (Barton and Pretty, 2010). Water may be one of those quality attributes that becomes subjective though; older adults considering water features as calming and reflective, whereas parents of young children may view them with suspicion.

More wild landscapes or those that look more wild do not always enhance the well-being benefits though. Martens *et al.* (2011) asked participants to walk through either a wild or a tended forest for 30 min. More pronounced changes in both positive and negative affect were actually associated with the well-tended forest landscape.

Biologically rich spaces and encounters with wildlife

A number of the theories outlined above suggest the health benefits of green space may be promoted when that green space is physically and biologically diverse (e.g. biophilia or human microbiome phenomena). At the continental scale, Methorst *et al.* (2021) using socio-economic data from 26,000 European citizens across 26 countries showed that bird species richness correlated positively with self-reported life-satisfaction. The relationship was comparable in strength to that associated with income, i.e. a relatively strong association. Two possible explanations were highlighted: (i) the multisensory experience and positive emotions associated with seeing birds and (ii) the landscape typology that supports more birds but also enhances human well-being. At more local levels, Mavoa *et al.* (2019) indicated subjective well-being positively correlated with faunal and floral species richness (number of species present). Higher plant species richness has also been positively linked with people's ability to recover from stress (Lindemann-Matthies and Matthies, 2018), while composite measures of multi-taxa species richness (plants, birds and bees/butterflies) are related to the restorative benefits of urban parks (Wood *et al.*, 2018). Other studies show positive relationships between bird and plant species richness and psychological well-being in urban parks (Cameron *et al.*, 2020), although results are not always consistent (Dallimer *et al.*, 2012). Further evaluations show positive effects due to bird biodiversity and watching birds (Wolf *et al.*, 2017; Wood *et al.*, 2018; Cameron *et al.*, 2020). Increase plant diversity and aesthetics associated with more natural planting designs in urban parks have been linked to enhanced well-being (Carrus *et al.*, 2015; Wolf *et al.*, 2017; Southon *et al.*, 2018; Wood *et al.*, 2018).

Wolf *et al.* (2017) showed videos comparing tree (one vs four species) and bird (one vs five species) species richness and found that videos with more species present reduced anxiety and increased positive affect and vitality. Some experiments show some, but not necessarily maximum, biodiversity promotes well-being (Douglas and Evans, 2022; Farris 2024b). Schebella *et al.* (2020) used 360-degree videos to both stress their participants and expose them to parks at different levels of species richness. Using a multisensory approach, species richness was controlled via visual (two, four or seven vegetation layers), audio (birdsong from more or fewer species) and olfactory stimuli (one to three smells from grasses species). Results showed that the low-biodiversity scenario (two vegetation layers, one bird and one smell) lowered anxiety and heart rate, compared against an urban control (i.e. little biodiversity) and also the treatments representing greater biodiversity. Farris *et al.* (2024a) showed that a green woodland setting could promote well-being, but adding additional (flowering) plants to the woodland did not enhance the effects further. Perceptions of biodiversity, however, did correlate better to well-being than actual biodiversity (Farris *et al.*, 2024a, 2024b), suggesting participants were not always aware how species-rich the landscape was. Ironically, if they thought the landscape was biologically rich, they felt better (even though that landscape itself did not necessarily have more species present than other options).

Aesthetically rich spaces – parks and gardens

Wilderness and even biologically rich places may not be that common in towns and cities. Yet, green space in urban areas is particularly valued for its capacity to counterbalance the stresses and strains associated with the hustle and bustle of city life. In their review of urban green space literature, Kondo *et al.* (2018) found exposure to green space was correlated with reduced levels of mortality, better heart rate data and reduced levels of violence. There were positive relations in terms of attention, mood and physical activity too. Results were mixed, or no association was found, however, on general health, weight status, depression and stress (via cortisol concentration). They also stated there were too few studies to give overall conclusions on the effects relating to blood pressure, heart rate variability, cancer, diabetes, respiratory problems and birth outcomes.

A number of designed landscapes, notably parks (Ayala-Azcárraga *et al.*, 2019; van Dinter *et al.*, 2022), private and public gardens (Cameron *et al.*, 2012; Cameron, 2023), green roofs (Williams *et al.*, 2019) and green walls (Ode Sang *et al.*, 2022) provide restorative value. Urban parks have been linked to a range of benefits including opportunities for physical activity (formal and informal sports at all skill levels and abilities) including walking, cycling, running and ball games. They also provide recreational activity (birdwatching, dog walking and photography) and foster social interactions (picnicking and meeting friends via music and art events). They provide green space and nature at a local level, thus helping with restoration from mental fatigue, providing opportunities for solitude and quiet, artistic inspiration and expression and educational development (e.g. natural and cultural history) (Romagosa *et al.*, 2015; Ma *et al.*, 2019).

Grilli *et al.* (2020) investigating park visits in Ireland showed a strong, statistically significant association between the number of park visits and health status (Fig. 4.7). Parks were more popular (visitor utility) when toilets, coffee shops and gym facilities were provided, and the park has standing or flowing water and higher tree densities. Proximity to your local park is important. Accessibility and distance to parks are crucial variables for their use and thus the benefits they offer the community at large (Sallis *et al.*, 2012; Evenson *et al.*, 2013; Ayala-Azcárraga *et al.*, 2019). English Nature (UK Government) indicates that everyone should have access to green areas at least 2 ha in size and less than 300 m from their homes. Adults who live less than 100 m from a park perform physical activities more regularly than those who live beyond that distance (Bonnefoy *et al.*, 2003). Quality of park also counts. A park full of litter, poor lighting or suffering from a lack of maintenance may be perceived as a location of crime, generating a sense of insecurity, regardless of any real relationship with actual crime.

Ornamental gardens and gardening improve health indices too. Chalmin-Pui *et al.* (2021b) introduced ornamental plants (a small tree and two planters) to 38 previously bare front gardens of terraced houses within an economically deprived region of northern England, UK. Over a 3-month period, residents reported significant decreases in perceived stress and normalised their daily cortisol patterns to a healthier profile (Fig. 4.8). Residents reported the new gardens were valued for their capacity to enhance relaxation, increase positive emotions, provide motivation to maintain the property and helped give a sense of pride around the neighbourhood. A follow-on study (Chalmin-Pui *et al.*, 2021a) which involved assessing questionnaire responses from gardeners (5766) and a non-gardening control group (250) indicated that gardening was linked with improvements in well-being, perceived stress and physical activity. Significant associations were found between health and more frequent gardening. Gardening two to three times a week corresponded with the greatest perceived health benefits. Such trends were apparent across cultures. In the Philippines, Diaz *et al.* (2020) reported that in a survey of 200 gardeners (and 200 non-gardeners), home gardening was associated with more frequent reports of good health, greater resilience and a greater capacity to cope with stress, both around the SARS COVID-19 pandemic and when people are faced with other distressing factors. The authors conclude those people who engage with gardening are better at adapting to change, have greater overall resilience and can adjust better to grief and the loss of a loved one. Cameron (2023) argues that home gardens should be given appropriate status and some form of protection from development within the UK, because they constitute a unique health resource. Others recommend gardening is incorporated into health intervention programmes, as an adaptive

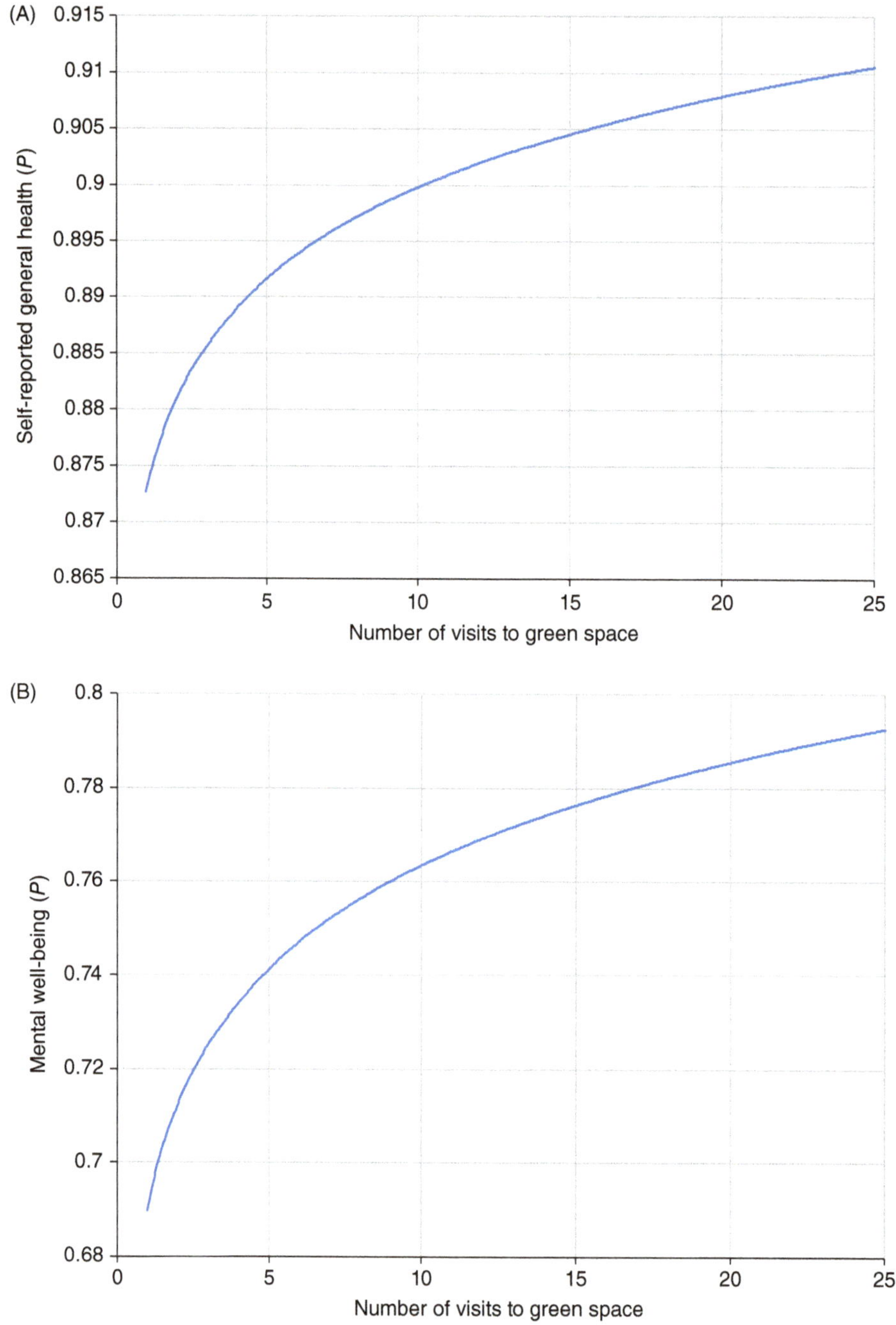

Fig. 4.7. The probability of health outcomes based on number of visits to green space: (A) Self-reported general health and (B) mental well-being. (Modified with permission from Grilli *et al.*, 2020.)

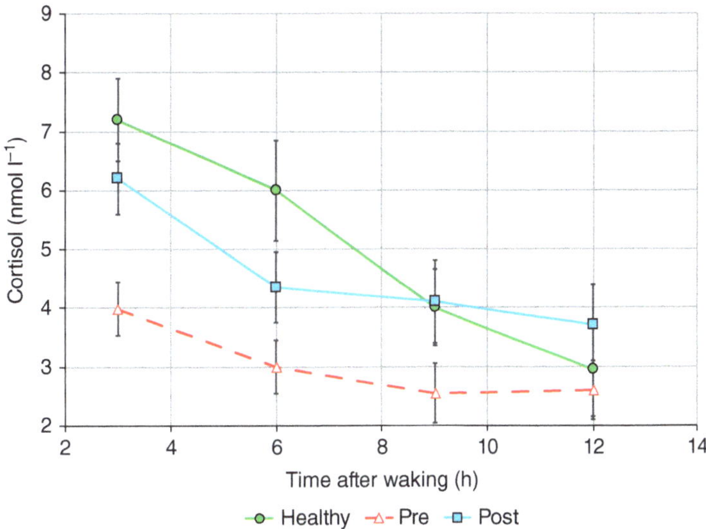

Fig. 4.8. Cortisol profiles of residents before (red dashed line) and after (blue solid line) the introduction of plants to the front of their house. The green line represents cortisol profiles of healthy, non-stressed individuals. (Used with permission from Chalmin-Pui *et al.*, 2021b.)

means to improve public health and well-being especially at times of health crises.

Gardens and other nearby green space were seen as a 'lifeline' by some during the COVID-19 pandemic – when movement outside the home was restricted in many countries. Life satisfaction was higher for those who owned a garden compared to those that did not during the pandemic (Lehberger *et al.*, 2021). Time outside was more important than the type of green space (own garden vs public green), but garden owners spent much more time in their garden than non-garden owners spent in public green spaces. Thus, private gardens appear important in facilitating time spent outdoors and maximising the health opportunities that confers.

Aspects of the aesthetic landscape have been linked to health (see sections relating to water – above – and trees – below). Flowers in the landscape (and in the home) have elicited attention for their effects on human emotions and thus links to health and well-being. Using digital images of daisy-like flowers in a range of different colours, Zhang *et al.* (2023; 2024) suggested that blue and white flowers promote relaxation/stress reduction, whereas warm flower colours – orange, yellow and red – evoke uplifted emotions and deliver better positive affect (Fig. 4.9). Interestingly, white was also considered an uplifting flower colour by some. Indeed, preference for specific colours could be linked to wider and intriguing emotional benefits. Red for example was usually seen by most people as an uplifting flower colour, but for those that cited red as their favourite colour (i.e. had a strong preference for it), then red flowers could also be relaxing. Li *et al.* (2012) indicated that respondents exposed to a range of natural images (images of flowering plant communities of various monochromatic colours, e.g. red *Papaver* spp. [poppy] vs purple/blue *Lavandula* [lavender]) exhibited physiological improvements, such as decreases in systolic and diastolic blood pressure, heart rate, electrocardiogram readings and fingertip pulse and increased galvanic skin response. This was irrespective of the colour they were exposed to. Overall, though, those who viewed green and purple/blue plantscapes had more positive psychological responses than those exposed to red, yellow or white flowers. This was evidenced by lower ratings of irritability, fatigue and anxiety and higher scores of vigour compared with the other groups. Other studies with office workers suggested viewing yellow and red flowers resulted in a significantly higher sense of relaxation, cheerfulness and comfort than viewing white flowers (Xie *et al.*, 2021). Electroencephalograms (EEGs) showed that the mean values of alpha relative power in the prefrontal lobe were significantly

Fig. 4.9. Colour can change the 'feel' of a landscape. Notice how the addition of warm yellow (bottom) draws the eye and proves greater vibrancy compared to the more muted and relaxing blues and purple (top).

higher when viewing yellow and red flowers vs white flowers. Another indoor study (Qin *et al.*, 2014) suggested that red flowers could be overstimulating and were fatiguing after a period of time. In this context, small, green, lightly scented plants were optimal for health and well-being. Flower/plant colours appear to affect human emotive states, but while there is some consistency in the literature about green, blue and purple acting to relax people, further work is required to clarify any universal response to the warmer colours such as red and yellow. The presence of flowers seems important though, at least for these 'cultural' green landscapes. Flowers and water often correlate better to relaxation and related health indicators than the presence of animals, trees, hills, natural aromas or sounds (Ogunseitan, 2005).

Plant form is another aspect to consider in parks and gardens. Plant and leaf shapes are cited as influencing human responses. Pine trees (*Pinus thunbergii*) clipped in the Japanese Sukashi technique (an important component of inferring a natural style in Japanese garden flora) promoted a greater relaxing effect as measured by decreased cerebral blood flow (CBF) compared to unpruned trees (Elsadek *et al.*, 2013). The positive effect due to the pruned trees was recorded in both genders of Japanese citizens. The fact that stylized trees (albeit to represent an elderly, wind-blown or weathered tree) was favoured in preference to untrained forms indicates that a cultural or aesthetic dimension may also be influencing the psychophysiological responses. This positive response and desire for nature or wilderness within garden style may account for the popularity of bonsai growing or alpine gardening which infer wilderness but actually involve highly contrived design and construction processes.

Gardening promotes physical health and provides opportunities for regular physical activity within a 'stimulating' environment. Gardening and like activities may be popular (e.g. with the elderly) because they provide exercise in a non-demanding way, with little stigma attached to factors such as age, gender or level of physical fitness. Exercise in the green space may motivate participants to undertake physical activities by increasing enjoyment and providing a better means to escape from 'everyday' problems and anxieties. In addition to physical activity, the benefits associated with home gardening include positive emotions through being creative (design and planting flowers, shrubs and trees), being close to nature on a regular basis, enjoying recreation time with family and friends and the garden being a source of pride and improving self-esteem. In the UK, almost 50% of the population do some form of home gardening, and this proportion increases in the elderly.

Public gardens are also therapeutic, again through aspects of design that promote biophilic responses or soft fascination. Carefully laid out plants and colourful borders create a sense of wonder. Weaving pathways prompt intrigue and exploration. Features such as rockeries or themed gardens (Japanese style, South African flora, Himalayan woodlands) elicit feelings of 'being away' and mental escapism.

Tree-dominated landscapes (urban woodlands, avenues and public parks)

Trees can define a city or neighbourhood location. They are the most prominent natural living features in most cities and towns. They provide a range of ecosystem services of which human health is just one. The urban forest comprises a diverse range of tree species and vegetation structures, orientated through areas of forest, groups of trees in parks, lined out along roadways, rivers and canals, or promoted in groves or 'mini' forests (e.g. Miyawaki schemes) as well as individual trees in public space or private gardens.

Landscapes with high tree cover have been associated with better health outcomes (Gathright *et al.*, 2006; Wolf *et al.*, 2020). Lower rates of lung, breast and uterine cancers in females, and lower prostate, kidney and colon cancers in males have been linked to tree cover in Japan (Li *et al.*, 2008a, 2008b, 2010). Walks in woodland correlate with lower blood glucose, HbA1c (glycated haemoglobin) and blood pressure in diabetic patients (Ohtsuka *et al.*, 1998). Tree-lined walkways have been associated with lower incidences of obesity, and people living in localities with a high proportion of tree-lined streets had lower body mass indices than those in areas without street trees (Lachowycz and Jones, 2011). The classical study by Ulrich (1984) on faster recovery rates and less pain in gall bladder patients after surgery, related to viewing trees from the hospital window. Viewing trees affects our psyche. A study in Uruguay indicated that people living in tree-lined streets considered themselves happier and to have an improved social life in comparison to those living in tree-less neighbourhoods (Gandelman *et al.*, 2012). Forest-based therapies are linked to lower symptoms of depression, remission rates, better mood and higher perceived mental restoration in those suffering depression or exhaustion (Sonntag-Öström *et al.*, 2014). Street trees (London, UK) (Taylor *et al.*, 2015) and trees in school playgrounds (Wu and Jackson, 2017) (California, USA) are reported to reduce the use of antidepressants and prevalence of autism, respectively. However, mixed results were found following forest-based therapy for children with attention deficit hyperactivity disorder (ADHD) (Van den Berg and Van den Berg, 2011) and patients with exhaustion syndrome (Nordh *et al.*, 2009). Removing trees from landscapes has been associated with increased stress and increased incidence of domestic violence (Kuo and Sullivan, 2001). Exposure to trees is thought to activate involuntary brain attention, a possible precursor to effective cognitive function (Martínez-Soto *et al.*, 2013).

Higher perceived stress restoration has been linked to woodlands compared to urban built-up areas, although cortisol profiles were not different (Tyrväinen *et al.*, 2014). When comparing outcomes between the urban streetscape and various urban forest settings – parkland, tended woodland and wild woods – the urban forest settings promoted stronger stress recovery, but no differences were found across the different natural settings (Van den Berg *et al.*, 2014). Videos showing different proportions of street tree density (2% to 62%) showed greater stress recovery with the higher densities (Jiang *et al.*, 2016). Though in another study using a similar approach, women showed no response to tree density while men did (Jiang *et al.*, 2014). People living near to forested areas have demonstrated MRI brain scans that indicate better capacity to cope with stress (Kühn *et al.*, 2017) and when asked to observe forest scenes, respondents have shown more relaxed brain patterns with near-infrared spectroscopy (NIRS) – this corresponding with lower self-reported 'anger and hostility' and 'total mood disturbance' (Joung *et al.*, 2015). Proximity to trees and forests have also been linked to fewer days of mental health complaints (Akpinar *et al.*, 2016). Moreover, lower cardiovascular disease has been reported for those neighbourhoods with higher densities of street trees (Kardan *et al.*, 2015), and heart disease in women is higher in counties of the USA which lost their trees to the pest *Agrilus planipennis* (emerald ash-borer) (Donovan *et al.*, 2015). Not all studies elicit positive effects though – Horiuchi *et al.* (2014) reported no change in cardi-vascular outcomes when participants viewed tree scenes and non-tree scenes. Other studies showed mixed responses on cardiovascular health and heart attacks. Sudden, unexpected cardiac death was less likely in counties with a higher percentage of forest in North Carolina (Wu *et al.*, 2018) but not in Texas, USA (Tarar *et al.*, 2015).

Excessive tree density, however, as might be encountered in a close-planted forest may not be relaxing. EEG response to simulations of forest tree stand density (30% to 100%) indicated more frontal brain activity when viewers observed the greater stand density, while lower stand density (from 30% to 50%) produced less brain activity and a more relaxed state, as well as reduced tension and fatigue (An *et al.*, 2004).

Walking, sitting and resting in woodlands ('forest bathing' shinrin-yoku) has shown positive psychological and physiological responses (Horiuchi *et al.*, 2013) (Fig. 4.10). Even short periods of forest bathing (15 min) resulted in enhanced subjective feelings of vigour, recovery and vitality (Takayama

Fig. 4.10. Japan was at the forefront of understanding how forest environments relate to citizen health (e.g. Forest bathing or Shinrin-yoku theory) with both the quiet peaceful ambience of these locations, but also the biochemical (phytoncide) influences on human physiological processes being cited.

et al., 2014). Other studies showed depression decreased and liveliness increased with forest activity, with greater effects for subjects having higher initial stress levels (Morita *et al.*, 2007); exposure to forest settings produced lower measures of anxiety, depression, anger, confusion and fatigue (Park *et al.*, 2007). For women, forest walking increased happiness more than walking in a gymnasium, with meditative walking in the forest being the most effective (Shin *et al.*, 2013). Forest school activities show benefits in adolescents, particularly for those who might struggle in a conventional classroom. Self-reported responses with forest bathing have been backed up with physiological tests. This includes exposure to forest settings resulting in reduced prefrontal cerebral activity, better cortisol profiles and suppressed sympathetic nervous activity (i.e., fight or flight response) accompanied by enhanced parasympathetic nervous activity (i.e., rest and digest state) (Lee *et al.*, 2011; Park *et al.*, 2010). Forest bathing is linked to phytoncides from trees, and these help induce NK cells and associated activity. It was found that increased NK activity can last more than 7 days after a forest trip (Li *et al.*, 2008a, 2008b, 2010). Spending time in forest settings, even short visits, may promote healthier human immune systems and regulate immuno-inflammatory responses, though the underlying pathways are not completely understood (Mao *et al.*, 2012).

Proximity to trees and woodlands may act as an encouragement to take physical exercise. Greater neighbourhood tree canopy cover is linked to lower prevalence of obesity (Ulmer *et al.*, 2016), including a 12% lower prevalence of obesity in preschool children (Lovasi *et al.*, 2013) Body mass index (BMI) was found to be significantly lower in those in counties in the USA that have a higher per capita of forestland cover compared to rangeland, pastureland or cropland (Ghimire *et al.*, 2017). Similarly, a higher percentage of normal BMI was found in neighbourhoods with greater forest edge density (Tsai *et al.*, 2016). So there is good evidence that trees (or the associated increased physical activity associated with them) help keep weight down!

There are some suggestions that the presence of trees reduces problems during pregnancy or improves infant heath at birth. A lower incidence of premature births was found in neighbourhoods with more street trees (Abelt and McLafferty, 2017) and birth weight was higher for babies of the same age when tree canopy was greater (Donovan *et al.*, 2011). Results are not always consistent however, and factors such as location and size of trees, proximity to the house, other urban factors (e.g. traffic volumes on nearby roads) and ethnicity can influence the results (Wolf *et al.*, 2020). As such, more studies are required in this area.

Green roofs, walls and façades

Green walls and façades tend to be visible to the public as they can be viewed from street level. In contrast, green roofs are only likely to be observed by people living/working on tower blocks nearby and who have a view down to a green roof. Interest is increasing though for these urban interventions to improve mood and well-being. In a comparison between viewing a brick wall and a green façade, Elsadek *et al.* (2019) showed that viewing the green façade improved cerebral activity (increase in alpha relative waves in the frontal and occipital lobes of the brain) and increased parasympathetic activity and decreased skin conductance. In essence, the viewers relaxed more and experienced improved mood and attention. Questionnaire data indicated that the green façade provided feelings of comfort, relaxation, cheerfulness, and vigour compared to the building wall. The experiment was well designed and controlled but did involve participants sitting 1.5 m directly in front of a wall or façade for 5 min, and further work is required to determine any effect from more incidental noticing of green walls and façades, as might be more representative of practice. Relatively limited studies have focused on such scenarios in the outdoor environment.

Some studies have assessed interior green walls though. Working with classroom green walls in schools, Van den Berg *et al.* (2017) showed that selective attention was improved in those pupils taught within the classrooms with the green walls. There were no measurable effects, however, of the green wall on children's self-reported well-being. In contrast, Gunn *et al.*, (2022) indicated that stress and anxiety levels decreased and mood improved when green walls were introduced to classrooms in their study. The positive effect did not continue indefinitely, however, and partially dissipated after 2–5 weeks. A study on university lecture rooms reported that green walls provided far better interior conditions, mainly in terms of lower carbon dioxide (CO_2) concentration and higher relative humidity, but also in improving students' and teachers' mood and health (Peterková, *et al.*, 2019). More data is warranted on green walls and façades (especially outdoors), but despite this, Elsadek *et al.* (2019) and others conclude that to improve public health in cities the integration of vertical greening in buildings in future compact cities should be considered.

Despite views of green roofs perhaps being limited to certain individuals working or living at a high vantage point, they are useful locations when readily accessed, for example, office workers who can use a roof garden or terrace at lunchtime. In such situations green roofs offer social, aesthetic and cultural aspects. Indeed, green roofs can be locations of relative peace and tranquillity being above traffic noise and immediate air pollution. Providing green space at rooftop level creates opportunities for community cohesion and improvements in health and well-being through both the planting of ornamental landscapes but also via community growing of fruit and vegetables (Kotzen, 2018).

Allotments, community gardens and urban farming

Not all gardeners work in isolation. Growing plants (usually food crops) in communal space is commonplace across a number of different cultures. In the UK, 'the allotment' has a long history, and relates to people owning, or renting small parcels of land for food cultivation. In other situations, community gardens help bring communities together. These communities may be associated with certain neighbourhoods but could also be linked to schools or even places of work – where employees will come together at lunchtime or after work and cultivate plants. Individuals attending a community garden may also have their own plots or may share growing beds, containers, roof garden space, etc. to grow food. The key component about this style of gardening is that it is communal and often highly social. Thus, additional benefits may arise compared to gardening on one's own.

Allotment and community gardens have been linked to the promotion of self-esteem, better health outcomes (both general and mental), life satisfaction, mood and social cohesion (Lampert *et al.*, 2021; Gregis *et al.*, 2021). The benefits are linked with a sense of achievement in growing one's food, potentially a healthier diet, physical activity, less isolation and opportunities for social interactions and camaraderie. Regular exercise in such landscapes improves muscular and upper body strength and dexterity. The fact though that these activities also involve a lot of soil cultivation also supports the notion that interactions with the human microbiome may also be playing a role in this and similar activities.

Wood *et al.* (2016) indicated that even a single session of gardening could enhance benefits; paired *t*-tests revealing a significant improvement in self-esteem ($P < 0.05$) and mood ($P < 0.001$) as a result of a single session. In comparison to non-gardeners, allotment gardeners also experience less depression and fatigue and greater levels of vigour. Similarly, a survey of 332 people in Tokyo, Japan reported that allotment gardeners, compared to non-gardeners, had better perceived general and mental health, less subjective health complaints and better social cohesion, although BMI did not differ between the two groups (Soga *et al.*, 2017). Some studies report a dose response. Mourão (2019) working on an organic allotment in Portugal showed that increased frequency of allotment gardening was related to greater perceptions of subjective happiness. They concluded that urban organic allotment gardening represents a means for enhancing citizen well-being, contributing positively to their feelings of happiness and life satisfaction, changing behaviours and developing personal capacities. Others support the idea of allotments and community gardens being used more extensively as a precaution against poor mental health and reducing the costs of healthcare in urban societies. A study from Singapore (Koay and Dillon, 2020) proposed that community gardening was even better for one's health than home gardening, with community gardeners reported significantly higher levels of subjective well-being than individual/home gardeners as well as non-gardeners. Community gardeners also reported higher levels of resilience and optimism than their non-gardening counterparts. The authors suggest that community garden interventions have implications for future research in clinical psychology, mental health promotion, and health policies in Singapore and elsewhere.

In a somewhat novel angle, allotment gardening has been linked to a better body image. In a study conducted in London, UK, a sample of 84 allotment gardeners were compared to a non-gardener control group of 81 individuals (Swami, 2020). Participants were asked to complete several measures related to positive body images - namely, body appreciation, functionality appreciation and body pride - selected to provide broad coverage of the positive body image construct. Results showed that allotment gardening significantly improved people's perceptions of body image and that longer periods spent on the allotment were correlated with larger-scale improvements. Gardeners showed significantly higher positive attitudes to their bodies across all the indices measured than non-gardeners.

Community gardening is popular in countries such as the USA and the UK and is growing elsewhere (Gregis *et al.*, 2021). Benefits can vary based on context and more research is required to understand impacts due to the type of intervention, climate and seasonality. Gregis *et al.* (2021) note that seasonality affects the overall development of community gardening and potential outcomes need to take this into account (presumably take-up is lower when cold and wet weather conditions prevail). This is particularly relevant, for instance, when outcomes such as physical activity are analysed.

School gardens

Encouraging school pupils to engage in garden activities is becoming more frequent. The motivations for this can be varied and include providing children opportunities to learn about food and a healthy diet, promoting wider pro-environmental knowledge and behaviours (e.g. on climate change and food security), creating a more constructive learning environment for some pupils and also encouraging activities that promote physical and mental health (Gregis *et al.*, 2021). The impetus of many school interventions is to directly increase fruit and vegetable consumption by the pupils and, through better knowledge and behaviours around diet, help combat childhood obesity (see section on diet above). A number of school projects have resulted in dietary improvements including increased fruit and vegetable consumption, dietary fibre and vitamins A and C intake and an improved BMI for many of the pupils who partake (Holloway *et al.*, 2023). Improvements in mental health have also been noted. In one study (Lam *et al.*, 2019), 16 secondary school pupils tracked their experiences about food and gardening through photography and writing. The youths explicitly associated growing food, their garden as 'a place', cooking, and food choices with positive mental health. They associated their school gardening activities with relaxation and wider themes of love and connectedness.

Holloway *et al.* (2023) suggest that the benefits of school gardening can be enhanced when the initiatives actively embed nutrition-based and garden-based education in the curriculum; provide experiential learning opportunities; and promote wider family engagement and participation. Successful projects often have an authority figure who drives the project forward and helps reinforce activities to maximise the benefits to the pupils. Appropriate engagement across different cultural contexts is also a key factor. In New Zealand, a study of school pupils who took part in home gardening demonstrated that most were likely to be male, of a Pacific Island ethnicity, of younger age and most likely to be living in a rural area (8500 students from 91 school responded). Two-thirds of students had a vegetable garden at home and one quarter of all students participated in home gardening (Van Lier *et al.*, 2017). The survey indicated that gardening was positively associated with healthy dietary habits, physical activity and improved mental health and well-being. Students who participated in gardening reported slightly lower levels of depressive symptoms and enhanced emotional well-being and experienced higher family connection compared to those students who did not participate in gardening.

The social and educational value of school gardening needs wider recognition. Horticultural educational programmes are used to reduce truancy from school for disaffected pupils. Qualitative data suggests that such programmes can not only reduce the incidence of truancy but aid academic performance (improved skills in English and mathematics), promote self-discipline, self-esteem and team skills. Green spaces around schools and in children's living environment strongly influence a child's development, particularly in terms of boosting capacity to focus and maintain attention. When children's cognitive functioning was compared before and after they moved from poor (low-volume green space) to better-quality housing (greater green space) differences emerged in attention capacity, even when the effects of the improved housing were taken into account. The impact of plants and greenery on attention, engagement and interest may also explain why children with attention deficit disorder (ADD) appear to respond better when playing in a green or natural environment. ADD children focused and performed better after activities in green settings, and increasing the proportion of vegetation resulted in less severe attention deficit symptoms.

Business and work environments

Exposure to greenery and green spaces may be advantageous in the workplace. The health benefits of interior plantscapes in offices and other workplaces has been frequently cited (e.g. Park and Mattson, 2008). Some companies provide gardens for their employees to sit out in at break times/lunch to encourage a sense of freedom and a 'coping mechanism' to seek diversion during the working day. Green views from workplace widows also reduce stress with employees feeling less uptight compared to views of urban scenes. Lottrup *et al.* (2013) indicated physical and visual access to workplace greenery can have a significant positive effect on employees, with physical access to outdoor patios, gardened areas and lawns particularly promoting positive attitudes. There were differences, however, between how male and female employees responded. For males, access and visual site of green space improved the workplace attitude and decreased perceived levels of stress, whereas with women, attitudes improved but there was no significant effect on stress levels. These gender variations may partially relate to the different pressures/stressors men and woman commonly experience at work. It was found that more men than women went outdoors during the working day, and that women often reported 'being too busy' as a reason not to go outdoors. Compared to men, women tend to report more interpersonal stressors, more stress due to multiple roles, lack of career progress, and discrimination and stereotyping. Coping mechanisms for such stresses may also vary – woman relying on social support from friends/colleagues rather than environmental influences.

4.4 Therapeutic Landscapes and Horticultural Therapy

Horticultural activities, including domestic gardening, food production on allotments and community gardening provide an opportunity for restorative processes. Local government organisations and charities may specifically use the existing green estate to help provide facilities for clients with health or social problems; closely supervised groups of individuals may help to maintain flower beds, propagate plants for sale or be involved in nature conservation work. Inherently, many individuals will engage with gardening and other similar activities to relax and 'be close to nature'. More formally, however, such activities are intentionally used as interventions in either social and therapeutic horticultural (S&TH) or horticultural therapy (HT) programmes. S&TH has a more general focus on well-being improvements and is not necessarily set against clinical objectives. These courses are well used with 21,000 clients engaging with S&HT every week in the UK.

HT, on the other hand, has predefined clinical goals and helps clients learn new skills or regain ones lost. It aims to help clients improve memory, initiate tasks, improve responsibility, enhance problem-solving skills, pay greater attention to detail or regain physical abilities: people with physical or mental disabilities or in rehabilitation from illness, injury, addiction or abuse (Harris and Trauth, 2020; Guglielmetti Mugion and Menicucci 2021; Yun *et al.*, 2024). Both forms of green care may utilise specially designed therapeutic gardens, but these are also closely linked with hospitals and care homes where much of the interactions may tend to be relatively passive (e.g. viewing and hearing nature or smelling and touching plants). Nevertheless, even passive engagement with gardens and other forms of green space has been associated with less agitation and aggressive behaviour with, for example, people suffering from Alzheimer's disease and other disorders that cause frustration (Uwajeh *et al.*, 2019).

There have been a number of discrete studies that highlight the benefits of horticultural programmes. Gonzalez *et al.* (2010, 2011) working with individuals who suffered from clinical depression, demonstrated that mental health scores (averaged across five different assessment criteria) increased significantly after HT, compared to the same assessments conducted before the programme. In this study, large reductions in the severity of depression were observed over the first 4 weeks, and these were still evident 3 months after the HT finished. The value of HT in relation to clinical depression is that it is thought to activate the 'being away' and 'fascination' components of Kaplan's ART as well as provide social interactions and cohesion. Yang *et al.* (2022) investigated the feasibility of a therapeutic gardening program during the COVID-19 pandemic. The program consisted of 30 sessions and was conducted at 10 nationwide sites in Korea over a 5-month period. Mental health and well-being were assessed using a number of research tools, namely the Mental Health Screening Tool for Depressive Disorders, Mental Health Screening Tool for Anxiety Disorders, Engagement in Daily Activity Scale, the World Health Organization Quality of Life Scale, and Mindful Attention Awareness Scale. Data showed that all five mental health variables improved significantly over time as the therapeutic gardening program progressed.

Positive responses have been recorded for dementia sufferers too; Gigliotti and Jarrott (2005) noted greater levels of engagement in patients when HT approaches were adopted (78%) compared to conventional activities (28%). Spring *et al.* (2013) reported that gardening was a constructive, outdoor activity that promoted social interaction, physical activity and provided stimulating cognitive challenges for patients with dementia. Interestingly, it also provided positive benefits for staff and visitors to the hospital. This point was reflected in earlier research involving a children's hospital garden. In this case, 54% of visitors reported feeling more relaxed and less stressed, 12% felt refreshed and rejuvenated with 18% feeling more positive and able to cope after visiting the garden (Whitehouse *et al.*, 2001). The garden being valued as a 'tonic' for the staff, almost as much as for the patients and their families. Even short visits were beneficial, as half of visitors spent less than 5 min in the garden at any one time.

HT is being promoted more often to deal with the 'explosion' in mental health problems over recent years. Depression is thought to affect 350 million people, but HT can decrease psychological symptoms, improve memory, reduce stress/anxiety and improve self-esteem/identity (Najjar et al., 2018; Kim and Park, 2018).

HT is now used across a wide range of settings and ailments (Harris and Trauth, 2020). It is applied as therapy for people of all ages (e.g. elderly with dementia, children with autism and middle-aged people with drug or alcohol dependencies) (Gigliotti *et al.*, 2004). It is used in hospice care, acute care, for people with sensory defects and for people with a wide range of physical and psychological disabilities (Boyle and Pryor, 2019). It is employed in prisons and in community groups to improve social cohesion, and to help instil feelings of self-worth. Horticulture can be a 'diversionary activity' – reducing chances of young people engaging with anti-social behaviour such as drug abuse and petty crime. Horticultural programmes with prison inmates have been linked with greater opportunity to take up vocational work after release and tend to reduce the chances of reoffending. Military veterans suffering post-traumatic stress disorder (PTSD) and other mental health problems can gain benefit through restorative experiences of working with plants, being close to nature and re-developing social interactions with other patients and staff (Lehmann *et al.*, 2018). It is a useful medium for refugees to cope with stress and trauma on leaving family and friends and entering a new culture. Capra *et al.* (2019) explained that 'horticultural therapists bring unique and multidisciplinary skills to the process of change, emphasize the strengths in those served, and offer an intimate connection with nature through engagement in gardening'. Söderback *et al.* (2004) summed it up by 'horticulture therapy mediates emotional, cognitive and/or sensory-motor functional improvement, increased social participation, health, well-being and life satisfaction'.

Despite the presence of a large number of qualitative studies on the effect of horticulture/gardening on psychosocial well-being, Spano *et al.*, (2020) suggest that quantitative studies are still lacking in this area. They argue there is a strong need to advance the number of high-quality studies on this research topic given that gardening has promising applied implications for human health, the community, and sustainable city management.

Green social prescribing

'Green social prescribing' or 'green prescribing' is a new element of social prescribing where medical practitioners can prescribe activities in the green space as medical interventions. Such activities include gardening, community conservation work, walking in nature, etc. Increasing numbers of general practitioner doctors are utilizing this approach to help patients deal with issues such as depression, isolation or even physical ailments. Often the links between a medical practice (doctor's surgery) and a green therapy group/organization is mediated by a practice nurse, social worker or other professional. A number of studies have positive results on medical outcomes (e.g. Thomson *et al.*, 2020; Sands *et al.*, 2023). In their review of 31 studies, Adewuyi *et al.* (2023) found positive health benefits for psychological health and well-being (16 out of 24 studies), cardiometabolic health (5 out of 9 studies), physical activity (8 out of 9 studies), and inflammation (2 out of 2 studies). The reviewed studies did not report any significant benefits in orthopaedic conditions, pain, and recovery from exhaustion, but the number of studies was too small to generalize the effect of green prescriptions on these outcomes. Robinson *et al.* (2020b) argue that green prescribing works best when there are effective cross-disciplinary communication pathways, and the locations for interventions are close to the medical practices.

4.5 Abstract Green

It is evident that many of the benefits of green space are stimulated by visual and other sensory responses associated with natural/semi-natural landscapes, even without the requirement for any physical activity. Indeed, just providing artificial representations of natural scenes or green space, namely virtual reality experiences, videos, photographs or paintings also elicit some positive psychological or psychophysiological responses (Calogiuri *et al.* 2021; Yen and Huang, 2024). The quality of the virtual experience can determine the level of the positive outcome: for example, Kahn *et al.* (2008) found green images on plasma screens not quite as beneficial as window views. However, as more sophisticated experiences become the norm, such as the use of visual reality with 'more immersive' experiences, then the benefits become more comparable. Yen and Huang, (2024) compared visits to real parks and virtual parks. Participants experiencing both kinds of parks had significant improvements in their self-rated health, physical and environmental quality of life, and sedentary time after the interventions. Those who experienced actual parks had significant increases in the social quality of life, moderate physical activity and decreased body weight, whereas those who experienced the virtual parks showed a significant increase in their mental quality of life. Interestingly, the authors conclude that because virtual parks have similar health benefits to real parks they can be recommended to citizens who lack opportunities and motivation to go to actual parks. In a number of research studies, including a number in the author's own research group, virtual green landscapes are used to elicit emotional responses and help determine well-being outcomes (e.g. Zhang *et al.*, 2023, 2024; Farris *et al.*, 2024a, 2024b). This would indicate that at least some of the mediators for well-being are purely psychological (e.g. not reliant on phytoncides or beneficial microbiomes).

4.6 Designing and Managing Restorative and Therapeutic Environments

The previous sections outline the theories explaining how humans may gain a health benefit from green space and demonstrate the value of different landscape types and activities. But how do you bring this information together to design and manage effective therapeutic or restorative landscapes? Who are these landscapes for and what elements are essential, preferable or even irrelevant in eliciting appropriate emotional responses and well-being benefits? Some of the theories linking nature to human well-being imply a degree of subjective engagement with the topic, 'compatibility' for example in the ART suggests that you should be familiar enough with certain elements of nature to be comfortable engaging with them or thinking about them. This does raise questions such as

- Do all people gain the same/similar benefits from the same landscape?
- Are some landscapes more restorative than others?
- Does preference for a landscape typology or feature affect the emotional response you receive and thus its restorative or therapeutic value to you?
- How much of the emotional responses are intrinsic and how many are learned?
- Are restorative landscapes subtly different for different cultures?

On the face of it these questions should not matter too much (there are lots of different landscapes out there!), but they do become important if your professional brief is to design or manage a therapeutic landscape – what do you include and what do you leave out? How important are scale, perceived quality and the heterogeneity of different features within the landscape? There is no point either in having a wonderfully designed landscape if it is inaccessible to most of the people who could benefit from it.

Proximity and access

Close proximity to green space has been seen as a factor in promoting citizen health and well-being. Relative closeness to green space has been shown to have both a positive (e.g. Richardson *et al.*, 2013; Cardinali *et al.*, 2024) or no (e.g. Potwarka *et al.* 2008) influence on physical activity/health. Results may vary with the type of health issue, with increasing green space within neighbourhoods improving mental health, but further research is required to verify links between proximity to green space and obesity (Luo *et al.*, 2020). Larger areas of green space may enhance health effects, but in reality these are often scarce in the more deprived areas where health problems are greatest. Such inconsistent findings may reflect variations in user groups (e.g. children vs adults), dietary behaviour, degree, type or frequency of physical activity or type and accessibility of green space.

There are positive links between living in green neighbourhoods and mental health (Orstad *et al.*, 2020; White *et al.*, 2021). Richardson *et al.* (2013) found that residents of the greenest urban neighbourhoods had significantly lower risks of mental health problems compared to areas with limited green space. Indeed, the relationship was linear; more green space equated to better mental health. A study of over 10,000 individuals in the UK indicated that people were happier when living in urban areas with greater amounts of green space. Compared with when they live in areas with less green space, they demonstrated higher levels of well-being and lower levels of mental distress (White *et al.*, 2013). Data were controlled for other factors that would affect these parameters: income, employment status, marital status, health, housing type, and local-area-level variables such as crime rates. Although at an individual level the effects appear small, the cumulative effect across the population was deemed significant. Residents living in locations with greater green space have been linked to better cortisol profiles (less physiological stress) than those deprived of green infrastructure (Roe *et al.*, 2013). Somewhat in contrast to these results, Francis *et al.* (2012), investigating the quantity and quality of public open space in Australian neighbourhoods, concluded that quantity of open space was not a factor influencing psychological stress but quality was. It should be noted that this study restricted itself to public open space, not the greenness of entire neighbourhoods.

Sleep may be affected too by proximity to trees and other vegetation cover (Astell-Burt *et al.*, 2013; Astell-Burt and Feng, 2020). Sleep deprivation contributes to a range of health issues, including longevity of life, cardiovascular disease, obesity, diabetes and poor self-rated health scores. Favourable mental health and active lifestyles are also considered to be drivers of a healthier duration of sleep (usually around 8 h per night). Increasing the proportion of green space correlated with increases in the proportion of people get a good night's sleep (e.g. greater than 7 h, Fig. 4.11). The cause and effect of this relationship is not known. Living in areas with large amounts of green space may encourage greater physical activity during daylight hours, thus improving sleep during the night period. The green space may be promoting relaxation and inducing less anxiety in residents, with positive consequences on sleep at night. Alternatively, it may be a more subtle relationship as areas with large proportions of green spaces are likely to be quieter or have less air pollution than more urbanised ones and thus could also be impacting on the quality of sleep.

Fig. 4.11. The amount of green space close to where one lives correlates with the quality and duration of sleep. The proportion of 'nearby' green space and percentage of the population with >7 h sleep per night.

Coombes *et al.* (2010) demonstrated that proximity to a green location was important in their study; people living farther from a park or similar green space were less likely to use it, were less likely to meet the minimum guidelines for physical activity and were more likely to be overweight. Frequency of use was increased by closer proximity. In light of this, one of the UK Government's conservation bodies – Natural England – claimed that no residence should be located more than 300 m from an area of green space, i.e. encouraging planners to embed green areas more frequently within the urban framework. The proximity question is open to debate however, as Schipperijn *et al.* (2010) found that only 56% of respondents used their nearest green space on a weekly basis.

The importance of local green space may vary with different user groups, for example having a higher priority for elderly people, the disabled or those with young children, where visiting green spaces further afield may require greater organisation. The convenience factor is also important for dog walkers, a high proportion being reliant on local spaces due to the requirement for regular use. Interestingly, factors such as the range of facilities available, the shape and diversity of the area, maintenance levels or dominant vegetation type did not seem to change the frequency of use of *local* green space. Likewise, preference for different activities within the space, preference for different green space elements, marital status or profession of visitors, the levels of stress people were experiencing or even people's view on nature did not affect the use of these local parks and spaces. Remarkably, those without access to a domestic garden did not use their local green spaces more frequently; people with gardens utilised local green space as, or more, frequently than those without. This may imply though that those who frequently use their garden are also more interested in spending time outside in general terms and perhaps enjoy engaging in a wide variety of green environments – the home garden being utilised for some activities but more extensive green spaces providing opportunities for alternative experiences (Maas *et al.*, 2008, 2009).

Access to green space is not equitable. Access is often determined by affluence, population density, available land and physical aspects such as a railway or river acting as a barrier (Wu and Kim, 2021). Psychological barriers can also limit access – for example, locations that do not reflect one's culture or where certain demographic groups feel unsafe or unwelcome (Williams *et al.* 2020). The greenest parts of cities are often the most affluent. For example, in Melbourne, Australia, the distribution of green space favours the more affluent communities, meaning that there are lower concentrations of low-income households in greener areas. A lack of green space and access to it exacerbates the health deprivation factors low-income households experience from aspects such as diet, poor housing stock and affordable medical care (Fig. 4.12).

Quality of landscape

Quality of landscape is often deemed an important factor in eliciting the health and well-being benefits (Phillips *et al.*, 2023). Although what is deemed 'quality' is open to interpretation. For some it means that the park has a coffee shop and a toilet – ideally well maintained. For others, high quality may mean some landscape that is reminiscent of near-wilderness.

Fig. 4.12. Access to nature. Is tree climbing a 'rite of passage' for every child? It is not without risk, but (low-level) climbing can stimulate brain function, arm and leg dexterity, exposure to beneficial phytoncides and microbiomes, provide an appropriate level of challenge and develops skills in assessing risk.

Van Dillen *et al.* (2012) defined quality in green space to mean the following:

- increased accessibility;
- high levels of maintenance;
- greater variation in landscape type and form;
- more natural and more colourful landscape forms;
- clear arrangement and harmonious design;
- provision of shelter;
- absence of litter;
- positive perceptions of safety; and
- aesthetic value and overall general impressions.

Similarly, improving quality aspects in local public open spaces (scored by assessments on pathways, lighting, sports and playgrounds, the provision of shade and well-maintained lawns, presence of water features or a close locality to water, and the relative abundance of birdlife) were shown to have a positive effect on mental health ratings (Francis *et al.*, 2012). Quality can relate to the vegetation. Trees and forests are linked with better cardiovascular and respiratory health than grasslands (Nguyen et al., 2021). In these cases, such physical factors were deemed more important than subjective aspects such as perceived friendliness, comfort or safety. Quality is determined by the landscape typology in view. Taking vegetated transport corridors as a context (greenways, i.e. access corridors that pedestrians or cyclists might take but in all cases provided with vegetation at the side), Cao *et al.* (2024) examined how those based around roads (urban) compared to park pathways or green/blue ways (i.e. those beside rivers) in terms of their health restorative values. Park greenways were somewhat superior to the river ones, and both were better than (vegetation-lined) urban roadways in most aspects evaluated (Fig. 4.13).

Cao *et al.* (2024) explained their results in that positive factors around the park greenways related to perhaps richer planting communities and more fascination, cooler microclimates, more open views and extent of space, and the balance of dense vegetation and open views that aids the prospect/refuge theories within biophilia. The rich trees and shrubs flanking urban park-type greenways created a sense of refuge for visitors by forming sheltered spaces, while interspersed open grasslands provided prospect. Urban green riverways on the other hand possess unique visual (reflection of light) and auditory (sound of running water) properties that are potentially attractive and restorative, helping to trigger positive associations and promote psychological recovery (Fig. 4.14). River corridors may also act as crucial 'air lanes' too, providing cool but fresh air to users, collectively increasing physical comfort. From an evolutionary perspective, urban river-types are located at the interface between land and water bodies. Such habitat intersections tend to have stronger biodiversity and are more likely to meet human life needs which have evolutionary significance. Although road greenways did not score so highly, they still provided opportunities for feelings around being away (on a journey?), compatibility and extent (a familiar process or event).

A study from Portugal (Madureira *et al.*, 2018) is insightful in how both health-related aspects (e.g. aspects relating to biophilia – presence of water, etc.) and pragmatic points (e.g. cleanliness) affect preference for different urban green spaces (Fig. 4.15).

4.7 Restorative Environments and the Individual

Although there may be some underlying universal relationships between humans and their environment, inevitably there are individual and personal aspects of an individual's psyche that affects their relationship with the natural world. These too may not be consistent. How people respond to a landscape varies with their own state of mental health and individual interests/perspectives. Although the

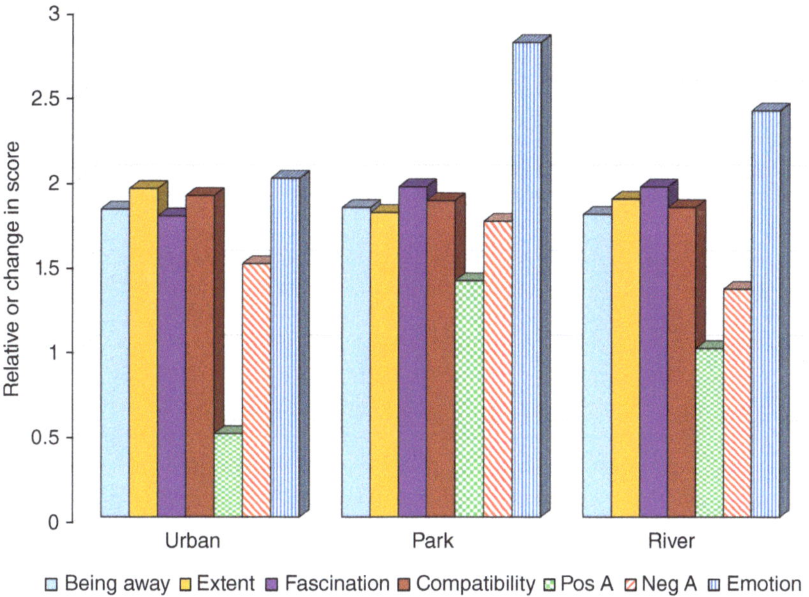

Fig. 4.13. How views of different types of greenways (transport corridors) influenced the components of the attention restoration theory (being away, extent, fascination and compatibility – solid colours) and changes in emotional state (positive affect, less negative affect and overall emotion – hatched colours) after viewing images of greenways. (Modified with permission from Cao *et al.*, 2024.)

Fig. 4.14. Water is considered restorative, with movement and sound being positive attributes, but the water feature does not need to be completely natural.

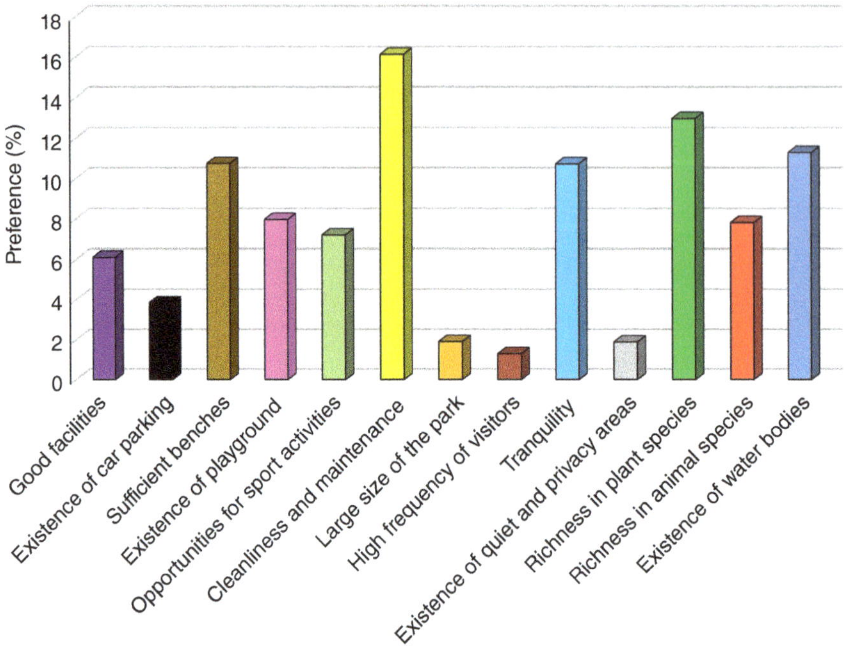

Fig. 4.15. Preference for urban green space is both based on aspects related to well-being (looking for restorative benefits, tranquillity, etc.) and more pragmatic points, such as the presence of a car park or sport facilities). (Modified with permission from Madureira *et al.*, 2018.)

benefits of green space are most often expressed with respect to individuals who have suffered some form of bereavement, mental breakdown or trauma, the benefits need not be exclusive to these circumstances, nor even work in the same way for people with similar experiences. Even individuals who have suffered trauma or prolonged stress may require different elements or components from the landscape as they proceed through progressive stages of recuperation (Stigsdotter and Grahn, 2002). Similarly, people looking for restoration may have preferences that differ from those not suffering any mental health problems at all. Such studies suggest that individuals suffering from physiological stress had a preference for landscapes that promoted concepts of 'refuge', 'nature' and 'prospect' (Akpinar, 2021). Indeed, landscapes that provided 'refuge', 'nature' and were deemed 'rich in species', but had few 'social' components, were considered optimal from a restorative perspective for those suffering stress. Non-stressed individuals, on the other hand, had an affinity for landscapes that promoted feelings of 'serenity' and offered 'space' and 'nature'. Attributes associated with 'rich in species' and 'refuge' were considered less important and 'culture', 'prospect' and 'social' were least preferred. Similar results were observed in follow-on studies investigating nine different small urban public green spaces in Copenhagen, Denmark (Peschardt and Stigsdotter, 2013) with the exception that 'social' ranked quite highly. 'Social' was particularly high for one city site that was characterised by cafes and restaurants but limited green infrastructure, suggesting that the restorative components may not be exclusive to green elements alone.

Much of the early work on the therapeutic aspects of green space and nature related to stress relief. Green space protects against stress or provides a restorative effect from mental fatigue and stress (Kaplan and Kaplan, 1989). Restorative experiences figure in human emotional and self-regulatory processes through which individuals develop an identity of place (Wilkie *et al.*, 2023). Individuals often go to a personal favourite place to help them relax, calm down, clear their mind of any negative thought or events or make themselves feel more positive (Korpela and Hartig. 1996; Korpela *et al.*, 2020). These favourite places frequently incorporate elements of vegetation, water or are landscapes of high scenic quality (Fig. 4.16).

Not all stimuli are necessarily visual. People respond positively to natural soundscapes, such as that produced by birdsong (Uebel *et al.*, 2021). Young residents of urbanised areas in Sweden were exposed to various combinations of birdsong:

1. Single species - *Passer domesticus* (house sparrow).
2. Single species - *Phylloscopus trochilus* (willow warbler).
3. Combined song from seven species - *Phylloscopus trochilus* (willow warbler), *Fringilla coelebs* (chaffinch), *Cyanistes caeruleus* (blue tit), *Parus major* (great tit), *Erithacus rubecula* (European robin), *Turdus merula* (blackbird) and *Dendrocopos major* (great spotted woodpecker).

Birdsong was played to the participants in different urban settings (three separate residential areas with varying amount of greenery). Results showed that urban settings combined with birdsong were more highly appreciated than the settings alone. Increasing the number of bird species singing within the 'chorus' further enhanced the feelings of appreciation. The authors of the report (Hedblom *et al.*, 2014) claim that birdsong contributes significantly to the positive values associated with urban green space. They conclude by urging planners to incorporate those habitats that maximize bird activity/diversity, so as to improve the recreational experience and positive emotions for citizens living within city zones.

The provision, location and type of urban green infrastructure has surprising effects on the range of benefits encountered. For example, expanding the amount of roadside vegetation has a positive influence

Fig. 4.16. Favourite places can provide a place of sanctuary or respite from stress and pressure. Such places can be created for those lucky enough to possess a large garden.

on car drivers, by allowing them to recover from stress more effectively, or to cope better with driving-related frustrations, anxieties and aggression (so-called 'road rage').

4.8 Social Cohesion

Another perceived benefit of green space is its role in aiding social cohesion. Well-designed spaces contribute to feelings of safety and familiarity and help develop a sense of community (Wan *et al.*, 2021). For individuals living in inner city apartment blocks, local green space provides an opportunity to develop stronger ties with neighbours and promotes a greater sense of safety in their local environment. In some, but not all residential areas, a greater proportion of green space correlates with enhanced 'social safety'. Oh *et al.* (2022) found that people who (i) strongly identified with nature, (ii) enjoyed being in nature or (iii) visited gardens on a regular basis tended to have a stronger sense of social cohesion.

Local green space encourages people to leave the confines of their home and promotes opportunities for social contact in a relaxed and informal way. These activities and improved social relations have further 'knock-on' positive effects particularly through increased informal surveillance of the neighbourhood, potentially reducing crime (Branas *et al.*, 2011; Stevens *et al.*, 2024). Removing trees in Chicago, USA (reducing the amount of green space in residents' views) was associated with increases in tension and domestic violence, with aggression against a partner increasing by 25–35% compared to residents that retained a green view (Kuo and Sullivan, 2001a, 2001b). Relationships may be subtle though (Wang *et al.*, 2024). In South Africa, for every 1% increase in total green space, there was a 1.2% decrease in violent crime and 1.3% decrease in property crime but no effect on sexual crime rates. Context is important, however, when situations are linked specifically with tree cover and park accessibility the direction of the association changed for property crimes; these increased by 0.4% with a percentage increase in tree cover and by 0.9% with every kilometre increase in proximity to a public park (Venter *et al.*, 2022).

Locally managed, familiar and welcoming green areas are valuable in that they help reduce the incidences of loneliness (and perceptions around a lack of social support) in elderly citizens and others.

Fig. 4.17. Green space and gardening provide opportunity for informal, low-level engagement. Being outdoors and people working with plants, etc. means that conversations are not forced and interactions tend to be relaxed and non-taxing.

Horticulture and the growing of plants is seen as an activity that brings different generations together too, breaking down some of the intergenerational divides (Fig. 4.17).

4.9 Health Risks

Despite the benefits being attributed to green space, the concept of some sort of 'Eden-like idyll' should be challenged of course. Engagement with green space is not risk-free. One percent of US citizens suffer some sort of injury in their garden every month. Horticultural activities including home gardening can result in injury due to certain pathogens, dermatitis, allergies and not least the misuse of tools. Powered machinery, particularly lawn mowers and hedge trimmers contribute to injuries commonly to the hand, foot or eye (Andrade Lima *et al.*, 2022). It is thought that 60% of lawn mower accidents relate to people cutting themselves on the blades or other sharp parts of the mower, which includes cases of users who touched the blades when the machine was operating (not recommended!), as well as during servicing of the machinery. Accidents occur with ride-on mowers, with these tipping over on steep, uneven ground. US studies suggest that stones and other debris flying out from the blades are a significant source of injury, especially when hitting vulnerable parts of the body such as the eyes (Yang *et al.*, 2024). Not surprising, children ≤15 years old had a proportionally higher risk of injury from mowers than adults (Jennissen *et al.*, 2023). Men were also more likely to be injured than woman, although this data was on a total, not a proportionate basis (i.e. fewer women may operate lawnmowers).

Green activities such as gardening and conservation work are recommended as a means to keep fit, but they can also be a source of muscle or skeletal injuries, most notably in the elderly or those who undertake excessively long, repetitive or heavy lifting tasks without being physically fit first. Excessive digging, lifting, twisting, stooping, kneeling and squatting are the reasons for injury, with lower back pain being the most commonly cited source of pain. If precautions are not taken, gardening can be associated with increased risk of arthritic pain, heat stress, skin cancer and carpal tunnel syndrome (numbness in the hand and wrist). Exposure to pollen induces allergenic reactions, both tree species and grasses can elicit hay fever (pollen allergy) and in some cases increased risk of asthma and rhinitis. Increased incidences of hay fever and asthma in urban areas have been blamed on enhanced levels of air pollutants (e.g. aerial particulate matter and ozone) but also the increased use of male trees that produce pollen (male trees being preferred over females in those species that readily drop fruit and result in unsightly fruit-littered pavements).

Green spaces – evenly highly managed spaces can house flora (and some fauna) that pose health risks, albeit often marginal ones. Numerous garden plant genotypes are poisonous if ingested, including some commonly occurring taxa (e.g. *Aconitum, Convallaria, Delphinium, Digitalis, Laburnum* and *Pieris* spp.) while others cause skin irritation and dermatitis on contact (e.g. *Chrysanthemum, Euphorbia, Ginkgo, Narcissus, Primula, Ruta* and *Tanacetum* spp.). In the less formal semi-natural landscapes, species such as *Urtica dioica* (stinging nettle) (Europe) and *Toxicodendron radicans* (poison ivy) (North America) can be commonplace – both species causing skin irritation and rashes. Fatalities due to plant poisoning are low however, with <5 p.a. in the USA. Children are most susceptible to poisoning, but even here an Irish survey suggested household medications contributed to 65% of ingested poisons, household or gardening products 34% and plants only 1%, with no fatalities reported (Rfidah *et al.* 1991). These relatively low levels relate to the fact that many poisonous plants are not visually enticing to eat, have a bitter taste or possess emetic properties (i.e. induce vomiting before toxins are absorbed through the stomach lining).

In public green space, effective management is required to ensure trees remain safe, with regular inspections of large specimen trees in publicly accessible locations. Increasing severity of weather events – strong winds, intense rainstorms, flooding but also drought – can all weaken the structural integrity of trees or their component parts. More than ever, trees need to be carefully planted and maintained through pruning and other activities to ensure, for example, roots are distributed effectively to optimise tree stability and that boughs remain strong, secure and pathogen-free. Inspections and activities need to be recorded and documented to ensure management plans are being implemented and that there is evidence that these are appropriate and proportional to the potential risks. Litigation has been brought against a number of land management bodies, where they have been seen to be negligent in the maintenance of tree stocks or even in their ability to logically document the maintenance schedules. Despite the range of health risks potentially posed by green space, health benefits are considered to outweigh the drawbacks.

For gardeners and professional horticulturists, being active close to city streets and poor-quality air is an increasing concern as particulate pollutants can be inhaled and exposure to gases such as NO_2, O_3 and CO increased. Those growing food in such locations also need to be aware of impacts of pollution. Heavy metals and polycyclic aromatic hydrocarbons (PAHs) sourced from vehicles are capable of bioaccumulation, depending on the element or compound characteristics and the vegetable species. Some leafy vegetables accumulate these toxic materials in significant quantities. As dietary uptake is considered the major exposure route of both heavy metals and PAHs, the consumption of impacted vegetables might pose human health risks (Hubai *et al.*, 2024).

Conclusions

- Humans are part of the natural world and have a dependence on it. Environmental horticulture has a key role to play in optimising the health and well-being benefits associated with green infrastructure.

- Green space promotes physical activity and encourages individuals to commit to more regular or prolonged exercise. This has implications for improved physical fitness and helps offset key sedentary lifestyle diseases such as heart disease and stroke.
- Similarly, views and activities within green space reduce psychophysiological stress and improve aspects of mental health. This includes aiding resilience to anxiety and depression. Positive thoughts (positive affect) induced by elements of the natural world are linked to better resilience against poor mental health.
- Positive effects on mental well-being have been closely linked with the attention restoration theory (ART), where key elements of 'being away', 'fascination', 'extent' and 'compatibility' are thought to influence recovery/recuperation from directed attention and reduce the likelihood of psychophysiological stress occurring.
- Benefits from green space include phytoncides and beneficial microbial communities associated with plants and soil. These act at a biochemical level within the human body and are considered to positively affect the immune system and human microbiome health and function, and in doing so impact on hormone regulation and mental health.
- Environmental horticultural activities that promote home or community food-growing schemes provide the potential for both nutritional and social benefits, as well as improving awareness about the natural world.
- Social benefits of green space include opportunities for community cohesion, social integration, crime reduction, educational attention and engagement.
- Horticultural therapy (HT) is one of a number of eco-therapy programmes available to people with special needs. These programmes are popular, with good anecdotal evidence of their success, but 'clinical' evidence is still being substantiated.
- Increments in the proportion and quality of vegetation enhances the work environment and can foster employee satisfaction.
- Green space does not necessarily provide universal benefits; there are 'down-sides' including risks associated with allergens, toxins, physical injury and repetitive strains. Not all demographic groups necessarily respond to green space in the same way, and what may be beneficial for one group may induce some disadvantages to another.
- Overall, urban green space provides a positive contribution to human health and well-being, but more information is required on how different sectors of society engage with green space and what the relative levels of these benefits are. Specifically, more information is required about what is meant by 'good design and management' in terms of promoting the health and well-being benefits, along with a range of other 'services'.

References

Abelt, K. and McLafferty, S. (2017) Green streets: Urban green and birth outcomes. *International Journal of Environmental Research and Public Health* 14, 771.

Adewuyi, F.A., Knobel, P., Gogna, P. and Dadvand, P. (2023) Health effects of green prescription: A systematic review of randomized controlled trials. *Environmental Research* 236, 116844.

Akpınar, A. (2021) How perceived sensory dimensions of urban green spaces are associated with teenagers' perceived restoration, stress, and mental health? *Landscape and Urban Planning* 214, 104185.

Akpinar, A., Barbosa-Leiker, C. and Brooks, K.R. (2016) Does green space matter? Exploring relationships between green space type and health indicators. *Urban Forestry and Urban Greening* 20, 407–418.

Amoly, E., Dadvand, P., Forns, J., López-Vicente, M., Basagaña, X. *et al.* (2014) Green and blue spaces and behavioral development in Barcelona schoolchildren: The BREATHE project. *Environmental Health Perspectives* 122, 1351–1358.

An, K.W., Kim, E.I., Jeon, K.S. and Setsu, T. (2004) Effects of forest stand density on human's physiopsychological changes. *Journal of the Faculty of Agriculture* 49, 283–291.

Andrade Lima, R.C., Soares Rocha, Q., Cazani, A.C., Guarnetti dos Santos, J.E. and Simões, D. (2022) Assessment of hand-arm vibration in semi-mechanized gardening activities. *Journal of Vibration Engineering and Technologies* 10, 2143–2149.

Antonelli, M., Donelli, D., Barbieri, G., Valussi, M., Maggini, V. and Firenzuoli, F. (2020) Forest volatile organic compounds and their effects on human health: A state-of-the-art review. *International Journal of Environmental Research and Public Health* 17, 6506.

Astell-Burt, T. and Feng, X. (2020) Does sleep grow on trees? A longitudinal study to investigate potential prevention of insufficient sleep with different types of urban green space. *SSM-Population Health* 10, 100497.

Astell-Burt, T., Feng, X. and Kolt, G. S. (2013) Does access to neighbourhood green space promote a healthy duration of sleep? Novel findings from a

cross-sectional study of 259 319 Australians. *British Medical Journal* open 3, p.e003094.
Ayala-Azcárraga, C., Diaz, D. and Zambrano, L. (2019) Characteristics of urban parks and their relation to user well-being. *Landscape and Urban Planning* 189, 27–35.
Barton, J. and Pretty, J. (2010) What is the best dose of nature and green exercise for improving mental health? A multi-study analysis. *Environmental Science and Technology* 44, 3947–3955.
Barton, J., Bragg, R., Pretty, J., Roberts, J. and Wood, C. (2016) The wilderness expedition: An effective life course intervention to improve young people's well-being and connectedness to nature. *Journal of Experiential Education* 39, 59–72.
Bonnefoy, X.R., Braubach, M., Moissonnier, B. Monolbaev K. and Röbbel N. (2003) Housing and health in Europe: Preliminary results of a Pan-European study. *American Journal of Public Health* 93, 1559–1563.
Boyle, C. L. and Pryor, J. (2019) Therapeutic horticulture: A mechanism for participate to learn in inpatient rehabilitation. *Journal of the Australasian Rehabilitation Nurses' Association* (JARNA) 22, 16–21.
Branas, C.C., Cheney, A and MacDonald J.M. (2011) A difference-in-differences analysis of health, safety and greening vacant urban space. *American Journal of Epidemiology* 174, 1296–1306.
Calogiuri, G., Litleskare, S. and Fröhlich, F. (2021) Physical activity and virtual nature: perspectives on the health and behavioral benefits of virtual green exercise. *Nature and Health* 2021 (29 Jul.), 127–146.
Cameron, R. (2023) "Do we need to see gardens in a new light?" Recommendations for policy and practice to improve the ecosystem services derived from domestic gardens. *Urban Forestry and Urban Greening* 80, 127820.
Cameron, R.W., Blanuša, T., Taylor, J.E., Salisbury, A., Halstead, A.J., Henricot, B. and Thompson, K. (2012) The domestic garden – Its contribution to urban green infrastructure. *Urban Forestry and Urban Greening* 11, 129–137.
Cameron, R.W., Brindley, P., Mears, M., McEwan, K., Ferguson, F. *et al.* (2020) Where the wild things are! Do urban green spaces with greater avian biodiversity promote more positive emotions in humans? *Urban Ecosystems* 23, 301–317.
Cao, S., Song, C., Jiang, S., Luo, H., Zhang, P. *et al.* (2024) Effects of urban greenway environmental types and landscape characteristics on physical and mental health restoration. *Forests* 15, 679.
Capra, C.L., Haller, R.L. and Kennedy, K.L. (2019) Introduction to the profession of Horticultural therapy. In: *The Profession and Practice of Horticultural Therapy* 3–22 CRC Press, Boca Raton, Florida, USA.
Cardinali, M., Beenackers, M.A., van Timmeren, A. and Pottgiesser, U. (2024) The relation between proximity to and characteristics of green spaces to physical activity and health: A multi-dimensional sensitivity analysis in four European cities. *Environmental Research* 241, 117605.
Carrus G, Scopelliti M, Lafortezza R, Colangelo G, Ferrini F, Salbitano F, *et al.* (2015) Go greener, feel better? The positive effects of biodiversity on the well-being of individuals visiting urban and peri-urban green areas. *Landscape and Urban Planning* 134, 221–228.
Chalmin-Pui, L.S., Griffiths, A., Roe, J., Heaton, T. and Cameron, R. (2021a) Why garden? – Attitudes and the perceived health benefits of home gardening. *Cities* 112, 103118.
Chalmin-Pui, L.S., Roe, J., Griffiths, A., Smyth, N., Heaton, T., Clayden, A. and Cameron, R. (2021b) "It made me feel brighter in myself"-The health and well-being impacts of a residential front garden horticultural intervention. *Landscape and Urban Planning* 205, 103958.
Christian, M.S., Evans, C.E., Nykjaer, C., Hancock, N. and Cade, J.E. (2014) Evaluation of the impact of a school gardening intervention on children's fruit and vegetable intake: A randomised controlled trial. *International Journal of Behavioral Nutrition and Physical Activity* 11, 1–15.
Coombes, E., Jones, A.P. and Hillsdon, M. (2010) The relationship of physical activity and overweight to objectively measured green space accessibility and use. *Social Science and Medicine* 70, 816–822.
Cruz-Piedrahita, C., Howe, C. and de Nazelle, A. (2020) Public health benefits from urban horticulture in the global north: A scoping review and framework. *Global Transitions* 2, 246–256.
Dadvand, P., Nieuwenhuijsen, M.J., Esnaola, M., Forns, J., Basagaña, X. *et al.* (2015) Green spaces and cognitive development in primary schoolchildren. *Proceedings of the National Academy of Sciences* 112, 7937–7942.
Daiz, B.G., Rosales, E.L., Diago, P. and De los Santos, J.A.A. (2022) Health and well-being benefits of gardening: A comparative study among gardeners and non-gardeners in the Philippines. *The Malaysian Journal of Nursing (MJN)* 13, 39–45.
Dallimer, M,. Irvine, K.N., Skinner, A.M.J.J., Davies, Z.G., Rouquette, J.R. *et al.* (2012) Biodiversity and the feel-good factor: understanding associations between self- reported human well-being and species richness. *BioScience* 62, 47–55.
Davis, J.N., Spaniol, M.R. and Somerset, S. (2015) Sustenance and sustainability: maximizing the impact of school gardens on health outcomes. *Public Health Nutrition* 18, 2358–2367.
Donovan, G.H., Michael, Y.L., Butry, D.T., Sullivan, A.D. and Chase, J.M. (2011) Urban trees and the risk of poor birth outcomes. *Health and Place* 17, 390–393.
Donovan, G.H., Butry, D.T., Michael, Y.L., Prestemon, J.P., Liebhold, A.M., Gatziolis, D. and Mao, M.Y. (2013) The relationship between trees and human health: Evidence from the spread of the emerald ash

borer. *American Journal of Preventive Medicine* 44, 139–145.

Donovan, G.H., Michael, Y.L., Gatziolis, D., Prestemon, J.P. and Whitsel, E.A. (2015) Is tree loss associated with cardiovascular-disease risk in the Women's Health Initiative? A natural experiment. *Health and Place* 36, 1–7.

Douglas, J.W.A. and Evans, K.L. (2022) An experimental test of the impact of avian diversity on attentional benefits and enjoyment of people experiencing urban green-space. *People and Nature* 4, 243–259.

Elsadek, M., Jo, H., Sun, M. and Fujii, E. (2013) Brain activity and emotional responses of the Japanese people toward trees pruned using sukashi technique. *International Journal of Agriculture, Environment and Biotechnology* 6, 465–470.

Elsadek, M., Liu, B. and Lian, Z. (2019) Green façades: Their contribution to stress recovery and well-being in high-density cities. *Urban Forestry and Urban Greening* 46, 126446.

Evenson, K.R, Wen, F., Hillier, A. and Cohen D.A. (2013) Assessing the contribution of parks to physical activity using global positioning system and accelerometry. *Medicine and Science in Sports and Exercise* 45, 1981–1987.

Farris, S., Dempsey, N., McEwan, K., Hoyle, H. and Cameron, R. (2024a) Does increasing biodiversity in an urban woodland setting promote positive emotional responses in humans? A stress recovery experiment using 360-degree videos of an urban woodland. *Plos One* 19, 0297179.

Farris, S., Zhang, L., Dempsey, N., McEwan, K., Hoyle, H. and Cameron, R. (2024b) 'The Elephant in the Room'– Does actual or perceived biodiversity elicit restorative responses in a virtual park? *Cities and Health* Aug 2024, 1–16.

Flies, E.J., Mavoa, S., Zosky, G.R., Mantzioris, E., Williams, C. *et al.* (2019) Urban-associated diseases: Candidate diseases, environmental risk factors, and a path forward. *Environment International* 133, 105187.

Francis, J., Wood, L.J., Knuiman, M. and Giles-Corti, B. (2012) Quality or quantity? Exploring the relationship between Public Open Space attributes and mental health in Perth, Western Australia. *Social Science and Medicine* 74, 1570–1577.

Gandelman, N., Piani, G. and Ferre, Z. (2012) Neighborhood determinants of quality of life. *Journal of Happiness Studies* 13, 547–563.

Gathright, J., Yamada, Y., Morita, M. (2006) Comparison of the physiological and psychological benefits of tree and tower climbing. *Urban Forestry and Urban Greening* 5, 141–149.

Ghimire, R., Ferreira, S., Green, G.T., Poudyal, N.C., Cordell, H.K. and Thapa, J.R. (2017) Green space and adult obesity in the United States. *Ecological Economics* 136, 201–212.

Gigliotti, C.M. and Jarrott, S.E. (2005) Effects of horticulture therapy on engagement and affect. *Canadian Journal on Aging / La Revue Canadienne du Vieillissement* 24, 367–377.

Gigliotti, C. M., Jarrott, S. E. and Yorgason, J. (2004) Harvesting health: Effects of three types of horticultural therapy activities for persons with dementia. *Dementia* 3, 161–180.

Gonzalez, M.T., Hartig, T., Patil, G.G., Martinsen, E.W. and Kirkevold, M. (2010) Therapeutic horticulture in clinical depression: A prospective study of active components. *Journal of Advanced Nursing* 66, 2002–2013.

Gonzalez, M.T., Hartig, T., Patil, G.G., Martinsen, E.W. and Kirkevold, M. (2011) A prospective study of existential issues in therapeutic horticulture for clinical depression. *Issues in Mental Health Nursing* 32, 73–81.

Grahn, P. and Stigsdotter, U.K. (2010) The relation between perceived sensory dimensions of urban green space and stress restoration. *Landscape and Urban Planning* 94, 264–275.

Gregis, A., Ghisalberti, C., Sciascia, S., Sottile, F. and Peano, C. (2021) Community garden initiatives addressing health and well-being outcomes: A systematic review of infodemiology aspects, outcomes, and target populations. *International Journal of Environmental Research and Public Health* 18, 1943.

Grilli, G., Mohan, G. and Curtis, J. (2020) Public park attributes, park visits, and associated health status. *Landscape and Urban Planning* 199, 103814.

Guglielmetti Mugion, R. and Menicucci, E. (2021) Understanding the benefits of horticultural therapy on paediatric patient's well-being during hospitalisation. *The TQM Journal* 33, 856–881.

Gunn, C., Vahdati,M. and Shahrestani, M. (2022) Green walls in schools – The potential well-being benefits. *Building and Environment* 224, 109560.

Harris, K. and Trauth, J. (2020) Horticulture therapy benefits: A report. International *Journal Current Scientific and Multidisciplinary Research* 3, 61–65.

Hartig, T., Evans, G.W., Jamner, L.D., Davis, D.S. and Gärling, T. (2003) Tracking restoration in natural and urban field settings. *Journal of Environmental Psychology* 23, 109–123.

Hedblom, M., Heyman, E., Antonsson, H. and Gunnarsson, B. (2014) Bird song diversity influences young people's appreciation of urban landscapes. *Urban Forestry and Urban Greening* 13, 469–474.

Holloway, T.P., Dalton, L., Hughes, R., Jayasinghe, S., Patterson, K.A. *et al.* (2023) School gardening and health and well-being of school-aged children: A realist synthesis. *Nutrients* 15, 1190.

Horiuchi, M., Endo, J., Akatsuka, S., Aaa, A., Hasegawa, T. and Seko, Y. (2013) Influence of forest walking on blood pressure, profile of mood states and stress markers from the viewpoint of aging. *Journal of Aging and Gerontology* 1, 9–17.

Horiuchi, M., Endo, J., Takayama, N., Murase, K., Nishiyama, N., Saito, H. and Fujiwara, A. (2014) Impact of viewing vs. not viewing a real forest on physiological and psychological responses in the same setting. *International Journal of Environmental Research and Public Health* 11, 10883–10901.

Hubai, K., Kováts, N. and Eck-Varanka, B. (2024) Urban gardening—How safe is it?. *Urban Science* 8, 91.

Hystad, P., Davies, H.W., Frank, L., Van Loon, J., Gehring, U., Tamburic, L. and Brauer, M. (2014) Residential greenness and birth outcomes: Evaluating the influence of spatially correlated built-environment factors. *Environmental Health Perspectives* 122, 1095–1102.

Jennissen, C.A., Krupp, T.D., Vakkalanka, J.P. and Hoogerwerf, P.J. (2023) Pediatric lawn mower-related injuries and contributing factors for bystander injuries. *Injury Epidemiology* 10, 51.

Jiang, B., Chang, C.-Y., Sullivan, W.C. (2014) A dose of nature: Tree cover, stress reduction, and gender differences. *Landscape and Urban Planning* 132, 26–36.

Jiang, B., Li, D., Larsen, L., Sullivan, W.C. (2016) A dose-response curve describing the relationship between urban tree cover density and self-reported stress recovery. *Environment and Behaviour* 48, 607–629.

Jong, M.C., Stub, T., Mulder, E. and Jong, M. (2022) The development and acceptability of a wilderness programme to support the health and well-being of adolescent and young adult cancer survivors: The WAYA programme. *International Journal of Environmental Research and Public Health* 19, 12012.

Joung, D., Kim, G., Choi, Y., Lim, H., Park, S., Woo, J.-M. and Park, B.-J. (2015) The prefrontal cortex activity and psychological effects of viewing forest landscapes in autumn season. *International Journal of Environmental Research and Public Health* 12, 7235–7243.

Kahn Jr, P.H., Friedman, B., Gill, B., Hagman, J., Severson, R.L., Freier, N.G. and Stolyar, A. (2008) A plasma display window?—The shifting baseline problem in a technologically mediated natural world. *Journal of Environmental Psychology* 28, 192–199.

Kaplan, S. (1995) The restorative benefits of nature: Toward an integrative framework. *Journal of Environmental Psychology* 15, 169–182.

Kaplan, R. and Kaplan, S. (1989) *The Experience of Nature: A Psychological Perspective*. CUP Archive.

Kardan, O., Gozdyra, P., Misic, B., Moola, F., Palmer, L.J., Paus, T. and Berman, M.G. (2015) Neighborhood greenspace and health in a large urban center. *Science Reports* 5, 11610.

Khan, M. and Bell, R. (2019) Effects of a school based intervention on children's physical activity and healthy eating: A mixed-methods study. *International Journal of Environmental Research and Public Health* 16, 4320.

Kim, K. and Park, S. (2018) Horticultural therapy program for middle-aged women's depression, anxiety, and self-identify. *Complementary Therapies in Medicine* 39, 154–159.

Kondo, M.C., Fluehr, J.M., McKeon, T. and Branas, C.C. (2018) Urban green space and its impact on human health. *International Journal of Environmental Research and Public Health* 15, 445.

Korpela, K. and Hartig, T. (1996) Restorative qualities of favorite places. *Journal of Environmental Psychology* 16, 221–233.

Korpela, K., Korhonen, M., Nummi, T., Martos, T. and Sallay, V. (2020) Environmental self-regulation in favourite places of Finnish and Hungarian adults. *Journal of Environmental Psychology* 67, 101384.

Kotzen, B. (2018) Green roofs social and aesthetic aspects. In: Pérez G. and Perini K. (eds) *Nature Based Strategies for Urban and Building Sustainability*. Butterworth-Heinemann, Elsevier, Kidlington, UK, pp. 273–281.

Kühn, S., Düzel, S., Eibich, P., Krekel, C., Wüstemann, H. *et al.* (2017) In search of features that constitute an "enriched environment" in humans: Associations between geographical properties and brain structure. *Science Reports* 7, 11920.64–72.

Kuo, F.E. and Sullivan, W.C. (2001a) Environment and crime in the inner city: does vegetation reduce crime? *Environment and Behavior* 33, 343–367.

Kuo, F.E. and Sullivan, W.C. (2001b) Aggression and violence in the inner city effects of environment via mental fatigue. *Environment and Behavior* 33, 543–571.

Lachowycz, K. and Jones, A.P. (2011) Greenspace and obesity: A systematic review of the evidence. *Obesity Reviews* 12 e183–189.

Lam, V., Romses, K. and Renwick, K. (2019) Exploring the relationship between school gardens, food literacy and mental well-being in youth using photovoice. *Nutrients* 11,1354.

Lampert, T., Costa, J., Santos, O., Sousa, J., Ribeiro, T. and Freire, E. (2021) Evidence on the contribution of community gardens to promote physical and mental health and well-being of non-institutionalized individuals: A systematic review. *Plos One* 16, e0255621.

Le Gear, K., Carlin, C. and Flaherty, G.T. (2023) Deep roots: Realising the public health benefits of exposure to forest environments. *Advances in Integrative Medicine* 10, 86–88.

Lee, J., Park, B.J., Tsunetsugu, Y., Ohira, T., Kagawa, T. and Miyazaki, Y. (2011) Effect of forest bathing on physiological and psychological responses in young Japanese male subjects. *Public Health* 125, 93–100.

Lehberger, M., Kleih, A.K. and Sparke, K. (2021) Self-reported well-being and the importance of green spaces–A comparison of garden owners and non-garden owners in times of COVID-19. *Landscape and Urban Planning* 212, 104108

Lehmann, L.P., Detweiler, J.G. and Detweiler, M.B. (2018) Veterans in substance abuse treatment program self-initiate box gardening as a stress reducing therapeutic modality. *Complementary Therapies in Medicine* 36, 50–53.

Lehtimäki, J., Karkman, A., Laatikainen, T., Paalanen, L., von Hertzen, L. *et al.* (2017) Patterns in the skin microbiota differ in children and teenagers between rural and urban environments. *Scientific Reports* 7, 45651.

Lehtimäki, J., Gupta, S., Hjelmsø, M., Shah, S., Thorsen, J. *et al.* (2023) Fungi and bacteria in the beds of rural and urban infants correlate with later risk of atopic diseases. *Clinical and Experimental Allergy* 53, 1268–1278.

Li, Q., Morimoto, K., Kobayashi, M., Inagaki, H., Katsumata, M. *et al.* (2008a) A forest bathing trip increases human natural killer activity and expression of anti-cancer proteins in female subjects. *Journal of Biological Regulatory Homeostatic Agents* 22, 45–55.

Li, Q., Morimoto, K., Kobayashi, M., Inagaki, H., Katsumata, M. *et al.* (2008b) Visiting a forest, but not a city, increases human natural killer activity and expression of anti-cancer proteins. *International Journal of Immunopathology and Pharmacology* 21, 117–127.

Li, Q., Kobayashi, M., Wakayama, Y., Inagaki, H., Katsumata, M. *et al.* (2009) Effect of phytoncide from trees on human natural killer cell function. *International Journal of Immunopathology and Pharmacology* 22, 951–959.

Li, Q., Kobayashi, M., Inagaki, H., Hirata, Y., Li, Y.J. *et al.* (2010) A day trip to a forest park increases human natural killer activity and the expression of anti-cancer proteins in male subjects. *Journal of Biological Regulatory Homeostatic Agents* 24, 157–165.

Li, X., Zhang, Z., Gu, M., Jiang, D.Y., Wang, J., Lv, Y.M. and Pan, H.T. (2012) Effects of plantscape colors on psycho-physiological responses of university students. *Journal of Food, Agriculture and Environment* 10, 702–708.

Liddicoat, C., Bi, P., Waycott, M., Glover, J., Lowe, A.J. and Weinstein, P. (2018) Landscape biodiversity correlates with respiratory health in Australia. *Journal of Environmental Management* 206, 113–122.

Lindemann-Matthies P and Matthies D. (2018) The influence of plant species richness on stress recovery of humans. *Web Ecology* 18, 121–128.

Lottrup, L., Grahn, P. and Stigsdotter, U.K. (2013) Workplace greenery and perceived level of stress: Benefits of access to a green outdoor environment at the workplace. *Landscape and Urban Planning* 110, 5–11.

Lovasi, G.S., Schwartz-Soicher, O., Quinn, J.W., Berger, D.K., Neckerman, K.M. *et al.* (2013) Neighborhood safety and green space as predictors of obesity among preschool children from low-income families in New York City. *Preventative Medicine* 57, 189–193.

Løvoll, H.S., Sæther, K.W. and Graves, M. (2020) Feeling at home in the wilderness: Environmental conditions, well-being and aesthetic experience. *Frontiers in Psychology* 11, 402.

Luo, Y.N., Huang, W.Z., Liu, X.X., Markevych, I., Bloom, M.S. *et al.* (2020) Greenspace with overweight and obesity: A systematic review and meta-analysis of epidemiological studies up to 2020. *Obesity Reviews* 21, e13078.

Ma, B., Zhou, T., Lei, S., Wen, Y. and Htun, T.T. (2019) Effects of urban green spaces on residents' well-being. *Environment, Development and Sustainability* 21, 2793–2809.

Maas, J., Verheij, R.A., Spreeuwenberg, P. and Groenewegen, P.P. (2008) Physical activity as a possible mechanism behind the relationship between green space and health: a multilevel analysis. *BMC Public Health* 8, 206.

Maas, J., Van Dillen, S.M., Verheij, R.A. and Groenewegen, P.P. (2009) Social contacts as a possible mechanism behind the relation between green space and health. *Health and Place* 15, 586–595.

Madureira, H., Nunes, F., Oliveira, J.V. and Madureira, T. (2018) Preferences for urban green space characteristics: A comparative study in three Portuguese cities. *Environments* 5, 23.

Mao, G.X., Lan, X.G., Cao, Y.B., Chen, Z.M., He, Z.H. *et al.* (2012) Effects of short-term forest bathing on human health in a broad-leaved evergreen forest in Zhejiang Province, China. *Biomedical Environmental Science* 25, 317–324.

Martens, D., Gutscher, H. and Bauer, N. (2011) Walking in "wild" and "tended" urban forests: The impact on psychological well-being. *Journal of Environmental Psychology* 31, 36–44.

Martínez-Soto, J., Gonzales-Santos, L., Pasaye, E. and Barrios, F.A. (2013) Exploration of neural correlates of restorative environment exposure through functional magnetic resonance. *Intelligent Buildings International* 5, 10–28.

Mavoa, S., Davern, M., Breed, M. and Hahs, A. (2019) Higher levels of greenness and biodiversity associate with greater subjective wellbeing in adults living in Melbourne, Australia. *Health and Place* 57, 321–329.

McDougall, C.W., Foley, R., Hanley, N., Quilliam, R.S. and Oliver, D.M. (2022) Freshwater wild swimming, health and well-being: Understanding the importance of place and risk. *Sustainability* 14, 6364.

Methorst, J., Rehdanz, K., Mueller, T., Hansjürgens, B., Bonn, A. and Böhning-Gaese, K. (2021) The importance of species diversity for human well-being in Europe. *Ecological Economics* 181, 106917.

Mills, J.G., Bissett, A., Gellie, N.J., Lowe, A.J., Selway, C.A. *et al.* (2020) Revegetation of urban green space rewilds soil microbiotas with implications for human health and urban design. *Restoration Ecology* 28, S322–S334.

Morita, E., Fukuda, S., Nagano, J., Hamajima, N., Yamamoto, H. *et al.* (2007) Psychological effects of forest environments on healthy adults: Shinrin-yoku (forest-air bathing, walking) as a possible method of stress reduction. *Public Health* 121, 54–63.

Mourão, I., Moreira, M.C., Almeida, T.C. and Brito, L.M. (2019) Perceived changes in well-being and happiness with gardening in urban organic allotments in Portugal. *International Journal of Sustainable Development and World Ecology* 26, 79–89.

Najjar, A.H., Foroozandeh, E. and Gharneh, A.A.H. (2018) Horticulture therapy. Effects on memory and psycho-logical symptoms of depressed male outpatients. *Iranian Rehabilitation Journal* 16, 147–154.

Nguyen, P.Y., Astell-Burt, T., Rahimi-Ardabili, H. and Feng, X. (2021) Green space quality and health: A systematic review. *International Journal of Environmental Research and Public Health* 18, 11028.

Nordh, H., Grahn, P. and Währborg, P. (2009) Meaningful activities in the forest, a way back from exhaustion and long-term sick leave. *Urban Forestry and Urban Greening* 8, 207–219.

Ode Sang, Å., Thorpert, P. and Fransson, A.M. (2022) Planning, designing, and managing green roofs and green walls for public health – An ecosystem services approach. *Frontiers in Ecology and Evolution* 10, 804500.

Ogunseitan, O.A. (2005) Topophilia and the quality of life. *Environmental Health Perspectives* 113, 143–148.

Oh, R.R., Zhang, Y., Nghiem, L.T., Chang, C.C., Tan, C.L. *et al.* (2022) Connection to nature and time spent in gardens predicts social cohesion. *Urban Forestry and Urban Greening* 74, 127655.

Ohly, H., Gentry, S., Wigglesworth, R., Bethel, A., Lovell, R. and Garside, R. (2016) A systematic review of the health and well-being impacts of school gardening: Synthesis of quantitative and qualitative evidence. *BMC Public Health* 16, 1–36.

Ohtsuka, Y., Yabunaka, N., Takayama, S. Shinrin-yoku (1998) Forest-air bathing and walking effectively decreases blood glucose levels in diabetic patients. *International Journal of Biometeorology* 41, 125–127.

Orstad, S.L., Szuhany, K., Tamura, K., Thorpe, L.E. and Jay, M. (2020) Park proximity and use for physical activity among urban residents: Associations with mental health. *International Journal of Environmental Research and Public Health* 17, 4885.

Park, S.H. and Mattson, R.H. (2008) Effects of flowering and foliage plants in hospital rooms on patients recovering from abdominal surgery. *HortTechnology* 18, 563–568.

Park, B.J., Tsunetsugu, Y., Kasetani, T., Hirano, H., Kagawa, T., Sato, M. and Miyazaki, Y. (2007) Physiological effects of Shinrin-yoku (taking in the atmosphere of the forest)-using salivary cortisol and cerebral activity as indicators. *Journal of Physiological Anthropology* 26, 123–128.

Park, B.J., Tsunetsugu, Y., Kasetani, T., Kagawa, T., Miyazaki, Y. (2010) The physiological effects of Shinrin-yoku (taking in the forest atmosphere or forest bathing): Evidence from field experiments in 24 forests across Japan. *Environmental Health and Preventative Medicine* 15, 18–26.

Peschardt, K.K. and Stigsdotter, U.K. (2013) Associations between park characteristics and perceived restorativeness of small public urban green spaces. *Landscape and Urban Planning* 112, 26–39.

Peterková, J., Michalčíková, M., Novák, V., Slávik, R., Zach, J. *et al.* (2019) The influence of green walls on interior climate conditions and human health. In: *MATEC Web of Conferences* 282, 02041. EDP Sciences.

Phillips, A., Plastara, D., Khan, A.Z. and Canters, F. (2023) Integrating public perceptions of proximity and quality in the modelling of urban green space access. *Landscape and Urban Planning* 240, 104875.

Potwarka, L.R., Kaczynski, A.T. and Flack, A.L. (2008) Places to play: association of park space and facilities with healthy weight status among children. *Journal of Community Health* 33, 344–350.

Qin, J., Sun, C., Zhou, X., Leng, H. and Lian, Z. (2014) The effect of indoor plants on human comfort. Indoor Building and Environment 23, 709–723.

Rfidah, E.I., Casey, P.B., Tracey, J.A. and Gill, D. (1991) Childhood poisoning in Dublin. *Irish Medical Journal* 84, 87–89.

Richardson, E.A., Pearce, J., Mitchell, R. and Kingham, S. (2013) Role of physical activity in the relationship between urban green space and health. *Public Health* 127, 318–324.

Robinson, J.M., Cando-Dumancela, C., Liddicoat, C., Weinstein, P., Cameron, R. and Breed, M.F. (2020a) Vertical stratification in urban green space aerobiomes. *Environmental Health Perspectives* 128, 117008.

Robinson, J.M., Jorgensen, A., Cameron, R. and Brindley, P. (2020b) Let nature be thy medicine: A socioecological exploration of green prescribing in the UK. *International Journal of Environmental Research and Public Health* 17, 3460.

Robinson, J.M., Watkins, H., Man, I., Liddicoat, C., Cameron, R. *et al.* (2021a) Microbiome-inspired green infrastructure: A bioscience roadmap for urban ecosystem health. *Architectural Research Quarterly* 25, 292–303.

Robinson, J.M., Cando-Dumancela, C., Antwis, R.E., Cameron, R., Liddicoat, C. *et al.* (2021b) Exposure to airborne bacteria depends upon vertical stratification and vegetation complexity. *Scientific Reports* 11, 9516.

Roe, J.J., Aspinall, P.A., Mavros, P. and Coyne, R. (2013) Engaging the brain: The impact of natural versus urban scenes using novel EEG methods in an experimental setting. *Environmental Science* 1, 93–104.

Romagosa, F., Eagles, P.F. and Lemieux, C.J. (2015) From the inside out to the outside in: exploring the role of parks and protected areas as providers of human health and well-being. *Journal of Outdoor Recreation and Tourism* 10, 70–77.

Roslund, M.I., Puhakka, R., Grönroos, M., Nurminen, N., Oikarinen, S. *et al.* (2020) Biodiversity intervention enhances immune regulation and health-associated commensal microbiota among daycare children. *Science Advances* 6, eaba2578.

Sallis, J.F., Floyd, M.F., Rodriguez, D.A. and Saelens, B.E. (2012) Role of built environments in physical activity, obesity, and cardiovascular disease. *Circulation* 125, 729–737.

Sands, G., Blake, H., Carter, T. and Spiby, H. (2023) Nature-based interventions in the UK: A mixed methods study exploring green prescribing for promoting the mental wellbeing of young pregnant women. *International Journal of Environmental Research and Public Health*, 20, 6921.

Sbihi, H., Boutin, R.C., Cutler, C., Suen, M., Finlay, B.B. and Turvey, S.E. (2019) Thinking bigger: How early-life environmental exposures shape the gut microbiome and influence the development of asthma and allergic disease. *Allergy* 74, 2103–2115.

Schebella, M.F., Weber, D., Schultz, L. and Weinstein, P. (2020) The nature of reality: Human stress recovery during exposure to biodiverse, multisensory virtual environments. *International Journal of Environmental Research and Public Health* 17. 17010056

Schipperijn, J., Ekholm, O., Stigsdotter, U.K., Toftager, M., Bentsen, P., Kamper-Jørgensen, F. and Randrup, T.B. (2010) Factors influencing the use of green space: Results from a Danish national representative survey. *Landscape and Urban Planning* 95, 130–137.

Schreinemachers, P., Rai, B.B., Dorji, D., Chen, H.P., Dukpa, T. *et al.* (2017) School gardening in Bhutan: Evaluating outcomes and impact. *Food Security* 9, 635–648.

Shin, Y.-K., Kim, D.J., Jung-Choi, K., Son, Y., Koo, J.-W., Min, J.-A. and Chae, J.-H. (2013) Differences of psychological effects between meditative and athletic walking in a forest and gymnasium. *Scandinavian Journal of Forest Research* 2013, 28, 64–72.

Söderback, I., Söderström, M. and Schälander, E. (2004) Horticultural therapy: The „healing garden" and gardening in rehabilitation measures at Danderyd Hospital Rehabilitation Clinic, Sweden. *Pediatric Rehabilitation* 7, 245–260.

Soga, M., Cox, D.T., Yamaura, Y., Gaston, K.J., Kurisu, K. and Hanaki, K. (2017) Health benefits of urban allotment gardening: Improved physical and psychological well-being and social integration. *International Journal of Environmental Research and Public Health* 14, 71.

Sonntag-Öström, E., Nordin, M., Lundell, Y., Dolling, A., Wiklund, U. *et al.* (2014) Restorative effects of visits to urban and forest environments in patients with exhaustion disorder. *Urban Forestry and Urban Greening* 13, 344–354.

Southon, G.E., Jorgensen, A., Dunnett, N., Hoyle, H. and Evans, K.L. (2018) Perceived species-richness in urban green spaces: Cues, accuracy and well-being impacts. *Landscape and Urban Planning* 172, 1–10.

Spano, G., D'Este, M., Giannico, V., Carrus, G., Elia, M. *et al.* (2020) Are community gardening and horticultural interventions beneficial for psychosocial well-being? A meta-analysis. *International Journal of Environmental Research and Public Health* 17, 3584.

Spring, J.A., Viera, M., Bowen, C. and Marsh, N. (2013) Is gardening a stimulating activity for people with advanced Huntington's disease? *Dementia* 13(6), 1471301213486661.

Stein, M.M., Hrusch, C.L., Gozdz, J., Igartua, C., Pivniouk, V. *et al.* (2016) Innate immunity and asthma risk in Amish and Hutterite farm children. *New England Journal of Medicine* 375, 411–421.

Stevens, H.R., Graham, P.L., Beggs, P.J. and Ossola, A. (2024) Associations between violent crime inside and outside, air temperature, urban heat island magnitude and urban green space. *International Journal of Biometeorology* 68, 661–673.

Stigsdotter, U. and Grahn, P. (2002) What makes a garden a healing garden. *Journal of Therapeutic Horticulture* 13, 60–69.

Stragà, M., Miani, C., Mäntylä, T., Bruine de Bruin, W., Mottica, M. and Del Missier, F. (2023) Into the wild or into the library? Perceived restorativeness of natural and built environments. *Journal of Environmental Psychology* 91, 1–11.

Swami, V. (2020) Body image benefits of allotment gardening. *Ecopsychology* 12, 19–23.

Takayama, N., Korpela, K., Lee, J., Morikawa, T., Tsunetsugu, Y. *et al.* (2014) Emotional, restorative and vitalizing effects of forest and urban environments at four sites in Japan. *International Journal of Environmental Research and Public Health* 11, 7207–7230.

Tarar, G., Etheredge, C.L., McFarland, A., Snelgrove, A., Waliczek, T.M. and Zajicek, J.M. (2015) The effect of urban tree canopy cover and vegetation levels on incidence of stress-related illnesses in humans in metropolitan statistical areas of Texas. *HortTechnology* 25, 76–84.

Taylor, R.P. (2021) The potential of biophilic fractal designs to promote health and performance: A review of experiments and applications. *Sustainability* 13, 823.

Taylor, M.S., Wheeler, B.W., White, M.P., Economou, T. and Osborne, N.J. (2015) Urban street tree density and antidepressant prescription rates. A cross-sectional study in London, UK. *Landscape and Urban Planning* 136, 174–179.

Thomson, L.J., Morse, N., Elsden, E. and Chatterjee, H.J. (2020) Art, nature and mental health: Assessing the biopsychosocial effects of a 'creative green prescription' museum programme involving horticulture, artmaking and collections. *Perspectives in Public Health* 140, 277–285.

Tsai, W.-L., Floyd, M.F., Leung, Y.-F., McHale, M.R. and Reich, B.J. (2016) Urban vegetative cover fragmentation in the U.S. *American Journal of Preventative Medicine* 50. 509–517.

Tyrväinen, L., Ojala, A., Korpela, K., Lanki, T., Tsunetsugu, Y. and Kagawa, T. (2014) The influence of urban green environments on stress relief measures: A field experiment. *Journal of Environmental Psychology* 38, 1–9.

Uebel, K., Marselle, M., Dean, A.J., Rhodes, J.R. and Bonn, A. (2021) Urban green space soundscapes and their perceived restorativeness. *People and Nature* 3, 756–769.

Ulmer, J.M., Wolf, K.L., Backman, D.R., Tretheway, R.L., Blain, C.J., O'Neil-Dunne, J.P. and Frank, L.D. (2016) Multiple health benefits of urban tree canopy: The mounting evidence for a green prescription. *Health Place* 42, 54–62.

Ulrich, R. (1984) View through a window may influence recovery from surgery. *Science* 224, 420–421.

Ulrich, R.S., Simons, R.F., Losito, B.D., Fiorito, E., Miles, M.A. and Zelson, M. (1991) Stress recovery during exposure to natural and urban environments. *Journal of Environmental Psychology* 11, 201–230.

Uwajeh, P.C., Iyendo, T.O. and Polay, M. (2019) Therapeutic gardens as a design approach for optimising the healing environment of patients with Alzheimer's disease and other dementias: A narrative review. *Explore* 15, 352–362.

Van den Berg, A.E. and Van den Berg, C.G. (2011) A comparison of children with ADHD in a natural and built setting. *Child Care and Health Development* 37, 430–439.

Van den Berg, A.E., Jorgensen, A. and Wilson, E.R. (2014) Evaluating restoration in urban green spaces: Does setting type make a difference? *Landscape and Urban Planning* 127, 173–181.

Van den Berg, A.E., Wesselius, J.E., Maas, J. and Tanja-Dijkstra, K. (2017) Green walls for a restorative classroom environment: a controlled evaluation study. *Environment and Behavior* 49, 791–813.

Van Dillen, S.M., De Vries, S., Groenewegen, P.P. and Spreeuwenberg, P. (2012) Greenspace in urban neighbourhoods and residents' health: Adding quality to quantity. *Journal of Epidemiology and Community Health* 66, e8–e8.

Van Dinter, M., Kools, M., Dane, G., Weijs-Perrée, M., Chamilothori, K. *et al.* (2022) Urban green parks for long-term subjective well-being: Empirical relationships between personal characteristics, park characteristics, park use, sense of place, and satisfaction with life in the Netherlands. *Sustainability* 14, 4911.

Van Lier, L.E., Utter, J., Denny, S., Lucassen, M., Dyson, B. and Clark, T. (2017) Home gardening and the health and well-being of adolescents. *Health Promotion Practice* 18, 34–43.

Venter, Z.S., Shackleton, C., Faull, A., Lancaster, L., Breetzke, G. and Edelstein, I. (2022) Is green space associated with reduced crime? A national-scale study from the Global South. *Science of the Total Environment* 825, 154005.

Vilcins, D., Sly, P.D., Scarth, P. and Mavoa, S. (2024) Green space in health research: An overview of common indicators of greenness. *Reviews on Environmental Health* 39, 221–231.

Von Hertzen, L., Hanski, I. and Haahtela, T. (2011) Natural immunity: Biodiversity loss and inflammatory diseases are two global megatrends that might be related. *EMBO Reports* 12, 1089–1093.

Wan, C., Shen, G.Q. and Choi, S. (2021) Underlying relationships between public urban green spaces and social cohesion: A systematic literature review. *City, Culture and Society* 24, 100383.

Wang, R., Cleland, C.L., Weir, R., McManus, S., Martire, A., Grekousis, G., Bryan, D. and Hunter, R.F. (2024) Rethinking the association between green space and crime using spatial quantile regression modelling: Do vegetation type, crime type, and crime rates matter? *Urban Forestry and Urban Greening* 101, 128523.

Warber, S.L., DeHudy, A.A., Bialko, M.F., Marselle, M.R. and Irvine, K.N. (2015) Addressing "nature-deficit disorder": A mixed methods pilot study of young adults attending a wilderness camp. *Evidence-Based Complementary and Alternative Medicine*, 2015.

White, M.P., Alcock, I., Wheeler, B.W. and Depledge, M.H. (2013) Would you be happier living in a greener urban area? A fixed-effects analysis of panel data. *Psychological Science*, 0956797612464659.

White, M.P., Elliott, L.R., Grellier, J., Economou, T., Bell, S., Bratman, G.N., Cirach, M., Gascon, M., Lima, M.L., Lõhmus, M. and Nieuwenhuijsen, M. (2021) Associations between green/blue spaces and mental health across 18 countries. *Scientific Reports* 11, 8903.

Whitehouse, S., Varni, J.W., Seid, M., Cooper-Marcus, C., Ensberg, M.J., Jacobs, J.R. and Mehlenbeck, R.S. (2001) Evaluating a children's hospital garden environment: Utilization and consumer satisfaction. *Journal of Environmental Psychology* 21, 301–314.

Wilkie, S., Platt, T. and Trotter, H. (2023) Does a brief virtual dose of an environment affect subjective wellbeing and judgements of perceived restorativeness? Considering the role of place preference. *Current Research in Ecological and Social Psychology* 4, 100127.

Williams, K.J., Lee, K.E., Sargent, L., Johnson, K.A., Rayner, J. *et al.* (2019) Appraising the psychological benefits of green roofs for city residents and workers. *Urban Forestry and Urban Greening* 44, 126399.

Williams, T.G., Logan, T.M., Zuo, C.T., Liberman, K.D. and Guikema, S.D. (2020) Parks and safety: A comparative study of green space access and inequity in five US cities. *Landscape and Urban Planning* 201, 103841.

Wilson, E.O. (1984) *Biophilia*, Harvard Press, Harvard, USA.

Wolf, L.J., Zu Ermgassen, S., Balmford, A., White, M. and Weinstein, N. (2017) Is variety the spice of life? An experimental investigation into the effects of species richness on self-reported mental well-being. *PloS One*,12: e0170225–e0170225.

Wolf, K.L., Lam, S.T., McKeen, J.K., Richardson, G.R.A., Van den Bosch, M. and Bardekjian, A.C. (2020) Urban trees and human health: A scoping review. *International Journal of Environmental Research and Public Health* 17, 4371.

Wood, C.J., Pretty, J. and Griffin, M. (2016) A case–control study of the health and well-being benefits of allotment gardening. *Journal of Public Health* 38, e336–e344.

Wood, E., Harsant, A., Dallimer, M., de Chavez, A.C., McEachan, R.R.C. and Hassall, C. (2018) Not all green space is created equal: Biodiversity predicts psychological restorative benefits from urban green space. *Frontiers in Psychology* 9.

Wu, J. and Jackson, L. (2017) Inverse relationship between urban green space and childhood autism in California elementary school districts. *Environmental International* 107, 140–146.

Wu, L. and Kim, S.K. (2021) Exploring the equality of accessing urban green spaces: A comparative study of 341 Chinese cities. *Ecological Indicators* 121, 107080.

Wu, J., Rappazzo, K.M., Simpson, R.J., Joodi, G., Pursell, I.W. *et al.* (2018) Exploring links between greenspace and sudden unexpected death: A spatial analysis. *Environmental International* 113, 114–121.

Xie, J., Liu, B. and Elsadek, M. (2021) How can flowers and their colors promote individuals' physiological and psychological states during the Covid-19 lockdown? *International Journal of Environmental Research and Public Health* 18, 10258.

Yang, Y., Ro, E., Lee, T.J., An, B.C., Hong, K.P. *et al.* (2022) The multi-sites trial on the effects of therapeutic gardening on mental health and well-being. *International Journal of Environmental Research and Public Health* 19, 8046.

Yang, Y., Chen, J., Hu, J., Shen, H., Chen, Q. *et al.* (2024) Ocular trauma from lawn mower accidents: Clinical insights, visual outcomes and microbial profiles. *The American Journal of Emergency Medicine* 80, 18–23.

Yen, H.Y. and Huang, H.Y. (2024) Actual and virtual parks benefit quality of life and physical activity: A cluster trial. *Journal of Urban Health* Apr 17, 1–10.

Yun, J., Yao, W., Meng, T. and Mu, Z. (2024) Effects of horticultural therapy on health in the elderly: A review and meta-analysis. *Journal of Public Health* 32, 1905–1931.

Zhang, L., Dempsey, N. and Cameron, R. (2023) Flowers–sunshine for the soul! How does floral colour influence preference, feelings of relaxation and positive up-lift? *Urban Forestry and Urban Greening* 79, 127795.

Zhang, L., Dempsey, N. and Cameron, R. (2024) 'Blossom Buddies'– How do flower colour combinations affect emotional response and influence therapeutic landscape design? *Landscape and Urban Planning* 248, 105099.

Zhu, S.X., Hu, F.F., He, S.Y., Qiu, Q., Su, Y., He, Q. and Li, J.Y. (2021) Comprehensive evaluation of healthcare benefits of different forest types: A case study in Shimen National Forest Park, China. *Forests* 12, 207.

5 Environmental Horticulture and the Conservation of Biodiversity

Abstract

The world's biodiversity is under threat from loss of habitat, over-exploitation, climate change, pollution and invasive species. Putting the 'environment' into environmental horticulture can do much to address these pressures on wildlife, at least within an urban environment, where there is capacity to retain/add green space to the built infrastructure. Well-planned, designed and managed green-blue cities can support an amazing level of wildlife, but understanding and acting on a number of ecological principles is required to achieve this. Increasing the variety of, and connections between, green spaces open up opportunities for wildlife and let species move through the city matrix with a degree of safety. Habitat quality can be improved through careful selection of plant taxa that optimises nectar, pollen, fruit and foliage which support wildlife. This chapter touches on how policy and practice relating to urban design and management can and needs to be improved to support biodiversity.

5.1 Introduction

The philosophy within environmental horticulture is to increase the opportunities for wildlife and promote greater biodiversity or at least maintain biodiversity at the existing level. Environmental horticulture should 'go with the grain' of nature, not undermine it. This is in contrast to traditional horticultural practices that tended to see cultivated plants predated on by pests and outcompeted by weeds. Environmental horticulture may still need to control plant populations through cultivation and weeding, and 'edge the ecological advantage' in favour of *desired* plants, but it should do this with the minimal impact to other organisms in the area. The answer to a rose plant covered in black spot (*Diplocarpon rosae*) and rust (*Phragmidium tuberculatum*) is not to reach for the chemical cupboard and apply fungicides but rather to improve the micro-environment to discourage the pathogens in the first place, to use a different, less susceptible variety of rose and/or indeed tolerate a degree of leaf blemish on the plant. Situations in commercial horticulture are not analogous to outdoor, environmental horticulture. In commercial horticulture there is often a monoculture of a single crop plant (one genotype) grown within a limited or enclosed area, and thus pathogen and pest populations can build up quickly and are subsequently difficult to manage using cultivation methods alone. Environmental horticulture in contrast tends to cultivate plants in mixed, multi-species communities, usually outdoors where natural food webs can be exploited to help keep pest and pathogen population in check. Thus, plant husbandry within environmental horticulture takes a holistic view. When trying to maintain populations of cultivated plants, this should be achieved by minimising the impact on other species that may use the landscape. This is not to be confused with the complete restoration of natural habitats and communities (although sometimes that is the objective), but rather the development of some form of semi-natural or artificial plant communities that still provide a range of ecological niches to local wildlife.

There needs to be a 'mind-shift' in how we view urban green spaces and how we can accommodate space for wildlife. Global biodiversity is under immense pressure from loss of habitat, climate change, competition from non-native invasive species, over-exploitation and pollution; and there are genuine worries that the planet will experience the sixth mass extinction of biodiversity within the coming decades. Better planned, designed and managed green space can do much to alleviate the pressure on urban biodiversity. Emphasizing the *environmental* component in horticulture is key to this. Urban green spaces can be good for people, but also good for wildlife. As indicated above management needs

DOI: 10.1079/9781800621763.0005

to take a more holistic approach and use resources sustainably. Key factors to consider include the following:

- Use more diverse plant communities. Plant diversity positively affects animal diversity as greater resources are provided for herbivores and in turn carnivores.
- Reduce excessively 'destructive' practices that damage vegetation dynamics or soil structure.
- Avoid (or at least reduce current) chemical use.
- Reduce/avoid the use of non-sustainably sourced materials or ones that cause environmental damage elsewhere.
- Provide more water features.
- Make physical (vegetated) links between the green spaces.
- Plant more trees (depending on site context).
- Keep dead wood on-site.
- Consider plants that act as food supplies for invertebrates and key bird and mammal species.
- Provide vegetation that enables whole invertebrate life cycles to be completed.

The focus of this chapter is on the urban environment, where horticultural land management is most prevalent, and where the pressures on wildlife are particularly challenging through the loss and fragmentation of green space. Specific management approaches discussed may be equally relevant in rural landscapes managed by horticulturists, for example in large country gardens and estates, but the wider landscape context may be somewhat different.

One key driver for environmental horticulture is to restore green landscapes and a degree of ecosystem function to areas that are currently highly degraded and where the biodiversity is impoverished. However, this does not automatically assume the use of native plant species exclusively in this process of 're-greening'. Indeed, environmental horticulture has been embroiled in the debates relating to 'is native always best?', for example, the arguments that non-native, highly floristic plant species can provide pollen and nectar resources to invertebrates just as well, or in some cases better, than native plant species communities (for example, by extending the periods pollen and nectar are available to invertebrates). This is not to advocate, of course, that native plant communities with high biodiversity indices should be replaced by non-native species, but that in the highly 'artificial' environment of our towns and cities non-native species may have a significant role in improving the opportunities for wildlife. A counterbalance to this, however, is that choice of species should be aimed at minimising the risk of introducing aggressive, invasive plant species that do threaten nearby native habitats.

5.2 What Is Biodiversity?

Biological diversity, more commonly termed 'biodiversity', is the range of taxonomic entities present in a given area. It is often defined as the 'totality of genes, species and ecosystems of a region' (Larsson, 2001) (Table 5.1). Most people simply understand it as the number of species present (biological richness), but actually variation within a species is also an important component. This is particularly relevant to environmental horticulture where the term needs to accommodate a given plant species and also the many and varied cultivated forms that may have been derived from that species. So, a location with a single taxon of *Sambucus nigra* (elder) would be considered to have a plant biodiversity richness of 1, but if we introduced specimens of *S. nigra* cv. Black Beauty, *S. nigra* cv. Black Lace, *S. nigra* cv. Aurea, *S. nigra* cv. Guincho Purple, *S. nigra* cv Golden Tower and *Sambucus racemosa* cv. Sutherland's

Table 5.1. Biodiversity arranged by ecological, organism and genetic levels

Ecological diversity	Organismal diversity	Genetic diversity
Biome	Kingdom	Population
Bio-region	Phylum	Individual
Landscape	Family	Chromosome
Ecosystem	Genus	Gene
Habitat	Species	Nucleotide
Niche	Subspecies	
Population	Population	
	Individual	

Gold the plant biodiversity richness increases to 7. In this case there are seven taxa present, even although there are all derived from the one species.

The term 'biodiversity' alludes to the fact that we somehow know all the species present, but science has probably only named about 16-30% of the total global biodiversity (and indeed significantly less if microbial communities are included). Nevertheless, the notion of biodiversity is a useful one as it allows some indication of the biological richness a given place, location or habitat may possess. The other term that is used commonly is biological 'abundance'; this relates to the number of individuals present within a taxonomic group. Other relevant terms commonly used in urban ecology are listed in Table 5.2.

Urban biodiversity is defined by 'the variety or richness and abundance of living organisms (including genetic variation) and habitats found in, and on the edge of, human settlements' (Müller *et al.*, 2013). Urban landscapes cover approximately 4% of the earth's land surface and 55% of the world's human population (4 billion people) now live in them (Ritchie and Roser, 2018). Urban biodiversity is the range of species associated with this landscape type and ecosystem, and essentially means the species found from the urban core out to the urban fringe of a human settlement.

Biodiversity as a term is often used to confer a notion of richness in *wild*life, at least to a lay audience. In reality, the life does not necessarily need to be 'wild'. In an urban context a highly decorative garden full of ornamental plants (e.g. *Begonia, Salvia, Petunia, Viola, Gladiolus* and *Lilium* species and cultivars) could be extremely rich in biodiversity, but there may be few *native* plant and animal species present at all. This point needs clarifying here, because in the subsequent sections where biodiversity is used it refers to all species, not just native ones.

5.3 The Urban Environment – Key Issues

The urban environment is often hostile to wildlife, although in reality the urban environment is a mosaic of different land forms, and some - the greener ones (parks, gardens etc. and indeed sometimes the brown ones – brownfield ex-industrial sites) can be wildlife rich. The challenges for animals and plants in an urban context are outlined in Table 5.3.

Urban ecosystems vary from 'natural' ecosystems in a number of key points. The high density of humans and human activities can deter many species from being present. Vegetation patches tend to be more fragmented (see below), and it is difficult for species to migrate from one patch to another, due to unsuitable landscapes in between (roads, buildings, etc.). Much of the vegetation is highly managed (disturbed) and may be composed of a high proportion of non-native flora. Aronson *et al.* (2014) claims that 75% of pre-urbanisation flora can be lost through the urban development process, although previous land use will affect this statistic (large tracks of intensely managed arable land can be species-poor too). There are high predation rates for wildlife due to the presence of domestic cats and dogs, as well as high incidences of road deaths. Natural food sources may be infrequently found, although human waste streams can for some species provide alternative source of nutrition. Urban *Vulpes vulpes* (red fox) are famous for 'checking out' park waste bins and removing the leftovers of passers-by 'fast-food' convenience dinners. People also intentionally put food out for wildlife – for example the seed and fat sources in bird feeders. Some animals (e.g. *Rattus norvegicus* – brown rat) are effective at feeding off such sources (and grain feed used for domestic poultry), even though they are not the intended recipient! Even when food is abundant, urban environments may provide little opportunity for nesting, roosting or breeding sites.

Although urban areas typically have lower biodiversity than rural areas (Aronson *et al.*, 2014), some urban green spaces can still act as important habitat for wildlife. For example, urban green spaces function as surrogate habitats for birds, refuge habitat for *Bombus* spp. (bumblebees) (McFrederick and LeBuhn, 2006) and dispersal and movement corridors for small birds and small mammals (Munshi-South, 2012). Ex-industrial brownfield sites are landscapes in transition and 'famous' for hosting rare plant (e.g. orchids) and invertebrate assemblages (e.g. *Bombus sylvarum* – shrill carder bee) (McCallum and Sardo, 2021; Cox and Rodway-Dyer, 2023). Green spaces are also environmentally important in they allow people to engage with nature and understand it better, as well as potentially gain health benefits from it. (Soga and Gaston, 2016; Cameron *et al.*, 2020). Given these benefits for both wildlife and humans, it is important that urban green spaces are located in the right places and designed well – including providing quiet refuges for animals.

Table 5.2. Common ecological terms and their definition

Ecological term or concept	Description
Abundance	The number of individuals representing a species in a particular ecosystem. Relative abundance is the ratio of individuals counted in one species compared to individuals in all species.
Arthropods	The group of invertebrate animals that includes insects, arachnids, crustaceans and myriapods such as centipedes.
Biome	A large naturally occurring community of flora (plants) and fauna (animals) occupying a major habitat, e.g. forest or savannah.
Biotope	Areas with uniform biological conditions (climate, soil, altitude, etc.).
Carnivore	A species that is reliant on predating and feeding on other animals.
Communities	Flora, fauna and microorganisms found in a location or habitat.
Competition	The simultaneous demand for an essential common resource by two or more organisms or species which is actually or potentially in limited supply. It may relate to food but also for the right to hold a territory or to mate. Competition can occur between individuals within a species (intraspecific) or between species (interspecific).
Competitor plant	Plants that thrive in areas of low-intensity stress and disturbance and outcompete other plants by efficiently exploiting available resources. They do this with traits such as rapid growth rate, high productivity and phenotypic plasticity (allocating resources where most needed at any given time).
Corridor	A strip of vegetation or waterway used by wildlife and potentially allowing movement of biotic factors between two areas. A number of intersection patches and corridors make up a network.
Depauperate	Areas poor in species number or diversity.
Dispersal	The ability of species to move, for example from one pond to another.
Disturbance	A change (often temporary) in environmental conditions that causes pronounced changes in an ecosystem or to species dynamics. Biodiversity is often promoted by moderate levels or infrequent occurrence of disturbance.
Diversity	Differences (usually in species or genes). Alpha diversity is the amount of diversity found within a community and beta diversity is the diversity resulting from differences amongst the various communities.
Ecocline	A gradual change in the ecosystem, for example change in environmental conditions as altitude increases
Ecotone	The area between two habitat types – edge or boundary communities, for example, 'woodland' edge where there may be species from the wood and from a neighbouring grassland community. Often more biodiverse than either habitat alone.
Ecotope	A small ecologically spatial unit where conditions tend to be relatively homogeneous.
Equitability	The 'evenness' of the community – how equally all species can be found in a habitat.
Food chain	A simple plot of energy flow in a community, e.g. primary producer–herbivore–carnivore.
Fragmentation	Break up of a habitat or ecosystem into smaller parcels. Affects the number of individuals located within an area.
Food web	Describes links between many species, i.e. the combination of different food chains.
Gastropods	A group of animals within the molluscs e.g. snails and slugs.
Generalist	Species that are both widespread and common because they can use many resources or are highly adaptable.
Guild	A collection of species that use similar resources in similar ways (e.g. fish-eating birds).
Habitat	The physical location/s or type of environment/s in which an organism or biological community lives or occurs. Habitat technically refers to all the types of environment a species requires to complete its lifestyle – not just a single vegetation patch or stand type. So, the habitat of the common frog (*Rana temporaria*) might include garden ponds, urban wetlands, rough grassland and woodland glades.
Herbivore	An animal that relies on plants for food.
Heterogeneity	The uneven distribution of various habitat patches, landforms or concentrations of each species within an area. A landscape with spatial heterogeneity has a mix of concentrations of multiple species (ecological), or geological formations or environmental characteristics (e.g. wind and sunlight) within it. A population showing spatial heterogeneity is one where various concentrations of individuals of this species are unevenly distributed across an area, i.e. patchily distributed.

Continued

Table 5.2. Continued.

Ecological term or concept	Description
Immigration	New species or individuals entering an area.
Island theory	This relates to the position of vegetation stands or other habitat types in relation to one another. For example, how areas of discrete oak woodland affect each other in determining their species composition. Larger areas of woodland (islands) will support more species. Areas of woodland further away from the others will be more isolated and will support fewer species. There is competition within an island and this can result in a turnover of the species present over time. Some species will become extinct, but other newcomers will arrive from time to time.
Matrix	Main or background ecological system, for example a forest matrix, i.e. a large area predominately covered with patches of forest.
Metapopulations	Local populations.
Mosaics	The patterns of patch habitats and corridors within the wider landscape matrix.
Networks	A number of interconnecting corridors allowing different patches to be linked. Connectivity relates to how well species can use these networks to cross the landscape. Species survival tends to be higher in patches that have higher connectivity.
Niche	The habitat, limited by environmental factors (fundamental niches) and/or competitors (realised niches) where individuals of a species can survive and reproduce.
Omnivore	A species that can eat either plant or animal material.
Parasitoid	An organism that in part of its life cycle lives in or on another organism and usually ends up killing the host.
Patch (habitat)	One of a number of areas depicting a type of habitat. A fragmented habitat may be one that has been broken up into a number of patches, by housing, road or rail networks, and where movement between patches may be difficult for wildlife.
Refuge	Area where a species is exposed to less environmental, predation or competitive pressures.
Remnant vegetation/habitat	Areas within the urban matrix that represent the type of habitat that was prominent before urban development took place – essentially 'left-over' patches.
Richness (species)	The number of different species present.
Ruderal plant	Plant species that are adapted to high-intensity disturbance and low-intensity stress. Such species tend to be fast-growing and have rapid, short life cycles, investing heavily in seed production to ensure the next generation. Ruderals often dominate the colonization of recently disturbed land.
Saturation	The concept when all the available niches are filled by one or more species.
Scale	The size of a landscape or habitat. Can vary markedly depending on the species being considered (puddle vs ocean). Larger areas will support more species and more individuals, as there are more resources available. Larger areas also allow more individuals to possess territories and avoid inbreeding in isolated populations.
Specialist	A species that may be very dependent on a single food source or habitat type.
Stand (vegetation)	See 'Patch' above.
Stress tolerator plant	Plants that compete by being adapted to intense abiotic stress and low-intensity disturbance. Found in adverse environments such as alpine or arid habitats, deep shade or soils which are nutrient deficient, contaminated or high/low pH. Typified by slow growth rates, long-lived leaves, high rates of nutrient retention, and low phenotypic plasticity.
Taxocene	Closely related species within a community (e.g. aphids).
Taxon (plural taxa)	A group of one or more populations of an organism or organisms seen by taxonomists to form a functional or discrete unit.
Trophic levels	Subsets of species that acquire energy in similar ways (e.g. herbivores, carnivores and detritivores).
Universal adaptive strategy theory	An evolutionary theory based on the trade-off that organisms (usually plants) face when the resources they gain from the environment are allocated between either growth, maintenance or regeneration – known as the universal three-way trade-off or C-S-R (competitor–stress tolerator–ruderal) theory.

Table 5.3. Key aspects of urbanisation that affect habitat provision

Change due to urbanisation	Effects on biodiversity
Loss of natural habitat or agricultural land.	Loss or degradation of previous ecosystems.
Configuration of buildings, technical infrastructure and open spaces.	Increase in non-permeable hard surfaces, roads, paving and roofs. In city centre 60% of land surface area can be covered.
Grey infrastructure, such as roadways and railway lines, that fragments habitats.	Inhibition of animal movement or high incidence of fatalities, causing population isolation and potential inbreeding.
Increase in built infrastructure.	Presents hazards for some species. More than 100 million birds are estimated to be killed each year by colliding into windows. Alternatively, some tall buildings provide habitat, e.g. for *Columba livia domestica* (feral pigeon) and *Falco peregrinus* (peregrine falcon).
New built structures.	Bridges, house roofs, underpasses, overpasses and culverts serve as nesting and roosting sites for a number of species, e.g. *Hirundo pyrrhonota* and *H. fulva,* cliff and cave swallows, respectively. In the USA, 50% of bat species use bridges as roosting sites.
Modification of the soil-moisture regimes, drier in temperate zones but wetter desert areas due to irrigation.	Changes plant species composition to drier-adapted species in the first instance and more nutrient-competitive species in the second.
High nutrient loads in some soils, due to atmospheric pollutants, eutrophication processes and active land use change.	As above, alters the balance with respect to plant colonisation strategies. Nutrient-richer soils promote competitive and ruderal 'weedy' species rather than stress tolerators.
Warmer temperature due to the urban heat islands.	Enhances range of plant and animal species present, with species from lower latitudes surviving where they would otherwise not. Higher temperatures through climate change, though may start to restrict diversity in those cities situated in warm, arid climates. Can alter plant growing periods and other phenological aspects (e.g. lack of winter chilling that some species require for seed germination or bud-break).
Higher productivity of plant biomass due to cultivation.	Can be exploited by some herbivore invertebrates but may cause 'pest' explosions if no natural predators.
Abundance of food and food wastes and other resources from human activities.	Allows high populations of some generalist species to be maintained.
Higher levels of disturbance through noise and human activities.	Can alter feeding and breeding patterns.
Soil contamination (nitrogen and calcium deposition and certain heavy metals), air pollution (elevated CO_2, NOx, aerosols, metals and ozone) and water pollution with particular impacts.	Impacts on species compositions, particularly soil organisms, lichens and aquatic species.
Disturbance such as removal of all vegetation, trampling, construction, mowing, radical soil change, light pollution, litter or illegal dumping, arson and vandalism.	Loss of species diversity.
Introduced non-native species of plants and animals.	Risk of invasion to surrounding semi-natural habitats.
High predation rates due to pet species.	Populations of vulnerable species such as lizards/snakes reduced. Impact on bird and small mammal breeding rates.
High proportion of habitat generalists and common plant and animal species.	Loss of biodiversity, but also cityscapes, flora and fauna become more uniform across the globe.

Urban green spaces also need protection. Most cities are still expanding and becoming more dense in terms of built infrastructure as a premium is placed on land for further 'development'. It is not always space for new buildings that is needed either; many cities are experiencing 'soil-sealing' as permeable soil is paved over to provide extra car-parking space. Cities such as Edinburgh (UK) are thought to have lost 50% of their (green) front gardens, as more people own cars and wish to park these off the street and adjacent to their house. In recent years, it has also become popular to cover garden area with synthetic (plastic) turf – as a labour-saving activity (no mowing required). Such changes are highly damaging to habitat provision for wildlife (as well as exacerbating other environmental problems). Change in housing styles – generally moves to smaller units and smaller gardens – means habitat provision also changes; such neighbourhoods are associated with less vegetation and smaller-stature tree species. Species such as *Strix aluco* (tawny owl) (Fig. 5.1) and *Dendrocopos major* (great spotted woodpecker) that can do well in older residential neighbourhoods with large gardens (due to the presence of mature trees with roosting and feeding sites, respectively) do less well in the more modern, tightly packed suburbs.

5.4 Important Urban Landscapes

The form of urban landscape dictates the opportunities for wildlife, with less densely built suburban and peri-urban areas holding more opportunities for wildlife, due to the presence of more green space than city centre locations (Fig. 5.2). Common green and blue landscape typologies found in the urban matrix are highlighted in Table 5.4. These have their own assemblages of plants and animals, and thus different approaches and management practices may be appropriate, based on these and their relative value to humans.

Fig. 5.1. *Strix aluco* (tawny owl) can be present in towns and cities but only when large park and garden trees persist. (Used with permission from P. Dixon)

5.5 Patch Size, Mosaics, Corridors, Stepping Stones and Networks

Key elements in ecology are patch (habitat) size and connectivity between different patches. In an urban context different green and blue spaces come together to create a mosaic, and species move across these through connections (green corridors) or ecological 'stepping stones' (small patches that may not be physically connected but are near enough for species to move between them). The entirety of these elements – patches, corridors and stepping stones – is known as the network.

Patch size and quality

As scale of patch increases, then the capacity to hold an entire population of a given species increases – especially if animals, for example, are territorial and each individual (or breeding pair) needs a space exclusive to them. Larger patches simply hold more individuals (or breeding pairs). Woodland patch size is often considered the best predictor of urban bird species richness and abundance (Kang *et al.*, 2015), and Kang's study showed that smaller forest plots (e.g. <3.5 ha to <5 ha in size depending on species) show much less bird diversity. Species such as woodpeckers need a certain size of woodland to be commonly present or to breed (Myczko *et al.*, 2014) (Fig. 5.3).

Towns and villages are less spatially heterogeneous than large cities, with the former matrices having been observed to favour native mammalian and native plant species. Both patch configuration and size (i.e. the mosaic) affect the richness of remnant vegetation, with larger and more closely aligned patches helping to retain plant populations. Not all taxa though are as dependent on patch size. Delaney *et al.* (2021) working with reptiles and amphibians indicated that patch quality was critical for these groups, and smaller patches of the right composition were supportive of species diversity.

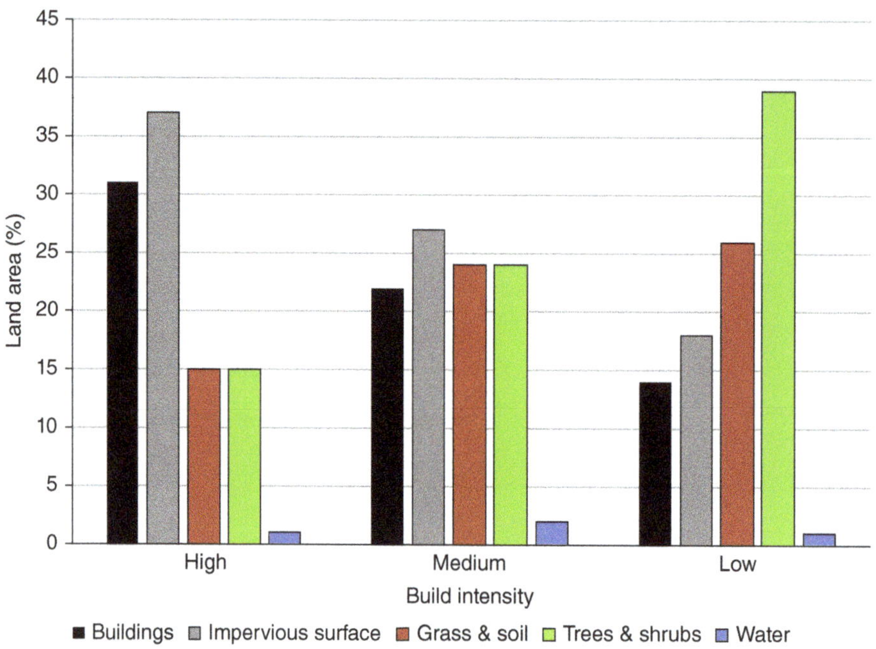

Fig. 5.2. The effect of building intensity in the city of Manchester, UK on land surface area type. (Modified with permission from Gill *et al.*, 2008.)

Small patches also maintain their value through acting as stepping stones for species movement. Indeed, Rudd *et al.* (2002) argue that small remnant patches of vegetation combined with new gardens form a habitat network in urban landscapes that is critical to the conservation of local species. Thus, preserving 'micro-reserves' could be as equally important for some species as trying to increase the overall habitat provision.

Connectivity

It is important populations do not get isolated (island effect) as this reduces the gene pool in any one area – leading to inbreeding and loss of population vigour (Lynch, 2019). If a species becomes extinct within a given patch, there is little opportunity for recolonisation from somewhere else – unless there is green corridor available to access the original site. As the human population in the landscape increases, both patch density (different types of land cover and land use per km^2) and edge density (total length of all edge segments per ha) are enhanced, but connectivity between patches is reduced. Trying to avoid this fragmentation of habitat has been an objective for urban ecologists and planners in recent years. They have advocated the use of green corridors and, to a less extent, increasing the number and proximity of green 'stepping-stone' spaces. Retaining species richness and abundance is strongly correlated with urban connectivity, and preserving/restoring connectivity across the urban matrix enhances the diversity of particularly sensitive taxa like birds (Kang *et al.*, 2015). Such corridors and links are increasingly important with climate change, as organisms migrate to seek out more suitable environments.

To understand how green space supports wildlife in highly fragmented urban landscapes, it is important to consider population dynamics, especially colonisation and persistence trends (Holyoak *et al.*, 2005). Persistence rates indicate habitat quality within a patch, and colonisation rates signify the permeability of the matrix. Using medium-sized mammals as a model, Gallo *et al.* (2017) reported that colonisation and persistence differed greatly between species and between the different types of green space within Chicago, USA. *Sylvilagus floridanus* (eastern cottontail rabbit) could colonise city parks effectively, and when present, then *Procyon lotor* (raccoon) and

Table 5.4. Urban green space typologies and value for biodiversity

Typology	Biodiversity features	Degree of horticultural interventions
Urban nature reserves	Prime purpose is to conserve local wildlife. The landscape may have originally been protected because of the presence of one or more rare species. Conversely, some are created precisely to allow citizens to access natural environments.	Structural and ecological management may take place, e.g. realigning of water levels, tree planting (natives) and grassland meadow management (e.g. to encourage invertebrate diversity and abundance).
Urban woodlands	May be present for a variety of reasons, including retainment of residual woodland present before the town/city enveloped them (remnants). Others may be newly planted, specifically to allow human access to green space and to be used for e.g. dog walking. Can be rich in plant and animal diversity.	Woodland management – tree maintenance – but with perhaps less emphasis on aesthetics and trunk management for safety than in other urban places, e.g. street trees.
River corridors	Important ecological corridors for urban species and other species migrating through the city landscape. Increasingly kept as a green space due to flooding risk and mitigation against flooding elsewhere in the city, i.e. flood 'sacrificial' land.	May include water meadow management. Clearing of fast-growing vegetation both within the river and on the banks may occur on a regular basis to allow river water to flow freely.
Parks	Vary from very informal 'semi-natural' landscapes to highly manicured green spaces (e.g. close-mown lawns).	Relatively high horticultural/landscape management input, especially for turf management, footpath maintenance and formal plantings.
Gardens	Again, design and management will dictate levels of biodiversity present – even formal gardens can be high in plant biodiversity – but it may be of non-native species. On the other hand, 'wild gardens' can host numerous native plant and animal species.	Often high levels of horticultural intervention – frequent grass cutting, hedge trimming, planting and watering.
Cemeteries	Old 'unmanaged' churchyards are a haven for wildlife and can have their open unique biodiversity, e.g. lichens on gravestones.	Older, full churchyards and cemeteries are relatively unmanaged. On the other hand, those still in regular use can be intensively maintained through frequent mowing. High maintenance (neat and tidy) often being associated with a sign of respect for the dead!
Allotments	Good for certain invertebrates and if managed organically – i.e. without pesticides – useful for wildlife.	High levels of maintenance on the growing beds themselves but surrounding spaces can be unmanaged and wildlife rich.
Urban farms and community food gardens	Similar to allotments but often found in denser built inner-city areas. Roof gardens and living 'edible' walls can come into this category. Provide habitat but may be disconnected from other green patches.	Cultural and ethical factors often mean wildlife conservation and wider ecological principles are to the fore.
Golf courses and other sports facilities	Many were once 'bland' landscapes, but increasingly the grassland 'rough' at either side of the fairways and tees are now partially managed for wildlife. Large specimen trees or blocks of woodland provide shelter and food for key species. Location may determine the species tolerated – for example, *Talpa europaea* (mole) are 'not welcome' on golf greens!	Contrasting levels of turf maintenance with daily grass cutting on golf greens, but perhaps only a single annual cut of the 'rough'.
Road verges	Road vehicles can be a threat to mammals and birds, but the verdant spaces at the side of the road often have limited human access and provide important vegetation types as habitat. Short-turf areas favour small mammals (e.g. *Myodes glareolus* – bank vole) and associated predators (e.g. *Falco tinnunculus* - kestrel).	Light horticultural interventions – grass mowing in some places. Increasing annual and perennial 'flower' meadows may be designed and implemented along roadsides.

Continued

Table 5.4. Continued.

Typology	Biodiversity features	Degree of horticultural interventions
Street trees	Discrete subjects and can host good numbers of invertebrates and avian species, e.g. *Sturnus vulgaris* (starling) in Europe and *Spinus psaltria* (lesser goldfinch) in North America. As ground flora is limited may not provide the whole woodland food web.	Once established, little maintenance other than annual health inspection and occasional removal of problematic (damaged/diseased) branches.
Brownfield sites	Unique habitats due to the transitional nature of the vegetation. Often unusual species (e.g. orchids) or those that thrive in open free-draining soil/gravel (e.g. tiger beetles) or are specialists that can tolerate residue chemicals in the soil such as heavy metals.	Actively unmanaged in most cases.
Railway corridors	The trackbeds and powerlines (especially those at ground level) and trains can be dangerous to animals. The ground at the side though provides important green corridors through the urban network.	Apart from keeping 'weeds' from the trackbed itself and occasional removal/cutting back of woody vegetation, management regimes tend to be light.
Rivers, lakes, ponds and canals	Aquatic and riparian habitats. Water quality can be sub-optimal due to pollutants from road runoff or sewage overflows.	River vegetation removed on an annual basis to avoid excessive build-up of nutrients and to reduce the height of water in the river (plant growth is biomass that adds volume to a river and increases flood risk).

Didelphis virginiana (Virginia opossum) could persist relatively well in the parks. These three species were also effective at colonising the open spaces of golf courses, as to some extent were *Canis latrans* (coyote) and *Mephitis mephitis* (stripped skunk), with coyotes, cottontails and raccoons having good persistence levels on some golf courses (coyotes perhaps due to limited human disturbance at night). Coyote, opossum and raccoon were good colonisers of cemetery sites, and if present coyote, *Odocoileus virginianus* (white-tailed deer), raccoon and to some extent striped skunk could persist in cemeteries (especially if there is good vegetation cover and low incidence of disturbance from humans). For more natural areas of vegetation within the city then raccoon, coyote and opossum had the capacity to colonise quickly and, along with white-tailed deer, also show good persistence levels. White-tailed deer do well too, when there is connectivity to large-scale natural areas beyond the urban greenways.

The degree of connectivity and type of mosaic is important and affects taxonomic groups differently. Fragmentation has been seen to be particularly damaging for toads (Hitchings and Beebee, 1998), prairie dogs (Magle and Crooks, 2009), carabid beetles (Angold *et al.*, 2006), bumblebees (Jha and Kremen, 2013) and other invertebrates (Vergnes *et al.*, 2012). Small mammals also seem to rely on effective green corridors, e.g. *Arvicola amphibius* (water vole), *Muscardinus avellanarius* (dormouse), *Sorex coronatus* (Millet's shrew), *Sorex minutus* (pygmy shrew) and *Crocidura russula* (greater white-toothed shrew) (Angold *et al.*, 2006; Vergnes *et al.*, 2013). In contrast, Angold *et al.* (2006) indicated that effective dispersal of species across the mosaic was not a limiting factor in maintaining populations of certain plant and butterfly taxa.

Quality of corridors

The quality of the green corridors is itself important, with many recommending the wider the better and that the green space is itself diverse and accommodating for wildlife – for example, higher, floral diverse grassland is better than 'open' short-mown turf. Green corridors are frequently a component of a transport network, e.g. road verge, railway embankment, cycle way, pedestrian walkway or canal towpath (Fig. 5.4), thus the degree of disturbance from

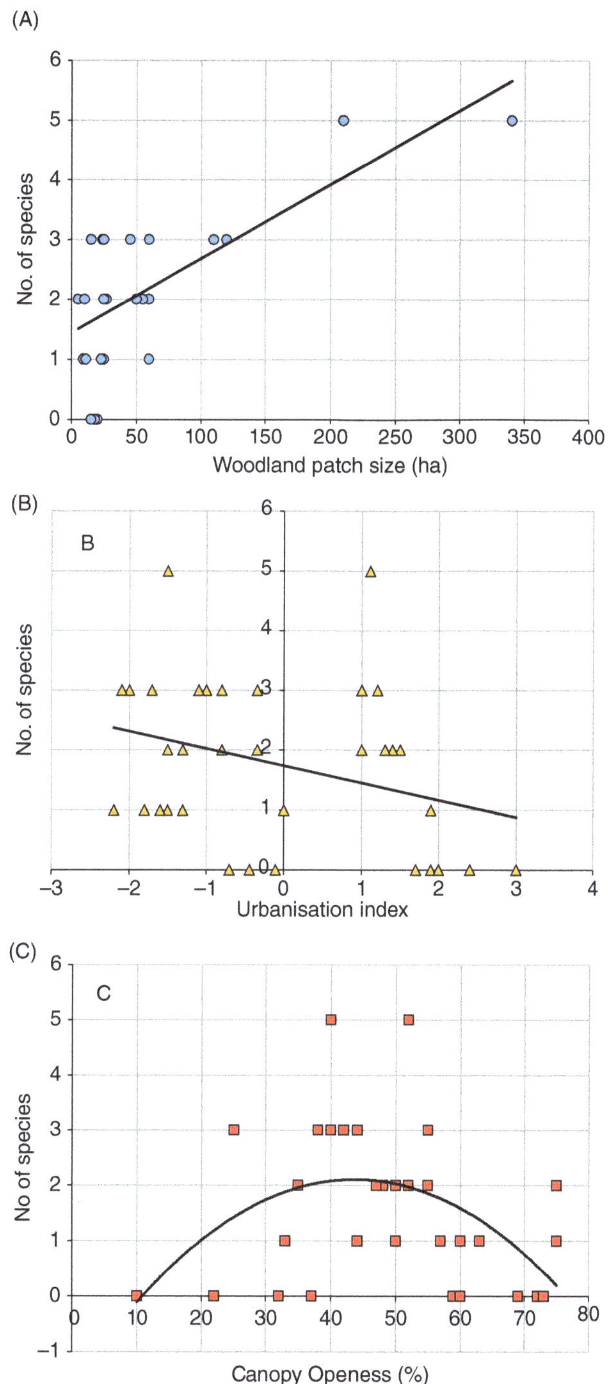

Fig. 5.3. Relationship between woodpecker species diversity and (A) woodland patch area, (B) urbanisation index and (C) canopy openness. The urbanisation index relates to distance from city edge, intensity of road networks and percentage cover of human settlement within 500 m. Species were *Dendrocopos major*, *Dryocopus martius*, *Picus viridis*, *Dendrocopos minor*, *Dendrocopos medius* and *Picus canus*. (Modified with permission from Myczko *et al.*, 2014.)

Fig. 5.4. Planting up this rainwater detention swale with native plant species helps biodiversity. The swale runs along an urban pedestrian/cycle route linking one of the suburbs of New Orleans (USA) to the city centre.

human activity is important too. Lynch (2019) advocates their value is enhanced when:

- Corridors are specifically linked to key habitat locations (i.e. used to connect the larger habitat patches).
- Corridors are managed as habitat, not just as a conduit between habitats.
- Land adjacent to the corridor is managed for wildlife, e.g. gardens butting up against the corridor.
- Transport routes within them are kept narrow and located at one edge of the corridor.
- Stepping-stone habitats are provided along the corridor (green necklace).

Wider corridors benefit more sensitive species. Woodland-interior bird species are rarely present in green corridors ≤50 m wide, and more sensitive species may require ≥300 m width. Compromise may be required due to the pressures on urban land, but planners are encouraged to provide ones 100–150 m or wider where possible (Kohut *et al.*, 2009). The presence of mature and decaying trees is beneficial, e.g. to vulnerable bird species (Zhang *et al.*, 2024), and mature trees combined with dense understorey vegetation provide opportunities for species often classed as 'urban avoiders'. Thick vegetation though concerns landscape architects, due to the loss of visual lines of sight and perceptions around security, for example on a public pathway.

Stepping stones

Ecological 'stepping stones' are small islands of habitat that serve as refuges between larger habitat patches (Carbó-Ramirez and Zuria, 2011; Xu *et al.*, 2024). To support species dispersal, stepping stones must be functionally – if not necessarily physically – connected, which means that organisms must be able to move through the landscape between the refuges. Corridors with their structural connections are more effective at supporting urban biodiversity than stepping stones, but the latter are still important in facilitating movement. Parks, gardens and garden networks host breeding birds and flying insects, but they also act as stepping stones allowing individuals to migrate; and newer features such as green roofs and rain gardens may have a role in supporting certain mobile taxa too. Urban and suburban gardens and other flower rich spaces can be effective at supporting some *Bombus* (bumblebee) species (Bjørn and Howe, 2023) and better than some core green spaces (McFrederick and LeBuhn, 2006). Studies of bats and arthropods conclude that green roofs are functionally connected to nearby ground-level green spaces and to each other (Braaker *et al.*, 2014). As one might assume, connections between green roofs and surrounding features are most pronounced for high-mobility flying species such as bees and bats (Berthon *et al.*, 2023), but taxa such as spiders are also functionally connected to ground-level vegetation (Braaker *et al.*, 2014). Results suggest that green roofs function as nodes in urban ecological networks.

Very small green spaces are predominately, ecologically speaking, 'edge' habitat, which even in optimal vegetation conditions can facilitate predation and disturbance. So while sensitive species may be attracted to small urban parks and gardens, this paradoxically might lead to increased mortality (Carbó-Ramirez and Zuria, 2011). Indeed, some stepping stones may be too small to serve as a refuge and emphasis should be on increasing size where feasible. Further research is required to understand better this aspect of scale – and to give greater confidence on the extent to which small green spaces support wildlife within this context of functional urban stepping stones.

Again, Lynch (2019) recommends that ecological stepping stones should:

- lead to other larger habitat areas;
- be in close proximity to one another (to facilitate easy movement);
- be as large as possible, or encourage communities to provide a larger asset, e.g. a number of gardens managed for wildlife next to one another;
- be structurally and ecologically diverse and complex; and
- be carefully (and minimally) managed.

5.6 Urban Ecosystems

The species that live within urban landscapes fall into four categories. Species that were present prior to, and survived, urbanisation; species native to the region that have migrated in because conditions are conducive; non-native species introduced by humans or which have colonised from somewhere else by their own means and finally, species which have no natural habitats but have evolved to adapt to agricultural, urban, and industrial landscapes (anecophytes). This includes a number of species that are common to many urban conurbations across the globe. For example, 'weedy' plant taxa such as *Taraxacum officinale* (dandelion), birds such as *Passer domestica* (house sparrow) or mammals including the ubiquitous *Mus musculus* (house mouse). Some species thrive precisely because they are adapted to co-habit human settlements (synanthropic species).

The terms urban 'adapter', 'exploiter' and 'avoider' are frequently used. 'Urban exploiters' often represent species that are generalists in terms of food and environmental needs (Table 5.5). Some species have some capacity to adapt to urban conditions and can be present in high numbers in the 'greener' suburbs, especially if there are stands of remnant vegetation (urban woodlands or wetlands) or the landscape mimics their natural habitat. Birds naturally found in the woodland edge can do very well in large gardens with trees present, e.g. *Erithacus rubecula*, robin; *Turdus merula*, blackbird; *Parus major*, great tit; and *Prunella modularis*, dunnock. Some species rarely enter urban environments (the urban 'avoiders') due to a lack of an ecological niche, additional pressures due to human presence, or simply they have specific habitat requirements that are not present in our towns and cities. Some species that have become adapted to 'open' arable landscapes, for example *Lepus europaeus*, brown hare and *Alauda arvensis*, Eurasian sky lark, do not enjoy the more congested and densely built environment of the city.

Animals that thrive in urban environments possess traits such as

- a catholic choice of food sources (omnivorous);
- tolerance to variation in the abiotic environment;
- adaptable in terms of the conditions required for shelter and breeding;
- high reproductive and survival rates;
- high rates of recruitment through immigration;
- limited pressure from competitors and/or predators; and
- habituation to human activities and an ability to cope with highly fragmented landscapes with abundant edges (Adams and Lindsay, 2009).

The effects of urbanisation do not impact on all species evenly. Often predators at the top of the food chain (apex predators) are the most vulnerable. These species are susceptible to population declines due to habitat fragmentation and loss, reduction of landscape connectivity and increases in road density and traffic volumes. More generalist and often smaller, better-adapted mesopredators on the other hand have adapted well to the highly fragmented urban landscape and have substantially increased in abundance in the absence of apex predators (a phenomenon known as mesopredator release) and due to an increase in food supply. Mesopredators adapted to urban conditions include *Procyon lotor* (raccoon), *Mephitis mephitis* (stripped skunk) and *Vulpes vulpes* (red fox), as well as some snake species. Enhancement of populations of these species, however, shift trophic structures and is frequently detrimental to smaller prey species, for example, specialised rodents and ground nesting birds. *Felis catus* (domestic cat) fit within this role too and it is estimated that in the USA alone cats kill as many as 1.3–4.0 billion birds and 6.3–22.3 billion mammals p.a. (Loss *et al.*, 2013).

Adaptation in urban animals

Urbanisation drives evolution! Animal morphology, physiology and behaviours are known to change due to urbanisation (synurbanization). Trends include a

Table 5.5. Examples of birds and mammals that have a range of tolerances to urban condition across three different continental areas.

		Europe	North America	Australia
Urban exploiters	Birds	*Columba livia* (feral pigeon) *Chroicocephalus ridibundus* (black-headed gull) *Passer domesticus* (house sparrow) *Pica pica* (magpie)	*Haemorhous mexicanus* (house finch) *Columba livia* (feral pigeon) *Larus californicus* (California gull) *Corvus brachyrhynchos* (American crow)	*Psephotus haematonotus* (red-rumped parrot) *Threskiornis molucca* (Australian white ibis) *Eolophus roseicapilla* (galah) *Cracticus tibicen* (Australian magpie)
	Mammals	*Mus musculus* (house mouse)* *Rattus norvegicus* (brown rat)* *Vulpes vulpes* (red fox) *Sciurus carolinensis* (grey squirrel)*	*Mus musculus (house mouse)** *Rattus norvegicus* (brown rat)* *Procyon lotor* (North American raccoon) *Didelphis virginiana* (Virginia opossum)	*Mus musculus (house mouse)** *Rattus norvegicus* (brown rat)* *Rattus rattus* (black rat)* *Isoodon obesulus* (Southern brown bandicoot) *Acrobates pygmaeus* (feathertail glider)
Urban adapters	Birds	*Fringilla coelebs* (chaffinch) *Turdus merula* (blackbird) *Cyanistes caeruleus* (blue tit) *Accipiter nisus* (sparrowhawk) *Milvus milvus* (red kite)	*Cardinalis cardinalis* (northern cardinal) *Spinus psaltria* (lesser goldfinch) *Buteo jamaicensis* (red-tailed hawk) *Turdus migratorius* (American robin)	*Trichoglossus moluccanus* (rainbow lorikeet) *Strepera graculina* (pied currawong) *Callocephalon fimbriatum* (gang-gang cockatoo) *Dacelo novaeguineae* (laughing kookaburra)
	Mammals	*Meles meles* (European badger) *Oryctolagus cuniculus* (rabbit) *Myodes glareolus* (bank vole) *Sus scrofa* (wild boar)	*Mephitis mephitis* (striped skunk) *Lynx rufus* (bobcat) *Canis latrans* (coyote) *Odocoileus virginianus* (white-tailed deer)	*Isoodon fusciventer* (quenda) *Macropus rufogriseus* (red-necked wallaby) *Trichosurus vulpecula* (common bushtail possum) *Pteropus poliocephalus* (grey-headed flying-fox)
Urban avoiders	Birds	*Perdix perdix* (grey partridge) *Tyto alba* (barn owl) *Aquila chrysaetos* (golden eagle) *Lyrurus tetrix* (black grouse) *Gavia immer* (northern diver)	*Scolopax minor* (American woodcock) *Melanerpes formicivorus* (acorn woodpecker) *Tympanuchus cupido* (greater prairie chicken) *Piranga flava* (hepatic tanager)	*Accipiter novaehollandiae* (grey goshawk) *Pedionomus torquatus* (plains-wanderer) *Dromaius novaehollandiae* (emu) *Cinclosoma castanotus* (chestnut quail-thrush)
	Mammals	*Martes martes* (pine marten) *Ursus arctos arctos* (Eurasian brown bear) *Mustela putorius* (European polecat) *Marmota marmot* (alpine marmot) *Lepus timidus* (mountain hare)	*Puma concolor* (mountain lion) *Urocyon cinereoargenteus* (grey fox) *Lepus californicus* (black-tailed jackrabbit) *Ovis canadensis* (bighorn sheep)	*Petrogale lateralis* (black-footed rock-wallaby) *Ornithorhynchus anatinus* (platypus) *Dendrolagus bennettianus* (Bennett's tree-kangaroo) *Thylogale stigmatica* (red-legged pademelon)

*Not native to the region

move to darker colours (melanic forms) (Leveau, 2021). This pattern has been long known as industrial melanism and is attributed to the effects of pollution on lichens on tree trunk and the resulting camouflage of several moth species (Walton and Stevens, 2018). Increased melanism has also be observed in *Columba livia domestica* (feral pigeon) and is thought to relate to fewer parasites in city populations (Jacquin *et al.,* 2013), better stress responses (Corbel *et al.,* 2016) and the detoxification of pollutants (Chatelain *et al.,* 2016). *Emydocephalus annulatus* (seasnake) living near urban and industrial areas are darker too, again as a mechanism to help with detoxification of pollutants from fishing boats (Goiran *et al.,* 2017). Other species show duller plumage/fur, e.g. *Parus major* (great tit) and *Junco hyemalis* (dark-eyed junco) as well as feral pigeons (Senar *et al.,* 2014). Duller colours are attributed to birds absorbing less carotenoids (via eating less fruit) or the prevalence of feather-degrading bacteria in urban environments (Giraudeau *et al.,* 2017). More abundant food has been put forward too as a reason behind the loss of more vibrant colours, as the requirement for these in signalling territorial displays and dominance becomes less relevant.

Feeding behaviours alter. Urban animals are more likely to explore and exploit novel food sources more quickly. Better feeding opportunities mean that species like *Meles meles* (Eurasian badger) require a smaller territorial home range. Regular food availability throughout the year has been cited as to why *Didelphis virginiana* (Virginia opossum) can now survive in colder, more northerly latitudes and why *Ursus americanus* (black bears) have shorter hibernation periods (Beckmann and Berger, 2003). Individual animals seem more tolerant of the presence of other individuals from the same or even other species, again possibly due to greater feeding opportunities reducing competition. Animals change their behaviour in line with threats or opportunities from humans. *Vulpes vulpes* (red fox) are less active and less likely to cross busy roads during the evening traffic 'rush-hour'. Conversely, *Corvus corone* (carrion crow) in Japan actually use traffic to yield a food source, crows being observed to place walnuts at spots on the road where cars are most likely to run over them and crush the hard external casing.

Due to the prevalence of urban noise, birds may change the pitch of their song, for example, by using a higher-frequency note or singing more frequently. Artificial street and house lightning encourages territorial birds to sing earlier in the morning during the breeding season. With *Cyanistes caeruleus* (blue tit), singing earlier in the morning has led to increased reproductive rates and opportunities for extra-pair copulations for males, because singing earlier is a sign of male quality (Kempenaers *et al.,* 2010).

The presence of humans is important. If humans are not perceived as a threat, then birds adopt a shorter flight initiation distance, i.e. they let people come closer before flying off. This is probably most notable in *Columba livia domestica* (feral pigeon), where birds can be almost in direct contact with humans before initiating a flight response – and even then, they do not fly very far away! Conversely, animals that may be persecuted/not tolerated change their diurnal patterns for foraging to help avoid humans in urban environments. In North America, *Lynx rufus* (bobcat), *Canis latrans* (coyote) and *Ursus americanus* (black bear) (Beckmann and Berger, 2003) change the timing of their activity in response to human presence. *Callithrix penicillata* (black-tufted marmosets) even adjust their behaviours to the day of the week in accordance to human activities (Duarte *et al.,* 2011). On Sundays, these monkeys retreat to the quieter areas of the park to avoid the large influx of humans that congregate near their usual feeding zones – areas they tend to exploit during the rest of the week.

The opposite may be true with *Macaca fascicularis* (long-tailed macaques), who actually rob temple visitors of inedible items (sunglasses, bowl, etc.) and then use these objects as tokens to barter for food (Brotcorne *et al.,* 2017). It is thought the macaques display a form of economic behaviour and basic arithmetic. They demonstrate social learning behaviours and can comprehend delays in gratification (they get a reward but not immediately).

Animals even use human dwellings for shelter. *Martes foina* (beech or stone marten) regularly use loft spaces as denning sites. *Osmia* spp. (mason bees) will hollow out holes in the masonry between brickwork to develop nest sites. *Passer domesticus* (house sparrow) and *Haemorhous mexicanus* (house finch) of Mexico City have learned to incorporate cellulose cigarette butts into nest structures to act as parasite repellents and act as an urban substitute to the green plant materials used in the wild.

Native and non-native species

Urbanisation not only changes the pressures on native wildlife and alters their habitat; it is also correlated with an increase in non-native species. These are often intentionally brought in to provide a greater variety of aesthetic flora (and sometimes fauna. e.g. *Aix galericulata*, mandarin duck in Europe). In terms of non-native plants, horticulture is a principal contributor to the non-native biodiversity of urban environments – plants being introduced from other parts of the world due to their aesthetic qualities and then widely used in gardens and parks. Since Neolithic times, approximately 12,000 plant species have been introduced into Europe for ornamental and cultural purposes, with about 10% of these becoming naturalised. Horticultural practices too have increased both intraspecific (within a species) and interspecific (between species) hybridization, thus swelling the numbers of non-natives further. Introduced plants seem to proliferate and become more variable too when introduced to new areas.

Urbanisation leads to changes in the proportions of native and non-native plants present. There are gradients across the urban matrix; in general, the richness of native species declines as one moves towards the city centre, whilst non-native species richness increases towards the centre of the city, or towards the older residential suburbs and away from the rural hinterland (Aronson *et al.*, 2015). In the urban core approximately 30–50% of the plant species are non-native, although Kowarik (2011) suggests that, collectively, native species can still comprise 50–70 % of total species richness in a city, albeit sometimes they are found as less frequently abundant species. In a study of 112 cities, Aronson *et al.* (2014) found that these supported 5% of the global plant species, with richness in cities overall still dominated by native species. Plant biodiversity overall peaks in the suburbs where there are still relatively adequate areas of green space and associated niches for native species, due largely to a heterogeneous landscape typology (parks, woodlands, gardens and waterways), whilst the increase in garden space correlates with an abundance of non-native garden plants. In other parts of the world, particularly vulnerable island ecosystems such as New Zealand, where European settlers retained a desire for the species common to their homeland, the proportion of non-natives is much greater, with negative consequences for many of the original endemic species.

In terms of cultivated plants there are concerns that many cities are dominated by the same species, i.e. there is a global homogenisation process at work. The popularity of aesthetically pleasing, strongly structured trees or shrubs or colourful flowering garden plants has resulted in garden and park landscapes being very similar irrespective of location around the world, or at least within similar climatic zones. The plants used in an ornamental park or garden in California (USA) may be very similar to ones used in the cities of Greece, Italy or Chile. Ignatieva (2010) analysed catalogues from a range of nurseries within the temperate zones of the USA, Russia, Germany and New Zealand and found a high degree of commonality on the genotypes used irrespective of the location. Favoured plants across temperate zones were European deciduous trees and shrubs and some 'fashion' conifers. Plants that appeared to have universal appeal include the following:

- deciduous trees species – *Acer* (maple), *Betula* (birch), *Fraxinus* (ash) *Prunus* (ornamental cherry), *Populus* (poplar), *Quercus* (oak), *Rhododendron, Salix* (willow) and *Ulmus* (elm);
- evergreen tree species - *Chamaecyparis lawsoniana* (Lawson cypress and related cultivars), *Juniperus* (junipers), *Picea* (spruce), *Pinus* (pine) and *Thuja* (cedar);
- annual and perennial flowering plant species and cultivars – *Impatiens* (busy lizzie), *Pelargonium* (geranium), *Petunia, Tagetes* (marigold) and *Viola* (pansy and viola);
- lawn grass species – *Agrostis capillaris* (common bent), *Festuca rubra* (red fescue), *Lolium perenne* (perennial ryegrass) and *Poa pratensis* (Kentucky blue grass); and
- ornamental grass species – *Cortaderia* (pampas grass), *Miscanthus*, *Pennisetum* and *Stipa* spp.

Parks and gardens in the tropics are less well documented but common 'global' species are likely to include *Acacia* (wattle), *Bougainvillea*, *Casuarina*, *Croton*, *Delonix regia* (flamboyant tree), *Hibiscus rosa-sinensis, Jacaranda, Plumeria* (frangipani tree) and *Strelitzia reginae* (bird of paradise) (Soderstrom, 2001).

Gradient trends are also apparent for animals. Work in Buenos Aires (Argentina) indicated that of the seven species observed, the four native rodents were dominant on sites with natural vegetation, often located at the edge of the city, whereas the three non-native species dominated in inner-city shanty towns, industrial sites and residential neighbourhoods (Cavia *et al.*, 2009). Overall, non-natives were faring better, as their species richness increased as new habitats were created with urbanisation. In contrast, the richness of native species declined as remnant natural habitats became further fragmented, isolated or destroyed.

Socio-economic factors though can play a role. High non-native animal presence can be linked to domestic animals such as cats and dogs, which also take the roles of mesopredators impacting on the populations of small rodents and birds. Conversely, more affluent citizens (tending to live in the suburbs) may spend money on bird food and help artificially boost populations of songbirds in such neighbourhoods. In the UK, populations of the native *Vulpes vulpes* (red fox) are probably highest in the more affluent suburbs of UK cities than any other 'habitat' in the country, due to ready sources of food (earthworms, fallen fruit, small mammals and discarded human food), less persecution and greater tolerance from the local human population. These suburbs are typified too by a useful balance of green open spaces to forage and secluded large gardens to rest up in and raise cubs. Certain bird and butterfly species follow a similar pattern across the urban matrix, being closely linked to those areas where parks and large gardens are most common. A linear gradient, of course, to some degree is a simplification and heterogeneity within the city often exists because of different building types and social contexts and this alters the opportunities available for different species.

The type of landscape affects the native/non-native balance. Remnant or planted urban woodlands are composed largely of native tree species, whereas street trees comprise a significant proportion of non-natives. Surveys from Geneva (Switzerland) suggest that of non-forest tree stocks 40% are composed of non-native genotypes (and comprise 90% of the species diversity) (Schlaepfer *et al.*, 2020). Non-natives accounted for 39% of street trees and 45% of park trees in Malmo (Sweden) and 33% of street trees in Copenhagen (Denmark), but only 8% of street trees in Stockholm (Sweden) and 3% in Tampere (Finland) (Sjöman *et al.*, 2012). Surveys in Flanders (Belgium), revealed that 15 urban parks contained about 30% of the total number of wild plant species held within the region, 50% of the breeding birds, 40% of butterflies and 60% of the region's amphibians (Cornelis and Hermy, 2004). Brownfield sites often have high ratios of spontaneously generating, non-native species, partially because the microclimate can be warmer – lots of concrete, tarmac and residual rubble. Those linked to landfill sites also have many non-native plants species thriving due to seed being imported from human food sources (e.g. in the UK, *Solanum lycopersicum* – tomato), animal feed (e.g. *Cannabis sativa* – hemp or cannabis) or the seed being resilient and long-living (e.g. *Datura stramonium* - thorn apple, or jimson weed).

Native and non-native plants and their capacity to support faunal wildlife

Are native or non-native plants best for supporting faunal wildlife? This raises a number of questions. Evolutionarily speaking, the answer should be clearly that native plant species are best as they evolved in tandem with the native animals. But these interrelations can be disrupted in an urban setting. Moreover, what is the definition of 'native'? Is the plant native to that city, that county, that 'political' country, or that biogeographical regional area? Studies on the use of native/non-native plants and their value to wider wildlife also often tend to focus on insects and birds, with intermediate numbers of studies on mammals and less on reptiles and amphibians. Few researchers study interactions at the microbial level, but increasingly people are interested in plant-fungal relationships. The matter is made more complex in that some of the urban mammal and bird species may themselves not be natives yet are popular with many citizens within the city.

Berthon *et al.* (2021) in a review of research in this area concluded that in 43% of studies native plants outperformed exotic plants in supporting faunal populations. Others indicated mixed (33%) or neutral effects (17%) of plant origin. A small number of studies showed a relative superiority from using non-native plants (8%). A couple of studies (Lessi *et al.*, 2016; Jasmani *et al.*, 2017) correlated both native and non-native plant biomass with increased biodiversity. Often, native plant origin was shown to benefit some taxa or species but had neutral or negative effects on others. Threlfall *et al.* (2017) found that increasing the proportion of both indigenous (to Melbourne) and native (to Australia) plants increased the richness of birds and bats. Sikora *et al.* (2020), on the other hand, found some *Bombus* (bumblebee) species prefer to forage on native plants, but others prefer exotic plants. Higher numbers of native trees were important in supporting 'bush-dependent' native species in Australia, but other native 'non-bush' species were unaffected. Although there were differences between taxa, native ornamental perennial plants were generally superior to non-native perennials and non-native annuals in attracting greater numbers of pollinating insects in Virginia (USA) (Palmersheim *et al.*, 2022). But insect diversity on non-native perennials was marginally greater than native perennials (Fig. 5.5).

It is important to note there are interactions with other factors. The composition and condition of urban vegetation drives species–species interactions, dictating which animals will succeed in urban environments. For example, the structure and composition of urban vegetation – tree, shrub, ground-layer composition and diversity – influence bird, bat and insect communities and this indicates that native vegetation supports higher levels of biodiversity partially through these means (Threlfall *et al.*, 2017). But plant and floral density, structure, nectar or pollen supply/quality and other aspects of resource provision are important irrespective of origin (French *et al.*, 2005; Salisbury *et al.*, 2017; Giovanetti *et al.*, 2020; Giuliani *et al.*, 2020). Positive responses relate to specific plant taxa that are especially supportive of other species (Parsons *et al.*, 2020; Wood and Esaian, 2020). In many cases plant origin does not matter as much as the resources provided by the plant. For example, certain taxa within the Asteraceae appear to be popular with insect pollinators, irrespective of whether the species is native to that region or not (Rollings and Goulson, 2019).

In some situations, native plants are shown to increase taxon abundance but have no effect on diversity (Lowenstein *et al.*, 2019) or vice versa (Rollings and Goulson, 2019; Parsons *et al.*, 2020). For example, Helden *et al.* (2012) found an increase in the abundance of both Hemiptera (true bugs) and birds with an increasing proportion of native trees, but there was no effect on species richness.

Non-natives species have been favoured in a limited number of studies (Jain *et al.*, 2016; Nagase *et al.*, 2019), and here attraction to individual plant species, or species-specific responses, seems important. Where studies show a positive influence of non-native plants for native animals, it is often also related to the habitat or resource value that they provide. For example, increased abundance of fruit from non-native *Lonicera* species increased the

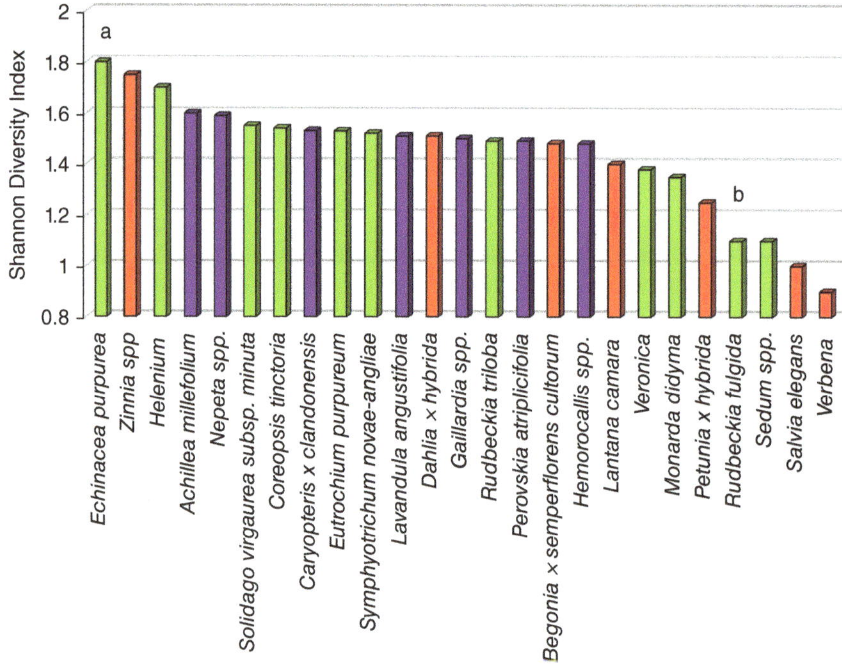

Fig. 5.5. Pollinating insect biodiversity (species richness – Shannon Diversity Index)) observed in native perennial plants (to the USA), shown in green; non-native perennials in purple; and non-native annuals in red. Letters note statistical differences i.e. no statistical difference between *Echinacea* and *Petunia*. (Modified with permission from Palemersheim *et al.*, 2022.)

abundance of frugivorous birds, and non-native *Sorbus* spp. tend to retain their fruit for longer than native *Sorbus*, thus supporting food provision for thrushes (Turdidae) and other birds, long into the winter months (Fig. 5.6). *Buddleia davidii* (from Asia) is often superior for nectar provision for certain butterflies (Lepidoptera) compared to native European shrubs, and the North American *Echinacea purpurea* is a 'staple' plant for many UK invertebrate pollinators. In one UK study (Hanley *et al.*, 2014), it was found that the most attractive plant to native bees was a *Hebe* from New Zealand. Papachristou (C. Papachristou, personal communication, 2015) found no difference in pollinating insect visits to six species of native and non-native Dipsaceae. Such observations support the work of Smith *et al.* (2006) and Owen (2010) in suggesting garden environments (with large numbers of non-native plants) can be important for pollinators. Factors other than nativeness may be affecting attractiveness to pollinators – these include pollen load, sugar concentration or monosaccharide type within the nectar, aromatic volatiles or access to the nectaries. Trials conducted by the Royal Horticultural Society (UK) evaluated the relative attractiveness of different species and cultivars to invertebrates. In the *Sedum* evaluation trial in 2006, surprisingly large differences were noticed in the attractiveness of different *S. autumnale* x *S. telephium* cultivars to insects. This has been mirrored in other plant taxa. Erickson *et al.* (2021) found significant variation in the abundance of pollinators attracted to herbaceous perennial cultivars of *Agastache* spp., *Rudbeckia* spp. and *Nepeta* spp. but not between cultivars of *Echinacea* sp. and *Salvia nemorosa* (Fig. 5.7). *Geranium* species have been cited as useful groundcover plants in Poland (Masierowska *et al.*, 2018) that accommodate the needs of bees, but different traits provided different opportunities. The native *Geranium sanguineum* provided an extended flowering period and continued supply of nectar/pollen, whilst the non-native *G. macrorrhizum* and *G. platypetalum* included large floral displays that attracted insects, with *G. macrorrhizum* providing the highest nectar volumes and sugar yields and pollen reward, marginally better than *Geranium sanguineum*.

Non-native plants can provide structural resources that may form important nesting sites or shelter from introduced predators in highly disturbed urban remnants. Ecological fit is important. Salisbury *et al.* (2017) found that 'near-native' plants from the northern hemisphere performed nearly as well as plants native to the UK, but that these were superior for supporting native biodiversity compared to plants derived from further away,

Fig. 5.6. Careful choice of non-native plants can help biodiversity in urban situations. Birds seem to prefer the orange berries of the native *Sorbus aucuparia* (top) early in the autumn in the UK, but as these are consumed, the palate changes to exploit berries from non-native *Sorbus* species and cultivars, e.g. *Sorbus hupehensis* (middle) and *Sorbus* cv. Joseph Rock (bottom).

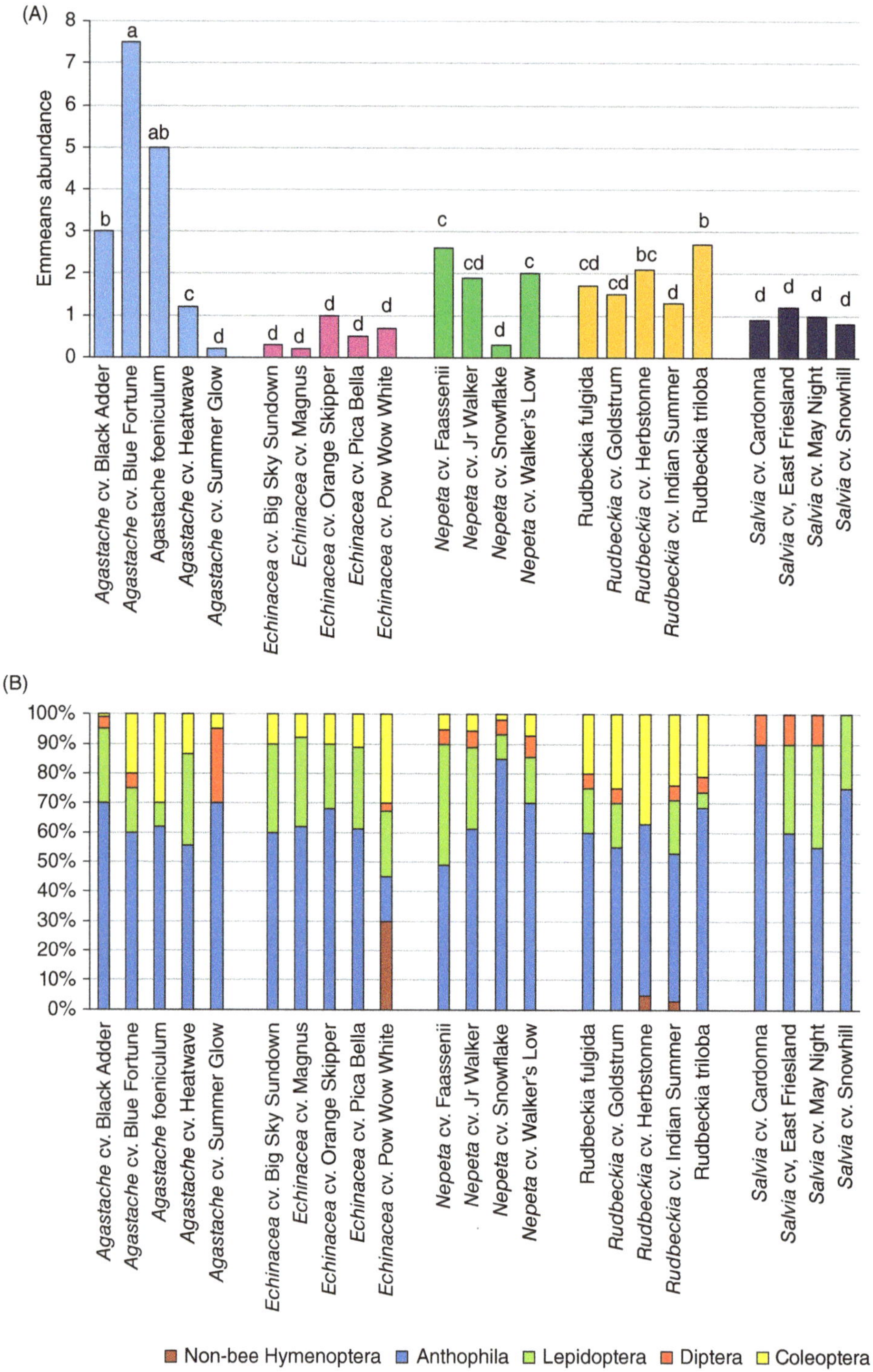

Fig. 5.7. Insect visits to different cultivars within five plant genera: (A) estimated marginal means (emmeans) of total visitor abundance (visits cultivar^{-1} 10 min^{-1}) and (B) mean proportional abundance of observed insect pollinator visitors by insect taxa. (Modified with permission from Erickson *et al.*, 2021.)

such as the southern hemisphere. Many insects within the Palearctic fauna, which range from Ireland to Japan and from the Arctic circle south to the Himalayas, share a long evolutionary overlap with the plants in this vast region, thus increasing likely suitability across it.

Tree choice at an urban scale is important. Comparisons between native and non-native trees in Bracknell (UK) of their potential to host members of the Hemiptera (true bugs) was conducted by Helden *et al.* (2012). In general, native trees proved more valuable for supporting a greater number of species within Hemiptera (Fig. 5.8) and increasing the numbers of individuals found on individual trees (Fig. 5.9), although due to the variation between tree species, there was no overall statistical difference between the two groups. Indeed, some non-native trees such as *Sorbus intermedia* (Swedish whitebeam) and *Quercus rubra* (red oak) were relatively useful in providing habitat and resources for a range of bug species. The non-native *Malus* (apple) also enhanced the number of individuals that could be found within the non-native tree group. These non-natives taxa often excelled in promoting Hemiptera richness and abundance compared to 'less conducive' native species such as *Salix fragilis* (crack willow) and *Ilex aquifolium* (holly). So even when restricted to using native tree species, environmental horticulturists need to consider functional choice carefully, although further research across other invertebrate taxa is warranted here.

In studies where a distinction was made between native and non-native animal taxa, native animals often favoured native plants, whereas non-native animals favoured non-native plants (Gray and van Heezik, 2016), but native plants often supported both native and non-native animals (e.g. Pardee and Philpott, 2014; Threlfall *et al.*, 2015). As mentioned previously, urbanisation tends to attract animals that are generalists and, for example, have a catholic diet (Ducatez *et al.*, 2018). This may partially explain why some urban fauna respond neutrally to plant origin (Nascimento *et al.*, 2020). In contrast, native animal species that are highly dependent on one or a limited number of native plant species (and have no substitute within the non-native flora) are often negatively impacted by

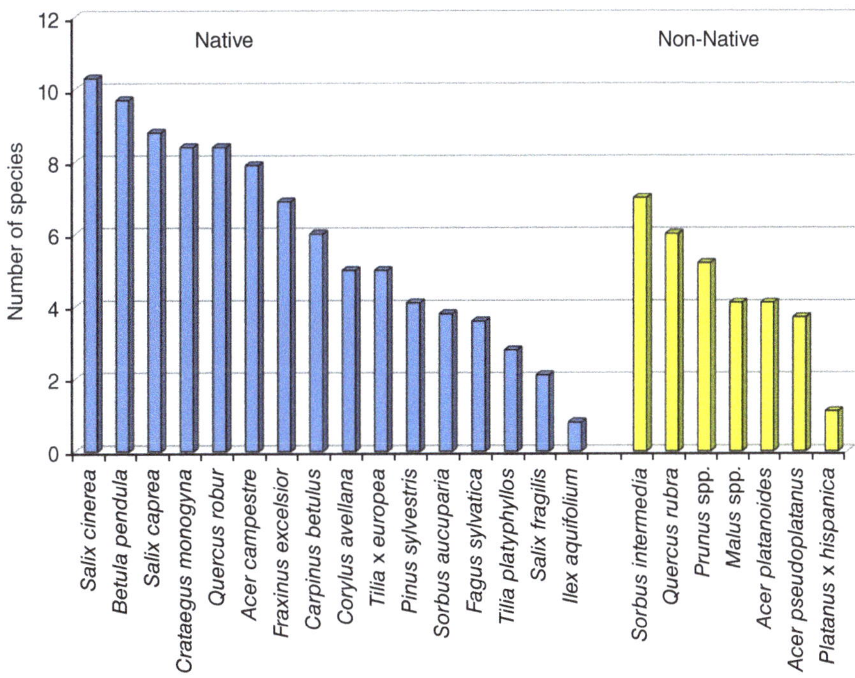

Fig. 5.8. The number of species of Hemiptera (true bugs) found in native and non-native tree species in the UK. (Modified with permission from Helden *et al.*, 2012.)

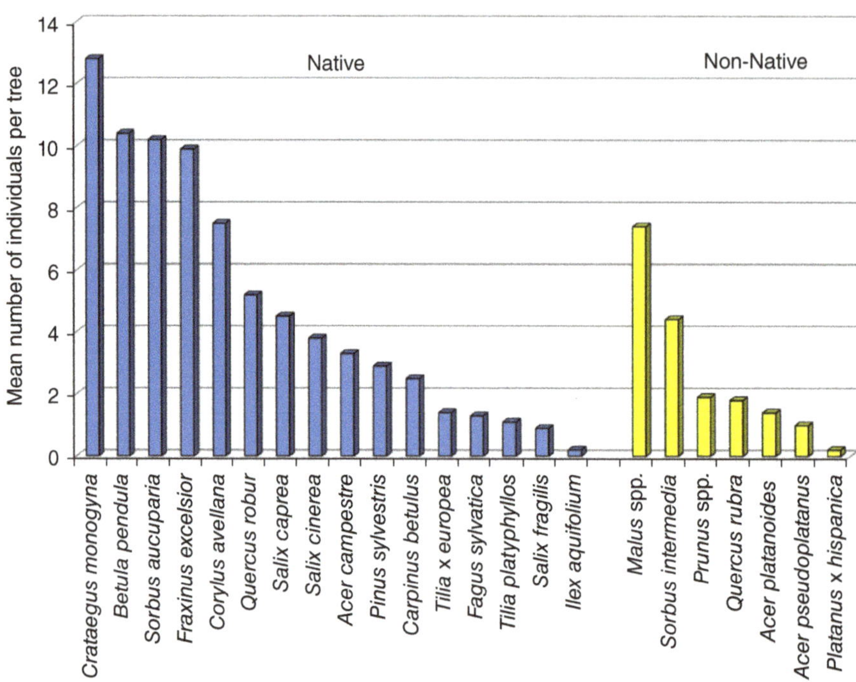

Fig. 5.9. The mean number of individual specimens (abundance) of Hemiptera (true bugs) found per tree across a range of native and non-native tree species in the UK. (Modified with permission from Helden *et al.*, 2012.)

urbanisation (Jain *et al.*, 2016). Actively promoting native plants in design might help, especially if a known animal taxon is dependent on key native plants. Interestingly, urban-avoiding species abundance and richness can increase on urban golf courses as more native vegetation is introduced.

To attract a specific species, group of animals or provide for the continuity of ecological communities, it is important to consider what specific plant–animal relationships need to be preserved. As outlined above, many animal species will not be present at a site without their specific plant hosts, and not all native plants provide resources that are useful for all native animals (Gray and van Heezik, 2016). For example, native grasses may not be beneficial plants in cities if the goal is to enhance the biodiversity of flower-dependent bees. Protocols already exist for target-based approaches to include habitat requirements for certain animal species (Apfelbeck *et al.*, 2020), but trait-based selection of plants for plant resources may also be informative.

Context is important in these discussions. The UK, for example has a relatively limited natural flora and has a long history of introducing non-native plants to its towns and cities. Invasive non-natives have been problematic on occasions but not nearly so damaging as the introduction of non-native flora and fauna to Australia and New Zealand. The fact that in UK cities (and many other European states) a large proportion of the urban flora is already non-native suggests flexibility is required, when considering what to include in future planting designs. Indeed, the highly modified nature of urban environments and the multiple social and ecological goals of urban green spaces tend to mean that a mixed planting strategy would appear prudent. Such a strategy may involve choosing plants of different origins or shifting the boundaries of native species definitions depending on the project goals. In their review Berthon *et al.* (2021) conclude that at least some non-native plants provide resources for urban fauna, but for native animals, native plants are more often useful.

Invasiveness

Whilst there seems merit in valuing the ecosystem services and value to native wildlife of non-native plant species, attention needs to be paid to the consequences of highly invasive non-native ('alien') plant species. One of ornamental horticulture's more notorious legacies has been that some introduced plant species have become invasive in climatic zones outwith their natural distribution. With reduced natural competition and less pressure from herbivores and pathogens, such plants can exploit new ecological niches, outcompeting native flora. Even though only a small proportion of introduced plants may become invasive (<10%), the ecological and cost implications are vast (Bigirimana *et al.,* 2012), with <1% contributing to the main costs (Haubrock *et al.,* 2021). Not only this but traded horticultural products are also a potential source of new invertebrate pests and pathogens to native plants. The costs of controlling or eradicating invasive species is significant. Global estimates (for 2017) are US$29,198M (Diagne *et al.,* 2021) to control both invasive plants and animals. Earlier data for the UK suggested plant invasions contributed about 16% of total costs, i.e. plants, US$455M p.a. (for 2008), and total, US$2,800M (Williams *et al.,* 2010). Some countries have been disproportionally affected by non-native species, e.g. weed control in Australia alone (in 2008) costing US$3,306M. As the climate warms, it is thought that some currently 'benign', introduced plants may become more invasive. It is also considered that a warmer climate may encourage new plant species to colonise naturally, adding to the pressure from human-activated introductions. The fact that indigenous, native species may be more stressed and prone to pathogens due to climate change also increases the threat posed by competitive 'newcomers'. Counterbalancing this notion of control though is the fact climate change itself may so radically alter the relationship between native plants and their ecological fit to any one location, so that new species migrations (invasions) from warmer geographical regions are inevitable.

5.7 Climate Change

Even without a changing climate, cities can be 3–11°C warmer than the surrounding countryside due to the urban heat island effect. This accommodates plants that come from somewhat warmer climates – particularly so if the city remains frost-free throughout the year. The timing of seasonal responses also changes; plants come into flower a few weeks earlier in the spring than their counterparts in rural locations. Climate change though is likely to warm our cities by a greater magnitude again. Bastin *et al.* (2019) indicates that a city like London (UK) by 2050 will experience a climate similar to that of present-day Barcelona (Spain). There is also an east–west 'drying shift' in some parts of the globe – for example, Nairobi (Kenya) may become as dry as Beirut (Lebanon) is currently. Such rapid shifts in climatic patterns are going to put further stress on native plant species, especially those at the southern edge of their natural distribution (in the northern hemisphere) and those at their northern edge in the southern hemisphere. It may be that many cities actually become reliant on tree and shrub species not native to the region but adapted to conditions 2000– 3000 miles closer to the equator. For example, in London, UK, native tree species that like mild, temperate conditions, such as *Alnus* and *Sorbus*, are likely to give way to non-native species better adapted to warmer continental climates, e.g. *Castanea* and *Juglans*. The author was involved with a project conserving trees in the mountains of Oman. It was concluded that for one keystone species – *Juniperus seravschanica* – the mountain landscape it evolved in was now too warm to ensure its survival, and the only course of action was to consider artificial introduction to cooler montane landscapes in central Asia (Al Farsi *et al.,* 2017), the ecological consequences of which are unknown. Yet this highlights the point that climatic shifts are likely to be too rapid to allow natural colonisation of new areas by slow-growing/slow-dispersing plant species. Humans will have to intervene. We are entering the era of human-generated AI – artificial introductions! A result of which will be new, human-determined and -constructed ecosystems.

Matching a current climate and ecosystem with one closer to the equator is challenging enough, but climate change is itself unpredictable in terms of the specific weather events that may predominate. The greater volatility predicted in the climate due to climate change will result in these new climatic regimes demonstrating greater oscillations in temperature and other weather phenomena. So, despite mean warmer temperatures, theoretically such 'newly warmed' areas may still experience periodically severe frosts for a few decades to come – a factor that could limit the movement or transplantation of species from further south (which have limited adaptation to frost, for example). Choosing street tree species to guarantee future success is extremely challenging, with a multitude of uncertainties likely to impact on what actually survives. In the UK, the climate change scenarios suggest warmer, wetter winters and drier summers in the south and east; and warmer, wetter winters and potentially warmer, wetter summers in the north-west of the region. So, a city like Sheffield (UK), being at the interface of these two zones, may have to deal with any of the broad scenarios outlined above. Add in variability (more turbulence in the climatic patterns) and this begins to focus the mind on what future tree species will need to tolerate. In essence they could require adaptation to

- drought;
- flooding (winter and summer; the latter tending to promote more anaerobic soil conditions);
- less stable soils during waterlogging;
- high-temperature stress (periods of temperature perhaps exceeding 50°C);
- occasional frosts;
- high wind speeds (as storms become more violent);
- alterations in phenological triggers that regulate plant development – lack of winter chilling, for example; and
- increased competition and pathogenic pressure from new invertebrate and microbial organisms colonising the region.

Understanding which (limited number?) of tree species possess all these tolerances/adaptations (and where in nature they might be found) will be important if a successful urban forest is to be maintained.

Identifying plant species that have wide distributions and can be present in different vegetation communities may help in this regard. When examining resilience to flooding and drought in *Primula*, Lewis *et al.* (2019) found that the widely distributed UK native *Primula vulgaris* (primrose) had higher survival rates, than more specialised native species (*Primula veris*, cowslip – dry-adapted, and *Primula elatior*, oxlip – wet-adapted) and *P. vulgaris* was also superior to any of the 'garden' cultivars of *Primula* tested in the study (Fig. 5.10).

Fig. 5.10. The widely distributed, native species *Primula vulgaris* (primrose – pale yellow flowers) showed more resilience against climate change-linked events (such as prolonged flooding) than more specialised *Primula* species and garden hybrids.

Cities themselves could be used more effectively as sanctuaries for plants as the climate changes and habitats are lost elsewhere. Blackmore (2019) argues that expanding the use of native plants within city green spaces, particularly the less frequently cultivated and, crucially, threatened plant species, is important, and could be a mechanism to help their conservation.

5.8 Rewilding

Rewilding is a concept that had been interpreted in a number of different ways. It is often defined as the restoration of land to its natural uncultivated state (with an implication that means the reintroduction of species of wild animal that have been driven out or exterminated in the past). It can mean, however, land that is simply released from intense human management and nature left to 'take its course', i.e. without necessarily actively reintroducing any species. Urban rewilding seems similarly open to a variety of interpretations. For many, urban rewilding seems to mean the acknowledgement that vacant or non-developed land should be left to develop its own ecologies, or that actively managed, existing green space is released from intensive management regimes. Again, there are differences as to what this means in practice – is litter blown in to the area no longer removed? Are people, e.g. dog walkers, who might – through the activities of their canine friends – have an impact on the ecology excluded? Are species, indeed, reintroduced? Genuine rewilding might be a concept difficult to deliver in the midst of strongly human-dominated cities. Nevertheless, the principle that areas of land can be left with low or minimal management is gaining traction. It also embraces the philosophy that non-human actions often deliver the most natural ecologies and arguably more effective ecosystem functionality.

Rewilding areas in cities has become a powerful strategy to encourage more diverse plant communities and help bring back insects, birds, mammals and reptiles (Lehmann, 2021). There is also a school of thought that naturally rewilding areas restore soil functionality and encourage a more diverse (and potentially healthier human) microbiome (Mills *et al.*, 2017). Active management – grass cutting, soil cultivation and the use of fertilisers and pesticides – is prohibited. The consequences of such initiatives are intriguing, because site-specific ecological processes and vegetation successions can be unpredictable or highly dependent on the presence or absence of certain herbivores (who in turn may have their activity levels determined by the presence of key predators). Rewilding initiatives, for example, that hope to aim for the restoration of wetland may end up with riparian forest systems instead – especially if the concept of no intervention is strictly adhered to throughout. Rewilding is still in its infancy and the public attitude to rewilding in the city remains undetermined.

5.9 Planning and Design for Wildlife

Urban planners are left with a historical legacy of urban development and so have only limited powers to ensure city layouts are completely logical and coherent. Nevertheless, as planning becomes a more important factor in future city development, then there is more scope to arrange city features and infrastructures that align better with natural process and features and that can aid urban biodiversity.

Connections and movement

As discussed above the ability to move through the urban landscape via corridors and mosaics of green space is vital for many species. City planning can help in this respect by ensuring the continuity of green and blue spaces across the urban network, River corridors should be linked to parks and urban woodlands. Urban meadows landscapes should be able to link up with brownfield sites and through green networks to the surrounding countryside. Planners can be more imaginative when dealing with difficult areas such as junctions where important ecological habitats or migration routes come up against major road or rail infrastructure. Facilities that provide safe passage such as underpasses that go under, or green bridges that pass over, roads and railways should be considered and invested in.

Terrestrial animal species are particularly vulnerable to roadkills. Many amphibian species require landscape types that complement each other at multiple scales; for example, they require access to ponds at certain times of year but low vegetation or even areas of forest at other times. Effective links between these distinct areas are important in the urban landscape as the advent of roads, rail networks, walls buildings and fences often disrupt natural migration routes. Reptile species are also susceptible to roadkills – especially for nocturnal species that use the heat radiating off a road to regulate their body temperatures before hunting at night. In addition, reptiles are intentionally killed because of human aversions (especially snakes, even when the species in question is harmless). Education and road signage can help in this regard, especially as some snake species are effective at keeping rat and mice populations in check.

Planning can work at small scale. Encouraging homeowners to have permeable boundaries to their gardens helps *Erinaceus europaeus* (hedgehog) populations. Hedges can be encouraged over the use of fences to allow movement of the animals from one garden to another. Where fences are used, special hedgehog 'doorways' can be cut into the base of the fence panels – again, to facilitate the roaming nature of nocturnal hedgehogs.

Quality habitat for different taxa

Planners also have a responsibility to ensure important habitat is not lost or isolated. Some urban habitat is of a high quality or hosts unusual plant and animal assemblages. There can be unusual trends. Barrett and Guyer (2008) noted the richness of reptile species actually increased with urbanisation, whilst amphibian species numbers declined. They hypothesised that this was because urbanisation altered the conditions from a closed-canopy, shallow-water habitat (generally conducive to salamanders and frogs) to a habitat characterised by open vegetation, warmer microclimate and deeper and more open water (conditions more favourable to snakes and turtles). Thus, the retention of some patches of appropriate habitat can help reptiles retain a foothold during urbanisation. This is manifest in locations such as Brisbane (Australia) where the viability of reptile populations in the greater city area is strongly dependent on local habitat composition, structure and the proximity and configuration of lowland remnant forests. Rapid changes in landscape structure and complexity tend to result in the decline of amphibian species diversity, and planners, designers and managers can do much to try and avoid this. Pro-environmental attitudes help and lead to corresponding positive activities. *Rana temporaria* (common frog) has benefited from an increase in popularity of garden ponds in the UK and elsewhere in Europe. Their breeding success though can be limited due to predation of the eggs and tadpoles by ornamental fish; but Wildlife Trusts and other conservation groups are doing much to educate people about such problems. A survey of home gardeners in the UK indicated a significant minority had 'designed in' a pond to their gardens, specifically for wildlife.

Retaining high-quality habitat is important. Water quality is an issue (as indeed is the case in agricultural landscapes), due to pollutant loads associated with fertilisers, pesticides, sediment and chemicals from storm water runoff. De-icing salts and petrochemical products are particularly common problems where ponds are located close to roads. Other implications of urbanisation include alterations in water flows, either through increased built infrastructure reducing water infiltration to the soil, or greater surface runoff or, alternatively, over abstraction for domestic and industrial uses.

Urbanisation also has more subtle effects; microclimates and alterations to hydrological flows can alter the period water is withheld within ponds, streams or wetlands, and this affects the dynamics between species. In Portland (USA), increased and prolonged water flows changed ephemeral wetlands to more stable permanent wetlands, with a consequential change in amphibian species present, namely a transition from those species with rapid larval (tadpole) development to those with prolonged larval development (Pearl *et al.*, 2005). Similarly, a change from temporary ponds to permanent ponds encourages fish populations. Permanent wetlands colonised by fish result in increased tadpole predation, and amphibian species that have become adapted to fish-free temporary ponds may suffer population crashes, e.g. in the USA, *Ambystoma macrodactylum* (long-toed salamander), *Ambystoma tigrinum* (tiger salamander), *Ambystoma laterale* (blue-spotted salamander) and *Rana sylvatica* (wood frog) have seen populations decline as more pools have become permanent features.

Scale of urban habitat is important for mammals. Mammals, in general, need greater spatial scales than many other taxa to maintain viable population sizes. Even so, although a relatively wide range of mammal species have been recorded in urban garden habitats and city parks, it tends to be the domain of a relatively select few. For example, in the UK, *Vulpes vulpes* (red fox) (Fig. 5.11), *Sciurus carolinensis* (grey squirrel) and *Erinaceus europaeus* (hedgehog), as well as certain bats, voles and mice. Where extensive areas of green space

Fig. 5.11. *Vulpes vulpes* (red fox) is one of the more successful urban mammals adapting to life in the city, where there are some parks and gardens to provide refuge and food. (Used with permission from P. Dixon.)

exist in the suburbs and urban fringes, then opportunities do exist for larger mammal species or those requiring greater territories. Indeed, the relative lack of persecution from hunting of larger predators in cities means some species may be more obvious, even if actual populations are not necessarily greater. Some selected species are becoming more adapted to urban environments too (synurbization – see above). This includes *Alces alces* (elk or moose) in Sweden or Canada, *Oryctolagus cuniculus* (rabbit), *Capreolus capreolus* (roe deer), *Meles meles* (badger), *Martes foina*, (stone marten) in western Europe and *Canis latrans* (coyote), *Lynx rufus* (bobcat), *Odocoileus virginianus clavium* (key deer) and *Tamias striatus* (chipmunk) in North America. In parts of Africa and India there is even evidence of *Panthera pardus* (leopard) entering urban situations and preying on feral dogs. Similarly, in Europe, a degree of relaxation in the persecution of the *Canis lupus* (wolf) has resulted in some individuals at least crossing urban environments, if not quite settling.

For bird populations, quality habitat relates to the coverage and arrangement of green spaces. This relates to all levels of vegetation, i.e. ground level, grasslands as well as shrubs and trees (Morelli *et al.,* 2021). More extensive and varied green spaces increases bird diversity. Significant secondary factors are light (and noise) pollution. Research on different taxa (e.g., insects, birds, reptiles and other wildlife species) shows that artificial light at night (ALAN) can alter behaviour, feeding locations and breeding cycles, mainly in the most densely populated urban areas (Van Geffen *et al.,* 2015; Gaynor *et al.,* 2018). Dominoni (2017) considers light pollution as one of the main factors affecting bird species living in urban areas.

Links have been made between the presence of bird species and the quality of the environment for humans, as well as other animals. In a European context, high urban environmental quality could be related to the presence of certain bird species, as indicator species (Morelli *et al.,* 2021). This list being topped by *Sylvia atricapilla* (Eurasian blackcap), as well as *Turdus merula* (blackbird), *Parus major* (great tit), *Cyanistes caeruleus* (blue tit), *Passer montanus* (tree sparrow) and *Pica pica* (magpie). Morelli *et al.* (2021) pointed out that those urban neighbourhoods with high environmental quality were also likely to be those best for human well-being. However, the issue of light pollution is controversial here, as increasing safety for humans (especially more vulnerable groups) is often linked to better lighting (e.g. along streets and within parks) (Portnov *et al.,* 2020).

Retaining fragments of remnant forests (and other habitats?) within urban areas helps bird populations. Older trees provide nesting habitat for some species and more diverse woodland structures support greater diversity. In Australia, remnant forests support species that otherwise would be excluded from the suburbs. A study of hummingbirds in the city of Campo Grande (Brazil) showed that species that were usually forest dwellers could survive in small urban fragments of cerrado (grass and tree) vegetation (Barbosa-Filho and Araujo, 2013). The hummingbird species relied both on plant species normally pollinated by hummingbirds (ornithophilous) but also other non-ornithophilous plant species for their source of nectar. The high instance of visits to the latter plant type was thought to relate to a number of contributing factors: the scarcity of the ornithophilous species that were the traditional food source, the limited period in the year they were in flower and that the non-ornithophilous species provide nectar with energy values equivalent to the ornithophilous species. In essence the hummingbirds were adapting to the reduced opportunities to feed in the urban environment, by relying on other types of flowering plants that would not normally be favoured. The number of hummingbird species recorded in the urban cerrado is comparable to that reported for well-preserved forested areas. This suggests that the small forest fragments contained within the cerrado are providing important refuge sites for those hummingbird species that manage to remain prevalent in urban areas. Similarly, the remnant vegetation may play a critical role in providing ecological stepping stones across the urban matrix for such small birds, as they attempt to reach other forested areas. Understanding the bird's requirements in detail will provide opportunities for the more cultivated 'horticultural' landscapes in Brazil to replicate these floral groups and thus afford year-round food supplies to the hummingbird species in questions, i.e. *Amazilia fimbriata* (glittering-throated emerald) *Anthracothorax nigricollis* (black-throated mango), *Chlorostilbon lucidus* (glittering-bellied emerald), *Eupetomena macroura* (swallow-tailed), *Heliomaster squamosus* (stripe-breasted starthroat), *Hylocharis chrysura* (gilded sapphire) and *Thalurania furcata* (fork-tailed woodnymph).

In New Zealand, the capacity to restore urban woodland by planting back native species and developing more complex canopy structures also aided bird conservation. Species richness increased with age of restoration planting, with plant community composition progressing towards that found in non-damaged forest areas (Elliot Noe *et al.*, 2022). The data revealed that the age of restoration (longer-term restoration being better) was a direct driver of bird species richness. Bird abundance was driven by the canopy structure, with more enclosed canopies proving useful. Better habitat structure may also be a reason why these bird populations seemed relatively resilient against the large number of non-native predators found in these urban sites (cats, rats and possums).

As with other taxa, the age of the habitat or the stage it is at within its successional process influences the population of invertebrates present (Gilbert, 1989). Comparing open, derelict spaces in cities, it is evident that those that have been left unmanaged for the longest time before they are 'redeveloped' have the greatest diversity of species, and species taxa and abundance shifts from younger to older plots. (Plots that are never redeveloped of course may eventually attain a 'stable' climax vegetation, which paradoxically may result in lower biodiversity.) Interestingly, McIntyre (2000) estimates that terrestrial arthropods (insects, spiders, mites, centipedes, millipedes and crustaceans) in urban landscapes tend to be more diverse than those in rural environments. In general, herbivore species within this grouping are more abundant in cities than rural sites. Conversely, parasitoids are more abundant in rural than urban sites. A number of parasitoids are reliant on butterfly and aphid species, and their abundance correlates closely with the presence of flowering plants; as such, the number of parasitoid wasps tends to decrease as flower diversity and abundance reduces. In general, as the area of impervious surfaces increases there is less space for gardens, and hence fewer parasitoid wasps present.

Other urban factors that affect invertebrates include access to water, air dynamics and thermal characteristics as well as extent of urban development. These influence the frequency of occurrence but also trophic structures. Air movement affects a spiderling's ability to colonise new areas through its 'ballooning' activities (where a young spider releases strands of silk into the air and these are caught by the wind, potentially pulling the spiderling into the high atmosphere before eventually returning to the earth's surface many kilometres away from its original location). Urban heat islands and the effects of climate change are raising the temperatures experienced within cities, thus permitting species to survive throughout the year in latitudes further north and south of the equator than has been the case in the past. Increasingly stressful urban environments impact on plants too, potentially increasing their susceptibility to herbivores and sucking insects.

In insect assemblies in the UK, urban areas are deemed richer in bee species (Fig. 5.12) and with greater numbers of individuals recorded (Fig. 5.13) than either agricultural land or even designated nature reserves (Baldock *et al.*, 2015). The trend, however, is not repeated for true flies or hoverflies. Turrini and Knop (2015), comparing agricultural, managed landscapes and urban landscapes, suggested the latter also supported a greater number of bug species and greater abundance of individuals for taxa such as bugs, beetles, leafhoppers and spiders as well as more variation within the bug and beetle species present (e.g. species adapted to different niches). Within the urban environment, increasing the number and size of vegetated areas present enhanced the species richness and abundance of most arthropod groups under observation. Somewhat in contrast to agriculturally productive landscapes, the isolation of vegetation stands and other habitats played only a relatively limited role in determining the diversity and abundance of species within the urban ecosystems. Conversely, river and other aquatic ecosystems tend to suffer more from pollution in urban areas than their rural

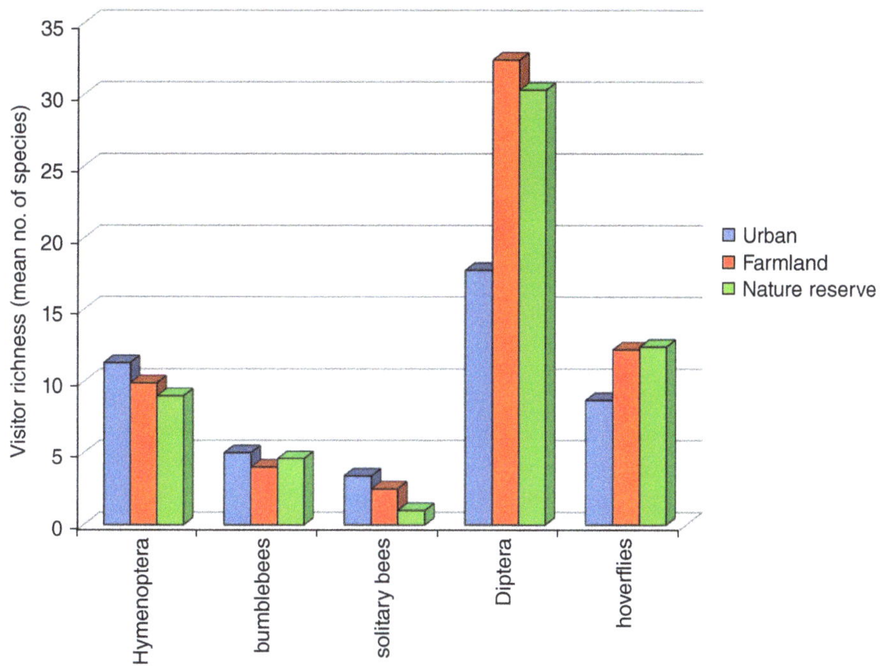

Fig. 5.12. Insect taxa richness in different landscapes of the UK. (Modified with permission from Baldock *et al.*, 2015.)

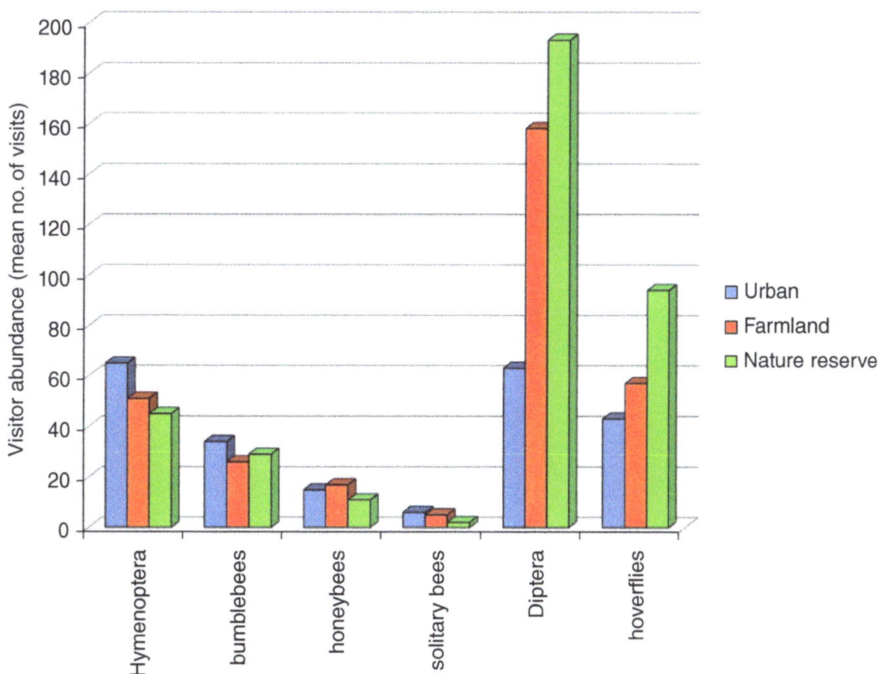

Fig. 5.13. Insect taxa abundance in different landscapes of the UK. (Modified with permission from Baldock *et al.*, 2015.)

counterparts, and this has a negative effect on the diversity of aquatic insects and crustaceans. Overall, McIntyre (2000) believes that arthropod diversity:

- decreases with increasing air and water pollution;
- increases with the age of an urbanised area;
- of non-native species increases with the age of an urban area;
- increases with a higher proportion of vegetation incorporated into the urban matrix: and
- is also affected by the juxtaposition to native habitats as these play an important role for recruitment and dispersal into new urban habitats.

5.10 Key Landscape Typologies and Their Management

A number of horticulturally important landscapes impact directly on wildlife, and through careful management of these landscapes, biodiversity can be enhanced. Historically, of course, the concept of horticulture has been to manage areas explicitly to favour the development of some plant species over others and to reduce the impact of herbivores. Thus, cultivated crops were encouraged at the expense of 'weeds' and invertebrate species considered pests, reduced or eliminated altogether. Even in environmental horticulture this may still be the case. Today, however, it may well be that non-native ornamental 'alien' plants are being eradicated in an attempt to help the establishment of native species, indeed precisely those indigenous plant species that once might have been seen to be the enemy of 'good horticultural practice'. The key areas where horticulture overlaps with biodiversity management are outlined below, although emphasis is given to those landscape typologies where the management is often led by an environmental horticulturist or other plant-focused professional, rather than professionals with a wider remit within nature conservation or an alternative discipline, such as ornithology.

Wetlands and ponds

It is frequently quoted in garden books and magazines that if you wish to introduce animal life to your garden, construct a pond or some other water feature. Open water is normally a rare resource in city business districts and even the outlying suburban belts may have few natural water features remaining. Therefore, the provision of new ponds and bog gardens, as well as careful management of existing natural water courses, are key components in drawing wildlife into cities. Despite the fact that many towns and cities developed precisely because they were located on a river, this rarely means the natural character of the river has been preserved. Rivers have been disconnected from their natural flood plains, channelised and culverted over with concrete: all factors which are highly detrimental to aquatic life. Thankfully, these approaches are beginning to be reversed in the more enlightened cities at least, as the value of blue and green infrastructure begins to become more widely recognised (see section on sustainable urban drainage systems in Chapter 3).

In a survey of the larger aquatic invertebrates (macroinvertebrates) it was shown that garden ponds tended to be less rich than the traditional field ponds in rural locations (Hill and Wood, 2014; Hill *et al.*, 2021). Most garden ponds (87%) were categorised as of low or moderate conservation value (compared to 35% of non-urban ponds). Garden ponds were found to support compositionally different macroinvertebrate communities compared to non-urban ponds, influenced by variation in water depth and conductivity. These researchers found 44 taxa from garden ponds (10 of which were unique to garden ponds) but over three times that number of taxa in field ponds. They, however, noted the value of garden ponds in providing habitat for certain taxa (e.g. Odonata – dragonflies and damselflies) due to their high mobility at the adult phase and thus their relative ease in colonising new ponds. Additional features such as fountains and artificial waterfalls allowed a number of taxa normally defined as specialists of natural fast-flowing water to colonise these ponds. In contrast, frequent cleaning and dredging of ponds was detrimental to some of the taxa under observation. Overall, urban garden ponds afford value at the landscape scale, due to their relative abundance and diversity, but the biodiversity that is held within them is still strongly determined by factors such as their size, depth of water and management.

Another area where ponds are increasing in frequency within towns and cities is the use of storm water ponds and rain gardens adjacent to roadways. These act as a sink for water running off hard road surfaces during intensive rainfall. Although primarily acting as a means to remove surface water and pollutants quickly from the highway, these features develop their own ecological features. Due to de-icing salts and pollution on the road surface, their water chemistry is often very different from natural ponds. In effect they have more salts present in the solution. Despite this, in a survey Meland *et al.* (2020) showed these water bodies could be almost twice the size of natural ponds and actually held more species, hence becoming an important component of the aquatic environment in built-up areas.

To enhance the value of urban ponds, again a number of practical activities should be encouraged:

- Larger, deeper ponds will have more ecological niches and be less prone to drying out or rapid changes in temperature. Balance the need for deeper ponds though with safety in the garden – especially with respect to access regarding children.
- Ponds can be isolated in the urban environment, especially if surrounded by concrete and tarmac – making it difficult for non-flying animals to colonise. Some studies have found urban density (greater numbers of buildings and grey infrastructure) around a pond had a negative effect on urban pond macroinvertebrate richness (Blicharska *et al.,* 2017). Mosaics of ponds and other wet areas across a number of gardens holds certain advantages.
- The creation of shallow edges allows terrestrial species to drink safely and for amphibious species to enter/exit the pond easily.
- Stones, logs, marginal plants or other 'refuge' features placed at the pond edge provide some protective cover for species that use the shallows.
- Ensure the middle of the pond is dug to a depth of at least 0.8–1 m as this will enable some water to remain unfrozen during cold winters (ensure such ponds, however, are not accessible to small children).
- Include substrate at the base of the pond as this provides a habitat for invertebrates and amphibians to burrow in and shelter from predators, but this needs to be formed from nutrient poor material, as excess nitrogen and phosphorus will quickly cause eutrophication problems by encouraging excessive algal blooms.
- Marginal plants provide cover and food sources – so use shallow edge areas to introduce plant species.
- As urban ponds will be slow to colonise naturally with aquatic plant species, these can be introduced, but care is required to avoid the use of invasive species, most notably in the UK: *Azolla filiculoides* (fairy moss), *Crassula helmsii* (New Zealand pygmyweed), *Eichhornia crassipes* (water hyacinth), *Hydrocotyl ranunculoides* (floating pennywort), *Lagarosiphon major* (curly waterweed), *Myriophyllum aquaticum* (parrot's feather) and *Pistacia stratiotes* (water lettuce). Balanced planting of marginal plants in the shallows, oxygenating species for the pond bottom and floating species such as *Nymphaea* spp. (water lilies) are advocated to help establish the pond quickly. Floating species exclude light, thus helping to counteract excessive algal growth.
- Avoid introducing ornamental fish, as most are superb predators of invertebrates and amphibian tadpoles.
- Encourage invertebrates such as *Daphnia* and *Cyclops* spp. as they are algal grazers and often important in the food chain of garden ponds.
- Remove sediment and sludge to stop excess silting of the pond but doing so too frequently can impair invertebrate populations. Pond sludge and dead vegetation can be left at the side of the pond to allow the more mobile invertebrates to migrate back to the pond, before being finally removed and composted at a location away from the aquatic environment. Cleaning ponds with detergents or other chemical agents is not encouraged.
- The design and construction of larger ponds or mosaics of ponds can encourage less abundant species, e.g. in the UK, amphibians such as *Triturus cristatus* (greater crested newt) or *Natrix natrix* (grass snake).
- Once a pond is established, it should not be lost or filled in. Urban meta-populations of amphibians may be dependent on a limited number, or even just the one pond, to breed in, which increases their vulnerability. Ponds can act as a habitat sink and the loss of a single pond or a pollution incident in it can eliminate an entire population (Battin, 2004).

Parks and gardens

Parks and gardens provide a novel set of habitats in the urban landscape. Depending on the socio-cultural factors and age of a city, private gardens can comprise a major component of urban green space.

In Europe, private residential gardens represent between 11% (Sweden, Hanson *et al.,* 2021) to 27% (UK, Smith *et al.,* 2006) of the total land area in a city. They deliver considerable opportunities for biodiversity, but fulfilment of this depends on design, scale and dominant management regimes. Their value is often dependent on the larger-scale 'landscape ecology' framework and how they are spatially organised and fit within the wider, interconnected green space network. Indeed, their value to biodiversity is often determined by the character of adjoining habitats. Gardens promote plant species richness (although frequently through non-native species or hybrids of native species) and seem to be particularly beneficial to insect and avian species diversity through the provision of shelter, nesting sites and food.

Parks particularly, and to some extent gardens, are legacies of preceding cultural values and trends. Many urban parks (and some of the larger gardens!) still retain an English landscape style promoted by the likes of Lancelot 'Capability' Brown (1716–1783) and retain large expanses of grass with clumps or avenues of trees. Urban gardens, on the other hand, retain elements of the gardenesque style, although they can also be influenced by minimalist or naturalistic styles of gardening too. The gardenesque style (John Claudius Loudon, 1783–1843) evolved during the Industrial Revolution in Europe and the Victorian era, and was characterised by the introduction of new, colourful, floristic plants, artificial designs and in some instances extravagant features like ornate summer houses and topiary. A common theme was that the plants were very much 'on show' and positioned and managed in such a way that each individual specimen could be displayed to its full potential in a scattered planting. The approach involved the creation of small-scale landscapes to promote beauty, variety and mystery. Typical features included island beds, winding paths, and tree-planted mounds as well as open areas of mown grass. Landscape gardeners of the day often preferred the use of non-native plants that were beginning to be introduced from other parts of the world including new genotypes from eastern Asia, and North and South America. Today, urban gardens may be on a smaller scale but still many retain the legacy characterised by 'pretty', 'tidy', 'colourful' and 'beautiful' homogeneous landscapes based on non-native plants. This has implications for the species that can utilise such landscapes.

In their assessment of garden biodiversity, Thompson *et al.* (2003) found 438 plant taxa in 60 gardens in Sheffield (UK) of which two-thirds were non-native and one third native to the UK. This was a richer collection of species compared to that found in derelict urban land (brownfield sites) or other plant communities typical of central/northern England (Fig. 5.14). Non-natives were mainly represented by species from Europe and Asia. The majority of the most frequently encountered species were weeds, not plants intentionally planted into the garden, e.g. *Geranium robertianum* (herb Robert) or *Digitalis purpurea* (foxglove), or which had self-seeded from introduced species but were now acting as weeds, e.g. *Alchemilla mollis* (lady's mantle), *Aquilegia vulgaris* (columbine) and *Meconopsis cambrica* (Welsh poppy). Despite the weed species being the most common, the authors noted that there was no saturation to the species count for the garden flora, in that the number of ornamental genotypes available to gardeners is vast (theoretically at least 70,000 species and cultivars in the UK). Overall, native species richness was not correlated with garden size but total species richness was. Garden design and management also influences the number of species present, and it was noted that through active management, more species were maintained in a given area than otherwise would have occurred naturally (Thompson *et al.,* 2003). Others agree that gardens promote biodiversity richness but argue that ecological functionality is not great, as they are developed largely for visual appeal and do not increase trophic diversity and, for the most part, are not self-sustaining (Quigley, 2011). Cameron (2023) argues gardens divide into three basic types, and their value to wildlife and wider ecosystem services is determined by these typologies, i.e. 'grey' gardens that are dominated by impermeable hard surfaces (e.g. concrete pavers) with limited vegetation and thus provide little habitat for wildlife; 'others' that are 'intensively managed' but have some vegetation although limited plant diversity and are of intermediate value to wildlife: and thirdly, 'green' gardens with a high proportion of vegetation, low-energy and other resource inputs and mimic natural vegetation stands or transitional ecotones that help provide ecological opportunities for some, if not all, wildlife.

In many tropical cities too, parks and gardens have a high proportion of non-native plants. In cities such as Bandung (Indonesia), Bangalore (India)

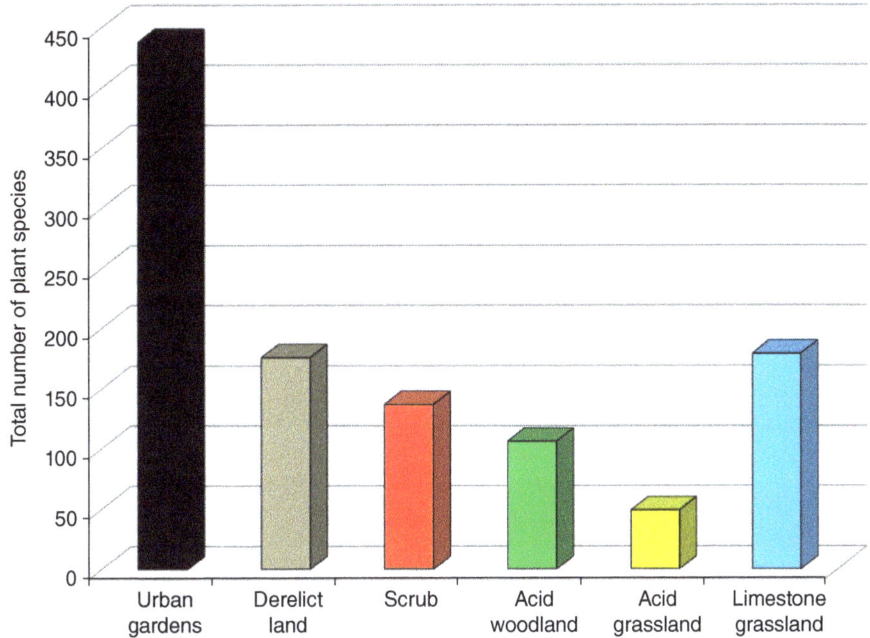

Fig. 5.14. The number of plant species found in 120 1 m^2 plots in a variety of habitats in the UK. (Modified with permission from Thompson *et al.*, 2003.)

and Rio de Janeiro (Brazil) between 70% and 80% of woody plants found in parks may be non-native. These parks host a diverse range of plant species, but large numbers of the genotypes have been imported into the region. Domestic gardens in Hong Kong (China) were evaluated for their diversity of tree species. The 72 species observed represented 19% of the species present in the province. Considering the relatively small area devoted to private gardens this was considered a fairly rich collection. In contrast to western European countries, however, (where tree species diversity is higher in parks and gardens than in native woodlands) the proportion was small compared to other locations surveyed. Species richness in domestic gardens (72 species) was comparable to the 'old and valuable' tree grouping (70 species), but considerably less than that noted for urban park trees (272 species), landscape trees around public housing (232 species) and roadside trees (149 species) in Hong Kong or, indeed, total urban trees in Guangzhou (254 species) or Taipei (164 species). Zhang and Jim (2014) put the arboreal richness of the gardens in Hong Kong down to two factors. The tropical humid climate is conducive for a wide range of tropical tree species to be grown, and once established garden trees have high survival and persistence rates at remarkably low population sizes, due solely to the high maintenance levels employed by gardeners and landscape professionals. Of the 72 species noted in gardens, some species were more common than others. *Juniperus chinensis* (18% of sampled trees) was dominant through its use as a garden hedge or border tree. Others were popular for their ability to provide shade, *Michelia alba* (10%) and *Dimocarpus longan* (5%); flower colour *Delonix regia* (flamboyant tree, 6%); or dramatic form, *Dypsis lutescens* (bamboo palm, 7%) and *Araucaria heterophylla* (Norfolk Island pine, 4%). Despite the overall diversity being considered good, as with other cities there is an overdependence on these relatively few popular species. This result was consistent with urban trees in Chicago, Mexico City and Bangkok where the top four dominant species comprised over 50% of the entire tree population.

Gardens provide particular opportunities for specific taxa and groups. The high propensity of garden ponds, for example, help maintain populations of amphibians in the UK (i.e. *Rana temporaria* common frog; *Bufo bufo,* common toad; and *Lissotriton vulgaris,* smooth newt) and help counteract the loss of wetlands and farm ponds over the last century – a point illustrated by the fact that populations of *Rana temporaria* have experienced declines in rural areas but increases in urban parks and gardens. As mentioned elsewhere, urban environments are important for bees, and gardens contribute to this. Colony expansion of bumblebee *Bombus terrestris* nests was greater in suburban gardens in southern England than neighbouring agricultural habitats, and the density of bumblebee nests recorded in UK suburban gardens (36 nests ha^{-1}) was comparable to that found in hedgerows and grassed pathways in rural habitats (20–37 nests ha^{-1}) (Osborne *et al.,* 2008). Similarly, in a comparison of public parks in San Francisco (USA), *Bombus* spp. were more abundant in a park within a central urban location compared to parks outside the city boundary.

The demise of bee species and other 'pollinators' is a growing concern, not only in relation to concerns for the species themselves but also for the pollinating services such taxa provide. Bees, butterflies, hoverflies and certain beetle and thrip species are important for ensuring the pollination of a wide range of food crops – the fruiting bodies of which, of course, are fundamental to human nutrition. Insect-dependent crops include *Malus* (apple), *Pyrus* (pear), *Fragaria* (strawberry), *Lycopersicum* (tomato), *Pisum* (pea) and other legumes to name but a few. Bees also provide honey directly.

A study in Basel (Switzerland) by Braschler *et al.* (2020) focussed on ground-living invertebrates in 35 gardens across a city centre–city edge gradient. Biodiversity levels were relatively high with species number (and % of known national totals) being 87 rove beetle (6%) and 26 ground beetle species (5%), 29 ant species (21%), 52 spider species (6%), 22 millipede species (17%), 10 woodlice species (23%) and 39 gastropod species (20%). Species richness varied with some gardens possessing two to eight times more species than others, although a garden that favoured a high number of species in one taxonomic group did not necessarily provide optimum habitat for other types of invertebrate. For example, one garden that was ideal for woodlice, gastropods, ants and beetles was poor in millipede diversity. Overall, species richness increased as one moved away from the city centre. Characteristics reflecting the naturalness and diversity of the gardens (native plant species richness, habitat richness and structural diversity) affected species richness of different groups to a varying degree. Notably, having more native plants present aided gastropods and ant diversity.

A number of initiatives have aimed to better understand pollinator populations and to encourage their viability, both in rural and urban environments. Within the UK, the Royal Horticultural Society (RHS) investigated the value of non-native plants to invertebrates, both as a source of pollen and nectar, but also as a general conducive habitat to a range of taxa. Their 'Plants for Bugs' project evaluated garden plant populations using species derived from either the UK alone, the northern hemisphere excluding the UK or from the southern hemisphere. These geographical 'communities' were used to monitor the habitat preferences of various invertebrate taxa. Each plant population was matched in their composition and form across the geographical populations, i.e. each group included similar numbers of shrubs and climbers, geophytes and herbaceous perennials as well as genotypes of ornamental grasses. Data from here and elsewhere (Salisbury *et al.,* 2025) has enabled the RHS to provide a comprehensive list of plants (RHS Plants for Pollinators) that support pollinating insects (Tables 5.6–5.11). This allows both gardeners to attain (and nurseries to promote) those taxa particularly useful to pollinating insects, irrespective of whether the plants are 'native' or not (Anon., 2025). Other studies have taken similar approaches, and information is now available to gardeners in North America as to what garden plants should be planted to encourage hummingbirds to visit (Table 5.12).

Although gardens are considered to be an opportunity for urban adapted or 'synurbic' mammals, a study by Van Helden *et al.* (2020) in Australia considered gardens could be used more often for conservation purposes for key or threatened species. Their results indicated that domestic gardens were a more positive asset for the critically endangered *Pseudocheirus occidentalis* (western ringtail possum) with its presence being reported in 65% of all gardens, whilst *Trichosurus vulpecula hypoleucus* (common brushtail possum) often termed an 'urban adapter' was present in only 175 individual gardens.

Many urban parks were criticised in the 1970s and again in the 2000s for being overly dominated by close-mown turf grass, which required excessive maintenance through gang mowing, resulting in a high use of fossil fuels. A consequence of which was

Table 5.6. Tree genotypes providing pollen and/or nectar resources to invertebrates within different seasons in the UK. (Modified with permission from the RHS 'Plants for Pollinators' plant list – Anon., 2025.)

Species	Common name	Native
Spring March–May		
Acer campestre	field maple	Y
Acer platanoides	Norway maple	
Acer pseudoplatanus	sycamore	
Acer saccharum	sugar maple	
Aesculus hippocastanum	horse chestnut	
Cercis siliquastrum	Judas tree	
Crataegus monogyna	hawthorn	Y
Ilex aquifolium	holly	Y
Malus species	apple/crab apple	
Mespilus germanica	medlar	
Prunus species	cherries/plums/peaches	
Pyrus communis	pear	
Salix species	willow	
Summer June–September		
Aesculus indica	Indian horse chestnut	
Arbutus unedo	strawberry tree	
Catalpa bignonioides	Indian bean tree	
Koelreuteria paniculata	pride of India	
Robinia pseudoacacia	false acacia	
Sorbus aria	common whitebeam	Y
Sorbus aucuparia	mountain ash/rowan	Y
Tetradium danielli	bee-bee tree	
Tilia species	lime	

that the term 'green deserts' was coined to describe park landscapes. Since then many park authorities have attempted to turn a higher proportion of their close-mown turf over to meadow-type landscapes with varying degrees of success (see Chapter 8). Increasing the length of the grass (height of cut) has been associated with increasing plant richness. Converting lawns to meadows by the reducing the frequency of mowing in Cambridge, UK resulted in the grasslands supporting three times more plant species, three times more spider and bug species and individuals and bats were observed to hunt three times more often over the meadow than the remaining lawn (Marshall *et al.*, 2023). Terrestrial invertebrate biomass was 25 times higher in the meadow compared with the lawn. Relaxed mowing or having mowing-free periods (e.g. 'No Mow May' in the UK) can allow a wide range of short-stature flowering broadleaf perennials to flower and complete their life cycle.

The land area of urban parks has been linked to bird diversity, with larger parks increasing richness. However, the number of sub-habitats, the complexity of these habitats and their ability to complement each other and hence fulfil the entire ecological needs of a species also play a key role. Biological factors such as size of home range and specific interactions between species are critical too. More intensively managed parks are thought to be a negative component, but the term 'intensively managed' needs to be carefully defined and designated here. Herbaceous flower borders are 'heavily managed' but are useful nectar 'hotspots' within a park, attracting many pollinating insects.

Bird populations in Israel were monitored in relation to landscapes within urban parks and the degree of management imposed (see Table 5.13 for definitions of management imposed and Fig. 5.15). Although intensively managed landscapes decreased the richness of the species present, areas that were managed at a moderate intensity were deemed good as bird habitats, and comparable to un-managed areas (Shwartz *et al.*, 2008). Overall, the moderately managed areas were richer habitats for all categories of birds (urban adapter, urban exploiter, alien and locally rare).

Landscape architects can enhance native plant species persistence in urban parks by providing a

Table 5.7. Shrub and climber genotypes providing pollen and/or nectar resources to invertebrates within different seasons in the UK. (Modified with permission from the RHS 'Plants for Pollinators' plant list – Anon., 2025.)

Species	Common name	Native
Winter November–February		
Clematis cirrhosa	Spanish traveller's joy	
Fatshedera lizei	tree ivy	
Lonicera × *purpusii*	honeysuckle	
Mahonia species	Oregon grape	
Sarcococca species	sweet box	
Viburnum tinus	laurustinus	
Spring March–May		
Berberis species	barberry	
Buxus sempervirens	box	Y
Ceanothus species	California lilac	
Chaenomeles species	Japanese quince	
Cornus mas	Cornelian cherry	
Cotoneaster conspicuus	Tibetan cotoneaster	
Crataegus monogyna	hawthorn	Y
Erica species	heath	
Hebe species	hebe	
Mahonia species	Oregon grape	
Pieris species	lily-of-the-valley bush	
Ribes nigrum	blackcurrant	
Ribes rubrum	redcurrant	
Ribes sanguineum	flowering currant	
Skimmia japonica	skimmia	
Stachyurus praecox	stachyurus	
Vaccinium corymbosum	blueberry	
Summer June–August		
Aesculus parviflora	bottlebrush buckeye	
Brachyglottis cv. Sunshine	brachyglottis cv. Sunshine	
Brachyglottis monroi	Monro's ragwort	
Buddleja davidii	butterfly bush	
Buddleja globosa	orange ball tree	
Bupleurum fruticosum	shrubby hare's ear	
Calluna vulgaris	heather	Y
Caryopteris × *clandonensis*	caryopteris	
Clematis heracleifolia	tube clematis	
Clematis vitalba	old man's beard	Y
Cornus alba	red-barked dogwood	
Elaeagnus species	oleaster	
Erysimum cv. Bowles's Mauve	wallflower cv. Bowles's Mauve	
Escallonia species	escallonia	
Fatsia japonica	Japanese aralia	
Fuchsia species– hardy types	fuchsia	
Hebe species	hebe	
Hedera species	ivy	
Hydrangea paniculata (with fertile flowers)	paniculate hydrangea	
Hydrangea anomala subsp. *petiolaris*	climbing hydrangea	
Hyssopus officinalis	hyssop	
Kalmia latifolia	mountain laurel	
Jasminum officinale	jasmine	
Laurus nobilis	bay tree	
Lavandula species	lavender	

Continued

Table 5.7. Continued.

Species	Common name	Native
Ligustrum species	privet	
Lonicera periclymenum	honeysuckle	Y
Olearia species	daisy bush	
Parthenocissus tricuspidata	Boston ivy	
Perovskia atriplicifolia	Russian sage	
Phlomis species	sage	
Potentilla species	cinquefoil	
Prostanthera cuneata	alpine mint bush	
Ptelea trifoliata	hop tree	
Pyracantha species	firethorn	
Rosa canina	dog rose	Y
Rosa rubiginosa	sweet briar	Y
Rosa rugosa	Japanese rose	
Rosmarinus officinalis	rosemary	
Rubus fruticosus	blackberry	Y
Rubus idaeus	raspberry	
Spiraea japonica	spiraea	
Symphoricarpos albus	snowberry	
Tamarix ramosissima	tamarisk	
Autumn September–October		
Thymus species	thyme	
Viburnum lantana	wayfaring tree	Y
Viburnum opulus	guelder rose	Y
Weigela florida	weigelia	
Zauschneria californica	Californian fuchsia	

diversity of habitat types and structures in their initial designs (Chang *et al.*, 2021). Using more varied elements (water, sand or gravel substrates, soils with different nutrient status, etc.) and promoting a higher abundance of native herbaceous species in the design encourages ecological function and supports associated animal communities. Chang *et al.* (2021) found that spontaneously occurring plant species respond to a variety of artificially created habitat structures. These approaches work particularly well when the parks are closely connected to natural populations on the outside of the city.

Practical aspects of garden and park management that environmental horticulturists can employ to aid wildlife include the following:

- Reduce lawn mowing to allow greater plant diversity and for more prostrate forbs (herbaceous, non-grass species) to flower.
- Avoid the use of pesticides, or use more benign, more targeted or organic alternatives.
- Provide water features, ponds and wet/damp areas. Damp mud from drying puddles is used by *Delichon urbicum* (house martin) to build their nests in the eaves of houses.
- Create heterogeneous plant groups to mimic natural habitats – canopy, shrub and ground layers of vegetation. Introduce features that provide shelter, e.g. hedges, compost heaps, log piles and areas of long grass. As most gardens are part of a mosaic, not all 'habitat' types will need to be incorporated within a single garden.
- Increase the number of trees and the amount of foliage.
- Encourage at least some trees/shrubs to grow to their natural height and promote a canopy structure 2 m in height.
- Although perhaps unsightly, dead tree trunks can be left and then used to support more attractive features such as climbing *Rosa* (rose), *Clematis* or *Lonicera* (honeysuckle).
- Leave cut logs and branches to support wood-boring invertebrates and fungal microorganisms.
- Provide an area of the garden/park that is not intensively managed.

Table 5.8. Herbaceous perennial genotypes providing pollen and/or nectar resources to invertebrates within different seasons in the UK. (Modified with permission from the RHS 'Plants for Pollinators' plant list – Anon., 2025.)

Species	Common name	Native
Winter November–February		
Helleborus species	hellebore (winter flowering)	
Spring March–May		
Ajuga reptans	bugle	Y
Aubrieta species	aubretia	
Aurinia saxatilis	gold dust	
Bergenia species	elephant ear	
Caltha palustris	marsh marigold	Y
Doronicum species	leopard's bane	
Erysimum cv. Bredon	wallflower cv. Bredon	
Geranium species	cranesbill	
Geum rivale	water avens	Y
Helleborus species	hellebore (spring flowering)	
Iberis sempervirens	perennial candytuft	
Lamium maculatum	spotted dead nettle	
Primula veris	cowslip	
Primula vulgaris	primrose	Y
Pulmonaria species	lungwort	
Summer June–August		
Achillea species	yarrow	Y
Actaea japonica	baneberry	
Agastache species	giant hyssop	
Anthemis tinctoria	dyer's chamomile	
Antirrhinum majus	snapdragon	
Aquilegia species	columbine	
Armeria maritima	thrift	Y
Aruncus dioicus	goat's beard (male)	
Asparagus officinalis	common asparagus	
Astrantia major	greater masterwort	
Buphthalmum salicifolium	yellow ox-eye	
Borago officinalis	borage	
Calamintha nepeta	lesser calamint	
Callistephus chinensis	China aster	
Campanula species	bell flower	Y
Centaurea species	knapweed	
Centranthus ruber	red valerian	
Cirsium rivulare cv. Atropurpureum	purple plume thistle	
Crambe cordifolia	greater sea kale	
Cynara cardunculus	globe artichoke/ cardoon	
Cynoglossum amabile	Chinese forget-me-knot	
Dahlia species	dahlia (open-centre flowers)	
Dictamnus albus	dittany	
Echinacea purpurea	purple coneflower	
Echinops species	globe thistle	
Erigeron species	fleabane	
Eriophyllum lanatum	golden yarrow	
Eryngium species	eryngo	
Erysimum × allionii	Siberian wallflower	
Eupatorium cannabinum	hemp agrimony	Y
Eupatorium maculatum	Joe Pye weed	
Euphorbia species	spurge	

Continued

Table 5.8. Continued.

Species	Common name	Native
Foeniculum vulgare	fennel	
Fragaria × ananassa	strawberry	
Gaillardia × grandiflora	blanket flower	
Gaura lindheimeri	white gaura	
Geranium pratense	meadow cranesbill	Y
Geum species	avens	
Helenium species	Helen's flower	
Heliopsis helianthoides	smooth ox-eye	
Hesperis matronalis	dame's violet	
Inula species	harvest daisy	
Knautia arvensis	field scabious	Y
Knautia macedonica	Macedonian scabious	
Lathyrus latifolius	broadleaved everlasting pea	
Leucanthemum × superbum	Shasta daisy (open-centre flowers)	
Leucanthemum vulgare	ox-eye daisy	Y
Liatris spicata	button snakewort	
Limonium platyphyllum	broadleaved statice	
Linaria purpurea	purple toadflax	
Lychnis coronaria	rose campion	
Lychnis flos-cuculi	ragged robin	Y
Lysimachia vulgaris	yellow loosestrife	Y
Lythrum salicaria	purple loosestrife	Y
Lythrum virgatum	wand loosestrife	
Malva alcea	greater musk mallow	
Malva moschata	musk mallow	Y
Mentha aquatica	water mint	Y
Mentha spicata	spearmint	
Monarda didyma	bergamot	
Nepeta × faassenii	garden catmint	
Origanum species	oregano	
Paeonia species	peony	
Papaver orientale	oriental poppy	
Penstemon species	penstemon	
Persicaria amplexicaulis	red bistort	
Persicaria bistorta	bistort	Y
Polemonium caeruleum	Jacob's ladder	Y
Rudbeckia species	coneflower (open-centre flowers)	
Salvia species	sage	
Scabiosa species	scabious	
Sedum spectabile	ice plant	
Sedum telephium	orpine	Y
Sidalcea malviflora	checkerbloom	
Solidago species	goldenrod	
Stachys byzantina	lamb's ear	
Stachys macrantha	big-sage	
Stokesia laevis	Stokes' aster	
Tanacetum vulgare	tansy	Y
Telekia speciosa	yellow ox-eye	
Teucrium chamaedrys	wall germander	
Verbena bonariensis	purple top	
Veronica longifolia	garden speedwell	
Veronicastrum virginicum	Culver's root	

Continued

Table 5.8. Continued.

Species	Common name	Native
Autumn September–October		
Actaea simplex	simple-stemmed bugbane	
Anemone hupehensis	Chinese anemone	
Anemone × *hybrida*	Japanese anemone	
Aster species	Michaelmas daisy	
Ceratostigma plumbaginoides	hardy blue-flowered leadwort	
Chrysanthemum species	chrysanthemum (open-centre flowers)	
Dahlia species	dahlia	
Helianthus × *laetiflorus*	perennial sunflower	
Salvia species	sage	
Symphyotrichum novae-angliae	New England aster	

Table 5.9. Geophyte genotypes providing pollen and/or nectar resources to invertebrates within different seasons in the UK. (Modified with permission from the RHS 'Plants for Pollinators' plant list – Anon., 2025.)

Species	Common name	Native
Winter November – February		
Crocus species	crocus (winter flowering)	
Eranthis hyemalis	winter aconite	
Galanthus nivalis	snowdrop	
Spring March–May		
Crocus species	crocus (spring flowering)	
Muscari armeniacum	Armenian grape hyacinth	
Ornithogalum umbellatum	common star of Bethlehem	
Summer June–August		
Allium species	Ornamental and edible onion	
Autumn September–November		
Colchicum species	autumn crocus	
Crocus species	crocus autumn-flowering types	

- Retain areas that are not excessively 'tidy'.
- Ensure the majority of the landscape is composed of soft rather than hard landscape materials (plants and exposed soil, rather than concrete and slabs).
- Introduce nectar- and pollen-rich flowering plants – create floral-rich plant communities using annuals, biennials or herbaceous perennials.
- Promote complementary flowering plant types, e.g. night-scented, rosette, panicle and tubular forms, continuity of flowering and long flowering periods, seasonal variety and flowers that produce edible seeds.
- Generally, single open-flower forms are more advantageous than doubles or multiples, because the nectaries and stamens will not have been sacrificed in the breeding process for more petals, or any nectaries that are present will be more easily accessible.
- Promote a proportion of plants that tolerate predation effectively, e.g. *Rosa* cultivars that are not unduly damaged by aphids. Use alternative species/genotypes if plants are highly susceptible to herbivores.
- Certainly for gardens in western Europe, plant provenance does not appear to be of paramount importance in supporting wildlife because many herbivores select species within a genus or family, regardless of original provenance.
- Use companion planting to protect crops and other desirable plants. These companion plants either produce volatiles that inhibit pest insect species from visiting the locality or, alternatively, attract predatory insects that prey on the pests.
- Provide habitat and nesting sites for birds, small mammals and invertebrates.
- Consider 'green' infrastructure solutions – rain gardens and swales rather than underground pipes to draw water off.

Table 5.10. Biennial plant genotypes providing pollen and/or nectar resources to invertebrates within different seasons in the UK. (Modified with permission from the RHS 'Plants for Pollinators' plant list – Anon., 2025.)

Species	Common name	Native
Spring March–May		
Lunaria annua	honesty	
Smyrnium olusatrum	alexanders	
Summer June–July		
Alcea rosea	hollyhock (single flowers)	
Angelica archangelica	angelica	
Angelica gigas purple	angelica	
Angelica sylvestris	wild angelica	Y
Campanula medium	Canterbury bells	
Dianthus barbatus	sweet William	
Digitalis species	foxglove	
Dipsacus fullonum	common teasel	Y
Eryngium giganteum	Miss Willmott's ghost	
Matthiola incana	Brompton stock	
Myosotis species	forget-me-not	
Oenothera species	evening primrose	
Onopordum acanthium	cotton thistle	
Verbascum species	mullein	

- Leave dead plant parts intact until late in the dormant season, as these provide overwintering shelter and potential food sources for wildlife.
- 'No tillage', i.e. avoiding or reducing soil digging is favourable to some but not necessarily all species. Although there is a wide range of responses among different species, evidence from agricultural systems would suggest most organism groups have greater abundance or biomass in 'no-till' than in conventional tillage systems. This is particularly so for the larger soil organisms as they are more sensitive to digging and other soil interventions than the microfauna. Population changes may not just be due to physical disturbance of the soil but also the addition and burial of organic matter such as crop residues or compost and the changes in soil chemistry, water and temperature this may bring on.
- Invest in bird and bat nest boxes and place these in suitable locations within the landscape.

Community gardens and allotments

Although still understudied, these urban landscapes seem valuable for wildlife, due to the heterogeneous nature of the landscape and variety of vegetation forms found within them. Many community gardens encompass areas of bare open ground but also vegetated areas through both the food crops present and rougher areas of ruderal/shrub cover. The food crops themselves mimic different tiers of natural vegetation: orchard fruit, tree layer; fruit bush, shrub layer; and vegetables and herbs, ground layer vegetation. These cropping plants are often augmented by herbaceous flowering plants, as well as many native, wild plants at the less managed edges of the gardens. Despite representing highly managed plant communities, Lin *et al.* (2015) consider urban agricultural sites to exhibit high levels of biodiversity, and that they often exceed the biodiversity of other green space areas within the city. They attribute this largely to the varied vegetation structure, increased diversity of native plants and the reduction in impervious surfaces compared to other urban land forms – features they note that also contribute to important urban ecosystem services such as pollination, pest control and climate resilience.

Two factors seem to drive plant biodiversity: the particular environment the allotment or garden finds itself in (scale, land form and proximity to other green areas) and the level of management imposed on the landscape. Cabral *et al.* (2017) studying allotments in Germany showed that moderately managed plots (221 species recorded, including 110 spontaneous species) exhibited a higher biodiversity than those intensively managed. Leaving plots vacant for periods of time was also very beneficial. Allotment gardens have high levels of native plants (higher than domestic ornamental gardens, for example) and are considered as 'biodiversity hotspots' for native species within urban green infrastructure (Borysiak *et al.*,

Table 5.11. Annual plant genotypes providing pollen and/or nectar resources to invertebrates within different seasons in the UK. (Modified with permission from the RHS 'Plants for Pollinators' plant list (Anon., 2025.)

Species	Common name	Native
Summer June–August		
Anchusa azurea	large blue alkanet	
Anchusa species	alkanet	
Centaurea cyanus	cornflower	Y
Cerinthe major cv. Purpurascens	honeywort	
Consolida ajacis	larkspur	
Convolvulus tricolor	dwarf morning glory	
Coreopsis species	tickseed	
Cosmos bipinnatus	cosmea	
Cosmos sulphureus	yellow cosmos	
Cucurbita pepo	marrow/courgette	
Echium vulgare	viper's bugloss	Y
Eschscholzia californica	California poppy	
Gilia capitata	blue thimble flower	
Glebionis segetum	corn marigold	Y
Gypsophila species	baby's breath	
Helianthus annuus	sunflower (not pollen-free cultivars)	
Helianthus debilis	cucumberleaf sunflower	
Heliotropium arborescens	common heliotrope	
Lavatera species	mallow	
Limnanthes douglasii	poached egg flower	
Lobularia maritima	sweet alyssum	
Malope trifida	large-flowered mallow wort	
Nemophila menziesii	baby blue-eyes	
Nicotiana alata	flowering tobacco	
Nigella damascena	love-in-a-mist	
Papaver rhoeas	poppy	Y
Phacelia campanularia	Californian bluebell	
Phacelia tanacetifolia	fiddleneck	
Phaseolus coccineus	scarlet runner bean	
Reseda odorata	garden mignonette	
Ridolfia segetum	false fennel	
Sanvitalia procumbens	creeping zinnia	
Scabiosa atropurpurea	sweet scabious	
Tagetes patula	French marigold	
Tithonia rotundifolia	Mexican sunflower	
Tropaeolum majus	nasturtium	
Verbena × hybrida	garden verbena	
Autumn September–October		
Machaeranthera tanacetifolia	tansy-leaf aster	
Verbena rigida	slender vervain	
Vicia faba	broad bean	
Zinnia elegans	youth and old age	

2017). Interestingly though, Cabral *et al.* (2017) noted that community gardens tended to provide more diverse microhabitats than allotments, this being due to the former having a wider range of landscape types, such as habitats for amphibians (ponds and stone walls), areas with woody perennial plants, beehives and livestock shelters.

Although allotments are managed for home food production, a high proportion of allotment holders in Yorkshire (UK) expressed a strong interest in urban biodiversity irrespective of the age or management style of their plots (Turnbull, 2012). Pitfall traps were used across 42 plots to assess the invertebrates living on the soil surface. The data showed a trend for abundance actually increasing from rural to urban plots, with Coleoptera (beetles), Oniscidea (woodlice) and Araneae (spiders) constituting almost 80% of species caught. To some extent, taxa abundance varied with management styles too, although not always in a consistent manner. When plots were split based on either traditional (highly managed) or a wildlife-friendly management style, molluscs (snails and slugs) and woodlice were noted to be more abundant on the wildlife-friendly plots, whereas beetles were more abundant on the traditional ones. Spiders, opilione (harvestmen) and myriapods (millipedes and centipedes) on the other hand showed no significant difference. The majority of species surveyed were generalists (flexible opportunists) when ranked within their taxa, and whilst spider populations fitted with the concept of intermediate levels of disturbance increasing biodiversity, the same was not true for the woodlice and beetle species present. Although there was a trend for increases in species diversity along the rural–urban gradient, individual sites also strongly influenced biodiversity, further highlighting the importance of site history and management (Turnbull, 2012). A key point to note from this data is that rural areas dominated by intensive agriculture practices may themselves be relatively poor in supporting a diverse range of species.

Recent increased interest in converting derelict, brownfield land and using it for food production led to one study examining the implications of these transitions on invertebrate biodiversity (specifically arthropod generalist predators) (Gardiner *et al.*, 2013). This study in Ohio (USA) showed that the abundance of ladybirds and hoverflies and the activity of predatory ground beetles, ants and wolf spiders were roughly equivalent among the vacant plots and the urban allotment sites. The conversion of brownfield sites to allotments, however, had a negative effect on abundance levels of long-legged flies and the density levels of small spiders and harvestmen. In contrast, the abundance of flower bugs and the number of active rove beetles were greater within the urban allotments compared to the derelict land.

Table 5.12. Examples of garden plants attractive to hummingbirds in the USA (Modified with permission from Anon., 2015a, 2015b.)

Species	Common name
Aesculus californica	California buckeye
Aesculus pavia	red buckeye
Aesculus x *carnea*	red horse chestnut
Aesculus x *carnea* cv. Briotii	ruby red horse chestnut
Agapanthus orientalis	African lily
Asclepias tuberosa	Indian paintbrush
Buddleia alternifolia	butterfly bush
Caesalpinia gilliesii	poinciana
Caesalpinia. pulcherrima	poinciana
Callistemon citrinus	red bottlebrush
Campsis radicans	trumpet vine
Cephalanthus occidentalis	buttonbush
Chilopsis	desert willow
Citrus species	citrus species – orange, lemon, etc.
Cuphea ignea	cigar plant
Erythrina herbacea	cardinal spear
Hamelia patens	firebush
Hibiscus species	swamp mallow
Jatropha integerrima	peregrina
Justicia californica	chuparosa
Malus floribunda cultivars	crabapple
Kalanchoe blossfeldiana	kalanchoe
Lantana species	lantana
Lonicera sempervirens	honeysuckle
Odontonema strictum	firespike
Rhaphiolepis indica	Indian hawthorn
Schlumbergera bridgesii	Christmas cactus
Strelitzia reginae	bird of paradise

Table 5.13. Description of the management regimes and habitat in Yarkon Park, Tel Aviv. (Modified with permission from Shwartz *et al.*, 2008.)

Management regime	Management Intensity[a]	Habitat description	Plant understorey	Irrigation	Human activity[b]
Intensive	++++	Open grass lawn, scattered trees	Mainly exotic	Yes	High
Moderate	+++	Gardens, orchards, agricultural farm	Partly exotic	Yes	Moderate
Light	++	Syrian ash, pine, Tamarisk coppices, Fallow fields with annuals	Mainly native	No	Low
Unmanaged	+	*Eucalyptus* groves, fallow fields, annual ground cover	Mainly native	No	Low

[a]Landscaping maintenance operations such as grass cutting, pruning, fertilising levels range from high (++++) to low (+).
[b]As estimated by the quantity and development of pathways (paved vs unpaved), number of people observed eating or playing within a 100-metre radius during five sampling minutes in each location (meaning 5 for low, 17 for moderate and 37 for high) and the number of waste bins (relative rankings of high, moderate or low).

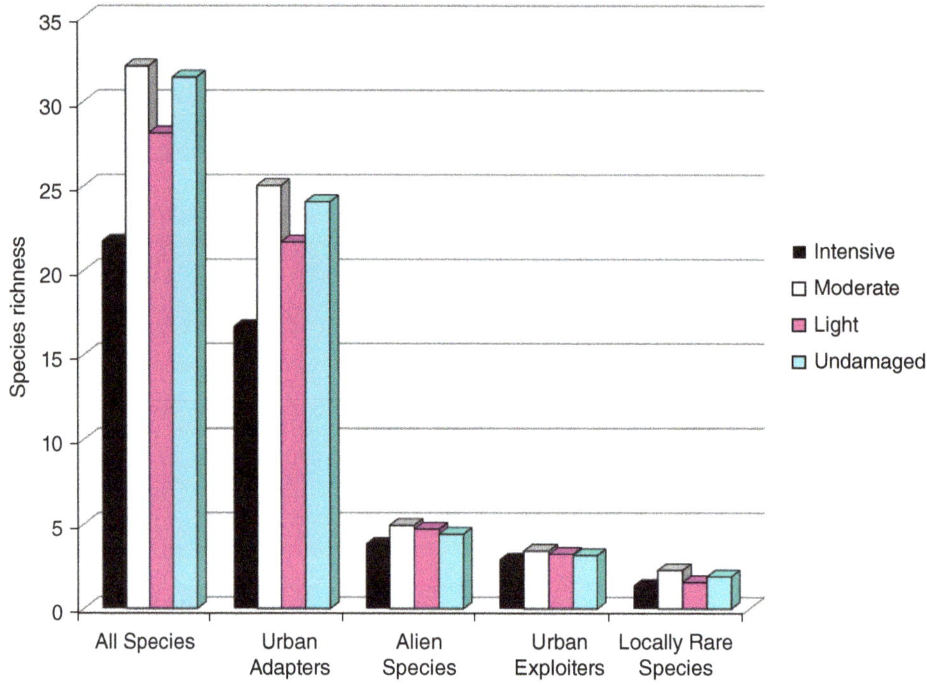

Fig. 5.15. Bird species richness in an urban park (Tel Aviv, Israel). (Modified with permission from Shwartz *et al.* 2008.)

Overall, although allotments and community gardens may impact on biodiversity, the response appears very taxa dependent, and as long as allotment sites are carefully managed, they can be positive locations for urban wildlife.

Urban trees and woodland

The urban forest is important for supporting urban biodiversity and indeed biodiversity across a wider context. The urban forest can comprise a significant percentage of a nation's tree canopy. In the USA, trees in urban counties account for nearly 25% of the country's total tree canopy cover (Dwyer *et al.*, 2000). In Flanders (Belgium), park trees provide a significant component of the region's canopy cover and this is largely composed of native species such as *Fagus sylvatica* (beech), *Acer pseudoplatanus* (sycamore), *Quercus robur* (English oak) and *Fraxinus excelsior* (ash) (Cornelis and Hermy, 2004). Trees either as individuals or as groups provide networks for species distribution and migration and act as refuge sites across the city landscape. Research suggests that urban forests contain a significant percentage of species that were endemic to an area before it was developed. In Guangzhou City (China), urban tree diversity was considered high (Jim and Liu, 2001). A survey of 115,000 trees in parks, university grounds and along city streets recorded over 250 different species. Native broadleaves dominated, with the top three abundant species recorded as *Ficus virens*, *Caryota mitis* and *Melaleuca leucadendra*. It is thought that plant species richness in Guangzhou City actually exceeds the degraded forests of the surrounding countryside (Jim and Liu, 2001). In Hong Kong (China), a review of the tree stock indicated 232 species present, but natives were the minority here, only contributing 69 of the species recorded. Urban forests also may be home to endangered or conservation priority species. In Sweden, Stockholm County contains two-thirds of the red-listed species despite being densely populated; and includes rare plant species such as the fern *Dryopteris cristata* and the moss *Buxbaumia viridis* which require damp woodlands (Colding *et al.*, 2003).

Trees are an essential element in city planning but the value of native and non-native tree species needs to be explored more deeply. Chong *et al.* (2014) suggest trees, irrespective of whether native or non-native, had a positive influence on the diversity of urban birds in Singapore by providing an extra dimension to an otherwise planar landscape. Even non-native trees provided more opportunities to local bird populations than alternative types of green space, such as grass, largely because many of these trees were tropical, rainforest-adapted species and thus could still provide a number of the services to the local bird population (although biodiversity was still optimised by stands of native trees, where these are left standing). Whilst evaluating tree population in commercial office areas of Washington State (USA), Dyson (2020) found bird species richness and community dynamics were positively influenced by the greater presence of native conifers, including the retention of pre-development tree stands with taller trees and denser groves. Measures of the native bird community were negatively influenced by higher non-native tree numbers.

Older, veteran trees provide greater opportunities than younger, 'still extending' trees. Thus, during urban expansion or densification priority should be given to conserving large veteran trees and mature stands of woodland. Work in Los Angeles (USA) (Clarke *et al.*, 2013) demonstrated that tree species diversity was correlated with older, high-income residential areas, and that to help extend the opportunities for wildlife, new tree plantings should, conversely, be focused on low-income neighbourhoods to help mitigate gaps in the urban forest.

Retaining existing woodland stands (patches) of remnant forest is key to encouraging bird biodiversity in most scenarios. However, as Catterall (2004) found the interrelationships between species can be exceedingly complex and the inadvertent responses to actions difficult to predict. She found that spatial scaling in urban design was important in conserving the small-bodied, foliage-feeding birds of the native *Eucalyptus* forests. She found that as urbanisation took place in Brisbane (Australia) the remaining vegetation was composed of small remnant vegetation patches (<5 ha) or well-vegetated suburbs through the planting up of gardens and parks. Local authorities and residents encouraged the retention of small, vegetated patches to provide sanctuaries for wildlife. One consequence of this fragmentation of the *Eucalyptus* forests and increase in plant diversity through the garden vegetation though was an increase of larger bird species, such as *Manorina melanocephala* (noisy miner). Increases in noisy miner bird numbers, however, correlated with decreases in the smaller foliage-feeders. Noisy miners drive other birds away that enter their territory using vigorous coordinated attacks and loud strident vocalisations. Indeed, targets are not limited to birds, with mammals and reptiles also sometimes wounded or killed. These were traits, Catterall suggested, that rather than protecting the native birds of the extensive *Eucalyptus* forests, meant the retention of smaller and mixed patches of vegetation (due to the fact that they included nectar-rich native plant species that were attractive to the noisy miner) had a negative effect on the desired species – a lesson that positive motives on promoting native biodiversity can have a different effect from that intended.

Invertebrates improve the health and functionality of urban forests. Their key roles and services include pollination of wild and cultivated plants, predation including on potentially pest species, herbivory, seed and microorganism dispersal and organic matter decomposition (Kotze *et al.*, 2022). Urban forests provide significant nesting resources for social and solitary wild bees. The queens of social bumblebees construct nests at the edge of forest clearings, and solitary bees nest in logs, snags and stumps. Leaf-cutter bee, collect leaves from a variety of trees and shrubs to partition their brood cells in the nest (MacIvor, 2016). Many bees, wasps and beetles depend on dead wood for nesting and laying eggs and are potentially limited by these resources in cities, largely because they exist almost exclusively in remnant urban forests.

Many invertebrate predators are associated with the woodland floor, including carabid and rove beetles (Carabidae and Staphylinidae), ants (Formicidae) and spiders (Araneae). Not only do these species play a crucial role in a traditional ecosystem's natural ecology but they can help pest management in urban forests. Warmer climates, disrupted habitats and greater food resources associated with the urban environment can encourage rapid populations of certain species that then reach pest proportions. Pest outbreaks in North America have includes the northward expansion of *Agrilus planipennis* (emerald ash borer) into Canadian cities and towns (Sun *et al.*, 2024) and *Adelges tsugae* (hemlock woolly adelgid) across the northeastern USA (Cornelsen *et al.*, 2024). Traditional pest species such as mosquitoes (Culicidae), beetles (Coleoptera) and cockroaches and termites (Blattodea) threaten stored products, public health or building structures. Yet robust woodland ecosystems with good predator populations help keep such pest species in check.

Urban forests too support large numbers of herbivorous insects and other invertebrates. These in turn provide food for birds, amphibians and mammals, as well as acting as hosts for invertebrate parasites and parasitoids. The key woodland herbivores include the juvenile and adult Lepidoptera (moths and butterflies), Coleoptera (beetles), Diptera (flies), Hemiptera (bugs) and Orthoptera flies (grasshoppers and crickets). Some are specialist feeders on certain woodland host plants, while others have the capacity to feed on a wide array of plant taxa.

Invertebrates are also ecological engineers, being important in the distribution and dispersal of plant seeds, fungal spores, bacteria and other microorganisms through either intentional (e.g. transporting fruiting bodies of fungi) or incidental means (e.g. through digestion and excretion or via surface adhesion) (Kotze *et al.*, 2022). A range of life forms recycle the nutrients from plants. The macro-detritivores (e.g. earthworms, woodlice and millipedes) break down leaf litter into smaller pieces (comminution) making it accessible to micro-detritivores (e.g. springtails and oribatid mites), bacteria and fungi (Ossola *et al.*, 2017). More complex decomposer communities can accelerate both carbon and nitrogen loss back to the atmosphere by on average 11% (Handa *et al.*, 2014; Meyer *et al.*, 2020).

Transport corridors

Road and rail traffic do not tend to mix well with wildlife! Road traffic can have a significant toll on the populations of some vertebrate species, particularly slow-moving creatures including toads and frogs. Wider and busier roads have been shown to be problematic for both bird and butterfly species. Birds appear to be disturbed by the noise and vehicle movement, and even the air turbulence causes problems for butterflies. Studies on flying insects suggest roadkills range from 21 to 27 roadkills km^{-1} day^{-1} depending on road type (and 0.5 to 10 km^{-1} day^{-1} for Lepidoptera alone) (Phillips *et al.*, 2020). Providing trees that overhang the edge of roads and building roads where trees can be incorporated into the middle reservation are deemed positive interventions to help aerial animals traverse roads and hence migrate across the wider urban matrix. There are some advantages too for humans, with vegetation planted down the middle of highways reducing glare from headlights from oncoming traffic. This needs to be balanced, however, with the potential for more serious injuries to drivers/passengers when cars collide directly with trees.

Low-growing woody vegetation, i.e. shrubs and hedgerows, immediately in the vicinity of roads (at least those with heavy traffic volumes) have a negative effect on bird populations, through increasing the number of fatalities due to vehicle impacts. Urban specialists such as *Turdus merula* (blackbird), *Passer domesticus* (house sparrow), and *Prunella modularis* (dunnock), which have low lines of flight, are placed in danger when flying from one patch of shrubbery to another across a roadway. Those that tend to patrol along hedgerows and ditches at 2–3 m above the ground such as *Tyto alba* (barn owl) are also particularly vulnerable where such features lead up to roadways. Careful design of roadside vegetation can offset some of these problems through avoidance of 'desire lines' that transect directly across a road. For example, staggering and extending the distance between patches of shrubbery at either side of a road will encourage some species to fly higher, and the placement of vegetation further away from the road will improve the line of vision, thus helping birds perceive any oncoming traffic.

Roadside verges are underutilised nature reserves. Through better management these could encourage more pollen- and nectar-rich flowering forbs that are beneficial to insects. Different mowing management treatments have been evaluated in the Netherlands (Noordijk *et al.*, 2009). Mowing of roadside verges twice per year with the grass cuttings (hay) being removed increased the number of flowering plant species (forbs) and the total number of inflorescences observed compared to other treatments imposed. This had positive effects on the abundance of insects and the frequency at which flowers were visited. Not all areas should be mowed down though, as leaving some patches of longer grass allow invertebrates to remain present until the following year.

The value and disadvantages of roadside locations for pollinating insects (Table 5.14) was reviewed by Phillips *et al.* (2020), with a range of management recommendations to improve insect conservation (Table 5.15).

Brownfield sites, vacant plots and wastelands

Brownfield sites, vacant plots and wastelands are not areas that would necessarily normally fall under the jurisdiction of a horticultural office, but the value of such brownfield sites needs to be

Table 5.14. Advantages and disadvantages for pollinators in roadside verges. (Modified with permission from Phillips *et al.*, 2020.)

Issue	Impacts
Advantages	
New habitat, especially in urban context where green space is limited	New opportunities for pollinators (and other wildlife).
Different location, soil type, aspect, etc. provide a range of ecological niches	Wider variety of species accommodated.
Apart from traffic, areas may have low human disturbance	Life cycles less likely to be disturbed.
Potential for long ecological corridors	Allows for species migration in parallel to the direction of the road.
Disadvantages	
Vehicle impacts	Insect mortalities occur when flying across roads.
Street lights	Attracts various types of flying insects, disrupts behaviour (foraging, breeding and feeding) and can increase predation (e.g. from bats). The type or light (wavelength and intensity) affects these factors (e.g. yellow light seems less attractive to some taxa than white).
Noise	May affect larval stages (higher heart rates in *Danaus plexippus*, monarch butterfly caterpillars), although these may become habituated to noise with time.
Turbulence	Slower traffic speeds allow for greater foraging time (insect is on the flower seeking nectar/pollen for longer).
Air pollution	Air pollution from traffic can affect pollinator behaviour and perception of plant odours. Fumes degrade floral perfumes making it more difficult for pollinators to find flowers. Poor-quality air may affect insect memory – exposure to diesel fumes reduced honeybees' ability to recall an odour after 72 h by 44%.
Heavy metals	The presence of roads has been linked to increased Na, Ca and Mn concentrations in the bodies of most insect studies and to higher As in beetles (Coleoptera). City centre sites and high traffic levels contributed to greater levels of Cd, Co, Mn, Cu, Ba, Fe, Ni, Sr and Pb in honeybees.
De-icing salts	Na from de-icing salt can accumulate in plants by the roadside. *Danaus plexippus* (monarch) butterflies reared on roadside milkweed plants had six times more Na in their abdomens and significantly lower survival rates compared to control plants. *Pieris brassicae* (large white) butterflies fed on an artificial diet that varied in Na had significantly lower survival on the high-Na diet.

considered in the wider context of urban green space and biodiversity conservation, not least as environmental horticulturists will often be called on to help in the 'development' of ex-industrial 'brownfield' sites. The reality is that these 'waste' areas are frequently quite species-rich and provide opportunities for a spectrum of species including those native to the area, or from agri-rural environments as well as ruderal, non-native species (Kelcey and Müller, 2011). Wastelands have been demonstrated to be a significant original habitat for species in cities. They are very atypical to other urban open/green spaces as they offer habitat for numerous rare and endangered species. Previous land activities and their legacy often dictate the vegetation dynamics and wider biological significance of such sites. They are characterised by a high diversity of anthropogenic substrates (nutrient-rich soil in places but also nutrient-poor rubble, ballast and brick), restricted public access and limited human activities, and spontaneous vegetation cover with irregular removal. Brownfield soils are largely, but not exclusively, neutral to alkaline, free-draining, well aerated and low in organic matter.

Brownfield sites are in a process of transition – with a key ecological driver being plant species succession. These areas are characterised by the abandonment of management, which leads to the

Table 5.15. Key management recommendations for roadside verges. (Modified with permission from Phillips *et al.*, 2020.)

Management recommendation	Description	Benefit Positive benefit (✓) Inconclusive or not fully established (?)	Practical implications
Create high-quality habitats	Follow best management practice. For example, species-rich grasslands can be created by ensuring low soil fertility (e.g. removing/not adding topsoil) and sowing native, local provenance wildflower seed.	Pollinator density ✓ Pollinator species richness ✓	Nutrient-poor habitats may require less mowing to maintain safe vegetation height. May have more bare ground which might affect aesthetics and soil stability.
Control/remove invasive, non-native plant species	Maintain high-quality habitats by ensuring that they do not become dominated by invasive, non-native plant species.	Pollinator density ✓? Pollinator species richness ✓?	May be costly.
Reduce mowing frequency to 0–2 cuts/year		Pollinator density ✓? Pollinator species richness ✓?	Long grass may not be popular in residential areas. Provide compromise with aesthetics and biodiversity by reducing number of cuts as much as possible.
Avoid mowing between spring and late summer	Avoid mowing between spring and late summer, when pollinators are most active.	Pollinator density ✓ ? Pollinator species richness ✓? Reduced mortality of pollinators, eggs and larvae ✓?	Not always plausible due to minimum safety requirements, for example, to retain driver visibility. A 1 m cutting strip alongside the road may be a compromise.
Remove cuttings		Flower species richness ✓ Flower density ✓? Pollinator density ✓?	May improve road verge aesthetics. Removal of nutrients may reduce subsequent vegetation growth and the need for verge cutting. Appropriate location is needed to dispose of cuttings.
Use mosaic mowing/ management	Apply different mowing/ management regimes along the length and/or width of a road verge. For example, split road verges into three sections:	Habitat diversity; balances the needs of different pollinator taxa; allows areas where pollinator eggs and larvae can persist. ✓?	Can be based around current number of cuts, so no increase in costs. More complex management requires greater training of ground staff. Mowing verge edges more frequently will increase visibility and safety for road users and perceived aesthetics around tidiness.

Continued

Table 5.15. Continued.

Management recommendation	Description	Benefit Positive benefit (✓) Inconclusive or not fully established (?)	Practical implications
	Section 1 (front of verge): mow twice per year (in early summer and from late summer onwards). Section 2 (middle of verge): mow once from late summer onwards. Section 3 (back of verge): leave unmown or cut on a multi-year rotational basis from late summer onwards.	Frequently mowing verge edges may reduce pollinator road crossing and roadkill. ✓ ?	
Create habitat diversity across the road network	Apply different mowing/ management regimes across different road verges (e.g. mow the opposite sides of roads at different times).	Habitat diversity; balances the needs of different pollinator taxa; allows areas where pollinator eggs and larvae can persist. ✓?	As for mosaic mowing/ management (above).
Identify pollinator roadkill hotspots and apply reduction measures	Use measures to reduce pollinator roadkills at roadkill hotspots. For example, improve road verge habitat quality and reduce traffic speeds both spatially (specific roads) and temporally (e.g. during key migration, for example *Danaus plexippus*, Monarch butterflies, in North America).	Butterfly roadkill reduced. ✓?	Initial investment to identify pollinator roadkill hotspots, improve verge habitat quality and enforce speed limits. Reducing speed limits is often impractical and probably socially contentious for the purpose of invertebrates, although education policies may help.
Reduce impacts of street lighting	Remove street lighting where possible by reducing the number of fixtures, the light intensity and the duration that street lighting is on, and use least harmful lighting technologies.	Reduction in pollinator predation. ✓ Reduces suboptimal behaviour ✓	Needs to be balanced against safety concerns.
Prioritise enhancement of road verges with the greatest capacity to benefit pollinators	Focus beneficial management on road verges with the greatest capacity to benefit pollinators, i.e. - wider road verges; - roads with lower traffic volumes; - road verges of particular conservation interest; - road verges in landscapes with few other suitable habitats for pollinators; - road verges that connect conservation areas and disproportionately increase habitat connectivity at a landscape scale.	Benefits for pollinators per unit investment ✓?	Requires initial investment to identify priority road verges and additional costs and training of ground staff to deviate from standard management.

coexistence of horticultural flora once cultivated with spontaneous flora. In young brownfield sites, the pioneer, early successional vegetation attracts bird species that prefer open habitat. It is also noted for its capacity to support ground beetles and polyphagous species (i.e. those with a catholic diet) that have reproduction strategies adapted to changing food availability. Rubble left over from demolished buildings results in patches denuded of vegetation and a prevalence of dry sandy or stony soils and their accompanying warm, localised microclimates; conditions conducive to ground beetle species such as *Cicindela campestris* (green tiger beetle) and *Cicindela sylvatica* (heath tiger beetle), in addition to solitary bees such as *Andrena barbilabris*. Brownfield sites also attract stress tolerator plant genotypes, such as orchids, due to the residual contamination from chemicals associated with the previous industrial processes. Nutrient-poor soils with high metal content enhance the competitiveness of plants with stress tolerator traits over ruderal and competitor species. Not all brownfield sites have nutrient-poor soil, however, and the heterogeneity of patches with different nutrient pools adds to the biodiversity. Indeed, soil fertility is often key in determining species dominance in these patches. In western Europe, woody plants such as *Populus tremula* (aspen) exploit nutrient-poor sites, whereas *Sambucus nigra* (elderberry) and *Salix caprea* (goat willow) dominate the moderately fertile locations. Del Tredici (2011) describes this suite of species occupying vacant plots rather protractedly as a 'cosmopolitan assemblage of early-successional, disturbance-tolerant species that are pre-adapted to the urban environment'!

In later stages of succession, perennial species gain dominance with tall grasses, meadow perennials and young, pioneering tree species becoming prevalent. These tend to increase the humidity of the site and start to lay down organic matter to the soils. Faunal communities may also change at this point, for example, with birds that prefer scrub and open woodland gaining traction. Machon (2021) claims the biodiversity of brownfield sites peaks when they are between 15–20 years into this successional process – biodiversity again being richest when sites are larger and well connected to other green spaces, and human interference remains low.

Brownfield sites are under socio-economic pressures due to redevelopment for housing and industrial uses. The impact of this on their biodiversity is poorly understood. For example, it is unknown how much open area needs to be preserved and indeed whether conservation is possible without completely excluding economic development (Kattwinkel *et al.*, 2011). Simulation studies using species distribution models at a city scale for plants, grasshoppers and leafhoppers demonstrated that overall dynamic land use (i.e. the 'cycle' of construction and development of new building on some sites but the demolition of buildings with plots left vacant on others) supports urban biodiversity in terms of species richness and rarity. From this modelling it was apparent that setting aside brownfields before redevelopment for a period of on average 15 years supported the highest conservation value. As such, Kattwinkel *et al.* (2011) recommend integrating the concept of 'temporary conservation' into urban planning when dealing with industrial and business areas. This concept requires habitat to be destroyed by redeveloping brownfield sites and converting them to built-up structures but simultaneously creating new open spaces due to abandonment of urban land uses at other locations. They claim this maintains a spatio-temporal mosaic of different successional vegetational stages ranging from pioneer to pre-forest communities. Such approaches enable urban populations of some unique species to remain viable, over the long term.

Overall though, the value of brown field sites is still not widely appreciated, except in the ecological science community. As a result, biodiversity issues around brownfield sites are often discounted in the redevelopment process. Harrison and Davies in 2002 stated that many conservation professionals involved in urban redevelopment were struggling to promote a pro-active approach to secure environmentally sensitive development. This is probably still largely true, although the concepts about (urban) rewilding may be changing attitudes – at least in the eyes of some of the public (Masood and Russo, 2023) if not necessarily developers. In a survey of attitudes to urban forests in Germany (Lippert *et al.*, 2022), the spontaneous ones that developed on wasteland were appreciated by the population (scored 0.4 for positive attitude and 1 for emotion, out of a range of –2 to +2), but less so than those that were natural forest remnants (scored 2 for positive attitude and 2 for emotion).

Conventional walls and roofs

Built infrastructure, although an adverse environment for most species, is not without life. Hard surfaces dominate due to the high density of buildings, but these vary in their construction and integrity. Modern city centre buildings constructed largely of glass or metal provide limited opportunities for wildlife, unless specific interventions relating to green roofs or walls are implemented. On more traditional brick and stone surfaces, on the other hand, non-vascular taxa such as lichens and moss can gain a foothold, even if the surface is relatively smooth. Non-mortared structures and those buildings, walls, pavement or roadways that are beginning to wear provide niches for vascular plants. Compared to newer walls, older walls and mortar tend to have more plant species because they have weathered more, have had time to neutralise alkaline salts and have accumulated some organic material in crevices. Nevertheless, these remain challenging environments due to the relative limited availability of soil and water and the extreme temperatures experienced. As temperature and moisture profiles strongly influence the potential for life, the predominate climate determines the range of species present. Maritime climates with high rainfall and low fluctuations in temperature promote a wider diversity and abundance of wall flora, and relatively large plant such as ferns and even *Buddleia davidii* (butterfly bush) can be commonplace on crumbling wall systems. Patterns of plant types found in walls and other hard surfaces have been classified (Lundholm and Richardson, 2010). In Atlantic and central Europe hemicryptophytes (herbaceous perennials with their buds at or near the soil surface) dominate, whilst in Mediterranean Europe, chamaephytes (woody species with resting buds at or near the soil surface) are more common. In India, on the other hand, therophytes (annual species) tend to be better adapted, whereas in Israel, phanerophytes (woody species with resting buds above the soil surface) were observed to be the dominate group colonising walls.

Green roofs and green walls

One of the key drivers for green roofs (see Chapter 10) has been habitat provision for wildlife. The opportunities for wildlife are influenced by factors such as the design of the green roof, substrate composition and depth, degree of exposure the roof experiences, water availability, the wet/dry cycles of the substrate and the location with respect to other nearby green spaces. For example, a study in Melbourne, Australia (Dromgold *et al.*, 2020) suggested invertebrate communities on green roofs were determined largely by their surrounding environment. So, their effectiveness as invertebrate habitat is highly dependent on their location (proximity to other green spaces) and their horizontal and vertical connection to other habitats. Green roofs can be 'extensive' with shallow substrates and limited plant diversity (often only using mats dominated by *Sedum* spp.); 'intensive' with deeper substrates that support more diverse plant communities (e.g. beginning to mimic dry steppe or wildflower meadow communities); roof gardens with significantly deep substrates (e.g. 0.5–1 m depth) and artificial irrigation-supporting herb, shrub and even tree species; and finally, bio-roof or brown roofs, which may be designed to replicate specific or unique brownfield wildlife habitat.

Most green roofs are effectively ecological islands, with continuity of habitat between them generally not feasible. Green walls may allow some connection with roofs and thus offer potential for green networks and corridors, but the vegetation type is often very different, and this still makes the assumption that the migrating animal species can climb! Species that can fly or are able to propel themselves via wind currents such as fungal spores, lightweight plant seeds, or young 'parachuting' spiders will be able to colonise new roofs. Human-assisted introductions are possible, as are inadvertent introductions, for example via substrate or soil imported on to the roof. In one of the small green roofs at the University of Sheffield, UK, a frog took up residence and was possibly brought in with a damp vegetation mat or suchlike. Wooster *et al.* (2022) studying green roofs in Australia indicated that these roofs supported four times the avian, more than seven times the arthropod and two times the gastropod diversity of that of a comparable conventional roof. Only green roofs attracted locally rare species including *Amegilla ingulata* (blue-banded bee) and *Scutiphora pedicellata* (metallic shield bug). Green roofs have been shown to support different arthropod (Durà *et al.*, 2023), including different beetle communities (Gonsalves *et al.*, 2022) compared to those present on ground-level green spaces. Moreover, green roofs designed intentionally to support wildlife (bio-roofs) can reliably increase insect diversity compared to simpler designs of green roof.

Brenneisen (2006) argues that natural soils and substrates are important in colonising a roof system. The adaptation of spider and beetle fauna to natural soil and other substrates such as sand and gravel from riverbanks seemed to be a key factor to success in a number of Swiss green roofs. This research showed that near-natural habitats can be established on roofs, with different roofs in Switzerland partially replicating certain microhabitats. This included the establishment of invertebrate populations associated with riverbanks, rock scree and debris, alpine and dry grassland habitats. Roofs with restricted drainage have been used to replicate the conditions typical of wet or dry meadows, heathland and acidic moorland. Even the inclusion of ponds and shallow moisture-retentive areas can encourage marsh and wetland species.

Local governments in different regions of both Switzerland and Germany have stipulated that new buildings of a certain size, and which have a flat roof, are legally obliged to put a green roof system in place. This has been a catalyst for the expansion of the green roof industries in these countries. In Switzerland, as a means to promote biodiversity, roofs exceeding 500 m^2 in area also need to be composed of appropriate natural soils from the surrounding region and must be contoured to provide varying depths of substrate to support different plant and invertebrate communities. One of the oldest green roof complexes in Switzerland is the Wollishofen water plant in Zurich, which was originally installed in 1914 to keep the water within the building cool. These roofs were composed of native soils and plant species and, unlike the surrounding agricultural land which has subsequently become intensively managed, have succeeded in retaining their complex and diverse plant communities. This green roof has now become famous due to its role as a refuge for 175 recorded plant species, including a number of Red Data book (i.e. endangered) terrestrial orchids. A number of orchid species have been found growing on this and other European green roofs (Table 5.16) (Fig. 5.16). Unlike many modern green roofs, the Wollishofen roof had local soil placed over a layer of gravel, resulting in a substrate that alternates between high water retention and much drier periods, thus providing conditions not dissimilar to those found in semi-natural habitats such as wet meadows and moorland. Such conditions were important factors in conserving the typical local and regional biodiversity on these green roofs (Brenneisen, 2003).

Table 5.16. European orchid species observed on green roofs (Brenneisen, 2003)

Orchid species	Common name
Dactylorhiza fuchsia	common spotted orchid
Dactylorhiza incarnate	early marsh orchid
Dactylorhiza majalis	western marsh orchid
Dactylorhiza sambucina	elder-flowered orchid
Epipactis palustris	marsh helleborine
Gymnadenia conopsea	fragrant orchid
Listeria ovata	twayblade
Ophrys apifera	bee orchid
Ophrys sphegodes	early spider orchid
Orchis mascula	early-purple orchid
Orchis militaris	military orchid
Orchis morio	green winged orchid
Platanthera bifolia	lesser butterfly orchid

Some researchers and land managers have advocated that green roofs can play an important role in preserving wildflower species that are under pressure in the wider rural environment, due to loss of habitat, soil nutrient enrichment and increased competition from more aggressive species or changes in land management practices – in the latter case, for example, the loss of traditional hay meadows to silage production. The paradox is that urban green roofs may facilitate the conservation of rare plant species normally associated with the traditional or rural, agrarian landscape. The fact that green roofs are a highly artificial system means that, where key factors such as soil nutrient levels or moisture can be manipulated, there are opportunities to recreate important ecological niches which have become rare in natural or agro-ecosystems. Working in the Mediterranean climate of Italy, Benvenuti (2014) monitored plant population dynamics on a semi-extensive 200 mm deep substrate roof, with irrigation being used during establishment and the more severe dry periods. He found that by planting communities of plants that flowered throughout the season continuous food supplies could be provided to a number of different insect taxa. The adverse environment encountered on the roof to some extent could be countered by a reliance on geophytes (bulb species) which survived from one year to the next, due to their dormant phase aligning with the driest and hottest mid-summer periods.

Fig. 5.16. A number of 'iconic' orchid species can appear on green roofs including left *Ophrys apifera* (bee orchid) and right *Dactylorhiza fuchsia* (common spotted orchid).

In conclusion, Benvenuti considers that wildflower roofs may contribute to ecological stepping stones across the city matrix and partially substitute for the loss of green space at ground level. The relatively 'unutilised' potential of roof space should be given greater priority in future to help address biodiversity loss in city centres. In contrast, Mayrand and Clergeau (2018) feel the ecological value of green roofs and walls is overstated. They believe their value as urban wildlife connectivity remains questionable due to common factors such as limited patch size, habitat quality at the building scale, and the limited numbers of green walls and roofs present across the city (see points above about ecological stepping stones).

One of the initiatives behind the advent of green roofs in the UK in the 1980s was the desire to recreate brownfield habitat that was being lost at ground level due to the economic 'boom' that took place during that decade. This resulted in ex-industrial vacant spaces within cities being developed for new office blocks and housing. Brownfield sites are important habitat for certain taxa of invertebrates (see above), but also one European bird species, namely *Phoenicurus ochruros* (black redstart). This species perhaps more than any other has become synonymous with the ecological green roof movement. Within the UK, the black redstart is at the edge of its northern distribution, and so having a resident breeding pair using your green roof as feeding habitat is seen as the ultimate symbol of success! As more brownfield sites were lost to development, the greater the pressure became to initiate green roofs as alternative habitat for black redstarts.

The key criteria that green roofs can provide for birds are water, food, shelter and roosting opportunities. Some species may also choose to nest on a green roof depending on their normal requirements, for example, the gravel substrates employed may replicate the stony beaches that species such as terns, gulls and ringed plovers normally utilise (Table 5.17). Food sources commonly include berries, seeds and invertebrates attracting in by the vegetation and substrates. As highlighted above, however, not all green roofs have the same ecological value for feeding or breeding birds. The opportunities presented for colonising bird species vary

Table 5.17. Bird species that have been reported to breed on green roofs. (Modified with permission from Fernandez-Canero and Gonzalez-Redondo, 2010.)

Species	Common name
Anas platyrhynchos	mallard
Anthus pratensis	meadow pipit
Alauda arvensis	sky lark
Branta canadensis	Canadian goose
Carduelis chloris	greenfinch
Charadrius dubius	little ringed plover
Charadrius hiaticula	ringed plover
Charadrius vociferus	kildeer
Columba livia domestica	feral pigeon
Columba palumbus	wood pigeon
Corvus cornone	carrion crow
Cyanistes caeruleus	blue tit
Falco peregrinus	peregrine falcon
Falco tinnunculus	kestrel
Fringilla coelebs	chaffinch
Galerida cristata	crested lark
Haematopus ostralegus	Eurasian oystercatcher
Laurus canus	common gull
Motacilla alba	pied wagtail
Muscipapa striata	spotted flycatcher
Oenanthe oenanthe	northern wheatear
Parus major	great tit
Passer domesticus	house sparrow
Passer montanus	tree sparrow
Phoenicurus ochruros	black redstart
Phylloscopus trochilus	willow warbler
Pica pica	magpie
Rissa tridactyla	kittiwake
Sterna hirundo	common tern
Turdus merula	blackbird
Vanellus vanellus	northern lapwing

depending on roof type, vegetation cover and plant species composition, presence of water, space (territory) available and levels of maintenance (potential disturbance).

Food is perhaps the main factor that attracts certain bird species to green roofs. This point may explain why the roofs in highly urbanised areas are more frequently visited (as there are few alternative sources of food) than those in rural areas or at the edge of the city (where there is more choice) (Dunnett and Kingsbury, 2004). Cantor (2008) observed that those bird species that relied on the plants or invertebrate species on roofs were present more frequently than more common urban species, which did not appear to have an ecological niche associated with the roofs. Green roofs may also provide a secure nesting or roosting site, and the fact that they are generally located in urban habitats may actually mean they are fewer predators around compared to more traditional rural nesting sites (Gering and Blair, 1999). Large numbers of young ground-nesting birds such as *Vanellus vanellus* (northern lapwing), *Sterna hirundo* (common tern) or *Larus canus* (common gull) are predated on by *Vulpes vulpes* (red fox), *Mustela nivalis* (weasel), *Mustela ermine* (stoat) and *Neovison vison* (American mink), yet these predators will be almost universally absent from roofs. Larger plants on roofs facilitate roosting by providing physical shelter against adverse weather, and roofs may have additional advantages such as enhanced night-time temperatures (due to heat loss from buildings) – extreme cold being a major cause of calorie consumption and thus a significant killer of small birds, or reducing their fitness to breed in the subsequent spring. Shrubs and small trees are particularly favoured by songbirds (passerines) and their presence on roofs can both widen the range of species visiting and enhance the numbers of individuals within a species found at a given location.

Although water is not an absolute requirement for adult birds, it is for chicks that cannot fly elsewhere until their primary wing feathers develop. So, green roofs without water supplies that attract nesting birds are in effect an ecological trap for young birds. Even for feeding or roosting adults the presence of water on a green roof may reduce the number of visitations a bird needs to make to water bodies at ground level, thereby reducing its chances of being predated upon and saving it energy. Surface or standing water is thus essential for those roofs that are designed to provide habitat for birds. Some bird species are adroit in attaining water from their food sources or even drinking dew; however, the provision of water baths and water stations may be an alternative means to ensure water is supplied effectively and regularly. Shade is another requirement that needs to be considered for ground-nesting birds. Even although these species naturally prefer open habitat, they may still rely on long grass and dwarf shrubs to shelter during the heat of the day – again, this is especially so for chicks that cannot fly off to cooler locations. Altering the structure of conventional green roofs may provide opportunities. Providing uneven topography, combined with hay spreading and seed sowing, significantly enhanced the reproductive performance of *Vanellus vanellus* (northern lapwing) (Baumann *et al.*, 2021).

It is thought better designed/managed green roofs could provide a network of green stepping stones across the city matrix and assist birds in migrating across urban conurbations. Potentially they could become important 'refuelling stations' for some species as they conduct their bi-annual migrations or even just move from one green space to another looking for new territories. Even if certain species do not nest on green roofs, appearances on them can become commonplace as they search out food. The green roof at the Ford Motor Company's Dearborn construction plant, Michigan (USA) is frequented by *Contopus cooperi* (olive-sided fly catcher) and *Charadrius vociferous* (killdeer). Keeping with the motor industry theme, Burgess (2004) observed on the green roof at the Rolls Royce factory in West Sussex (UK) bird species as diverse as *Corvus frugilegus* (rook), *Larus canus* (common gull) *Charadrius hiaticula* (ringed plover), *Motacilla alba* (pied wagtail), *Turdus philomelos* (song thrush) and *Carduelis cannabina* (linnet). Green roofs in Portland, Oregon (USA) have been recorded hosting *Cyanocitta cristata* (blue jay), *Passer domesticus* (house sparrow), *Selasphorus rufus* (rufous hummingbird), *Corvus brachyrhynchos* (American crow) and *Hirundo rustica* (barn swallow). Green roofs may provide these generic services of food, shelter and water but can also be designed to mimic habitats within the urban area to specifically benefit relatively rare or endangered species.

Green walls, particularly green façade types, provide nesting and foraging habitat for urban-adapted birds. Research in north Staffordshire (UK) compared bird activity on or within a 10 m radius of green walls and compared this to similarly unclad walls. Birds were not encountered on bare walls, although they were observed on nearby roofs and vegetated areas. Birds, however, were more abundant in areas with green walls. On the green walls themselves, the birds' activity was always restricted to the upper half of the wall vegetation. Cladding green walls with dense foliage appears to offer some habitat opportunities to birds and may supplement urban green infrastructure in those locations where urban density restricts the opportunities for trees and other ground-covering vegetation (Chiquet *et al.*, 2013). In suburban situations the use of wall climbers such as *Hedera helix* (ivy), *Clematis montana*, *Passiflora, caerulea* (Passion flower), *Hydrangea petiolaris* and *Parthenocissus quinquefolia* (Virginia creeper) on houses or garden structures are known for their ability to afford secluded nesting habitat.

Birds may be drawn to green façades as a resource for food. Observational studies by Matt (2012) suggested green façades had a higher abundance and diversity of arthropods than non-vegetated building façades. Vegetated walls were shown to contain 16× to 39× more arthropods per m^2 than equivalent bare walls. Invertebrate species found included herbivores, predators, parasitoids and detritivores, indicating a range of niche opportunities within the wall systems. Arthropod abundance and richness were most strongly correlated with habitat availability and increased with the density of the foliar canopy.

Green façades in Poland were linked to significantly enhances species' biodiversity compared to non-vegetated walls (Oloś, 2023). Synanthropic bird species found nesting within the green façades included *Turdus merula* (blackbird), *Passer domesticus* (house sparrow), *Columba palumbus* (wood-pigeon) and *Streptopelia decaocto* (collared dove). Preying on nesting bird and their eggs was the predator *Martes foina* (beech marten). Morton (2022) found that green walls that were located in busier, noisier areas hosted fewer birds. Height and area of the green wall were positively correlated with bird species richness, abundance and behaviour while plant diversity and ecological value did not appear to influence these factors. However, Morton found that greater bird activity and wider behaviours occurred on street trees than on green walls. Factors associated with the green walls seems important, and design aspects need further investigation to improve the opportunities for wildlife. Thicker, more established vegetation can aid invertebrate communities (Salisbury *et al.*, 2023). A semi-controlled study in the UK showed that the abundance of invertebrates increased with wall vegetation depth and cover, where considerably more invertebrates were collected from vigorous/deeper leaf wall cover by *Hedera helix* (ivy) compared to the other treatments. There was an indication that increasing plant diversity helps; the addition of another *Hedera* cultivar (*Hedera helix* cv. Glacier) resulted in higher invertebrate abundance compared to *Hedera helix* alone. Long-established and more varied vegetation types are likely to increase opportunities for invertebrates.

Although green roofs provide better opportunities for wildlife than bare roofs, some have criticised the implications that green roofs, particularly extensive roofs with shallow substrates, are an alternative to habitat provision at ground level. Therefore, the justification that a new building can be placed on a greenfield site and a replacement habitat simple placed on the roof of the building should be challenged. The disadvantages associated with green roofs are not solely focussed on their accessibility for some species (such sites being difficult for numerous terrestrial species) but also the thermal tolerances of the thin substrates are limited. Soil temperatures may be too high in summer and too low in winter. Moreover, they may retain inadequate moisture levels to let some plant and animal species survive. Earthworms, for example, struggle to thrive on many green roofs, as there is not the depth of soil required to retreat to cooler, moister conditions in summer. Each green roof may be too small to support individual populations of any one species, and it is difficult to replicate the scale of habitat that some species require. In comparing green roofs and ground habitats in Basel, Switzerland, Brenneisen (2006) noted that the areas available for colonisation at ground level covered several hectares and were thus of a different order of magnitude to that of green roofs, which may only cover between a hundred and a few thousand square meters.

5.11 Urban Biodiversity and Humans

Urbanisation is a significant threat to biodiversity. It is not that nature is not valued in the city development processes, but that it is just not valued enough. It often becomes behind 'imperatives' for economic development, the need for new housing and jobs and transport infrastructure. As Heymans *et al.* (2019) state

> A prevailing notion of the mechanistic, reductionist worldview underpinning the modernist urban planning paradigm is that humans are separate from, and superior to, nature. One outcome of this has been a human value system that assumes the right to use ecological resources and change ecological processes for maximum human benefit without limitation.

Even human attitudes to green space can be problematic for 'nature'; urban green spaces 'need to look neat and tidy' – but that paradigm can result in loss or degradation of habitat. When green spaces are valued too, they are often valued in a simplistic way. Green infrastructure interventions are often justified based of a single benefit (e.g. storm water management or air quality improvement) but then miss the opportunity to provide wider, multiple benefits including habitat provision. The 'nature-based solution' approach is to be welcomed as, in blunt terms, this substitutes natural process for physical or engineered ones (e.g. using a reedbed to clean wastewater, rather than an expensive water purification plant), but even here, the ways to optimise a greater number of ecosystem services and improve habitat potential may be given limited consideration. This can be partially due to a lack of knowledge and understanding of the dynamic nature of urban ecosystems including their social and ecological interactions, spatially and temporally. This then results in planning models where the factors are not well integrated and thus fail to achieve multiple synergies and benefits. Wildlife can lose out if green space planning does not embrace the green network approach, i.e. effective connectivity.

Landscape architects and environmental horticulturists need to better understand ecological process more fully too and educate the public about them. Many iconic, attractive floral landscapes, highly popular with the public, are now providing resources for pollinating insects, but those invertebrate species also need food plants for their larval stages to facilitate their complete life cycle. *Urtica dioica* (stinging nettle), *Cirsium* spp. (thistles) and *Rubus fruticosus* (bramble) often meet with less approval from the public than the colourful arrays of *Papaver* spp. (poppy), *Linaria maroccana* (toadflax) and *Glebionis segetum* (corn marigold), yet ideally these plants may need to be in the location too to help strengthen local populations of Lepidoptera (butterflies and moths).

Progressive policies can help wildlife, but they need to be upheld with rigour. The UK has just introduced a 'Biodiversity Net Gain' policy, where any building development needs to enhance the biodiversity on site by at least 10% or mitigate the impact by purchasing land elsewhere and improving the biodiversity there. This latter solution, of course, does nothing for the plants and animals directly impacted by the loss of habitat/resource within the city. Similar moves can be deemed 'half-hearted' elsewhere. Statutory requirements to include green roofs on new buildings, thus theoretically replacing the 'green stamp of land' with like for like, does nothing of the sort if the green roof has minimal soil and is dominated by a monoculture of *Sedum* species.

The other problem with urbanisation is that it divorces people from nature and natural processes. Living in cities contributes to the physical, geographical and emotional separation of people from nature (Lehmann, 2023). Potentially, half the world's population now have at best irregular interactions with the natural world. The separation of people from nature is likely to be an increasingly important environmental issue, as it fundamentally influences the way people value nature and their willingness to conserve it. A lack of knowledge leads to a lack of empathy. The value of biologically rich urban green space is that it can help mitigate this. A vibrant, observable and easily accessed urban biota helps address this trend and improves people's understanding of the natural world, although perhaps not always providing a comprehensive awareness.

The ability of citizens to understand and connect with urban biodiversity was studied by Shwartz *et al.* (2014), where they intentionally enhanced the diversity of flowers, birds and insect pollinators in small public gardens in Paris (France) and observed people's responses. Species diversity and abundance was increased by providing flower meadows and introducing bird nest boxes. Residents were interviewed before and after the interventions. Results showed respondents expressed a strong preference for a rich diversity of species (excluding insects) and interestingly related this diversity to feelings of well-being in the gardens. There was no indication, however, that they actually noted the change in the diversity of species before and after the habitat interventions. Respondents underestimated species richness and only noticed the changes in native flower richness in those locations where the interventions were explicitly advertised and where public engagement activities were organised. Other studies note similar trends. Perceptions of biodiversity and actual biodiversity can often be misaligned by the public (Cameron *et al.,* 2020; Farris *et al.,* 2024). In a computer-generated photomontage of a flowering park landscape, Zhang (L. Zhang, 2023, unpublished data) introduced an image of a mouse at the forefront of the image, but less than 5% of participants noticed it. Shwartz *et al.* (2014) claim that their results highlight a 'people–biodiversity paradox' as there is a mismatch between people's perceptions of biodiversity and actual awareness of it. Further studies are needed to explore the role that urban biodiversity plays in people's daily life and the importance of this interaction for raising public support for general conservation policies.

Conclusions

- Global biodiversity is under threat and environmental horticulturists need to play their role in supporting wildlife. This often translates to better and more benign management of urban green spaces.
- Urban biodiversity is 'the variety or richness and abundance of living organisms (including genetic variation) and habitats found in and on the edge of human settlements.
- Urbanisation tends to diminish and fragment green space, resulting in limited resources for remaining populations of plants and animals, and inhibiting movement (reduced connectivity) between one area and the next for some species.
- Other problems associated with urban environments include direct injury due to vehicles and buildings, limited access to soil and altering hydrological patterns, temperature profiles and soil nutrients, more disturbance through physical means and noise, greater predation due to pet species and competition from introduced 'alien' species.
- Animal species can be divided based on how well adapted they are to urban living, i.e. urban exploiters, urban adapters and urban avoiders.
- A key driver for environmental horticulture is to restore green landscapes and a degree of ecosystem function to areas that are currently highly degraded and where the biodiversity is impoverished. This may include utilising non-native plant species as well as native, where appropriate to do so. A counterbalance to this, however, is that choice of species should be aimed at minimising the risk of introducing aggressive, invasive plant species that do threaten native habitats.
- Occurrence of species within urban habitats depends on factors such as a species' key traits and adaptability, population size, the history of a site or habitat, the availability of appropriate habitat and the quality and spatial arrangement of habitats.
- Typically, in city centres approximately 30–50% of the plant species are non-native. Plant biodiversity though tends to peak in the suburbs where there are still relatively adequate areas of green

space and associated niches for native species, due largely to a heterogeneous landscape typology (parks, woodlands, gardens and waterways), whilst the increase in garden space correlates with an abundance of non-native garden plants.

- Environmental horticulturists need to champion the urban forest concept as urban forests (garden, street and park trees as well as areas of dedicated woodland) are important for supporting urban biodiversity and allowing species to move across the urban landscape.
- The design and management of garden space will become increasingly important as the effects of urbanisation and climate change place pressure on native animal and plant populations.
- Carefully managed gardens, allotments parks and even brownfield sites are important refuge sites for urban wildlife and have the potential to support greater biodiversity and numbers of individuals of certain species than agriculturally managed areas in the rural environment.
- Green roofs and green walls can help certain taxa in the urban environment but will not necessarily replace terrestrial green space lost to urbanisation on a like for like basis. Soils and substrates may be limited in volume and in their ability to remain sufficiently moist, the vegetation mantle may be relatively thin and some species may have difficulty accessing such 'habitats'. Future research should evaluate the mechanisms by which green roofs and walls can be better planned, designed and managed to promote biodiversity and other ecosystem services.
- Urban biodiversity will become increasingly important in enabling a largely urban human population to access and understand nature. Environmental horticulturists can help 'bridge the gap' between traditional, highly manicured green space and more ecologically robust spaces, whilst ensuring such places remain highly relevant to the human population.

References

Adams, E. and Lindsey, K.J. (2009) *Urban Wildlife Management*, 2nd edn. CRC Press, Georgia. USA.

Al Farsi, K.A., Lupton, D., Hitchmough, J.D. and Cameron, R.W. (2017) How fast can conifers climb mountains? Investigating the effects of a changing climate on the viability of *Juniperus seravschanica* within the mountains of Oman, and developing a conservation strategy for this tree species. *Journal of Arid Environments* 147, 40–53.

Angold, P.G., Sadler, J.P., Hill, M.O., Pullin, A., Rushton, S., Austin, K. and Thompson, K. (2006) Biodiversity in urban habitat patches. *Science of the Total Environment* 360, 196–204.

Anon. (2015a) Universities of Florida Extension Services. *Hummingbirds of Florida* //edis.ifas.ufl.edu/uw059 (accessed 25 January 2024).

Anon. (2015b) Oregon State University Extension Services. *Plants for Hummingbirds* //extension.oregonstate.edu/gardening/plants-hummingbirds (accessed 25 January 2024).

Anon. (2025) RHS Perfect for pollinators plant list – Royal Horticultural Society. Available at: https://www.rhs.org.uk/science/research/plants-for-pollinators (accessed 4 August 2025)

Apfelbeck, B., Snep, R.P., Hauck, T.E., Ferguson, J., Holy, M. *et al.* (2020) Designing wildlife-inclusive cities that support human-animal co-existence. *Landscape and Urban Planning* 200, 103817.

Aronson, M.F., La Sorte, F.A., Nilon, C.H., Katti, M., Goddard, M.A. *et al.* (2014) A global analysis of the impacts of urbanization on bird and plant diversity reveals key anthropogenic drivers. *Proceedings of the Royal Society B: Biological Sciences* 281, p. 20133330.

Aronson, M.F., Handel, S.N., La Puma, I.P. and Clemants, S.E. (2015) Urbanization promotes non-native woody species and diverse plant assemblages in the New York metropolitan region. *Urban Ecosystems* 18, 31–45.

Baldock, K.C., Goddard, M.A., Hicks, D.M., Kunin, W.E., Mitschunas, N., Osgathorpe, L.M. and Memmott, J. (2015) Where is the UK's pollinator biodiversity? The importance of urban areas for flower-visiting insects. *Proceedings of the Royal Society of London B: Biological Sciences* 282, 20142849.

Barbosa-Filho, W.G. and Araujo, A.C.D. (2013) Flowers visited by hummingbirds in an urban Cerrado fragment, Mato Grosso do Sul, Brazil. *Biota Neotropica* 13, 21–27.

Barrett, K. and Guyer, C. (2008) Differential responses of amphibians and reptiles in riparian and stream habitats to land use disturbances in western Georgia, USA. *Biological Conservation* 141, 2290–2300.

Bastin, J.F., Clark, E., Elliott, T., Hart, S., Van Den Hoogen, J. *et al.* (2019) Understanding climate change from a global analysis of city analogues. *PloS One* 14, e0217592.

Battin, J. (2004) When good animals love bad habitats: Ecological traps and the conservation of animal populations. *Conservation Biology* 18, 1482–1491.

Baumann, N., Catalano, C., Pasta, S. and Brenneisen, S. (2021) Improving extensive green roofs for endangered ground-nesting birds. In: *Urban Services to Ecosystems: Green Infrastructure Benefits from the Landscape to the Urban Scale*. Springer International Publishing, Cham, pp. 13–29.

Beckmann, J.P. and Berger, J. (2003) Rapid ecological and behavioural changes in carnivores: the responses of black bears (*Ursus americanus*) to altered food. *Journal of Zoology* 261, 207–212.

Benvenuti, S. (2014) Wildflower green roofs for urban landscaping, ecological sustainability and biodiversity. *Landscape and Urban Planning* 124, 151–161.

Berthon, K., Thomas, F. and Bekessy, S. (2021) The role of 'nativeness' in urban greening to support animal biodiversity. *Landscape and Urban Planning* 205, 103959.

Berthon, K., Thomas, F., Baumann, J., White, R., Bekessy, S. and Encinas-Viso, F. (2023) Floral resources encourage colonisation and use of green roofs by invertebrates. *Urban Ecosystems* 26, 1517–1534.

Bigirimana, J., Bogaert, J., De Cannière, C., Bigendako, M.J. and Parmentier, I. (2012) Domestic garden plant diversity in Bujumbura, Burundi: Role of the socio-economical status of the neighborhood and alien species invasion risk. *Landscape and Urban Planning* 107, 118–126.

Bjørn, M.C. and Howe, A.G. (2023) Multifunctional bioretention basins as urban stepping stone habitats for wildflowers and pollinators. *Urban Forestry and Urban Greening* 90, 128133.

Blackmore, S. (2019) Cities: The final frontier for endangered plants? *Sibbaldia: The International Journal of Botanic Garden Horticulture* 17, 3–10.

Blicharska, M., Andersson, J., Bergsten, J., Bjelke, U., Hilding-Rydevik, T. *et al.* (2017) Is there a relationship between socio-economic factors and biodiversity in urban ponds? A study in the city of Stockholm. *Urban Ecosystems* 20, 1209–1220.

Borysiak, J., Mizgajski, A. and Speak, A. (2017) Floral biodiversity of allotment gardens and its contribution to urban green infrastructure. *Urban Ecosystems* 20, 323–335.

Braaker, S., Ghazoul, J., Obrist, M.K. and Moretti, M. (2014) Habitat connectivity shapes urban arthropod communities: The key role of green roofs. *Ecology* 95, 1010–1021.

Braschler, B., Gilgado, J.D., Zwahlen, V., Rusterholz, H.P., Buchholz, S. and Baur, B. (2020) Ground-dwelling invertebrate diversity in domestic gardens along a rural-urban gradient: Landscape characteristics are more important than garden characteristics. *PloS One* 15, e0240061.

Brenneisen, S. (2003) *Ökologisches Ausgleichspotenzial von extensiven Dachbegrünungen—Bedeutung für den Arten- und Naturschutz und die Stadtentwicklungsplanung*. Doctoral dissertation, Institute of Geography, University of Basel, Switzerland.

Brenneisen, S. (2006) Space for urban wildlife: Designing green roofs as habitats in Switzerland. *Urban Habitats* 4, 27–36.

Brotcorne, F., Giraud, G., Gunst, N., Fuentes, A., Wandia, I.N. *et al.* (2017) Intergroup variation in robbing and bartering by long-tailed macaques at Uluwatu Temple (Bali, Indonesia). *Primates* 58, 505–516.

Burgess, H. (2004) An Assessment of the Potential of Green Roofs for Bird Conservation in the UK (Masters research report). University of Sussex, Brighton, UK.

Cabral, I., Keim, J., Engelmann, R., Kraemer, R., Siebert, J. and Bonn, A. (2017) Ecosystem services of allotment and community gardens: A Leipzig, Germany case study. *Urban Forestry and Urban Greening* 23, 44–53.

Cameron, R. (2023) "Do we need to see gardens in a new light?" Recommendations for policy and practice to improve the ecosystem services derived from domestic gardens. *Urban Forestry and Urban Greening* 80, 127820.

Cameron, R.W., Brindley, P., Mears, M., McEwan, K., Ferguson, F. *et al.* (2020) Where the wild things are! Do urban green spaces with greater avian biodiversity promote more positive emotions in humans? *Urban Ecosystems* 23, 301–317.

Cantor, S.L. (2008) *Green Roofs in Sustainable Landscape Design*. Norton and Company. New York, USA.

Carbó-Ramírez, P. and Zuria, I. (2011) The value of small urban greenspaces for birds in a Mexican city. *Landscape and Urban Planning* 100, 213–222.

Catterall, C.P. (2004) Birds, garden plants and suburban bushlots: Where good intentions meet unexpected outcomes. In: Lunney, D. and Burgin, S. (eds) *Urban Wildlife: More Than Meets the Eye*. Royal Zoological Society of New South Wales, Mosman, pp. 21–31.

Cavia, R., Cueto, G.R. and Suárez, O.V. (2009) Changes in rodent communities according to the landscape structure in an urban ecosystem. *Landscape and Urban Planning* 90, 11–19.

Chang, C.R., Chen, M.C. and Su, M.H. (2021) Natural versus human drivers of plant diversity in urban parks and the anthropogenic species-area hypotheses. *Landscape and Urban Planning* 208, 104023.

Chatelain M., Gasparini J. and Frantz A. (2016) Trace metals, melanin-based pigmentation and their interaction influence immune parameters in feral pigeons (*Columba livia*). *Ecotoxicology* 25, 521–529.

Chiquet, C., Dover, J.W. and Mitchell, P. (2013) Birds and the urban environment: The value of green walls. *Urban Ecosystems* 16, 453–462.

Chong, K.Y., Teo, S., Kurukulasuriya, B., Chung, Y.F., Rajathurai, S. and Tan, H.T.W. (2014) Not all green is as good: Different effects of the natural and cultivated components of urban vegetation on bird and butterfly diversity. *Biological Conservation* 171, 299–309.

Clarke, L.W., Jenerette, G.D. and Davila, A. (2013) The luxury of vegetation and the legacy of tree biodiversity in Los Angeles, CA. *Landscape and Urban Planning* 116, 48–59.

Colding, J., Elmqvist, T., Lundberg, J., Ahrne´, K. Andersson, E. *et al.* (2003) The Stockholm Urban Assessment (SUASweden). In: *Millennium Ecosystem Assessment Sub-Global Summary Report*, Stockholm, Sweden,

The Royal Swedish Academy of Sciences, Millennium Ecosystem Assessment, United Nations.

Corbel H., Legros A., Haussy C. *et al.* (2016) Stress response varies with plumage colour and local habitat in feral pigeons. *Journal of Ornithology* 157, 825–37.

Cornelis, J. and Hermy, M. (2004) Biodiversity relationships in urban and suburban parks in Flanders. *Landscape and Urban Planning* 69, 385–401.

Cornelsen, C.T., MacQuarrie, C.J. and Lee, S.I. (2024) Modeling the distribution of hemlock woolly adelgid under several climate change scenarios. *Canadian Journal of Forest Research* 54, 1458–1470.

Cox, L. and Rodway-Dyer, S. (2023) The underappreciated value of brownfield sites: motivations and challenges associated with maintaining biodiversity. *Journal of Environmental Planning and Management* 66, 2009–2027.

Del Tredici, P. (2011) Spontaneous urban vegetation: Reflection of change in a globalized world. *Nature and Culture* 5, 299–315.

Delaney, K.S., Busteed, G., Fisher, R.N. and Riley, S.P. (2021) Reptile and amphibian diversity and abundance in an urban landscape: Impacts of fragmentation and the conservation value of small patches. *Ichthyology and Herpetology* 109, 424–435.

Diagne, C., Leroy, B., Vaissière, A.C., Gozlan, R.E., Roiz, D. *et al.* (2021) High and rising economic costs of biological invasions worldwide. *Nature* 592, 571–576.

Dominoni, D.M. (2017) Ecological effects of light pollution: How can we improve our understanding using light loggers on individual animals? In: *Ecology and Conservation of Birds in Urban Environments*. Springer International Publishing, Cham, pp. 251–270.

Dromgold, J.R., Threlfall, C.G., Norton, B.A. and Williams, N.S. (2020) Green roof and ground-level invertebrate communities are similar and are driven by building height and landscape context. *Journal of Urban Ecology* 6, p. juz024.

Duarte, M.H., Vecci, M.A., Hirsch, A. and Young, R.J. (2011) Noisy human neighbours affect where urban monkeys live. *Biology Letters* 7, 840–842.

Ducatez, S., Sayol, F., Sol, D. and Lefebvre, L. (2018) Are urban vertebrates city specialists, artificial habitat exploiters, or environmental generalists? *Integrative and Comparative Biology* 58, 929–938.

Dunnett, N. and Kingsbury, N. (2004) *Planting Green Roofs and Living Walls*. Timber Press, Portland, USA.

Durà, V.B., Meseguer, E., Crespo, C.H., Monerris, M.M., Doménech, I.A. and Santamalia, M.E.R. (2023) Contribution of green roofs to urban arthropod biodiversity in a Mediterranean climate: A case study in València, Spain. *Building and Environment* 228, 109865.

Dwyer, J.F., Nowak, D.J., Noble, M.H. and Sisinni, S.M. (2000) Connecting people with ecosystems in the 21st century: An assessment of our nation's urban forests. *USDA Forest Service. General Technical Report* PNW-GTR-490.

Dyson, K. (2020) Conserving native trees increases native bird diversity and community composition on commercial office developments. *Journal of Urban Ecology* 6, p. juaa033.

Elliot Noe, E., Innes, J., Barnes, A.D., Joshi, C. and Clarkson, B.D. (2022) Habitat provision is a major driver of native bird communities in restored urban forests. *Journal of Animal Ecology* 91, 1444–1457.

Erickson, E., Patch, H.M. and Grozinger, C.M. (2021) Herbaceous perennial ornamental plants can support complex pollinator communities. *Scientific Reports* 11, 17352.

Farris, S., Zhang, L., Dempsey, N., McEwan, K., Hoyle, H. and Cameron, R. (2024) 'The Elephant in the Room'–does actual or perceived biodiversity elicit restorative responses in a virtual park? *Cities and Health* 2014, 1–16.

Fernandez-Canero, R. and Gonzalez-Redondo, P. (2010) Green roofs as a habitat for birds: A review. *Journal of Animal and Veterinary Advances* 9, 2041–2052.

French, K. Major R. and Hely K. (2005) Use of native and exotic garden plants by suburban nectivorous birds. *Biological Conservation* 121, 545–559.

Gallo, T., Fidino, M., Lehrer, E.W. and Magle, S.B. (2017) Mammal diversity and metacommunity dynamics in urban green spaces: implications for urban wildlife conservation. *Ecological Applications* 27, 2330–2341.

Gardiner, M.M., Burkman, C.E. and Prajzner, S.P. (2013) The value of urban vacant land to support arthropod biodiversity and ecosystem services. *Environmental Entomology* 42, 1123–1136.

Gaynor, K.M., Hojnowski, C.E., Carter, N.H. and Brashares, J.S. (2018) The influence of human disturbance on wildlife nocturnality. Science 360, 1232–1235.

Gering, J. and Blair, R. (1999) Predation on artificial bird nests along an urban gradient: Predatory risk or relaxation in urban environments? *Ecography* 22, 532–541.

Gilbert, O.L. (1989) Allotments and leisure gardens. In: *The Ecology of Urban Habitats*. Springer, Netherlands. pp. 206–217.

Gill, S.E., Handley, J.F., Ennos, A.R., Pauleit, S., Theuray, N. and Lindley, S.J. (2008) Characterising the urban environment of UK cities and towns: A template for landscape planning. *Landscape and Urban Planning* 87, 210–222.

Giovanetti, M., Giuliani, C., Boff, S., Fico, G. and Lupi, D. (2020) A botanic garden as a tool to combine public perception of nature and life-science investigations on native/exotic plants interactions with local pollinators. *PLoS One* 15, e0228965.

Giraudeau M., Stikeleather R., McKenna J., Hutton P. and McGraw K.J. (2017) Plumage micro-organisms and preen gland size in an urbanizing context. *Science of the Total Environment* 580, 425–9.

Giuliani, C., Giovanetti, M., Lupi, D., Mesiano, M.P., Barilli, R. *et al.* (2020) Tools to tie: Flower characteristics, VOC emission profile, and glandular trichomes of two Mexican *Salvia* species to attract bees. *Plants* 9, 1645.

Goiran C., Bustamante P. and Shine R. (2017) Industrial melanism in the seasnake *Emydocephalus annulatus*. *Current Biology* 27, 2510–2513.

Gonsalves, S., Starry, O., Szallies, A. and Brenneisen, S. (2022) The effect of urban green roof design on beetle biodiversity. *Urban Ecosystems* 25, 205–219.

Gray, E.R. and van Heezik, Y. (2016) Exotic trees can sustain native birds in urban woodlands. *Urban Ecosystems* 19, 315–329.

Handa, I.T., Aerts, R., Berendse, F., Berg, M.P., Bruder, A. *et al.* (2014) Consequences of biodiversity loss for litter decomposition across biomes. *Nature* 509, 218–221.

Hanley, M.E., Awbi, A.J. and Franco, M. (2014) Going native? Flower use by bumblebees in English urban gardens. *Annals of Botany* 113, 799–806.

Hanson, H.I., Eckberg, E., Widenberg, M. and Olsson, J.A. (2021) Gardens' contribution to people and urban green space. *Urban Forestry and Urban Greening* 63, 127198.

Harrison, C. and Davies, G. (2002) Conserving biodiversity that matters: Practitioners' perspectives on brownfield development and urban nature conservation in London. *Journal of Environmental Management* 65, 95–108.

Haubrock, P.J., Cuthbert, R.N., Sundermann, A., Diagne, C., Golivets, M. and Courchamp, F. (2021) Economic costs of invasive species in Germany. *NeoBiota* 67, 225–246.

Helden, A.J., Stamp, G.C. and Leather, S.R. (2012) Urban biodiversity: Comparison of insect assemblages on native and non-native trees. *Urban Ecosystems* 15, 611–624.

Heymans, A., Breadsell, J., Morrison, G.M., Byrne, J.J. and Eon, C. (2019) Ecological urban planning and design: A systematic literature review. *Sustainability* 11, 3723.

Hill, M.J. and Wood, P.J. (2014) The macroinvertebrate biodiversity and conservation value of garden and field ponds along a rural-urban gradient. *Fundamental and Applied Limnology/Archiv für Hydrobiologie* 185, 107–119.

Hill, M.J., Wood, P.J., Fairchild, W., Williams, P., Nicolet, P. and Biggs, J. (2021) Garden pond diversity: Opportunities for urban freshwater conservation. *Basic and Applied Ecology* 57, 28–40.

Hitchings S. and Beebee T. (1998) "Loss of genetic diversity and fitness in common toad (*Bufo bufo*) populations isolated by inimical habitat." *Journal of Evolutionary Biology* 11, 269–83.

Holyoak, M., Leibold, M.A. and Holt, R.D. (2005) *Metacommunities: Spatial Dynamics and Ecological Communities*. University of Chicago Press, Chicago, Illinois, USA.

Ignatieva, M. (2010) Design and future of urban biodiversity. In: Müller, N., Werner, P. and Kelcey, J.C. *Urban Biodiversity and Design*, 1st edn. Blackwell Publishing Ltd. Oxford, UK, pp 118–144.

Jacquin, L., Récapet, C., Prévot-Julliard, A.C., Leboucher, G. *et al.* (2013) A potential role for parasites in the maintenance of color polymorphism in urban birds. *Oecologia* 173, 1089–1099.

Jain, A., Kunte, K. and Webb, E.L. (2016) Flower specialization of butterflies and impacts of non-native flower use in a transformed tropical landscape. *Biological Conservation* 201, 184–191.

Jasmani, Z., Ravn, H.P. and van den Bosch, C.C.K. (2017) The influence of small urban parks characteristics on bird diversity. A case study of Petaling Jaya, Malaysia. *Urban Ecosystems* 20, 227–243.

Jha, S. and Kremen, C. (2013) Urban land use limits regional bumble bee gene flow. *Molecular Ecology* 22, 2483–2495.

Jim, C.Y. and Liu. H.T. (2001) Patterns and dynamics of urban forests in relation to land use and development history in Guangzhou City, China. *The Geographical Journal* 167, 358–375.

Kang, W., Minor, E.S., Park, C.R. and Lee, D. (2015) Effects of habitat structure, human disturbance, and habitat connectivity on urban forest bird communities. *Urban Ecosystems* 18, 857–870.

Kattwinkel, M., Biedermann, R. and Kleyer, M. (2011) Temporary conservation for urban biodiversity. *Biological Conservation* 144, 2335–2343.

Kelcey, J.G. and Müller, N. (2011) *Plants and Habitats of European Cities*. Springer Science and Business Media, Berlin/Heidelberg, Germany.

Kempenaers, B., Borgström, P., Loës, P., Schlicht, E. and Valcu, M. (2010) Artificial night lighting affects dawn song, extra-pair siring success, and lay date in songbirds. *Current Biology* 20, 1735–1739.

Kohut, S.M., Hess, G.R. and Moorman, C.E. (2009) Avian use of suburban greenways as stopover habitat. *Urban Ecosystems* 12, 487–502.

Kotze, D.J., Lowe, E.C., MacIvor, J.S., Ossola, A., Norton, B.A. *et al.* (2022) Urban forest invertebrates: How they shape and respond to the urban environment. *Urban Ecosystems* 25, 1589–1609.

Kowarik, I. (2011) Novel urban ecosystems, biodiversity and conservation. *Environmental Pollution* 159, 1974–1983.

Larsson, T.-B. (2001) Biodiversity evaluation tools for European forests. *Ecological Bulletin* 50. Wiley-Blackwell, New Jersey. USA.

Lehmann, S. (2021) Growing biodiverse urban futures: Renaturalization and rewilding as strategies to strengthen urban resilience. *Sustainability* 13, 2932.

Lehmann, S. (2023) Reconnecting with nature: Developing urban spaces in the age of climate change, *Emerald Open Research* 1, 5.

Lessi, B.F., Pires, J.S.R, Batisteli, A.F. and Macgregor-Fors, I. (2016) Vegetation, urbanization, and bird richness in a Brazilian peri-urban area. *Ornitologia Neotropical* 27, 203–210.

Leveau, L. (2021) United colours of the city: A review about urbanisation impact on animal colours. *Austral Ecology* 46, 670–679.

Lewis, E., Phoenix, G.K., Alexander, P., David, J. and Cameron, R.W. (2019) Rewilding in the Garden: Are garden hybrid plants (cultivars) less resilient to the effects of hydrological extremes than their parent species? A case study with *Primula*. *Urban Ecosystems* 22, 841–854.

Lin, B.B., Philpott, S.M. and Jha, S. (2015) The future of urban agriculture and biodiversity-ecosystem services: Challenges and next steps. *Basic and Applied Ecology*. 16, 189–201.

Lippert, H., Kowarik, I. and Straka, T.M. (2022) People's attitudes and emotions towards different urban forest types in the Berlin region, Germany. *Land* 11, 701.

Loss, S.R., Will, T. and Marra P.P. (2013) The impact of free-ranging domestic cats on wildlife of the United States. *Nature Communications* 4, 1396.

Lowenstein, D.M., Matteson, K.C. and Minor, E.S. (2019) Evaluating the dependence of urban pollinators on ornamental, non-native, and 'weedy'floral resources. *Urban Ecosystems* 22, 293–302.

Lundholm, J.T. and Richardson, P.J. (2010) MINI-REVIEW: Habitat analogues for reconciliation ecology in urban and industrial environments. *Journal of Applied Ecology* 47, 966–975.

Lynch, A.J. (2019) Creating effective urban greenways and stepping-stones: Four critical gaps in habitat connectivity planning research. *Journal of Planning Literature* 34, 131–155.

McCallum, R. and Sardo, A.M. (2021) "Britain's rainforests": Engaging the public with brownfield sites for conservation in the UK. *Journal of Science Communication* 20, A07.

Machon, N. (2021) Urban wastelands can be amazing reservoirs of biodiversity for cities. *Urban Wastelands: A Form of Urban Nature?* Springer Nature, London, UK, pp 3–18.

MacIvor, J.S. (2016) DNA barcoding to identify leaf preference of leafcutting bees. *Royal Society Open Science* 3, 150623.

Magle, S.B. and Crooks, K.R. (2009) Investigating the distribution of prairie dogs in an urban landscape. *Animal Conservation* 12, 192–203.

Marshall, C.A., Wilkinson, M.T., Hadfield, P.M., Rogers, S.M., Shanklin, J.D. *et al.* (2023) Urban wildflower meadow planting for biodiversity, climate and society: An evaluation at King's College, Cambridge. *Ecological Solutions and Evidence* 4, e12243.

Masierowska, M., Stawiarz, E. and Rozwałka, R. (2018) Perennial ground cover plants as floral resources for urban pollinators: A case of *Geranium* species. *Urban Forestry and Urban Greening* 32, 185–194.

Masood, N. and Russo, A. (2023) Community perception of brownfield regeneration through urban rewilding. *Sustainability* 15, 3842.

Matt, S. (2012) Green façades provide habitat for arthropods on buildings in the Washington, DC metro area Doctoral dissertation, The University of Maryland, Maryland, USA.

Mayrand, F. and Clergeau, P. (2018) Green roofs and green walls for biodiversity conservation: A contribution to urban connectivity? *Sustainability* 10, 985.

McFrederick, Q. S., and G. LeBuhn (2006) Are urban parks refuges for bumble bees *Bombus* spp. (Hymenoptera: Apidae)? *Biological Conservation* 129, 372–382.

McIntyre, N.E. (2000) Ecology of urban arthropods: A review and a call to action. *Annals of the Entomological Society of America* 93, 825–835.

Meland, S., Sun, Z., Sokolova, E., Rauch, S. and Brittain, J.E. (2020) A comparative study of macroinvertebrate biodiversity in highway stormwater ponds and natural ponds. *Science of the Total Environment* 740, 140029.

Meyer, S., Rusterholz, H.-P., Salamon, J.-A. and Baur, B. (2020) Leaf litter decomposition and litter fauna in urban forests: Effect of the degree of urbanisation and forest size. *Pedobiologia* 78,150609.

Mills, J.G., Weinstein, P., Gellie, N.J., Weyrich, L.S., Lowe, A.J. and Breed, M.F. (2017) Urban habitat restoration provides a human health benefit through microbiome rewilding: The Microbiome Rewilding Hypothesis. *Restoration Ecology* 25, 866–872.

Morelli, F., Reif, J., Díaz, M., Tryjanowski, P., Ibáñez-Álamo, J.D. *et al.* (2021) Top ten birds indicators of high environmental quality in European cities. *Ecological Indicators* 133, 108397.

Morton, L. (2022) Urban Green Infrastructures: An Assessment of Urban Bird Diversity, Abundance and Behaviour Associated with Green Walls and Street Trees in Manchester and Salford, UK (Doctoral Dissertation).

Müller, N., Ignatieva, M., Nilon, C.H., Werner, P. and Zipperer, W.C. (2013) Patterns and trends in urban biodiversity and landscape design. In: *Urbanization, Biodiversity and Ecosystem Services: Challenges and Opportunities*. Springer, Netherlands, pp 123–174.

Munshi-South, J. (2012) Urban landscape genetics: Canopy cover predicts gene flow between white-footed mouse (*Peromyscus leucopus*) populations in New York City. *Molecular Ecology* 21, 1360–137.

Myczko, Ł., Rosin, Z.M., Skórka, P. and Tryjanowski, P. (2014) Urbanization level and woodland size are major drivers of woodpecker species richness and abundance. *PLoS One* 9, e94218.

Nagase, A., Kurashina, M., Nomura, M., MacIvor, J.S. (2019) Patterns in urban butterflies and spontaneous plants across a university campus in Japan. Pan-Pacific *Entomology* 94, 195–215.

Nascimento, V.T., Agostini, K., Souza, C.S. and Maruyama, P.K. (2020) Tropical urban areas support highly diverse plant-pollinator interactions: An assessment from Brazil. *Landscape and Urban Planning* 198, 103801.

Noordijk, J., Delille, K., Schaffers, A. P. and Sýkora, K.V. (2009) Optimizing grassland management for flower-visiting insects in roadside verges. *Biological Conservation* 142, 2097–2103.
Oloś, G. (2023) Green facades support biodiversity in urban environment: A case study from Poland. *Journal of Water and Land Development* 59, 257–266.
Osborne, J.L., Martin, A.P., Shortall, C.R., Todd, A.D., Goulson, D., Knight, M.E. and Sanderson, R.A. (2008) Quantifying and comparing bumblebee nest densities in gardens and countryside habitats. *Journal of Applied Ecology* 45, 784–792.
Ossola, A., Aponte, C., Hahs, A.K. and Livesley, S.J. (2017) Contrasting effects of urban habitat complexity on metabolic functional diversity and composition of litter and soil bacterial communities. *Urban Ecosystems* 20, 595–607.
Owen, J. (2010) *Wildlife of a Garden, A 30 year study*. RHS publications, London, UK.
Palmersheim, M.C., Schürch, R., O'Rourke, M.E., Slezak, J. and Couvillon, M.J. (2022) If you grow it, they will come: Ornamental plants impact the abundance and diversity of pollinators and other flower-visiting insects in gardens. *Horticulturae* 8, 1068.
Pardee, G.L. and Philpott, S.M. (2014) Native plants are the bee's knees: Local and landscape predictors of bee richness and abundance in backyard gardens. *Urban Ecosystems* 17, 641–659.
Parsons, S.E., Kerner, L.M. and Frank, S.D. (2020) Effects of native and exotic congeners on diversity of invertebrate natural enemies, available spider biomass, and pest control services in residential landscapes. *Biodiversity and Conservation* 29, 1241–1262.
Pearl, C.A., Adams, M.J., Leuthold, N. and Bury, R.B. (2005) Amphibian occurrence and aquatic invaders in a changing landscape: Implications for wetland mitigation in the Willamette Valley, Oregon, USA. *Wetlands* 25, 76–88.
Phillips, B.B., Wallace, C., Roberts, B.R., Whitehouse, A.T., Gaston, K.J. *et al.* (2020) Enhancing road verges to aid pollinator conservation: A review. *Biological Conservation* 250, 108687.
Portnov, B.A., Saad, R., Trop, T., Kliger, D. and Svechkina, A. (2020) Linking nighttime outdoor lighting attributes to pedestrians' feeling of safety: An interactive survey approach. *PloS One* 15, e0242172.
Quigley, M.F. (2011) Potemkin gardens: Biodiversity in small designed landscapes. In: Niemelä J. (ed.) *Urban Ecology: Patterns, Processes and Applications*. Oxford University Press, New York. pp. 85–91.
Ritchie, H. and Roser, M. (2018) "Urbanization". Published online at OurWorldInData.org. Available at: https://ourworldindata.org/urbanization (accessed 12 October 2024).
Rollings, R. and Goulson, D. (2019) Quantifying the attractiveness of garden flowers for pollinators. *Journal of Insect Conservation* 23, 803–817.
Rudd, H., Vala, J. and Schaefer, V. (2002) Importance of backyard habitat in a comprehensive biodiversity conservation strategy: A connectivity analysis of urban green spaces. *Restoration Ecology* 10, 368–375.
Salisbury, A., Al-Beidh, S., Armitage, J., Bird, S., Bostock, H., Platoni, A., Tatchell, M., Thompson, K. and Perry, J. (2017) Enhancing gardens as habitats for plant-associated invertebrates: Should we plant native or exotic species? *Biodiversity and Conservation* 26, 2657–2673.
Salisbury, A., Bird, S., Bostock, H., Barrie, C., Harris, S., Marshall, R., Sanford, R., Tuson, M. and Tew, N. (2025). Choosing to be picky: An evidence-based system for selecting pollinator-friendly garden plants to support biodiversity. *BioScience*, biaf090.
Salisbury, A., Blanusa, T., Bostock, H. and Perry, J.N. (2023) Careful plant choice can deliver more biodiverse vertical greening (green façades). *Urban Forestry and Urban Greening* 89, 128118.
Schlaepfer, M.A., Guinaudeau, B.P., Martin, P. and Wyler, N. (2020) Quantifying the contributions of native and non-native trees to a city's biodiversity and ecosystem services. *Urban Forestry and Urban Greening* 56, 126861.
Senar J.C., Conroy M.J., Quesada J. and Mateos-Gonzalez, F. (2014) Selection based on the size of the black tie of the great tit may be reversed in urban habitats. *Ecological Evolution* 4, 2625–32.
Shwartz, A., Shirley, S. and Kark, S. (2008) How do habitat variability and management regime shape the spatial heterogeneity of birds within a large Mediterranean urban park? *Landscape and Urban Planning* 84, 219–229.
Shwartz, A., Turbé, A., Simon, L. and Julliard, R. (2014) Enhancing urban biodiversity and its influence on city-dwellers: An experiment. *Biological Conservation*, 171, 82–90.
Sikora, A., Michołap, P. and Sikora, M. (2020) What kind of flowering plants are attractive for bumblebees in urban green areas? Urban Forestry and Urban Greening, 48, 126546.
Sjöman, H., Östberg, J. and Bühler, O. (2012) Diversity and distribution of the urban tree population in ten major Nordic cities. *Urban Forestry and Urban Greening* 11, 31–39.
Smith, R.M., Warren, P.H., Thompson, K. and Gaston, K.J. (2006) Urban domestic gardens (VI): Environmental correlates of invertebrate species richness. *Biodiversity and Conservation* 15, 2415–2438.
Soderstrom, M. (2001) *Recreating Eden: A Natural History of Botanical Gardens*. V´ehicule Press, Montreal, Canada.
Soga, M. and Gaston K.J. (2016) Extinction of experience: The loss of human-nature interactions. *Frontiers in Ecology and the Environment* 14, 94–101.
Sun, J., Koski, T.M., Wickham, J.D., Baranchikov, Y.N. and Bushley, K.E. (2024) Emerald ash borer management

and research: Decades of damage and still expanding. *Annual Review of Entomology* 69, 239–258.

Thompson, K., Austin, K.C., Smith, R.M., Warren, P.H., Angold, P.G. and Gaston, K.J. (2003) Urban domestic gardens (I): Putting small-scale plant diversity in context. *Journal of Vegetation Science* 14, 71–78.

Threlfall, C.G., Walker, K., Williams, N.S., Hahs, A.K., Mata, L., Stork, N. and Livesley, S.J. (2015) The conservation value of urban green space habitats for Australian native bee communities. *Biological Conservation* 187, 240–248.

Threlfall, C.G., Mata, L., Mackie, J.A., Hahs, A.K., Stork, N.E., Williams, N.S.G. and Livesley, S.J. (2017) Increasing biodiversity in urban green spaces through simple vegetation interventions. *Journal of Applied Ecology* 54, 1874–1883.

Turnbull, S. (2012) Epigeal Invertebrates of Yorkshire Allotments: The Influence of Urban-Rural Gradient and Management Style. Doctoral dissertation. The University of Hull, Hull. UK.

Turrini, T. and Knop, E. (2015) A landscape ecology approach identifies important drivers of urban biodiversity. *Global Change Biology* 21, 1652–1667.

Van Geffen, K.G., Groot, A.T., Van Grunsven, R.H.A., Donners, M., Berendse, F. and Veenendaal, E.M. (2015) Artificial night lighting disrupts sex pheromone in a noctuid moth. *Ecological Entomology* 40, 401–408.

Van Helden, B.E., Close, P.G. and Steven, R. (2020) Mammal conservation in a changing world: Can urban gardens play a role? *Urban Ecosystems* 23, 555–567.

Vergnes, A., Le Viol, I. and Clergeau, P. (2012) Green corridors in urban landscapes affect the arthropod communities of domestic gardens. *Biological Conservation* 145, 171–178.

Vergnes, A., Kerbiriou, C. and Clergeau, P. (2013) Ecological corridors also operate in an urban matrix: a test case with garden shrews. *Urban Ecosystems* 16, 511–525.

Walton, O.C. and Stevens, M. (2018) Avian vision models and field experiments determine the survival value of peppered moth camouflage. *Communications Biology* 1, 1–7.

Williams, F., Eschen, R., Harris, A., Djeddour, D., Pratt, C. *et al.* (2010) The economic cost of invasive non-native species on Great Britain. *CABI* VM10066, CABI, Wallingford, UK. 199pp.

Wood, E.M. and Esaian, S. (2020) The importance of street trees to urban avifauna. *Ecological Applications* 30, e02149.

Wooster, E.I.F., Fleck, R., Torpy, F., Ramp, D. and Irga, P.J. (2022) Urban green roofs promote metropolitan biodiversity: A comparative case study. *Building and Environment* 207, 108458.

Xu, D., Peng, J., Jiang, H., Dong, J., Liu, M. *et al.* (2024) Incorporating barriers restoration and stepping stones establishment to enhance the connectivity of watershed ecological security patterns. *Applied Geography* 170, 103347.

Zhang, H. and Jim, C.Y. (2014) Contributions of landscape trees in public housing estates to urban biodiversity in Hong Kong. *Urban Forestry and Urban Greening* 13, 272–284.

Zhang, P., Fahey, R.T. and Park, S. (2024) The importance of current and potential tree canopy on urban vacant lots for landscape connectivity. *Urban Forestry and Urban Greening* 94, 128235.

6 Woody Plants (Trees, Shrubs and Climbers)

Abstract

Woody plants tends to divide into three main groups: trees, shrubs and climbers (vines). Trees are defined by having a single main stem (trunk), whereas shrubs are more likely to branch from/near the base and are generally smaller in stature. Climbers require a physical structure to support vertical growth but can also scramble across the ground where necessary. Woody plants are grown for their flowers, foliage and sometimes stem traits, such as bark colour or texture. Woody plants provide three-dimensional geometry to the landscape and a strong 'sense of place'. Trees, for example, can be important in giving 'character' to a city as well as providing important ecosystem services and supporting wildlife. This chapter deals with the propagation, distribution and management of woody plants within the urban landscape. Environmental horticulturists should understand how dry soils, waterlogged soils, soils with a low pH, etc. determine taxonomic choice.

6.1 Introduction

The woody plants tend to be divided into three groups based on their growth characteristics. Trees are predominately of a larger scale (mostly but not exclusively over 5 m tall when mature) and are characterised by a single stem (trunk) growing up from the ground surface level (Fig. 6.1). Shrubs are smaller (generally less than 5 m in height) and often have a greater number of stems, originating low down in the hierarchy of the plant; shrubs may be described as multi-stemmed or multi-branched, with no single shoot (leader) dominating the geometry. Both these types of woody plants build up rigid structure over time and are self-supporting. In contrast, the climbers or vines, although also having woody stems, usually require a structure to support their growth. In essence, they are designed to climb over other plants, cliff faces or 'scramble' across the surface of the ground. Their cells possess lignin (that provides structural strength), but the stems are typically not as rigid as that of trees and shrubs. As such they are sometimes denoted as semi-woody plants. Many climbing plant species also have specific ways of helping themselves climb and reach sunlight, so will utilise specialised organs such as thorns, tendrils, adhesion pads or simply twist their stem growth 'corkscrew'-like to twine up another vertical object such as a tree or pole.

Trees are a key component in many landscapes and provide the dominant feature in many of the world's biomes, being the climax vegetation type where appropriate temperature and water availability facilitate their growth (Table 6.1). Trees are dominated by dicotyledonous species (gymnosperms – the conifers and angiosperms – the broadleaves), although there are a number of notable monocotyledenous species too, some with important economic value, e.g. *Phoenix dactylifera* (date palm). Shrubs and climbers are mostly angiosperms. Tree growth tends to be dictated by light (solar irradiance) and water availability. Trees will compete for light in forest situations, with not all specimens surviving to maturity, as competition means some small or slower-growing individuals are 'shaded-out'. Many tree species have a strategy to grow fast and tall to capture the available sunlight, then spread their upper branches wide to capitalise on this light fully, but also to shade out trees growing below them to reduce the competition from them. A consequence of this is that young woodlands have a very high density of actively growing young trees, but mature woodland has fewer, larger and more spaced-out specimens. In warm climates, water availability also dictates the density of the tree stand – drier savannahs for example have their trees well-spaced apart, whereas in wetter parts the tree density is greater for any given area. In Boreal conifer forests, the tree density can be intermediate, because each tree needs sufficient light, but there is also an advantage in having neighbouring trees nearby to reduce injury from severe, cold winds.

DOI: 10.1079/9781800621763.0006

Fig. 6.1. Trees have a defined main stem (trunk) (left), whereas shrubs have a mosaic of branches, developing from the base (middle). Climbers twine and twist their way over other objects, in this case metal frames (right). From left to right: tree, *Quercus robor*; shrub, *Ribes sanguinium* cv. Elkington's White; and climbers *Clematis* cv. Nubia (red flowers) and *Clematis* cv. Amethyst Beauty (purple flowers).

Table 6.1. Forest biomes and their key characteristics

Forest biome	Annual precipitation range (mm)	Range of mean annual temperature (°C)	Features
Boreal (Taiga)	100 to 2000	–7 to 5	Dominated by slow-growing gymnosperms particularly conifers.
Temperate	350 to 2500	0 to 18	Dominated by deciduous 'hardwood' species. Leaves abscised before winter.
Temperate rainforest	2000 to 3400	5 to 20	Has both fast-growing gymnosperms/conifers and deciduous hardwood species.
Tropical seasonal	1300 to 2750	18 to 28	Dominated by deciduous 'hardwood' species. Leaves abscised before dry season.
Tropical rainforest	2400 to 4500	18 to 27	Dominated by evergreen 'hardwood' species.

Ecologically, shrubs tend to play a role as understorey plants (the shrub layer) in woodland but also as effective pioneering species when habitats are in transition, e.g. 'scrub' vegetation on 'unmanaged' chalk downland. Shrubs can dominate too, where ecological pressures such as limited water availability, frequent fires or browsing by herbivores inhibits true tree development (e.g. Mediterranean biomes such as the chaparral of North America or the Maquis of southern Europe have a high preponderance of woody shrub and woody sub-shrub species). Woody climbing plants or vines are found in woodland settings or in rocky escarpments, where their growth habit allows them to compete for light by growing up other plants or cliff faces.

The number and variety of woody plants used within urban public space can vary significantly. Within Europe, most cities have 50–80 street trees per 1000 inhabitants, but density can be as low as 20 street trees per 1000. Although there are geographical trends, the strategies and priorities within different local authorities also has a strong influence; Grenoble in France had 90 street trees per 1000 inhabitants, compared to only 20 in Nice despite the cities only being 200 km apart (Pauleit *et al.*, 2002). Where the trees are is quite important too. Smart *et al.* (2020) showed that street trees were planted at greater frequency in cities within warmer regions (e.g. Buenos Aires), compared to cooler ones (e.g. Stockholm, Fig. 6.2).

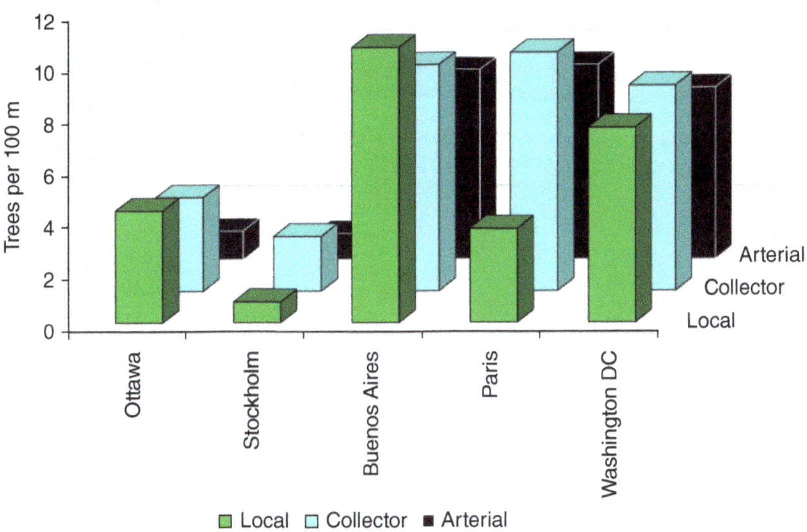

Fig. 6.2. Street tree frequency per 100 m of road length. Data determined by road type (local, collector or arterial) across five cities: Ottawa, Canada; Stockholm, Sweden; Buenos Aires, Argentina; Paris, France; and Washington DC, USA. (Modified with permission from Smart *et al.*, 2020.)

There is also a broad correlation between forest cover within a country and the prevalence of urban trees. Tree cover in cities within the UK, the Netherlands, Denmark and Iceland is relatively low (10 m^2 canopy cover per person) compared to Finnish cities (140 m^2 cover per person), reflecting the total woodland in each country – a similar pattern also being observed between different states of the USA. Climate determines species richness, with a trend for more natural diversity of plant (and animal) species as latitude decreases. This is not necessarily reflected in the use of landscape plants, however, as this is dictated by climate but also cultural factors, such as private ownership of gardens, interest in gardening and the strength of garden industries in different countries. In mild temperate climates such as western Europe and the north-western USA, which are conducive to the successful cultivation of plants from a wide range of different biomes, the range of garden plants can be extensive. This is due to the use of a large proportion of non-native, imported species and the development of new genotypes (taxa) bred/selected for their aesthetic traits. In the UK, the Royal Horticultural Society lists 70,000 ornamental plants available to the garden trade; these include woody and non-woody genotypes but exclude varieties of bedding plants (due to their extensive range and relative limited longevity in the marketplace). The range of plants used in public open spaces tends to be much less than that of private gardens (see section on Urban Forests).

6.2 Function in the Landscape

Within the landscape, woody plants not only determine much of the characteristics of the ecosystem, but from a human perspective are also valued for providing 'a sense of place', possessing functional and also cultural or spiritual value (Fig. 6.3.).

Although trees, shrubs and woody climbers were originally exploited as a source of food, firewood, timber, fibres and herbal medicines, they have also been catalysts for improving the aesthetics of landscapes, both rural and urban. They possess a range of aesthetic qualities; these being defined as form, colour (hue), texture and line within the landscape. They are used architecturally to provide 'canopies' or 'walls' of vegetation and screen off unsightly objects or provide privacy. In addition to performing as living sculpture in their own right, woody plants 'soften' the surrounding architecture by acting as a

Fig. 6.3. The wishing-well tree at St Helen's Well, Market Weighton, East Riding of Yorkshire, UK. The tree is festooned with ribbons as a token of good fortune and remembrance.

foil to adjacent buildings or improving the texture of the building surface (e.g. green walls/façades). Trees and shrubs are used to define boundaries, roadways and other access routes, and specimen trees have a long tradition as landmark icons.

Increasingly, woody plants are being valued not just for their aesthetics but also for a range of other functional purposes (see Chapters 3 and 4). Trees and shrubs modify microclimates through shading, evaporative cooling and reducing air movement by acting as wind breaks. Landscape architects now advocate their wider use in regulating building temperature. They can be used to divert sound waves, filter sunlight, reduce glare or block sources of light pollution, such as street lights or car headlights. Woodlands, forests, parklands and even street and garden trees contribute to reducing precipitation intensity and rates of runoff, thereby reducing the likelihood of flash flooding. As some societies become more conscious of their environmental footprint, trees are increasingly valued as a habitat and food resource for wildlife – standing dead specimens often providing as valuable a resource as living ones.

In urban locations, trees are used to line streets and provide avenues. They are employed as individual specimens or as small groups in parkland or gardens and in mass plantings within urban forests. Taxonomic selection will depend on aspects such as final size, natural tree form (e.g. 'fastigiate' or 'compact') or the ability of the taxa to respond to management techniques such as stem training, coppicing or pollarding. As pressure for space increases within high-density urban environments, there is a trend to plant species with a small final height and breadth. As a consequence, planting of smaller species and cultivars has become widespread in the most densely built parts of a city. For example, in some suburbs within Europe, *Prunus* (ornamental cherries and almonds, typical final height 8–12 m), *Malus* (crab apple, 10–14 m) *Sorbus* (rowans, 8–16 m) and *Pyrus* (ornamental pears, 10–16 m) are being preferred to larger growing species such as *Platanus* × *acerifolia* (London plane, 40–44 m), *Quercus* spp. (oaks, 25–35 m), *Fagus* (beech, 35–38 m) or *Acer platanoides* (Norway maple, 25–28 m). Conversely though, more local authorities are increasingly concerned about flood risk, poor air quality and urban heat islands at a city scale, and there are aspirations to increase tree canopy cover overall (Walters and Sinnett, 2021). Thus, wide arterial roads, parks and other areas with adequate space are being planted/replanted with the larger tree species. Attitudes are important here. In cultures where housing development originally took place in woodland settings, or where the shade of large trees is appreciated in warm, sunny climates, then larger specimen trees tend to be accepted more readily, even when grown in close proximity to a building.

Shrubs too reflect their natural ecological niche by being utilised as the understorey of new woodland plantings, but they are also exploited in roadside plantings, used as screens, barriers, and ground cover, and to stabilise soil on slopes. This is in addition to providing structure and interest in shrubberies and mixed ornamental borders. Aesthetic attributes largely rely on foliage colour and form, as well a multitude of taxa grown notably for their flowers (e.g. *Rosa, Rhododendron, Hydrangea, Hypericum, Hibiscus, Callistemon, Lagerstroemia*, *Ceanothus,* and *Helianthemum*) fruiting berries (e.g. *Cotoneaster, Berberis,* and *Viburnum*) or scent (e.g. *Philadelphus, Syringa,* and *Daphne*). Some of these genera are popular due to the wide range of species used in cultivation; in the case of *Rhododendron* species, they range from prostrate alpine garden plants, e.g. *Rhododendron campylogynum* (0.025–0.1 m tall) to semi-tree-like *R. arboreum* (20 m tall). Crossing different species (hybridisation) has increased the range of garden/landscape taxa further. This no more so than for *Rosa* (roses) of which there are in excess of 3000 cultivars currently in commerce – covering every hue of colour in their flowers, with the exception of a genuine true blue or the 'fabled' black.

Woody climbing plants tend to fall into two categories: large vigorous types that can cover a building wall or act as ground cover over a slope or embankment (*Hedera*, ivy; *Parthenocissus*, Virginia creeper; *Wisteria*; *Monstera* (Swiss cheese plant); *Bougainvillea*; *Rosa filipes* cv. Kiftsgate; *Rosa* cv. Rambling Rector; and *Rosa* cv. Paul's Himalayan Musk) and climbers that are more limited in their extension growth and are often associated with container planting or better-quality cultivated soil, e.g. *Jasminum* (jasmine), *Clematis* hybrids and *Lonicera* (honeysuckle).

6.3 Propagation

Trees, shrubs and woody climbers are propagated by seed, cuttings and grafting onto rootstocks. A few commercially important species are also propagated via micropropagation (tissue culture), due to being difficult to produce successfully from the first three methods. A large component of the ornamental woody plant genotypes (taxa) available is derived from either selected mutations (sports) or cross-bred 'cultivated' genotypes, and these will not reproduce 'true to type' via seed (i.e. sexual) propagation. Clonal (asexual) propagation via cuttings, grafting or micropropagation is then utilised as a means to build up stock of these genotypes.

Sexual propagation

This is used where genetic variability is desirable and a lack of uniformity does not impact on the plant performance or design.

Seed

The use of seed for the propagation of woody perennials relates most to the production of woodland or hedgerow specimens, particularly where native species are being exploited, or even demanded. For woodland grants in the UK for example, landowners may be required to plant young trees produced from local provenances of seed (genotypic forms common to that geographical area). These are predominately the specimens used for woodland planting or large landscapes contracts, such as mass roadside plantings. Seed is used much less as a means of propagation for plants in more formalised urban areas such as street boulevards, parks, or even for those genotypes that are retailed through garden centres, largely because the latter are composed of a selected clone of a particular species (desired for either its uniformity of growth or enhanced aesthetic characteristics such as unusual flower/foliage colour or form).

Seed is collected from known 'superior' specimen trees or, where large volumes are required, from 'seed orchards'. These are collections of parent trees planted together for the specific purposes of producing high-volume/high-quality progeny. Such orchards may include selections that are of a particular form of phenotype or have enhanced disease tolerance characteristics.

For seeds to germinate successfully, a number of factors need to come together:

- Individual seeds must possess enough carbohydrate, fat and protein reserves (for this reason, large seeds frequently have better germination rates compared to smaller seeds of the same taxon).
- The dormancy-breaking requirements have been satisfied:
  - sufficient moisture has been imbibed by the endosperm to activate growth;
  - the seed coat has been weakened enough to allow the radicle (first root) and hypocotyl (first 'storage' leaves) to emerge.
- Temperature and irradiance spectra are optimal for the given species.
- The moisture content of the soil is appropriate for further development, without impeding oxygen diffusion to the rapidly respiring tissues.

Dormancy covers a number of factors constraining seed germination. Physical dormancy may be due to too hard a seed coat. This is overcome by physical abrasion with sandpaper or a file, or exposure to acidic or warm water solutions: a process known as scarification. These techniques mimic the processes that would occur in nature before a seed coat was weakened, i.e. freezing and thawing stress, action of soil microorganisms, passage through the digestive tract of mammals/birds or action due to fire. Even if the integrity of the seed coat is compromised, the embryo itself may not germinate. This is due to endogenous dormancy controlled by phytohormones that act as inhibitors of growth. Exposure to moisture and chilling (stratification) is required to deactivate the action of these hormones (e.g. the degradation of endogenous abscisic acid) and thus facilitate normal cell division and expansion.

In contrast, ectodormancy usually refers to the situation where seeds remain viable, but external environmental conditions are not conducive for germination (temperature, irradiance and humidity). As a further challenge, some woody plants enter secondary dormancy if water has been imbibed but external conditions do not promote growth. This is again controlled hormonally, and further treatment with chilling (or in some species, irradiance) is essential. Artificial exposure of seed to treatment with gibberellic acid (GA_3 – a phytohormone that promotes cell expansion and growth) can also overcome secondary dormancy.

Controlled germination is feasible for many woody plants through the use of temperature schedules, scarification or the application of hormonal treatments. Various protocols exist that encompass the requirements of different species. These are widely used in commercial production and facilitated by the use of refrigerated cabinets and glasshouses. In nature of course germination is much more sporadic, but natural re-seeding is still a valid technique in certain landscape redevelopments – for example, allowing woodland regeneration via a residual 'seed bank' left in the soil, including after all the mature trees have been felled.

Vegetative propagation

This allows specific taxa to be 'cloned' and ensures their 'trueness to type', thus providing a degree of predictability on final size, form and colour of foliage or flower. Various techniques exist for landscape plants (Table 6.2).

Cuttings

The technology involved in rooting cuttings of woody plant species varies considerably. The simplest method is to insert cut stem sections in the ground and encourage latent, pre-formed root initials to expand and develop. This occurs in *Salix* (willow), *Populus* (poplar), *Ribes* (currants) and *Forsythia* (propagated as hardwood leafless cuttings in winter). For species that do not naturally develop preformed roots in the stem tissue then the synthesis of new adventitious roots is required. These are generated from undifferentiated cells usually formed after wounding (such as occurs at the base of an excised stem section). In commerce, this is most commonly achieved by propagating new plants from softwood, leafy cuttings, although the levels of success are very much dictated by genotypic and environmental factors.

With these softwood, leafy cuttings, creating a propagation environment that supports the cutting long enough to allow roots to form is critical. In practical terms this means providing an environment that facilitates leaf photosynthesis whilst minimising water loss from the cutting *yet* also ensuring the medium the cuttings are inserted into is free-draining and does not become anaerobic. Tolerance to the stresses experienced during propagation and the inherent rhizogenic potential of different taxa affects the ability of cuttings to form

Table 6.2. Vegetative propagation techniques

Technique/facility	Tissue	Methods
Hardwood cuttings	Semi- or fully lignified stem sections, with no leaves present.	
Field hardwood cuttings		Stems inserted into field soil – often used for species with pre-formed root initials that generate new root systems readily, e.g. *Salix* (willow) spp.
Protected hardwood cuttings		Basal heat applied to encourage callus and root formation at base of cut stems. Systems such as the 'Malling Bin' used, where stems are bundled and placed on heated sand beds covered with bark. Bark/sand strata keep cuttings moist whilst allowing extra water to drain. Basal temperatures of 18–21°C encourage rooting, but aerial temperatures are kept cool and light excluded to avoid premature bud development and shoot extension.
Softwood cuttings	Un- or semi-lignified stems, with leaves present.	
Sun tunnel		This is a 'mini-polytunnel' with polythene placed over low hoops and cuttings inserted into the ground or raised bed. The polythene helps retain humidity, with cuttings hand-watered intermittently.
Shade box		Box placed over soil, pit or raised bed and covered with palm leaves or open hessian. Used in the tropics where natural humidity is high, and shading material helps reduce the high irradiance intensity and minimises stress on the leafy cuttings.
Contact polythene		Cuttings inserted into sand beds or rooting media held within shallow trays. Polythene sheet draped over the cuttings is in direct contact with the upper leaves. Contact with polythene reduces transpirational water loss from the leaf. Sheets pulled back temporarily at times of watering. Often used in combination with glasshouse shading to avoid 'leaf scorch'.
Open mist		Cuttings in sand beds, trays or pots and placed under mist nozzles. Mist frequency controlled by either computer (e.g. based on irradiance integrals), timer or 'wet-leaf' system. In the latter system, an electrical signal is maintained by moisture on the artificial leaf and when this dries out it triggers the nozzles to spray, thus re-wetting the cuttings but also reinstating the electrical circuit on the artificial leaf. Growing medium needs to be free draining.
Enclosed mist		As above but enclosed within polythene tents that also help raise and maintain humidity, whilst still providing intermittent leaf wetting.
Fog		Controlled in a similar manner to mist. Finer water particles, though, help in the maintenance of humidity, without excessively wetting the rooting media. Fog also disperses incoming solar irradiance, indirectly providing a form of shade.
Aeroponics/Hydroponics		Roots can be induced in some woody plants by placing the base of the cutting in a (well-oxygenated) water bath (hydroponics) or high humidity/misted air (aeroponics).
Root cuttings	Roots	
		Procedures vary depending on the diameter of the roots being harvested. For species with thin roots, sections 25–50 mm long are cut and placed longitudinally on an 'open' medium within trays. These are covered with polythene or glass to provide heat and humidity. For species with thicker, fleshier roots, sections are 50–75 mm long and placed vertically into the growing medium, ensuring polarity is correct. The upper part of the root cutting is conventionally given a horizontal cut, while the basal end is cut at an angle to bestow a point at the base, thus aiding insertion into the medium. New shoots regenerate from the cut root sections.
Layering	Stems	
		This involves the bending or laying down of a shoot, pegging it at ground level and covering with earth. The act of bending the shoot alters phytohormone flow, and 'earthing-up' with soil encourages roots to form on the stem. Once these develop the original shoot can be cut, severing it from the mother plant. In due time, the resultant plant is lifted from the soil and potted on.

Continued

Table 6.2. Continued.

Technique/facility	Tissue	Methods
Air-layering		Air-layering follows a similar principle to above, but in this case the rooting medium is packed around a stem and held within a polythene bag. Again, exposure to the dark, moist medium encourages root formation and the rooted stem section can be cut from the original plant.
Micropropagation/tissue culture	Apical meristems, but also other tissues such as pollen or ovaries.	
		This involves the use of aseptic *in vitro* systems. Small pieces of plant tissue (explants) are sterilised and then placed on a culture medium. Explants are placed in liquid or agar-based media containing a cocktail of chemical components (nutrients, growth regulators, vitamins and carbohydrates), the composition of which determines how the explants develop subsequently. For example, an agar medium may be used to proliferate new shoots from explants, or form an undifferentiated cell, 'callus' culture. Transfer of these tissues to a new medium encourages further cell transformations, for example inducing root formation on shoot sections, or the formation of somatic embryos (seed-like structures). These final tissue cultures are subsequently weaned off the *in vitro* systems and form young plantlets. In this way many thousands of new plants can be generated from just a single explant.
Grafting	Stems	
		This is the fusion of two separate tissue sections, normally from two distinct genotypes. Most commonly shoot sections (scion) are grafted on to another plant which has a developed root system (rootstock). The shoots of the rootstock are then removed (headed back) allowing the scion wood to develop and form the new branch framework. Grafting is used on woody plants to improve or control the vigour of the scion growth. For example, *Rosa* (roses) are grafted onto rootstocks that impart vigour to the flowering scions. In contrast, fruit such as *Malus* (apple) may be grafted on rootstocks that constrain growth and encourage earlier fruiting in the life of the plant (precocity).
Chip budding	Buds and stems	
		Used when the plant is not actively growing as in spring before bud-burst or in late summer when extension growth has ceased. The scion bud is removed with about 20–30 mm of stem attached. This is placed into a notch made on the rootstock of equivalent size and tied in with budding tape. The rootstock is not normally cut back until the union is complete, e.g. the following spring.
T-budding		Used on active growing stem sections 5–25 mm diameter. A T-shaped cut is made on the rootstock. Existing budwood from the scion (bud with approximately 40 mm of cambial wood) is inserted into the 'T' and tied in.
Patch budding		A regular section of bark is removed from the rootstock and an equivalent-sized piece of bud wood excised from the scion. This is then 'patched-on' and tied in.
Whip graft	Stems	
		This is used for small diameter material (e.g. 5–15 mm diameter) and involves the rootstock being given a complete slanting cut, of which a further cut is inserted in the middle, leaving a small flap (tongue) of tissue. Similarly, a slanting cut is made in the scion, with a further insertion to provide a corresponding flap. The stock and scion are slipped together with the flaps interlocking. The intersection is then tied tightly and sealed in warm wax to minimise moisture loss
Splice graft		As above but without the flaps (tongues).
Side graft		There are various forms of these, but all involve a vertical cut in the rootstock, with a wedge-shaped piece of scion wood slipped into the aperture. This is then tied and waxed, with the rest of the rootstock above the union being cut off. Side veneer grafts involve longer sections of scion wood (e.g. 40–50 mm). These are often used for grafting young conifers and other evergreens.

Continued

Table 6.2. Continued.

Technique/facility	Tissue	Methods
Cleft graph		Cleft graphing involves the horizontal cutting of the rootstock and then inserting a vertical split across the cut section. Small sections of scion budwood are then slipped into the aperture, ensuring cambium contact between scion and rootstock. The whole is then sealed with wax.
Wedge graft		Similar to the cleft graft, but vertical notches are cut in at right angles to the cut horizontal section of the rootstock. Scion sections are then gently tapped into place to allow a tight fit between cambial tissues.
Bark graft		Similar to the wedge graft but lose flaps of bark around the vertical cuts are nailed back in place to help apply pressure around the scion/rootstock interface.
Topworking	Stems	A special type of grafting where the crown is reduced on mature trees and complete cross-sectional cuts made across the branches leaving stumps. Scion sections are then grafted on to these stumps and develop into branches of the new scion genotype. Opportunities exist for more than one scion genotype to be carried by the rootstock, so increasing the varieties of, for example, cherries or apples derived from the 'one tree'.

roots. Some taxa have reputations as being 'difficult to root', whereas others are categorised by nursery managers as 'easy to root'. In reality, the levels of stress a cutting experiences depends on the interactions between the different environmental components. So, propagation environments that have relatively high irradiance may need to provide higher humidity to minimise moisture loss from the leaves. This is accomplished by enclosing cuttings within polythene tunnels or alternatively fogging the cuttings with fine water droplets (10 to 50 μm diameter). High irradiance levels in a protected structure of glass or polythene corresponds to enhanced temperatures, therefore additional leaf cooling and wetting through misting (water droplets of 50 to 100 μm diameter) may be warranted.

Propagation benches that are more heavily shaded, on the other hand, may be able to keep cuttings viable simply by draping a polythene sheet over the canopy of the cuttings (contact polythene) to help minimise water loss. Systems that retain high humidity and keep leaves cool through the deposition of water droplets often enhance rooting success. Fog and mist systems enclosed in a polythene tent, for example, are useful for optimising rooting in challenging subjects by providing both these key factors. In contrast, open mist has more variable humidity levels with negative effects for some species (Fig. 6.4). The majority of shrubs and woody climbers are propagated via softwood cuttings, and although cutting preparation procedures are similar throughout, idiosyncrasies do arise; *Clematis* cuttings, for example, are prepared with the base derived from the former internodal section of the parent stem, whereas in most other species the base of the cutting conventionally comprises a former leaf/bud node. The nodal section is considered to have greater cell regeneration potential.

A secondary factor affecting 'ease of rooting', however, is the level and action of phytohormones either within the cutting or applied to the cutting (i.e. auxin-based rooting powders and solutions). For cuttings of difficult-to-root species the generation/movement of endogenous auxins can be critical: auxins are often synthesised in regions of active cell division (developing shoot tips, new leaves, etc.) and then exported to the base of the cutting where root initiation occurs – a process known as polar auxin transport. For this reason, rooting potential in difficult species is optimised when cuttings are derived from the actively growing shoots in early summer, with a very rapid drop-off in the ability to root as shoot extension rates slow and resting apical buds begin to form. In *Cotinus coggygria*, stock plants were manipulated to encourage branching on the developing shoots (Cameron *et al*., 2001). These were then excised as cuttings to demonstrate the importance of the developing branches on root generation within the cuttings. Rooting was highest when the developing branches were left intact on the cutting as these were providing a source of endogenous auxin. Placing a chemical inhibitor to auxin movement within the cutting, either on the branches or the stem, reduced rooting. Conversely, where the developing branches were removed and subsequent rooting percentages were low, adding exogenous

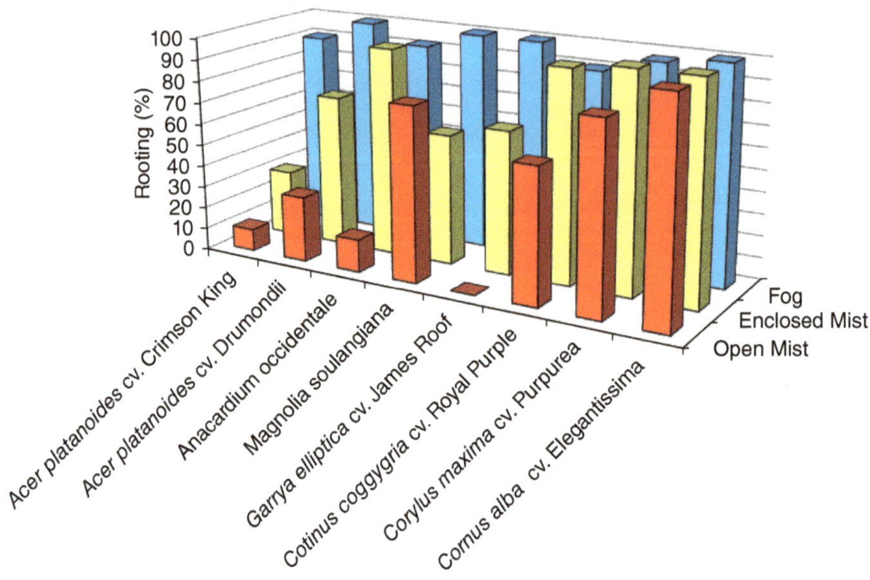

Fig. 6.4. The rooting percentage of cuttings from a selection of woody plants species and cultivars when propagated in either open mist (red), polythene-enclosed mist (yellow) or fog (blue) propagation environments. Note the variation in tolerance to open mist between taxa. (R. Harrison-Murray and J. Saranga, 2002, personal communication)

auxin to either the cut branch stumps or the base of the cutting tended to increase rooting once again.

As cuttings are prone to stress, nursery managers will apply fungicides to the crop, particularly to control *Botrytis cinerea*, although their use is not always absolutely necessary. Very heavy wetting through frequent fogging/misting for example may inhibit fungal spore germination and development.

The rooted cutting is known as a 'liner' or 'steckling' and will require weaning from the high-humidity environment it was rooted under. This is easily accomplished by rolling up the polythene covers around the propagation bed and/or reducing the frequency of misting/fogging by a small proportion each day for about 1 week. Once weaned, liners are left in their propagation modules to establish an effective root system, and once a small intact root system is formed (rootball) they can be potted on into small pots (e.g. 90 mm diameter). These are then grown on under protection or placed on container beds outdoors to reach marketable size. Landscape plants typically produced via softwood, leafy cuttings include trees such as *Chamaecyparis, Magnolia, Paulownia* (foxglove tree), *Thuja* cvs; shrubs, *Buddleia* (butterfly bush), *Buxus* (box), *Cistus, Cotoneaster, Euonymus, Forsythia* and *Viburnum* cvs; and climbers, *Clematis, Hedera* (ivy) and *Lonicera* (honeysuckle) cvs.

Grafting

Grafting is the union of a favourable taxon (scion) onto another plant (rootstock). Once a successful union between the two plants has taken place, the stem and branches of the rootstock plant are removed (headed-back), allowing the scion taxon solely to exploit the resources derived from the rootsystem. There are various types of graft-union and mechanisms for grafting (Table 6.2), but ornamental deciduous trees are commonly produced under field conditions when a single bud from a scion is grafted (budded) onto the stem of the rootstock at about 100 to 150 mm above ground level. Common landscape trees such as *Acer platanoides* cv. Crimson King, *Robina pseudoacacia* cv. Frisia, *Sorbus* cvs (rowans/whitebeams), e.g. cv. Joseph Rock, *Prunus* cvs. (flowering cherries and almonds), e.g. cv. Fragrant Cloud, *Malus* cvs. (crab apples), e.g. cv. Red Jade, and *Betula* cvs. (birches), e.g. cv. Grayswood Ghost are all produced this way. Budding normally takes place in August in the northern hemisphere, when the rootstock stem has started to lignify and there is sufficient warmth for the bud cells to fuse with the cambium of the rootstock.

In contrast, a side veneer graft carried out in winter is the preferred method for grafting selected cultivars of conifers. Rootstocks are grown in pots and allowed to go dormant in winter. After exposure to low temperature (5 to 10°C) for at least 6 weeks the rootstock is brought into a cool greenhouse. These are left for a few days to activate new root growth before grafting takes place. The plant should not be watered at this time though, as excessive sap rise through the xylem can push the graft apart after 'tying-in'. Notable ornamental conifers including *Picea pungens* cv. Hoopsii, *Picea omorika* cv. Pendula, *Pinus strobus* cv. Blue Shag, *Pinus sylvestris* cv. Glauca Fastigiata and *Pinus radiata* cv. Aureum are usually propagated this way.

6.4 Production

Growing-on

Production involves the propagation and the growing-on stages. Some nurseries may only specialise in one of these stages, i.e. there are specialised propagation nurseries and nurseries that just buy in already rooted or grafted material. Some 'traditional' nurseries will do both. Once plants have been successfully propagated, they are grown-on for a period before being sold as a 'finished' plant. Retailers specify the size and shape of the finished plants they require, as well as stipulating other parameters (the crop specification), e.g. minimum number of flower buds present, pot colour, and a peat-free growing medium. These specifications need to be achieved and a relatively uniform crop provided by the nursery in as speedy and as economical manner as possible. Profit margins for commercial nurseries rely strongly not only on volume of sales but also by minimising wastage and reducing production times. Hence, the newly propagated liners are often grown-on under protection (glasshouse or polytunnels) to optimise growth and aid root establishment, with plants being frequently pruned to improve their branching habit and retain crop uniformity (Fig. 6.5). With woody climbers this stage of production is associated with tying and training to a framework of bamboo canes or the like. Not only does this make the plants more presentable, but it also aids handling and avoids the stems/flowers being damaged. Chemical growth regulators have traditionally been applied during production for similar objectives or to encourage flower formation on the young stock. (Depending on species, a premium is paid for plants just coming into flower as this enhances sales impact, or it may be part of the pre-agreed specification.) A number of chemicals have been reduced in recent years, due

Fig. 6.5. *Cotinus coggygria* cv. Royal Purple (smoke bush) is a shrub with a strongly apically dominant growth habit (top), so frequent pinching or pruning of the shoot tips is required during production to get a better branched and 'balanced' specimen more suitable for retail sale (bottom).

to environmental or human exposure concerns, and agronomic means to retain compact growth or induce more flowers have been investigated. These include plant breeding to use more compact genotypes and the management of temperature, fertilizer application, irrigation and light in the commercial production of ornamental plants (Bergstrand, 2017). Reducing irrigation supply and altering the wavelength of irradiation plants are exposed to affects stem extension rates, and plants can be kept more compact, by applying regulatory techniques at certain critical stages in the growth cycle. Regulated deficit irrigation (RDI) applies water by using daily estimates of plant evapotranspiration. Applying only a percentage of the water (e.g. 80%) that a plant could potentially use induces physiological responses within the plant as it attempts to regulate its water use and save water. The plant will prioritise cell hydration but may curtail cell expansion for example, thus reducing the amount of new shoot growth it invests in. Thus, plants stay shorter and more compact in stature (Cameron *et al.,* 2006, 2008; Álvarez and Sánchez-Blanco, 2021). Another technique – partial root drying (PRD) – irrigates a part of the root system, but not the entire system (Davies *et al.,* 2002). This then induces endogenous biochemical signals in the dry portion of the roots that inhibit stomatal opening and growth. Comparison of the two systems with container-grown ornamentals through drip irrigation showed that both RDI and PRD reduced growth in *Cotinus coggygria* cv. Royal Purple, but in *Forsythia x intermedia* cv. Lynwood significant reductions were recorded only with RDI (Cameron *et al.,* 2008). RDI was also deemed successful with evaluations of *Cornus alba* cv. Elegantissima, *Lonicera periclymenum* cv. Belgica, *Choisya ternata* cv. Sundance, *Lavandula angustifolia* cv. Munstead and *Hydrangea macrophylla* cv. Blue Wave. Aspects of plant growth regulation are also covered in Chapters 7 (bedding plants) and 11 (interior plants).

During nursery production plants may be potted on 2–3 times and the final pot/plant size reflects the labour and other resources involved and the space/time taken to grow the crop; large stock is considerably more expensive than smaller specimens of the same species. Depending on geographical location, the growing media that is used during production can vary. In countries such as the UK, peat has traditionally been the favoured medium due to its capacity to hold high volumes of both oxygen and water, its good cation exchange capacity and it being relatively light to use and transport. Concerns over carbon dioxide emissions from peat extraction and the loss of peat bog habitat, however, have put pressure on the commercial use of peat in horticulture and alternative media are being sought and researched. Shredded bark, composted woodchips or sawdust, green (waste) compost, coir, sand and loam and many other locally sourced organic-based media are also used across the globe as components in potting substrates.

Shrubs and woody climbers tend to be grown in containers throughout (container or pot grown), but trees (and roses) are more frequently budded and grown-on in the field and either sold as bare root stock or put into containers before selling (containerised). Field-grown specimens that are larger than these are lifted from the soil and their root systems wrapped up in burlap (hessian or similar material) in a process known as 'rootball and burlap'. A large proportion of the root system can be lost during field-lifting and nursery teams will often undercut their stock some months prior to lifting. This is accomplished to encourage a fibrous root system to form in close proximity to the stem. Undercutting with a sub-soil blade, severs the primary roots – a process that initiates secondary (side) roots to form. During the lifting operations, bare root stocks are particularly susceptible to root desiccation, and they should be kept damp and placed in a cold store, before further packing and shipping.

Trees are either sold based on their height (younger specimens) or stem girth (older specimens where the circumference is measured at breast height). Within the UK there is a size classification (BS 3936 - Part 1: Nursery stock specification for trees and shrubs). In addition, certain terminology helps describe the tree age and form (Table 6.3).

6.5 Retail

Ornamental trees, shrubs and woody climbers tend to have three main market arenas: garden centres/nurseries which sell plants to the gardening public, nurseries which sell plants via the internet (both termed 'retail') and the landscape (or amenity) sector which provides nursery stock lines to landscape architects or their contractors (via wholesale transactions). Within both retail and wholesale sectors product price can vary considerably depending on more subtle divisions within each. Garden plants, for example, may be retailed through 'state-of-the-art'

Table 6.3. Nursery tree classification within the UK

Terminology	Description
Seedling	A 1-year-old or younger tree, sold by height, e.g. 40–60 cm.
Transplant	Trees started in a seed bed and undercut (primary roots severed by a sub-soil blade) at either 1 or 2 years of age. These can be left *in situ* to grow-on or be transplanted to a new bed. Sold with information relating to undercutting or transplanting and plant height.
Whip	Trees produced from seed or cuttings, with a central stem and little or no side branching. Sold by height.
Feathered whip	Larger than a whip, with side branches (feathers). Sold by height.
Maiden	Similar to a whip but produced via budding or grafting.
Feathered maiden	A maiden tree with side branches, usually 1 sometimes 2 years older than a maiden.
Standard trees	Tree with girth (circumference) measured at breast height (gbh).
Light standard	6–8 cm gbh.
Standard	8–10 cm gbh.
Select standard	10–12 cm gbh.
Heavy standard	12–14 cm gbh.
Extra heavy standard	14–16cm gbh.
Advanced heavy standard	16–18 cm gbh.
Semi-mature	18 cm+ gbh.

garden centres as a component of an extensive range of garden products on sale or alternatively may be sold by large multiple retailers such as DIY (do-it-yourself) stores and supermarkets. Those nurseries which specialise in supplying the landscape/amenity sector may grow plants to a specific contract or provide a more limited and predictable range of plants. Conversely, they are more likely to produce greater volumes of any single line to supply the needs of their customers for large landscaping contracts, such as a new housing development. A number of these nurseries also set up their own 'cash and carry' outlets which have a greater range of plants, but they are still retailed at wholesale prices to landscape or garden designers. In addition to these two main retail routes, the traditional outlet, and still a significant component of the market, are those nurseries that grow their own lines and sell direct to the public. This may be either through their own retail area or via the internet. Sales on the internet were already growing but became a lifeline for the ornamental plant sector during the SARS Cov-2 (COVID-19) pandemic, when physical access to garden centres and nurseries were restricted. This corresponded to a period when people had spare time and could not leave their home properties for more than 1 h a day (e.g. in the UK). Gardening boomed in popularity but only largely because internet sales were able to meet the demand for garden products, including plants. There is still a partial legacy from this, with internet sales remaining popular with some customers.

Many nurseries specialise in particular groups of plants, e.g. *Rhododendron* (Fig. 6.6), *Rosa*, *Clematis* or conifers, to develop their own unique market niche. Some of these may grow many hundreds of varieties within the one plant genus.

Sustainable production

As with other sectors, concerns over the environmental footprint of woody plant production are driving research and new initiatives to improve performance and sustainability. This includes investigations in to new growing media, including the use of recycled products, a reduction in pesticide use and a drive for greater integrated pest management and biocontrol approaches, more efficient use of water and the development of water recycling technologies, applying fertilizers to meet crop requirements rather than routine applications, irrespective of need, and finally attempting to reduce energy use, through alternative heat sources and low-energy crop lighting, e.g. the use of low-energy LED systems. (see also Chapter 11).

6.6 Establishment

Mature trees are large significant landscape features, not readily moved, so site selection for planting

Fig. 6.6. Some nurseries specialise in a single genus for production, such as *Rhododendron.* Even so, a single nursery may have as many as 500–600 taxa in their catalogue covering a vast range of flower colours, sizes, and plant forms. Top: *Rhododendron* cv. April Showers; bottom: *Rhododendron* cv. Polarnacht.

young trees is critical. Provision of adequate space is key, but consideration also needs to be made for proximity to buildings, roads and other infrastructure, the impact of the trees on sight lines, future shading traits and potential change in land use. A number of factors require consideration in selecting the appropriate species, particularly climatic conditions (temperature, wind exposure, including coastal 'salt-laden' winds, and humidity), soil structure and chemistry (drainage and water availability, gas exchange, nutrient levels, pH, organic matter content and the presence of phytotoxic elements) and biotic factors such as the likely presence of pests or pathogens (Table 6.4). Susceptibility to pests and pathogens has resulted in rapid declines in popularity for certain tree species. *Aesculus hippocastanum* (horse chestnut) is no longer recommended for planting in western Europe due to its susceptibility to *Pseudomonas syringae* pv. *aesculi* (bleeding canker) and *Cameraria ohridella* (horse chestnut leaf minor). *Hymenoscyphus fraxineus* (ash dieback or Chalara) has caused an almost complete collapse in the market for ash trees (*Fraxinus excelsior*) in the UK, and even crops with phytosanitary certificates are unsaleable due to the perceived threats. Shrubs too are affected. *Buxus sempervirens* (box) was once the 'go-to' choice for gardeners who wanted an easy-to-trim plant for low hedging or topiary purposes, but planting is now inhibited by the presence of *Cydalima perspectalis* (box tree moth) (Zemek and Pastirčáková, 2023).

In North America, the invasive jewel beetle *Agrilus planipennis* (emerald ash borer) has imperilled populations of *Fraxinus pennsylvanica* (green ash), *Fraxinus nigra* (black ash) and *Fraxinus americana* (white ash). *Fagus grandifolia* (American beech) trees are also under threat from beech bark disease: a combination of *Cryptococcus fagisuga*, (a scale insect) that helps introduce *Neonectria ditissima* and *Neonectria faginata*, two fungal species that produce cankers on the tree's bark. These lesions eventually grow into one another, girdling the tree which cuts off water and nutrients to the tree's canopy and ultimately kills the tree over time. *Tsuga canadensis* (eastern hemlock) is impacted by *Adelges tsugae* (hemlock woolly adelgid), a sap-sucking insect that reduces the plant's capacity to acquire/transport photosynthates, resulting in needle stunting and defoliation.

Even in circumstances where a whole species is not under threat, horticultural knowledge is important in making prudent decisions. Rosaceous species should not be planted on soils that previously held the same/similar species due to a phenomenon known as 'rose replant disease'. The precise cause of this is unknown but may be a combination of a build-up of pathogens in the soil and a depletion of certain micronutrients. Populations of *Streptomyces* bacteria and *Nectria* fungal species have been implicated with the possible causes (Yim *et al.*, 2020). Replacement plantings of the same/similar species often demonstrate signs of poor vigour and the dieback of roots.

Trees and shrubs are categorised by their preferences or tolerances to climatic and other environmental factors which aids in appropriate plant selection (Table 6.4). Climate change and associated changes in biotic populations, however, is beginning to alter what is suitable for specific locations (including the concept that native species may not always represent the optimal or best-adapted species choice).

Table 6.4. Factors to consider when choosing tree and shrub species

Climatic	Temperature	Summer temperatures too low to support appropriate shoot growth and development. Frost and freezing mid-winter temperature below that which genotype can tolerate – 'hardiness zone' information provides preliminary guidance on species selection. 'Unseasonal' frost and damage to tissues such as developing shoots or flowers. Sufficient winter chilling is required to release buds from endodormancy – poor and uneven bud development can occur if winter temperatures are too high.
	Wind	Physical damage due to strong wind, e.g. newly emergent leaves, or stem/branch structure. Exposure due to strong or excessively cold wind – tissue desiccation. Coastal, salt laden winds may induce osmotic and phytotoxic stress, e.g. tip burn on leaves.
	Precipitation/soil moisture availability	Frequency of drought or flooding events (see points on soil structure).
	Solar irradiance	Supraoptimal irradiance may cause photoinhibition, with UV wavelengths inducing direct cell damage. Tree canopy cover or planting in the shade of a building or wall may be a prerequisite for some shade-adapted species. Cultivars with gold/yellow foliage are particularly susceptible to excessive irradiance. Conversely, consider affects due to excessive shade. Some woody plants are well-adapted to shady woodland understorey conditions (e.g. *Acer palmatum* and *Camellia japonica*), others less so (e.g. *Cistus* spp. and *Genista* spp.).
Soil	Particle size and chemical composition	Determines soil type (sand, silt or clay) and degree of organic matter content, which will in turn affect physical, chemical, hydraulic and biotic factors.
	Structure	Defines drainage, water-holding and aeration characteristics – roots require free-draining soils, with pores to retain moisture and provide root aeration.Soils with high bulk density will impede root extension.
	Fertility	Provide macronutrients – nitrogen, phosphorus and potassium – as well as essential minor nutrients.
	pH	Needs to be suitable to avoid direct damage to roots but also allow availability of key nutrients. If plants are 'acid-loving' such as ericaceous species, then these should avoid chalky or limestone soils. Even non-ericaceous species such as certain *Quercus* or *Acer* provide more intense autumnal foliage colour if grown on lower pH acidic soils.
	Toxicity	Soils/substrates reclaimed from ex-industrial sites may contain chemical contaminants that are phytotoxic to plants, e.g. mercury, cadmium, chromium and residues of oil.
Biotic	Pests or pathogens	Likelihood on damaging infestations of pests or the presence of pathogenic fungi, bacteria and viruses.
	Weeds	If management of site is to be limited, then species need to be competitive and resilient against weed competition.
	Mycorrhizae	The presence of these fungi may aid establishment of certain woody plants through their symbiotic action.
	Wildlife	Wildlife conservation may be a consideration and species chosen for their ability to support invertebrates or provide food for birds and mammals.
Aesthetic		Meet the requirements outlined with respect to colour, form, line and texture. Provide interest and beauty at specific, or for prolonged, periods. This may include spring blossom, autumnal foliage colour or interesting bark colour/texture in winter.
Physical		Ability to provide shade, wind abatement, shelter from rain or solar glare, cooling influence, noise and aerial pollutant barrier, stabilise soil, etc.
Social and economic	Functional qualities	Privacy and seclusion, complement buildings, provide scale to landscape and sense of place, provide timber or food.
Management		Heavy and long-term investment required, training, staking pruning, etc.

Failure of young trees and other woody plants to establish successfully in the landscape is a common problem. This is due to a variety of reasons including the following:

- Lack of water during the establishment phase: not watered immediately after planting or during dry periods in the first year.
- Competition from weeds for (largely) water but also nutrients and light.
- Transplant shock (technically, a check in growth due to poor plant acclimation when moved from one environment to another). In practical terms, most land managers define it as an imbalance of the root-to-shoot ratio. Young trees may lose a high proportion of the fine roots due to lifting from the field, root trimming or desiccation when bare root stocks are left in warm or low-humidity conditions. Once planted, such plants struggle to acquire sufficient water to support top-growth (leaf and stem development). Typical symptoms are that the tree does not grow or growth is slow over a number of years, as the roots and shoots re-establish homeostasis, a period where the tree is susceptible to competition.
- Herbivore damage (rabbits/voles/deer, etc. eating young trees or stripping their bark).
- Vandalism.
- Poor maintenance – particularly injury to the bark of young trees due to strimmer damage (used to keep weed growth down around the base of the tree).
- The use of de-icing salts for roads and pedestrian areas.
- Limited 'root-run' due to restricted space – small containers or planting pits.
- Trenching of the soil by utility companies, cutting roots and causing soil compaction.

Where trees and other woody plant are located has a strong influence on their establishment rates and subsequent success, with trees in urban environments probably experiencing the greatest range of stress factors. As such, urban trees are considered to have slower growth rates than their rural counterparts (Quigley, 2004), but final size is strongly influence by their ecological characteristics. Early- and mid-successional-type species (often light demanding, e.g. *Liriodendron* [tulip tree] and *Gleditsia* [honey locust]) can reach their final height and girth dimensions in urban situations, whereas late-successional types (often shade-tolerant species such as *Fagus* [beech] or *Taxus* [yew]) regularly fail to attain their final potential size. As well as environmental and social influences, establishment rates vary strongly based on the level of management at and after planting. Regional variances are apparent in this respect, with 220 out of every 1000 Danish urban street trees needing to be replaced within 10 years of planting, compared to only 100 in Finland and Iceland. In general, park trees have higher survival rates than street trees, although local problems, e.g. vandalism in some parks, still induce significant losses. Costs associated with planting and establishing trees also vary between countries with some spending less than €250 per tree and others in excess of €2000. Within Europe there was a gradient from north to south in terms of the number of genera or even species used in streetscapes, with northern cities often relying on limited taxa. In contrast, Mediterranean cities and towns utilise a wide range of species, including frost-sensitive subjects such as *Citrus* and palms (Pauleit *et al.*, 2002).

Size/age of transplant

In more prestigious or iconic planting schemes there is a preference to plant large specimens to provide an instant effect. The downside is that larger plants may take longer to establish successfully, especially if there has been root damage due to lifting (e.g. via the ball and burlap method). In temperate regions (e.g. northern USA) it may take 5 years for a 100 mm diameter tree to regenerate its root system entirely, whereas after transplanting a larger tree (250 mm diameter) the root volume may not be fully restored for up to 13 years (Watson and Himelick, 1997). Root regeneration rates, however, tends to be faster in warmer climates.

It is often quoted that young trees, once successfully established, will outgrow more mature specimens, but there are conflicting reports as to whether this is true (Watson, 2005). So, although, many reports indicate planting younger trees is advantageous, others such as Struve *et al.* (2000) point out that in reality young specimens are still unlikely to outgrow their larger counterparts. Larger (older tree) transplants of *Tilia cordata* (small-leaved lime) were shown in one study to be better at resisting weed pressure and gaining height and full establishment than conventional, smaller-sized transplanted trees (Gallo *et al.*, 2020). A key element is how much root has been lost in the transplanting process; for example, large-container-grown material (with minimal root damage?) may establish and grow more effectively than smaller plants (Thetford *et al.*, 2005). Other factors too can influence the relative benefits between larger or smaller transplanted trees, including the species chosen, provenance of the genotypes, type of root system (predominant tap root vs fibrous roots), seasonal and other environmental conditions at time of planting, previous husbandry and the physiological state of the tree, as well as root-to-leaf canopy ratios and relative rootball to soil backfill volume (size of hole).

Soil cultivation and amelioration techniques vary significantly depending on the size of tree being planted. For young whips, planting into the parent soil simply involves inserting a slit or notch into the soil and carefully teasing in the plant roots, whereas for planting a large standard tree, major soil excavations and backfilling with quality topsoil are a requirement. For planting large trees, the planting 'pit' is recommended to be 2–3 times the diameter of the rootball and backfilled with a good quality soil of a known standard (e.g. in the UK complying with British Standard BS 3882). It is important the backfilled soil integrates with the parent soil to ensure movement of moisture through the profile and to avoid proliferating roots potentially having to bridge an air gap where the two substrates meet (for example, if there is soil/substrate shrinkage due to dry periods). More mature plants usually require greater investment in staking, tying and irrigation compared to smaller, younger stock. For some of the large grades of trees, tying and securing may involve both above and below ground techniques.

Soil conditions

Droughty soils and soils prone to compaction and waterlogging influence survival and establishment. For environmental horticulturists there is a trade-off between how often woody plants should be watered post-planting and the labour costs associated with doing so. Similarly, in arid areas, there may be considerations with respect to conserving water whilst meeting the needs of newly planted specimens. Some species can be classified by their water use characteristics, with, for example, *Salix matsudana* (corkscrew willow) and *Tilia cordata* (small-leaved lime) having high water use, *Platanus hispanica* (London plane) and *Fraxinus pensylvanica* (green ash) moderate use and *Acer platanoides* (Norway maple) low water use traits (Montague *et al.*, 2004). Providing water may not only be important in ensuring the survival of recently planted specimens but also has longer-term influences. Differences in the root systems of *Acer rubrum* (red maple) due to irrigation regimes imposed over the first 24 weeks after planting were still evident 5 years after the treatments terminated (Gilman *et al.*, 2003). Regular irrigation during establishment pays dividends in terms of growth. Trees that were irrigated frequently during the post-planting stage had larger trunk cross sectional areas (175 cm^2) compared to those watered less frequently (137 cm^2). This was attributed to greater root numbers in the surface soil layers. In contrast, the infrequently watered trees had a higher proportion of their roots at a deeper level (30% of roots were at least at 300 mm below the surface). Reducing irrigation *once plants have become established* is often considered a positive influence as it encourages root proliferation at deeper depths and hence can help secure water supply, should surface layers dry out during drought events. However, as drought events are likely to increase with climate change, irrigation may need to be supplied for longer periods post-planting than has been the case up to now. Schütt *et al.* (2022), studying water availability to young trees in Hamburg, Germany, showed that soil water availability became critical for 3–5 months per year, and trees were exposed to prolonged periods of critical soil-water availability for two consecutive years.

Preconditioning plants in the nursery through controlled moderate drought stress has been considered as a technique to improve their resilience once planted out in the landscape. (They should, however, still be well watered in the container before planting as well as watered in on planting). Drought preconditioning can be carried out in the nursery 3–4 weeks preceding sale, once plants have reached their specified height. For woody plants destined to be planted out in the autumn, reduced irrigation can be introduced in August or September. It is important that irrigation is not switched off completely, but rather plants are weaned onto lower moisture availability, thereby induced physiological changes that enhance acclimation to drought. Alterations include more effective or prolonged closure of the stomata to minimise water loss, reductions in cell water content and increases in solute (osmotic) potential, induction of cells that are smaller in size and increases in cell wall elasticity, alterations in leaf and xylem water potential and modification to endogenous hormone concentrations, e.g. increased concentrations of abscisic acid.

In an attempt to minimise failure rates with bare root seedlings after planting out, roots can be dipped into clay slurry or hydrogels or packed in damp peat to help minimise root desiccation. Even relatively short periods of exposure to air can reduce viability, i.e. as measured in minutes. *Pseudotsuga menziesii* (Douglas fir) seedlings exposed to warm air (32°C) dropped from 100% survival after 15 min exposure to only 60% and 50% survival after 60 min and 120 min exposure, respectively (Hermann, 1967). Overall, the effect of hydrogels used either as a root dip or at the time of planting appears mixed. Apostol *et al.* (2009) indicated applying hydrogels to roots of *Quercus rubra* (red oak) reduced cell leakiness and promoted more rapid budbreak in the spring after planting (both evidence of reduced physiological stress on the roots). After planting though, there was no apparent advantage associated with this treatment as there was no difference with non-treated plants in terms of shoot length, dry weight, root volume, net photosynthesis or stomatal conductance. Similarly, others have found no advantage (e.g. Heiskanen, 1995), whereas when Specht and Harvey-Jones (2000) incorporated gel into the growing medium, they noted increased water uptake, stomatal conductance and increased biomass in *Quercus coccinea* (scarlet oak).

Drought is not the only stress to reduce transplant survival or delay effective establishment. Tolerance to waterlogging, wind, heavy metals, low nutrient availability, salinity, etc. are also strongly influenced by the taxa selected.

Urban soils

Urban soils are frequently impoverished and particularly problematic to trees and other woody plant species (see Chapter 2). The reduced lifespan of urban trees compared to trees in more rural settings has been attributed to poor soil structure, lack of water and pollutants prevalent in the urban environment. In surveying urban soils in Hong Kong, Jim (1998) identified that they were a difficult medium for plant growth due to limited space for root development and interference from the abundant underground utility lines, pipes and cables, and their management. Surface and subsurface soil layers are commonly compacted (both unintentionally, e.g. by vehicle movements, and intentionally in attempts to stabilise building foundations) resulting in restricted root spread and limiting the movement of air and water to roots. Soil 'sealing' through the use of impermeable roadways and pavements similarly impairs the movement of water and gases from the atmosphere to the rhizosphere. The presence of stones, boulders, construction rubbles, mortar, bricks, old paving, old and extant foundations and other obstructions often frustrate planting efforts and render many sites unusable. To add to the problems most soils are depleted of organic matter with insufficient quantities of the major nutrients required for vigorous plant growth. Gardeners often improve their soil texture by adding organic matter, but it is not always cost-effective to do this in public green space. Nevertheless, Somerville *et al.* (2020) suggest that adding organic matter can help poorly structured, compacted urban soils, under some circumstances. They found that organic matter improves plant available water in sand but not in clay soils; incorporation of compost and biochar in soil reduced tree water stress in times of drought and that compost and/or biochar increased tree growth compared to just tilling the soil before planting.

Other problems with urban conditions are that carbonate is released from calcareous construction waste and tends to increase pH, making soils more alkaline. This in turn makes it more difficult for plants to absorb phosphorus and micronutrients such as iron, copper, zinc, boron, cobalt and manganate. There may be other soil contaminants such as lead that directly impair growth. Additional injury to planted specimens can be induced by trenching and tunnelling through the rhizosphere via the actions of utility companies. Many urban soils have lost their natural soil horizons and instead have artificial horizons composed of poor-quality infill material.

To some degree, soils can be 'designed' to help street and plaza trees perform well, whilst minimising the damage to the physical infrastructure such as pavements and roadside kerbs. 'Structured' soils are an aggregate mix of gravel and soil that encourage deeper and more prolific tree root development. They tend to have a high porosity that aids aeration and supplies an effective water-holding capacity after rain, with large pores draining while small pores can hold water against gravity. Such materials are becoming popular when backfilling tree planting pits (Fig. 6.7). Structural soils not only allow for tree root development but are also structurally

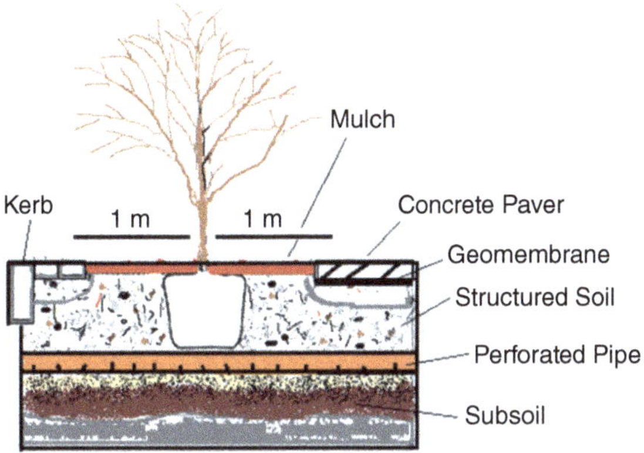

Fig. 6.7. Tree pit with structured soil incorporated.

strong enough to enable paving stones and other surface coverings to be positioned firmly on top. Planting pits themselves are useful to designate a specific volume for root growth and impair roots from causing displacement of pavement slabs or kerbstones. The downsides of structured soils, however, may relate to the open structure and high aggregate composition not retaining enough moisture during very dry periods or, indeed, not retaining nutrients as effectively as organic or clay-based soils (which tend to have a higher cation exchange capacity).

Polypropylene cells systems are sometimes used in the root zone, again to strengthen soil weight-bearing capacity and protect roots from compressive forces. Rather than being applied in isolation around individual trees, the combination of structured soils and cells can facilitate the formation of 'root zone corridors' placed linearly along roadways and pavements. This process thereby enhances the total volume of soil accessible to the roots of any one tree. Such systems can be linked to storm water management process – where excess rainwater runoff is fed in from buildings and roads and stored as irrigation water for the trees (Siering and Grüning, 2023).

In other locations where planting into true soils is the option, then soil compaction can be a problem. High soil bulk density inhibits root penetration and development, potentially limiting tree stability in the early stages of establishment. In such situations, pneumatic tools are used to expand the soil with compressed air. This creates fractures and fissures in the soil which allows roots to explore further through the soil profile and thereby expands the overall area for root development.

Mycorrhizae

Mycorrhizae are soil fungi that have a symbiotic relationship with plants and, through attachment to roots (Ectomycorrhizae) or even integration into the plant root cells (Endomycorrhizae), effectively extend the plant's ability to access nutrients and water from the soil (the surface absorbing area can be increased 100- to 1000-fold). In return the mycorrhizae derive energy from photosynthates directed towards the plant roots. Through the interchange of nutrients, water and hormones, the presence of mycorrhizae stimulates plant growth and accelerates further root development. They also improve resilience against soil-borne pathogens and to environmental stresses such as drought, salinity and high pH. The Endomycorrhizae fungi exchange nutrients with plants via finely branched arbuscules that form within the plant cell and are known as arbuscular mycorrhizal or AM-fungi (Glomeromycetes). Rather than arbuscules, other species form sac-like vesicles within the host cell and are denoted as vesicular arbuscular fungi or VAM-fungi. Arbuscular mycorrhizae are the oldest in evolutionary terms and the most widespread mycorrhizal fungi. It is thought that more than 90% of all plant species form symbiotic associations with these fungi. In contrast, Ectomycorrhiza are most commonly associated with tree interactions, forming relationships with species in the Betulaceae, Fagaceae, Pinaceae, Rosaceae and Salicaceae. These fungi (mainly Basidiomycetes) form a 'Hartig net': a network of hyphae held within the extracellular spaces of the root that in time provide a protective sheath around the root.

Plant/mycorrhizae relations are a normal part of soil ecology, but in soil that has been disturbed by human activity, the quantity of mycorrhizae decreases markedly. Following disturbance of soils, infection rates or levels of colonisation from arbuscular mycorrhizal are reduced by 30–50% compared to undisturbed soil (Oehl and Koch, 2018). So, typically, damaged urban soils may be bereft of the appropriate mycorrhizal partners. Indeed, Stabler *et al.* (2001) found lower mycorrhizal colonisation in residential areas compared to nearby undisturbed natural desert landscapes, even 10 years after such sites had been landscaped.

Low fungal propagule numbers in damaged soils may result in poor colonisation by mycorrhizae and present subsequent difficulties in establishing mycorrhizal-dependent woody plant species. Even when colonisation occurs, rates of proliferation are reduced. Trees growing in urban areas have lower arbuscular mycorrhizal and ectomycorrhizal colonisation levels compared to those in more rural environments (Bainard *et al.*, 2011). A number of factors could account for this: the highly disturbed and modified nature (pH, nutrient content, pollutants and lack of aeration) of urban soils, and the relatively lower proportions of host plant species present in urban environments and greater competition from non-arbuscular mycorrhizal species. Previous studies have shown that mycorrhizal associations with plants can be negatively influenced by environmental stresses, regardless of mycorrhizal abundance (Klironomos and Allen, 1995).

Compounding the fact that urban/degraded or disturbed soils have lower levels of mycorrhizae present, or reduced ability to propagate and colonise woody plants, plants grown in nurseries may themselves have limited mycorrhizal associations. Container plants grown-on with the 'luxury' of supraoptimal levels of nutrients and water seem to provide little opportunity for their symbiotic mycorrhizae to colonise. Moreover, those mycorrhizae that do occur naturally in the nursery are not capable of surviving in urban soils (Weber and Claus, 2000). This then leads to the scenario of plants which have been grown under ideal conditions being transplanted out to harsh field conditions but with little opportunity for adaptation (acclimation) through the action of their 'natural' mycorrhizal 'support' network.

Mycorrhizae can be introduced artificially at two points: to tree containers during production or via inoculation of the planting hole at the time of transplanting. Both approaches may result in limited success, however, if conditions are still not conducive to the fungi. As such, results from research are mixed. Drought tolerance was increased in young specimens of *Acer*, *Quercus* and *Tilia* during the nursery phase when mycorrhizae were introduced (Fini *et al.*, 2011). Although inoculation did not enhance shoot growth, it provided several physiological benefits such as the maintenance of less negative leaf water potential, higher apparent carboxylation rate, higher ribulose 1,5-bisphosphate regeneration (the compound that accepts a carbon dioxide, CO_2,molecule in photosynthesis) and higher quantum yield of photosystem II under water shortage. In essence, inoculation with mycorrhizae appeared to enhance the trees' ability to extract water, use it more efficiently and prioritise its use in photosynthesis.

Metzler *et al.* (2024) advocate that artificial inoculation with appropriate mycorrhizal species to urban situations can be successful. Using artificial structured substrates they found that active inoculation stimulated more distinct and diverse fungal communities, an effect that intensified over time. In the inoculated substrates, 45% of arbuscular mycorrhizal fungal taxa were subsequently detected to be present, compared to only 15% in non-inoculated areas. In other studies, the benefits of inoculation, especially during the nursery stage have been less tangible, e.g. in oak species such as *Quercus palustris Q. coccinea* and *Q. virginiana* (Gilman, 2001; Martin *et al.*, 2003).

Planting depth and protection

One area often overlooked in practice is planting depth. It is vital that the tree is planted so that its root collar is aligned to the surface of the soil. Amazingly, it has been estimated that between 75% and 93% of professionally planted trees may have their root collars buried by soil or mulch (Wells *et al.*, 2006). Burying a plant too deep interferes with the roots' capacity to uptake water, nutrients and oxygen, thereby delaying establishment. In the worst cases it may induce transplant failure. Although replication rates were relatively low (n = 6), Arnold *et al.* (2007) observed that survival rates were reduced to between 66% and 50% in *Lagerstoemia*, *Fraxinus*, *Nerium* and *Platanus* when root collars were planted 76 mm below soil level. Similarly, *Prunus x yedoensis* suffered 50% failure when planted at either 150 mm or 310 mm below the soil surface, although 100% survival was recorded in *Acer rubrum* (red maple). *Acer*, however, had increased incidence of root girdling (roots crossing and constraining the stem and other roots) when planted at the deeper depths. In contrast, deeper planting (200 mm compared to 100 mm) with *Eucalyptus pellita* increased plant volume, biomass and carbon storage by 22–32%, 18 months after planting (Wirabuana *et al.*, 2020).

Young trees need to be protected from the local wildlife too. Tree guards are commonly used to protect the stems of young trees from herbivores such as rabbits and deer and are considered to provide a favourable microclimate to the young tree, although the extent to which the latter is true has been challenged. Strapping and other physical barriers are also utilised to protect bark being stripped. Tree shelters are widely used to protect trees and there is strong evidence of improved survival rates (Abe, 2022).

6.7 Maintenance

Weed control

Weed control is required for effective establishment of newly planted trees and shrubs. Weeds compete with landscape plants for moisture, nutrients and, where growth is excessive, for light. Competition for water is greatest in drier climates and where soils have limited water-holding capacity. Removing weeds improves tree establishment under such circumstances (Fig. 6.8), although factors that aid weed growth, such as the incorporation of fertilizer, can result in even greater competition (Davis *et al.*, 1999). Davies (1985) states that to be effective weed control must eliminate competition below the soil surface, i.e. favouring tree root development over the root growth of the weeds. Control methods such as cultivation, herbicides or mulching are most effective, whereas cutting weeds above ground level is ineffectual over the long term. Studies on pine (*Pinus taeda* [loblolly pine] and *P. elliottii* [slash pine]) demonstrated that removing weeds stimulated new growth in recently planted specimens just as effectively as any fertilizer treatment. In *Quercus ilex* (holm oak), optimum establishment was achieved when weed control was combined with ventilated tree shelters in areas where weed pressure was high. The importance of removing weed competition to aid young trees cannot be overstated.

Traditionally weed control in amenity areas would involve herbicides; however, legislation and the cost of producing new agrichemicals for relatively niche roles is now limiting the number of active ingredients available for use in urban areas. Some countries have banned completely the use of herbicides in amenity

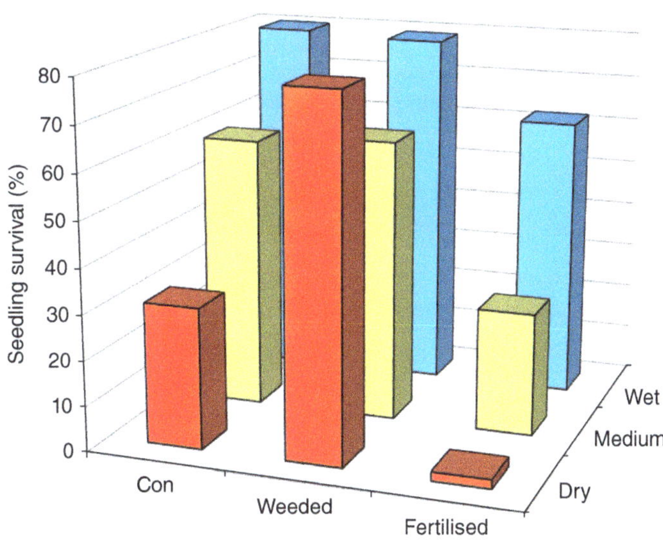

Fig. 6.8. Seedling survival of *Quercus ellipsoidalis* (northern pin oak) grown in dry (red), medium (yellow) or wet (blue) soils with weeds present ('Con'), removed ('Weeded') or present but with nitrogen fertilizer ('Fertilized'). (Modified with permission from Davis *et al.*, 1999.)

settings (e.g. Denmark and Sweden, Kristoffersen *et al.*, 2008) due to concerns relating to ground water contamination.

Mulches

Mulches are ground coverings that suppress weed growth as well as restrict moisture evaporation from the soil surface. Some mulch materials also help increase water infiltration and reduce soil surface temperature (e.g. from 2.2 to 3.3°C, Skroch *et al.*, 1992; Machado *et al.*, 2024). Mulches are classified as organic (pine needles, bark, chipped wood, etc.) and inorganic (usually polyethylene or polypropylene woven fabrics, although pebbles and other aggregates are employed in some high-value landscapes). Choice depends on aesthetic values, cost, durability and availability of the materials as well as effectiveness. Polyethylene and polypropylene mulches are considered more effective than organic materials in suppressing robust weeds but are unattractive and other mulch materials are often placed on top. Of the organic mulches, pine bark is popular due to its attractive surface as well as possessing a high durability. Organic mulches reduced total weed counts by 50% compared to control plots, and underlying organic mulches with polyethylene resulted in complete control (Skroch *et al.*, 1992). Billeaud and Zajicek (1989), evaluating a range of organic mulches, showed that deeper mulches are more effective at controlling weed growth, but they also tend to lock up soil nitrogen, decrease soil pH and reduce the visual quality of the landscape plants (Table 6.5). A good compromise was shallow organic mulches (e.g. 50 mm) laid over a polypropylene weed barrier. Others believe the role of mulches in immobilising nitrogen is overstated, and any effect of immobilisation is very localised to a few mm around the mulch/soil interface. There are concerns around plastic mulch materials leaving residue microplastics in the soil as they degrade (Liu *et al.*, 2024).

Pruning

Pruning in trees tends to fall into two categories. As a young tree, judicious pruning helps form the basic framework of the tree, and in later life it is used to remove problematic branches or alter the balance/weight of the trees canopy (crown). '*Structural pruning*' is encouraged on relatively young trees to ensure the continued development of a leading shoot (the 'leader' that eventually forms the main trunk), an 'open' branch structure with even spacing of the lateral branches initiated from the main stem and to remove any poorly angled branches that may later misshape the tree or cause splitting of the trunk. This structural pruning is important as it improves the integrity of the whole structure, enabling trees to have greater resilience against storms. It is important to remember that young trees may be pruned for different purposes, not solely for aesthetics, for example to create specific shapes in topiary, or if the tree provides a food crop, to encourage the appropriate amount and quality of fruit and easy access (harvesting of that fruit).

In older trees, pruning tends to be for structural issues or to aid management of the location, for example avoiding interference with overhead power or communication cables. Four techniques are used. '*Crown cleaning*' is employed to remove dead, diseased or overcrowded branches, and '*crown thinning*' aims to increase air movement and irradiance penetration within the tree canopy. Where lower branches may interfere with vehicle or pedestrian movement, or where there is a requirement to provide more room for buildings and other urban infrastructure, these may be removed via '*crown raising*'. '*Crown reduction*', on the other hand, reduces the overall size of the canopy by shortening the length of lateral branches. This can be done to ensure the trees remain in scale to surrounding buildings or to avoid entanglement in facility power or telephone wires. Pruning of large, mature trees is often limited to the removal of dead or hazardous limbs, and even here care is required to avoid pathogens and insects penetrating the wound site. Branches and limbs should not be cut flush to the main trunk, as a small section of remaining branch stub (collar) ensures cambial tissues can help form a protection covering around the heartwood, thus helping to protect the wound site.

The philosophy around the pruning of shrubs and climbers varies somewhat to that of trees. It is a subject that baffles many amateur gardeners and the occasional professional environmental horticulturist. Primarily though, most shrubs are grown for their flowers, and pruning is aligned to maximise the flowering display. This requires some knowledge, however, on how and where flower buds form on a given species or even cultivar. This is not always easy – in the *Clematis* group for example, some cultivars flower on the 'new' growth, i.e. the stems

Table 6.5. Potential benefits and drawbacks associated with the use of mulch materials

Benefits	Drawbacks
Weed suppression. Can eliminate solar irradiance thereby reducing weed germination, but some may also form physical barriers to weed growth.	Additional costs, including labour associated with placement of the mulch.
Some mulches are considered to have allelopathic properties, inhibiting weed germination by phytochemical means.	Cost–benefit ratios can be positive if plant survival is higher, or there is a lower requirement for subsequent irrigation and there are lower costs associated with weeding and pest/pathogen control.
Reduced evaporation.	Organic mulches may have contaminants present such as arsenic or formaldehyde, e.g. if wood materials have been previously treated with preservatives.
Temperature modification, e.g. insulation against frost. Inorganic mulches such as gravel and pebbles can help keep soil cool in summer.	Wood chips and other organic mulches if not properly composted before use can potentially contain pathogen spores or weed seeds.
Resistance against compaction/compression forces.	Black plastic mulches may heat the soil – a possible advantage in winter, but a disadvantage in summer. Even dark organic mulches may raise local temperatures, resulting in greater evapotranspiration and water loss in the adjacent plants.
Reducing precipitation runoff (e.g. by 28–80% (Wang *et al.*, 2021) and improving infiltration to soil.	Some plastic or fabric mulches interfere with water infiltration and movement.
Protect soils from wind and water erosion.	Wind may blow mulch materials off the soil.
Rapid root growth and hence stabilisation of new plants.	Subsequent removal of fabric mulches can cause root damage, as fine surface roots are removed too.
Increased survival rates in young transplants.	Bark, wood and certain other organic mulches may 'immobilise' soil nitrogen due to the microbial process associated with their degradation. Research suggests though that this may be very localised in the soil/mulch interface, and would, for example, not interfere with the ability of deeper roots of woody plants to access nitrogen (Chalker-Scott, 2007).
Reduce 'rainsplash' back into the plant foliage, thereby reducing potential for pathogens (*Pythium*/*Phytophthora* spp. to infect leaves).	Mulches may accommodate non-pathogenic microorganisms, such as the dog's vomit slime mould (*Fuligo septica*), which detracts from the aesthetic value of the mulch.
Microbial populations associated with organic mulches can be competitive or predatory on plant pathogens, thus reducing disease incidence. Other mulches may possess chemicals that are toxic to pathogenic organisms or inhibit microbial activity, e.g. thujaplicin from *Thuja* spp. wood or bark chips.	Certain mulch materials may be flammable and add to the risks in fire sensitive locations. Plastic mulches may be linked to nano- and microplastic pollution.
Such chemicals may also repel invertebrate pests.	Mammals, particularly rodent species utilise plastic and fabric sheets for shelter. Some of these species may damage tree/shrub stems.
Some mulch materials have a positive aesthetic value.	Plastic/fabric materials are often considered unsightly
Market for recycled wood, rubber and glass products.	The sustainability of some materials may be questionable, either derived from petrochemical products, or sourced and transported from distant quarries.

that have just grown that summer. Others though flower on the growth that was produced the previous season (i.e. the summer before). Thus, in practice those species (e.g. *Clematis orientalis*, *C. texensis* and *C. viticella* and cultivars (e.g. *C.* cv. Duchess of Cornwall, *C.* cv. Hagley Hybrid, *C.* cv. Nubia and *C.* cv. Star of India) that flower on the current season's growth (Group 3 *Clematis*) can be pruned hard back to ground level every winter. On the other hand, *Clematis* that flower on the 'old' wood are lightly pruned (Group 2) – removing weak and dead stems only, in late winter. Although after flowering, a few shoots may also be hard pruned to encourage some regrowth/rejuvenation from the base of the plant. *Clematis* that occur in Group 2 include *C.* cv. Sunset, *C.* cv. Bees Jubilee, *C.* cv. Guernsey Cream and *C.* cv. Fuyu-no-tabi and most double flower forms, e.g. *C.* cv. Vyvyan Pennell. The missing group so far (Group 1) requires no pruning at all and include early-blooming species and cultivars, including the *C. montana* and *C. alpina* types, e.g. *C. montana* cv. Elizabeth or *C. alpina* cv. Frankie.

Similar approaches are true for other climbing plants and shrubs. For taxa that flower on the current season's growth, hard pruning can be implemented without detriment to flowering capabilities, e.g. *Buddleia davidii* and bush (hybrid-tea) roses. Hard pruning can also be employed to spring flowering taxa such as *Forsythia* but only immediately after flowering, so any new shoots have time to develop enough maturity to form flowers for the subsequent spring. Other taxa are also pruned immediately after flowering, but here only the very old mature 'unproductive' stems or those shoots that have just had flowers are removed (e.g. *Philadelphus*, mock orange), leaving younger shoots to flower the next year.

Foliage and stem colour are the other attributes that shrubs are grown for. Hard pruning of colourful-stemmed *Cornus alba* (dogwood) and *Salix* (willow) varieties results in more vivid colouring associated with the new, young shoots, and these can be a wonderful focal feature of the winter garden. Likewise, large bold swathes of shoots and leaves can be induced by hard pruning of *Cotinus coggygria* (smoke bush), *Corylus avellana* (hazel), *Corylus maxima* (filbert) and *Pawlonia tomentosa* (foxglove tree). Other shrubs just need pruning infrequently to keep their size manageable.

Frequently, landscape shrubs are pruned to regular shapes such as squares and rectangles; where this fits well with the geometrical patterns of nearby buildings/other urban infrastructure or is in keeping with historical landscapes such as parterres, this is appropriate. All too often though it is used within a mindset to keep things 'neat and tidy', which results in landscapes that look constrained or incongruous with the taxa that have been chosen. Moreover, such approaches limit the geometrical heterogeneity within the planted landscape, thus reducing the opportunity to provide much needed 'habitat' for urban birds and invertebrates.

6.8 Plants in the Landscape – Right Plant, Right Place

Horticulture, and gardening in particular, has always had an element of 'challenge' in it. This is one of the attractions of the discipline, to grow something that others can't, or to create unique designs that express different ideas or concepts. There is also a strong desire for novelty, a consequence of which is that many new ornamental plants are bred every year. Whether it is the immense effort that goes into growing the biggest, the latest and the best for the local flower show, or the challenge of developing a new colour break or form of flower, these aspects add much of the intrigue and fascination involved with the subject. A garden, almost by definition, is an area where a range of plants grow that would not normally occur there naturally, or at least where the local species are actively manipulated to favour certain plant taxa over others. So, in essence, the idea of using plant species that might not quite fit the local conditions has always been a driving force in horticulture, with the skill of the cultivator being to adapt the conditions to suit the plant. As a consequence, tropical plant species are grown in expensive to construct and run glasshouses at 50° north and 40° south of the equator, *Rhododendron* and other acid-loving ericaceous species are plunged into peat beds in gardens over alkaline chalk, bog gardens are created on sandy free-draining sites and heavy clay soils are backfilled with gravel to facilitate the growing of species that require 'sharp' drainage!

Life is made much easier, however, if the plant species used in a set location have some adaptation to the conditions prevalent on the site. As such, a number of the concepts within environmental horticulture relate to using, on a more frequent basis, species well fitted to the conditions presented and not necessarily attempting to alter radically the conditions themselves. This does not automatically mean that only local-adapted species need be used, but that consideration should be given to the types of plant that are likely to be successful, including species from other continents or biomes (Fig. 6.9).

An understanding of a species ecophysiology (where it comes from and what is it adapted to), or even just a quick inspection of its morphology, can help provide some indications of the conditions it will tolerate.

Fig. 6.9. Gardens mimic nature, rather than replicate it, Here, 'temperate' plants are grown in the author's garden in the UK, but the genotypes are based on taxa from a wide range of geographical locations including North America (e.g. *Acer negundo, Thuja occidentalis* and *Picea sitchensis*), New Zealand (*Phormium tenax* and *Cordyline australis*), Mediterranean basin (e.g. *Olea europea*), Eurasia (e.g. *Cotinus coggygria*), South Africa (e.g. *Crocosmia* spp.) and western and central Europe (*Taxus baccata*).

Wind tolerance

Trees and shrubs that tolerate strong winds or can resist desiccation from drying winds often have identifiable traits. This includes waxy or needle-like leaves that help restrict water loss from the leaf. Pine and spruce trees are useful for exposed windy sites due to these characteristics, for example, *Pinus nigra maritima* (Corsican pine), *Pinus sylvestris* (Scots pine) and *Picea sitchensis* (Sitka spruce). Similarly, broadleaf species may include *Quercus ilex* (holm oak) or *Escallonia* spp. (particularly good on more temperate coastal situations) due to the thick cuticles on their leaves. At a lower scale, the strap-like leaves of sub-shrubs such as *Calluna vulgaris* (heather), *Erica* spp. (heath), *Lavandula* (lavender) and *Rosmarinus* (rosemary) help retain moisture. Some species avoid the wind by growing close to the ground in a hummocky (*Pinus mugo* and its cultivars) or prostrate form (e.g. *Salix arctica* [arctic willow] or *Juniperus squamata* 'Blue Star'). Other species may grow tall but can resist the wind by either possessing thin branches which allow the wind to filter through or have an ability to bend and be flexible in the wind (e.g. *Betula* spp. [birches], *Salix* spp. [willows] and *Pinus flexilis* [limber pine]). In addition, species as typified by *Acer pseudoplatanus* (sycamore) are expert at tolerating wind, cold and even salt spray in combination.

Wet soils and flooding tolerance

Some locations are more prone to water accumulation that others and planting should take account of this. Swales and rain gardens are designed to capture runoff water and hold it temporarily in an attempt to avoid urban drainage systems being overloaded. Other planted areas may just naturally be at the base of a hill or slope and receive all the natural runoff and percolation of water from the surrounding landscape. Alternatively, many urban green spaces are actually located along riverbanks, on floodplains or on other areas prone to natural flooding. These locations being seen by planners as 'sacrificial land', i.e. areas that should not be developed on precisely because they are a relief space to allow flood water to accumulate rather than inundate nearby buildings and houses. Finally, certain clay soils may also be prone to wetness, even when on higher land due simply to poor drainage capacity.

Many species in the plant kingdom are well adapted to wet soils and periods of flooding, but many others are not and careful selection is required for such sites. Riparian (riverside) woody plants usually do well in such locations, and species such as *Alnus glutinosa* (alder), *Populus nigra,* (black popular), *Populus deltoides* (cottonwood), *Betula nigra* (river birch), *Salix* spp., *Taxodium distichum* (swamp or bald cypress) and *Taxodium ascendens* are ideal for their ability to tolerate 'out-and-out' flooding. Other species are more likely to be damp woodland species in terms of their ecophysiology and are good at coping with wet clay soils rather than prolonged periods of inundation per se. These include small trees such as *Amelanchier canadensis* (service berry), *Aronia arbutifolia* (red chokeberry), *Styrax japonicus* (Japanese snowbell), *Halesia carolina* (little silverbell), *Carpinus betulus* (hornbeam), *Corylus avellana* (hazel), *Sambucus nigra* (elder) and *Populous tremula* (aspen); and larger trees, *Fraxinus excelsior* (ash), *Cercidiphyllum japonicum* (katsura), *Nyssa sylvatica* (black tupelo), *Davidia involucrata* (handkerchief tree) and *Liquidamber styraciflua* (sweetgum). Shrubs that do well in damp, if not waterlogged, conditions include *Ribes sanguineum* (flowering currant), *Euonymus europaeus* (spindle), *Cornus stolonifera* (dogwood), *Rubus hispidus*, *Viburnum opulus* (guelder-rose), *Hydrangea* spp., *Vaccinum corymbosum*, (highbush blueberry), *Kalmia polifolia* and *Myrica gale*.

The severity of effects due to waterlogging depends on the extent to which plant species themselves can adapt to the prevalent conditions and also on aspects such as season of flooding (temperature and growth phase), duration of flooding, depth of waterlogging and soil type and degree of microbial activity, as well as the condition of the flood water (e.g. degree of deoxygenation). Most plant species are more prone to waterlogging injury when temperatures are higher (i.e. summer) as this accelerates respiration and oxygen deficiency within the soil (Fig. 6.10).

So how do some species seem to tolerate waterlogging better than others? This may be due partially to different biochemical responses (see Chapter 2 for responses to waterlogging), but other physiological factors/alterations help some species cope with anaerobism. This includes the formation of airways (aerenchyma) through the stem to the roots that allow oxygen to diffuse down the plant internally and/or the development of surface roots that can access oxygen from the atmosphere more readily. Indeed, genera such as *Salix* have preformed root initials present in their stems, allowing them to send out new roots quickly when water levels change. These new 'surface' roots then provide plants with the water and nutrients they need and compensate for the older, now dysfunctional, roots located deeper down the soil profile in the anaerobic zone.

Some of the differences in tolerance between species are explained by such factors. When reviewing those trees suitable for urban soils prone to waterlogging, Smith *et al.* (2001) noted that *Corymbia maculata* and *Platanus orientalis* were able to induce new roots when waterlogged and to recover quickly after waterlogging. In contrast, *Platanus hispanica* did not induce roots whilst under stress but recovered well on draining, whereas the least well-adapted *Lophostemon confertus* was unable to initiate roots both during and sometime after waterlogging. Although riparian species generally have the best adaptations to waterlogging, it can be surprising which other species possess favourable characteristics and tolerances. *Corymbia*, for exam-

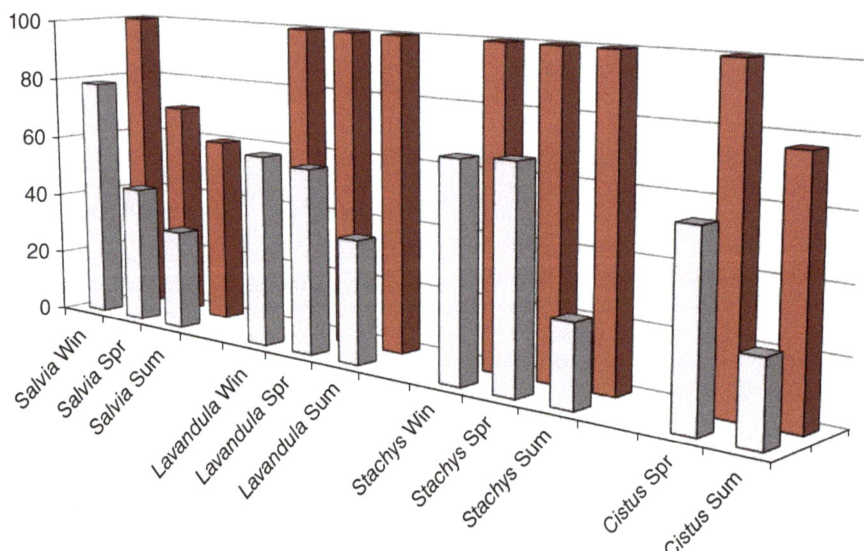

Fig. 6.10. Waterlogging and seasonal (temperature) effects. The effect of 18-day waterlogging in four Mediterranean plant species on survival (red) and biomass (grey) when flooded in semi-controlled, glasshouse conditions in winter ('Win') spring ('Spr') or summer ('Sum'). Survival rates decrease in *Salvia officinalis* and *Cistus* x *hybridus* during warmer periods. *Lavandula angustifolia* and *Stachys byzantina* survive all waterlogging events but biomass is reduced during flooding in summer. (Modified with permission from King *et al.*, 2012.)

ple, grows in well-drained, rocky locations in eastern Australia, not habitats normally prone to flooding. Similarly, certain drought-adapted 'Mediterranean' sub-shrubs such as *Lavandula* and *Salvia* possess some adaptation to waterlogging (King *et al.*, 2012). As indicated, anaerobism is a critical stress within waterlogging and providing low oxygen conditions (hypoxia) in this study, prior to the full removal of oxygen (anoxia), increased root survival in *Salvia,* indicating the potential to adapt to low oxygen environments.

Drought tolerance and xeriscaping

Drought-tolerant plants tend to be easier to identify than those with tolerance to flooding. Typical clues include the presence of succulent, pubescent (hairy), small or narrow leaves, or leaves that are coloured to reflect the incoming irradiance (blue, grey or silver). The absence of leaves altogether may be a feature! Some tropical species shed their leaves during the dry season to minimise water loss. In a similar manner, species such as *Eucalyptus*, adapted for warm, dry climates, have chlorophyll in their stems and bark which contributes to photosynthesis when the leaves' ability to fix CO_2 is compromised. Other adaptations may include deep tap roots or conversely fine surface roots (the latter being used in desert situations to capture moisture from light rains or dew that forms overnight, and where the water does not penetrate deep into the soil profile). There are exceptions to these rules too, of course. The conventionally shaped, broadleaved, rapid-growing shrub *Cotinus coggygria* is surprisingly drought tolerant.

Xeriscaping is a special form of landscape design, where landscaping and horticultural strategies are designed to minimise water use. It is used in those situations where soil moisture supplies are insufficient, and/or irrigation water is in short supply or too expensive to be used to support non-drought-adapted landscape plants. Depending on location, soil type and choice of plant species, water use can be reduced by one- to two-thirds if xeriscaping principles are adopted when designing, planting and maintaining a landscape. Obviously, there are variations within this theme, ranging from entire desert-scapes with cacti and other succulents, to plants representing Mediterranean biomes including trees, seasonal flowering shrubs, sub-shrubs, geophytes and herbaceous perennial species (forbs). Within continental and even temperate climatic zones too, there are some woody species better adapted to drier, low-water conditions. Typical woody plant species useful in xeriscapes are outlined in Tables 6.6 and 6.7.

6.9 Pests and Pathogens

Many woody plants are now planted to encourage biodiversity in urban landscapes (see Chapter 5), and in a robust ecosystem, plants are armed with a range of defences to tolerate pest and pathogen pressure. This combined with the more complex food chains that are associated with natural/semi-natural landscapes ensures that herbivorous invertebrates themselves are predated upon, thus regulating numbers attacking the plants. As such, plant populations for the most part are resilient enough to tolerate a degree of pressure from herbivorous animals, pathogenic fungi, bacteria and viruses, although individual plants may succumb from time to time. In natural woodland settings a large percentage of dead wood actually represents a 'healthy' ecosystem.

Table 6.6. Trees with some tolerance of dry soils and potentially useful for xeriscapes (depending on climatic zone)

Acer grandidentatum	*Fraxinus pennsylvanica*	*Pinus ponderosa*
Acer negundo	*Ginko biloba*	*Prunus armeniaca*
Acer tataricum	*Gleditsia triacanthos*	*Pyrus ussuriensis*
Ailianthus altissima	*Juniperus* spp.	*Quercus gambelii*
Catalpa speciosa	*Ilex* spp.	*Quercus ilex*
Cedrus deodar	*Pinus aristata*	*Quercus macrocarpa*
Celtis occidentalis	*Pinus edulis*	*Robinia pseudoacacia*
Crataegus ambigua	*Pinus mugo*	*Styphnolobium japonicum*
Crataegus crus-galli	*Pinus pinea*	*Zelkova* spp.

Table 6.7. Shrubs and sub-shrubs with some tolerance of dry soils and potentially useful for xeriscapes (depending on climatic zone)

Amorpha canescens	*Cowania mexicana*	*Prunus besseyi*
Artemisia spp.	*Elaeagnus* spp.	*Rhus glabra cismontana*
Atriplex canescens	*Ephedra equisetina*	*Rhus trilobata*
Berberis spp.	*Fallugia paradoxa*	*Rosa woodsii*
Buddleia alternifolia	*Fendlera rupicola*	*Rosmarinus officinalis*
Caragana arborescens	*Forestiera neomexicana*	*Rubus deliciosus*
Caryopteris x *clandonensis*	*Genista* spp.	*Shepherdia argentea*
Ceanothus spp.	*Hippophae rhamnoides*	*Symphoricarpos* spp.
Cercocarpus spp.	*Holodiscus dumosus*	*Salvia* spp.
Cistus spp.	*Ligustrum vulgare*	*Syringa vulgaris*
Chamaebatiaria millefolium	*Lavandula angustifolia*	*Viburnum lantana*
Chrysothamnus spp.	*Potentilla* spp.	

In ornamental and other urban situations, the story is more complex. Unsightly, damaged trees and shrubs are not often appreciated for their aesthetic value. Dead trees and particularly dead branches (limbs) that might fall are considered a danger to the public and so need managing; ideally individual trees should not become so stressed that they die prematurely anyway. Urban environments do not represent entire natural ecosystems and limited choice of trees and shrubs, including the use of monocultures (e.g. an avenue composed of one tree species alone), actually encourages pest and pathogen pressure on that particular species. The use of clonal genotypes, with no variation in genetic resistance to herbivores or pathogens, also increases the susceptibility of the tree and shrub stock. Abiotic (environmental stress) exacerbated by urban conditions and climate change can reduce plant defences and increase susceptibility to pest/pathogen attack. The movement of plant material through increased global travel, the import of planting stock from other regions/countries and perhaps climatic changes encouraging wider species distributions are also resulting in trees/shrubs being exposed to new species of pathogens or pests, or new pathovars (races) of existing pathogens. In these scenarios, tree and shrub stock may have limited defences and whole populations can be, and are from time to time, decimated.

Possibly the most infamous case of trees succumbing to a new pathogen is that of the elm. Populations of elms (of different species and hybrids) in both Europe and North America have been killed by the ascomycete pathogens of the genus *Ophiostoma*, e.g. *Ophiostoma novo-ulmi*, (the activities of which are generally known as Dutch elm disease). These ascomycete fungi are spread through the activity of bark-burrowing beetles such as *Hylurgopinus rufipes*, *Scolytus multistriatus* and *Scolytus schevyrewi*. In the UK, the numbers of mature elms killed over the last century exceeds 25 million, resulting in radical changes to the character of the UK landscape, notable through the loss of large iconic hedgerow trees. Similarly, cities such as New Haven, USA, which relied strongly on elms for street and park trees, lost much of their character within a few years.

New epidemics following the pattern of Dutch elm disease are a concern for environmental horticulturists. The threats vary depending on the dominant tree species used in different parts of the world and potential exposure to new pathogens (Table 6.8) or pests (Table 6.9). 'New' pathogens such as *Xylella fastidiosa* are particularly concerning for their ability to damage a wide range of host plants and to spread quickly from one region to another. *Xylella fastidiosa* is a Gram-negative bacterium of the Xanthomonadaceae family that colonises the xylem vessels of its host plants and is transmitted by insects that feed by sucking xylem sap (Uceda-Campos *et al.*, 2022). Partially due to the way it spreads it is extremely polyphagous and thus can enter a large number of taxonomically different host plant species. It potentially can cause great economic damage to key crops such as *Olea europea* (olive), *Vitis vinifera* (grape) and *Coffea arabica* (coffee) and also has significant implications for the viability of certain landscape trees and shrubs. The spread of *Xylella* is currently being controlled primarily by the quarantining of transported plants. Overall, in terms of landscape trees, most documentation relates to Europe or North America, although Australia and New Zealand also document threats (and impose strict regulations on plant imports to help protect native spp.).

Table 6.8. Examples of potentially damaging pathogens of woody plants in temperate climates

Ceratocystis facgaearum (oak wilt)	Spreads through the movement of infected wood or through insect activity and between one tree and its neighbour via natural root grafts (fusion of two separate roots).
Ceratocystis platani (plane wilt)	Very damaging to *Platanus* spp. (up to 80% of trees affected in some parts of France). Enters through wounds, causing cankers and eventual death. Originating in south-eastern USA, it has spread throughout urban populations of *P. hispanica* (London plane) in the cities of the eastern USA.
Chalara fraxinea (ash dieback)	An aggressive fungal disease of *Fraxinus* spp. (ash) which causes crown death, leaf wilting and dieback of branches.
Cryphonectria parasitica (chestnut blight)	Currently, a localized but highly damaging disease of *Castanea sativa* (sweet chestnut).
Dothistroma septosporum (Dothistroma needle blight or red-band needle blight)	A fungal pathogen of pine that can cause significant defoliation and subsequent death in species such *Pinus nigra maritima* (Corsican pine), *Pinus contorta* (Lodgepole pine) and *Pinus sylvestris* (Scots pine).
Erwinia amylovora (fireblight)	Affects species in the Rosaceae. Affected areas appear blackened, shrunken and cracked, as though scorched by fire. Primary infections are established in open blossoms and tender new shoots and leaves in the spring when blossoms are open.
Fusarium circinatum (pine pitch canker)	Affects *Pinus* spp. and *Pseudotsuga menziesii* (Douglas fir). Spores are spread by wind/rain and can enter trees or seedlings through existing wounds or the roots. Originally found in Mexico but has spread to North, Central and South America, Europe and parts of Asia and Africa. Causes bleeding infections that can encircle branches, exposed roots and trunks. The wood beneath the infection site is saturated with resin and becomes a characteristic honey colour.
Mycosphaerella dearnessi (brown-spot needle blight)	In North America, it causes serious growth check to seedlings and young trees and has rendered Christmas tree plantations unsalable. In central Europe, it has recently been found affecting several pine species.
Phytophthora austrocedrae	Confirmed as the cause of dieback and deaths of *Juniperus* (juniper, a priority conservation spp.) in northern England, UK in 2011. This pathogen had previously been almost solely associated with *Austrocedrus chilensis* (Chilean cedar) trees in South America.
Phytophthora kernoviae	Confirmed only in UK, Ireland and New Zealand, and only in a very few trees, so far. The fact that it seems to be able to infect a number of trees such as *Fagus* (beech) and *Quercus* spp. (oak), as well as woodland understorey species such as *Vaccinium* (bilberry) and *Rhododendron* spp., makes it a concern, particularly for historically important gardens with rare or unusual species.
Phytophthora lateralis	Usually kills *Chamaecyparis lawsoniana* (Lawson cypress) trees that it infects.
Phytophthora ramorum	A fungus-like organism which attacks a large range of woody plants. *Larix decidua* (European larch), an economically important silvaculture species, is a key host, and large numbers have had to be felled.
Puccina psidii (eucalyptus/guava rust)	A pathogen with a wide host range in the Myrtaceae, which includes about 3000 tree and shrub species – many of which are of great economic and conservation significance. It is native to parts of South America but also now occurs in parts of North and Central America. The pathogen causes disease symptoms in young shoots, flower buds and developing fruit depending on the host plant. Highly susceptible trees may be malformed or killed outright. Growth rates of infected trees are diminished.
Acute oak decline	A condition affecting *Quercus* in parts of the UK, in which bacteria, including one species previously unknown to science, are believed to be involved.

Continued

Table 6.8. Continued.

Xylella fastidiosa (Xylella disease/Pierce's disease in grapevine/olive quick decline syndrome)	*Xylella fastidiosa* is xylem-limited bacterium capable of infecting a wide range of woody host plants (estimated to be nearly 700 plant hosts) (Trkulja *et al.*, 2022). First identified in Europe in 2013 on *Olea europaea* (olive) it has spread widely and caused significant economic damage in parts of Mediterranean Europe. Insects that puncture plants and feed off xylem or phloem sap are primarily responsible for the spread and include *Philaenus spumarius* (meadow spittlebug) in Europe as well as various species within the cicadas (families *Aphrophoridae, Cicadellidae,* and *Membracidae*).

Table 6.9. Examples of potentially damaging pests of woody plants in temperate climates

Agrilus anxius (birch borer)	European and Asian *Betula* (birch) spp. are highly susceptible. Characteristic D-shaped exit holes but only once adults have left.
Agrilus plannipennis (emerald Ash borer beetle)	A major pest in North America able to infect *Fraxinus pennsylvanica* (green ash), *F. mericana* (white ash), *F. nigra* (black ash) *and F. quadrangulata* (blue ash). Larvae feed on inner bark leading to girdling, leaf yellowing, branch dieback and eventual tree mortality. Symptoms of infestation similar to birch borer.
Anoplophora chinensis (Chinese longhorn beetle)	Can spread long distance via movement of infested living (particularly potted plants) or sawn wood. Attacks base of trunk of suitable hosts which include wide range of broadleaved trees, e.g. *Acer, Cotoneaster Platanus, Malus, Prunus, Salix, Rosa* spp. First sign is usually a round exit hole. Small indentation from female chewing and oviposition may be the only other external sign. Dieback of foliage during early attack phase with sustained attacks resulting in tree mortality.
Anoplophora glabripennis (Asian longhorn beetle)	A wood-boring insect that can cause extensive damage across a range of broadleaf tree spp. Characteristic dieback of foliage during early phases. Beetles develop galleries within the wood and exit holes weaken the integrity of infested trees. Larvae though are considered to be the most damaging as they tunnel through the cambium tissues. Now one of the most destructive non-native insects in the USA; it and other wood-boring pests cause an estimated US$3.5 billion in annual damages.
Cameraria ohridella (horse chestnut leaf miner)	Originating in Macedonia the species has spread rapidly throughout Europe. The larvae mine within the leaves of *Aesculus hippocastanum* (horse chestnut) and up to 700 visible tunnels have been recorded on a single leaf under favourable conditions. Severely damaged leaves shrivel and turn brown by late summer, well before natural abscission.
Dendroctonus micans (great spruce bark beetle)	Present throughout much of the Eurasian region. In addition to spruce such as *Picea abies* (Norway spruce), it also attacks *Pinus sylvestris* (Scots pine) and *Abies* spp. (firs). Beetles infect the root and stems of trees and breed in the bark, where the larvae then feed creating tunnel galleries and undermine the viability of the whole tree.
Dendrolimus pini (pine tree lappet moth)	This moth can cause extensive defoliation of pines and other conifer trees in parts of its native range in Europe and western Asia.
Thaumetopoea processionea (oak processionary moth)	Capable of defoliating *Quercus* spp. over several seasons, leading to dieback, long-term decline and mortality. Larval hairs cause severe health problems for people and animals (skin rash, eye and throat irritation and allergic reactions).

Over and above these potentially devastating pests and pathogens, many woody plants will host a range of invertebrates and microorganisms that will cause infestations or disease symptoms when conditions are conducive. For example, these includes the powdery mildews (e.g. *Erysiphe, Podosphaera, Microsphaera* and *Sawadaea* spp.), root rots (e.g. *Pythium* and *Phytopthora* spp.), bootlace fungus (*Armillaria* spp.), bacterial cankers

(*Pseudomonas* spp.) and pests such as aphids (e.g. *Myzus, Metopolophium, Phorodon, Rhopalosiphum, Aphis* and *Elatobium* spp.), capsid bugs (e.g. *Plesiocoris* and *Corythucha* spp.) scale insects (e.g. *Lecanium, Aulacaspis* and *Saissetia* spp.), spider mites (e.g. *Tetranychus, Panonychus* and *Oligonychus* spp.) and caterpillars of various moth and butterfly species (Lepidoptera) amongst many others.

The most lethal or economically damaging diseases are not necessarily the most common. The most common may also change depending on location and weather, as well as inoculum levels from season to season. The Royal Horticultural Society (RHS) in the UK offers an advisory service to its members and other gardeners. The Plant Health Team at the RHS document the top ten most commonly occurring diseases reported to their advisory service each year (RHS, 2023). The most common in 2023 were *Armillaria* spp. (honey fungus), *Venturia inaequalis*/*Venturia pyrina* (apple and pear scab), *Diplocarpon rosae* (rose black spot), *Gymnosporangium sabinae* (pear rust), *Monilinia laxa* and *M. fructigena* (blossom wilt), *Pseudomonas syringae* pv. *morsprunorum*/*P. syringae* pv. *syringae* (bacterial leaf spot/canker on *Prunus* spp.), *Phytophthora* spp. (root rots), *Taphrina pruni* (pocket plum), *Botrytis tulipae* (tulip fire) and a 'cocktail' of different bacterial and fungal species known as 'slime flux' (Henrici et al., 2006). A number of these pathogens are disfiguring but not lethal to the host plant.

A number of strategies can be employed to minimise infection of woody plants in the landscape or control infections/infestations depending on the pathogen/pest in question. Careful vigilance can help offset problems associated with new diseases/pests to a specific area. This can include restricting imports of timber or live plants, using 'plant passports' (certificates of plant health) and imposing quarantine conditions to any imported material. Disease/pest outbreaks can be monitored through raising awareness within horticultural/landscape professional circles or with the general public, e.g. notices in entrances to urban woodlands.

Once a pathogen is present, then cultivation regimes can help avoid disease. This includes avoiding drought stress in landscape plants or minimising insect damage (for example, viruses can be transmitted from plant to plant via aphids or other sap-sucking invertebrates). A number of soil pathogens build up high spore levels or other propagules when their host plant has been present for a long time. Subsequently planting the same species again in the parent soil can result in poor growth and disease symptoms (specific replant disease or disorder).

Chemical pesticides have been the traditional means of controlling pests and diseases and are still used to a high degree. In landscape situations, however, there is increasing reluctance to spray chemicals where the public may be present or where more than the target pest species is harmed (non-selective pesticides). Nevertheless, in some countries copper and other metal-based chemicals are used, especially in nursery situations, e.g. fungicides such as chlorothalonil, Bordeaux mixture, benomyl and maneb. In mature trees, trunk injections are used to control pests and pathogens, e.g. imidacloprid or thiamethoxam have been used to help control Asian and Chinese longhorn beetles (*Anoplophora glabripennis* and *A. chinensis*). Where a number of individuals of the same species are grown together, it may be worth investing in pheromone traps; these emit insect sex pheromones, attracting attacking insects to the traps rather than the food source trees. In a similar manner, physical barriers can be put in place to stop insects accessing the vulnerable part/s of the plants, e.g. these may be sticky or cloth bands around the trunks of trees to stop flightless insects from climbing up into the canopy. Where labour allows, manual removal of pests is feasible; tent-forming caterpillars can be removed via hand or even vacuuming. For localised infections, selective removal of infected branches may help stop the spread of a pathogen. Similarly, radical action such as the felling and removal of entire stands of trees may be necessary for stopping the spread of the more virulent pathogens or invasive pest species. When managing infected specimens, tools and machinery need to be cleaned and sterilised afterwards to avoid cross-contamination onto healthy stock. There may be some benefit from treating wounded areas and cut stumps with fungicide, but the value of bitumen-based wound sealants is unproven and may even inhibit the natural callus and wound sealing processes.

Biocontrol or integrated pest management (IPM) techniques are attempted in woody plant populations prone to specific disease or pests. To some extent in outdoor situations, this is accomplished by encouraging natural predators of the pest species through habitat management, for example, by providing more ground-cover vegetation or companion plants that offer habitat or alternative food sources to predatory species. In some circumstances the use of specific biocontrol agents is justified, for example, spraying inoculations of bacteria (*Bacillus thuringiensis*) or nematodes (e.g. *Steinernema kraussei*) onto insect larvae, where these parasitic organisms invade the host's body and kill them. Parasitoid wasps do something similar, via laying their eggs within the body of the pest species. A longer-term strategy for woody plant biosecurity, of course, is to select or breed for taxa that have greater tolerance to the specific problematic pest or pathogen. Likewise, for some genera, now avoiding clonally produced stock and relying on the natural variations that seed material confers (and hence more likely to include some individuals with enhanced pathogen/pest tolerance) would seem prudent.

6.10 Urban Forests

Urban forest is a term used to denote the tree stock in towns and cities and incorporates street, park and garden trees as well as areas of woodland (natural, semi-natural or intentionally planted) within urban conurbations. In turn, urban forestry was defined by Johnston and Hirons (2014) as "the planned, systematic and integrated approach to the planting, maintenance and management of trees and woodland in and around urban areas". Traditionally, urban trees were valued for their aesthetic qualities, although wider recognition of their ecosystem service role is increasingly used as the rationale underpinning careful maintenance and new plantings (see Chapters 3 and 4). Urban forests can be a fundamental component of promoting a city's character or a 'sense of place' (Fig. 6.11). Many of the world's great cities and conurbations can be identified with their tree stocks, e.g. the use of iconic London plane (*Platanus hispanica*) in London (UK) city squares, or large imposing specimens of *Dipterocarpus alalus* in the streetscapes of tropical cities such as Hoi Chi Minh in Vietnam. Such tree stocks, however, are under pressure from environmental factors or continued densification of the urban environment. Street trees

Fig. 6.11. Urban trees are important in defining a place. *Liquidamber styraciflua* (sweetgum) creates colour and vibrancy around late autumn in the civic squares of Sheffield, UK.

particularly can be exposed to a range of environmental stresses (see above and Chapter 2).

Species that form large specimens are particularly under threat as perceptions associated with a lack of city centre space mean that they may not be selected during the planning of inner-city redevelopment, with smaller-stature species being selected in preference. Large and older specimens, however, are thought to provide more benefits (Armour *et al.*, 2012) (Fig. 6.12). Well-maintained trees in appropriate locations may live for 100 years or more, but life expectancy of many street trees post-establishment is thought to be only 19–28 years (Roman and Scatena, 2011), with failure at the establishment phase perhaps reducing this mean figure to as little as 12 years in some localities. In Philadelphia, USA, street tree attrition rates were on average 3.7% per year, with younger trees being most vulnerable (Bigelow *et al.*, 2024). Here, multivariate logistic regression revealed significant associations between street tree survival and tree crown vigour, species traits (expected mature stature and drought tolerance), location type, land use type, the percentage of impervious surface in a 15 m buffer around the planting site, other activities going on in the location such as site demolition/renovation, per cent of neighbourhood residents in poverty, per cent unoccupied households and per cent renter-occupied households.

Diversity in city trees is also a concern, especially in the context of promoting resilience against future climatic changes and biotic factors. In central and northern Europe, at least 250 landscape plant species are thought to be used in city locations, but actually less than five genera may account for 50–70% of all street trees in many of these cities. Similarly, surveys in Argentina demonstrated that five species alone accounted for 86% of city tree stocks. In Shanghai (China), Liu *et al.* (2021) indicated there was often a mismatch with tree species that were most commonly grown in the city and those that they considered would have greatest tolerance against a changing climate (Figs 6.13–6.17). These results suggest that warm subtropical species with low usage frequency, such as *Quercus virginiana*, *Carya illinoinensis* and *Q. texana*, could be planted more frequently due to their relatively higher climate adaptability and stress tolerance. Rarely used species from other climatic zones also have potential for Shanghai, such as *Ormosia henryi*, *Cyclobalanopsis glauca* (from the cool subtropics), *Ormosia hosiei*, *Dalbergia hupeana*, *Q. fabri* and *Cornus wilsoniana* (from the moist subtropics) and *Koelreuteria paniculata* (from the temperate zone).

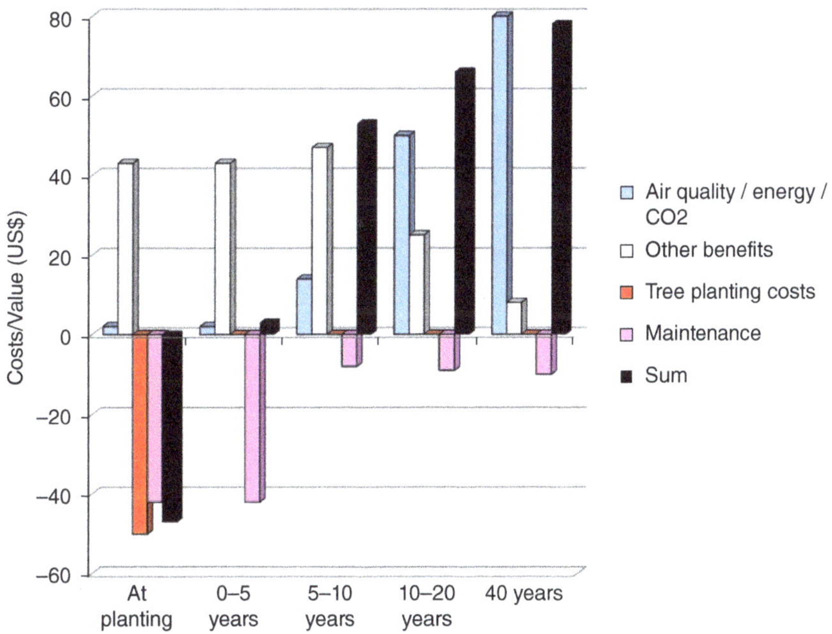

Fig. 6.12. Trees cost money for planting and maintenance over the first 2–3 years but rapidly provide net positive value via their ecosystem services as they grow over subsequent years.

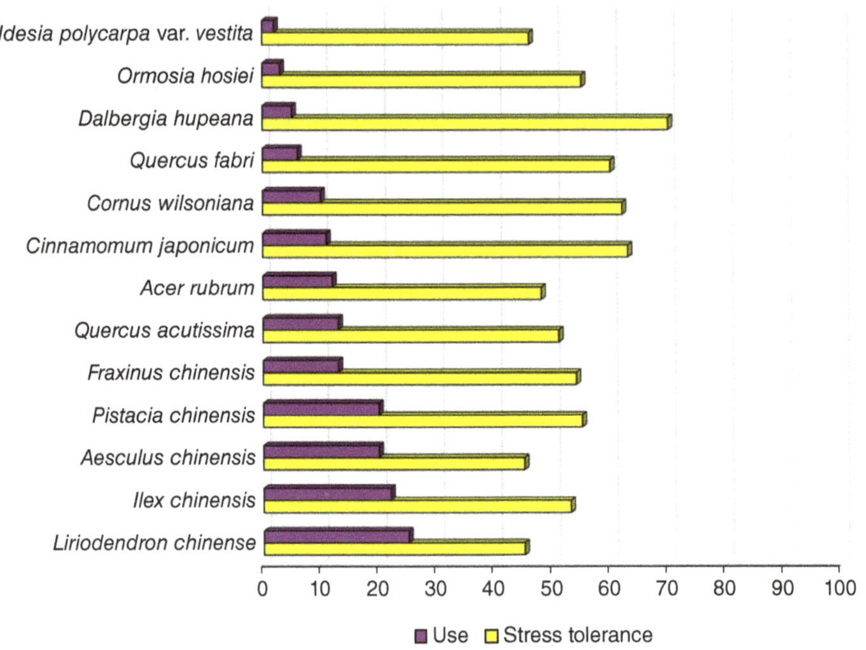

Fig. 6.13. Predicted stress tolerance (relative score) to a changing climate and current frequency of use (%) of tree species across different locations in Shanghai: Trees from moist subtropical climates used less frequently in the city. (Modified with permission from Liu *et al.*, 2021.)

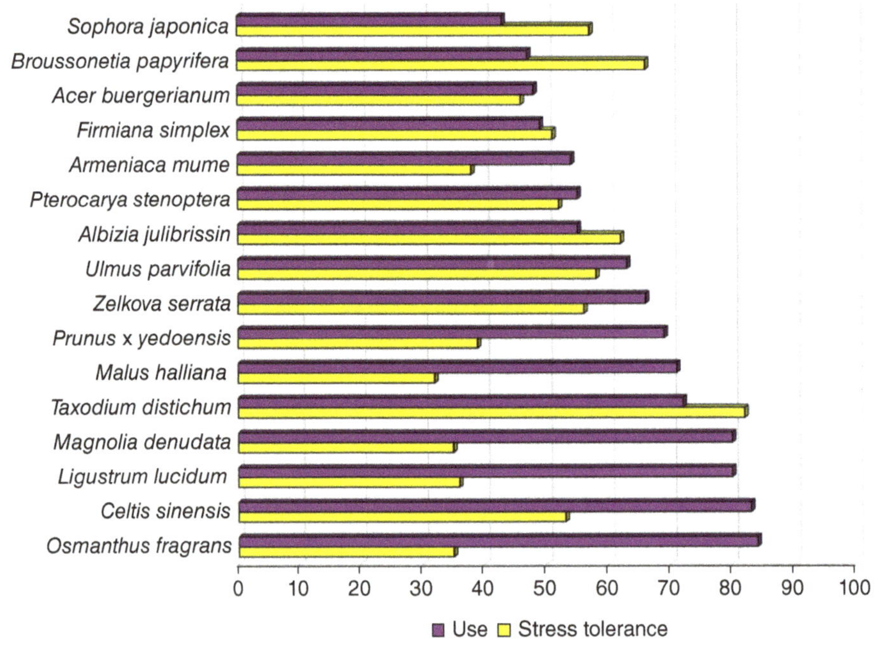

Fig. 6.14. Predicted stress tolerance (relative score) to a changing climate and current frequency of use (%) of tree species across different locations in Shanghai: Trees from moist subtropical climates used more frequently in the city. (Modified with permission from Liu *et al.*, 2021.)

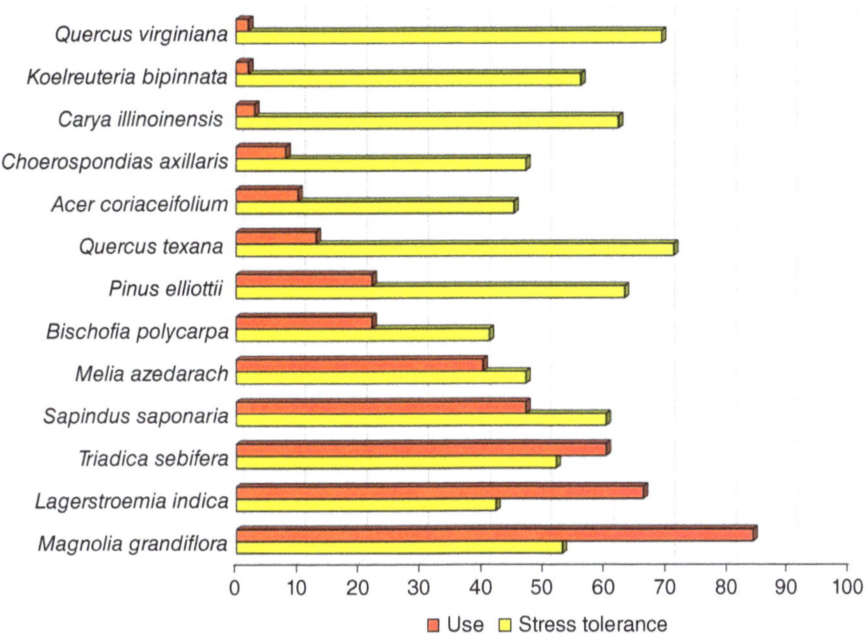

Fig. 6.15. Predicted stress tolerance (relative score) to a changing climate and current frequency of use (%) of tree species across different locations in Shanghai: Trees from warm subtropical climates. (Modified with permission from Liu *et al.*, 2021.)

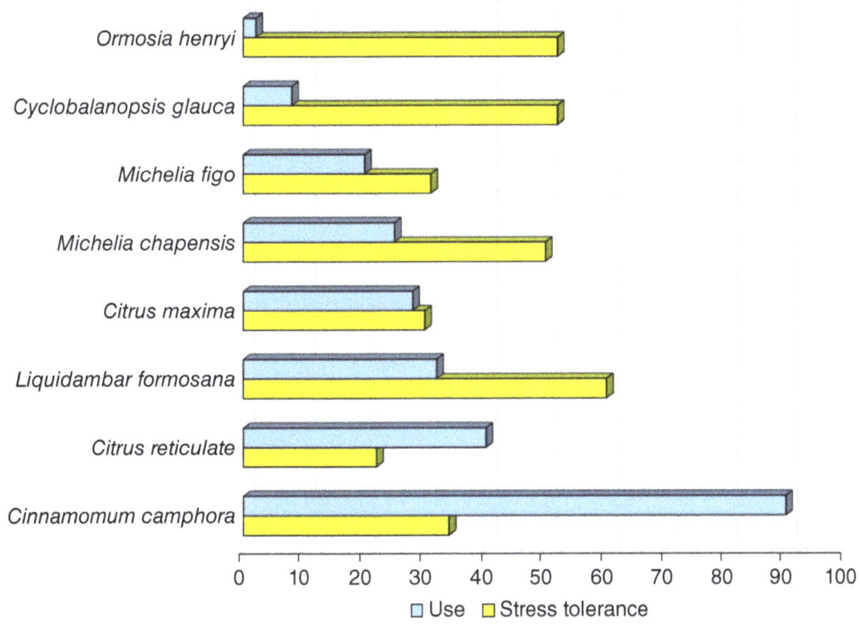

Fig. 6.16. Predicted stress tolerance (relative score) to a changing climate and current frequency of use (%) of tree species across different locations in Shanghai: Trees from cool subtropical climates. (Modified with permission from Liu *et al.*, 2021.)

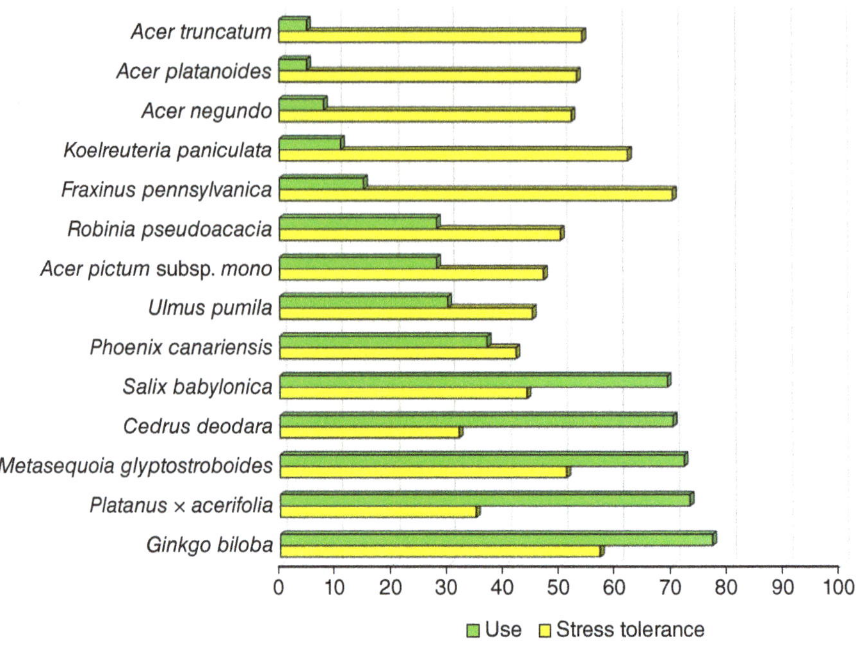

Fig. 6.17. Predicted stress tolerance (relative score) to a changing climate and current frequency of use (%) of tree species across different locations in Shanghai: Trees from temperate climates. (Modified with permission from Liu *et al.*, 2021.)

Paradoxically, in temperate climates the urban heat island effect is becoming more common (where city centres can be up to 11°C warmer than surrounding rural districts), and this can increase the range of species grown, as well as encouraging better rates of growth and earlier flowering/fruiting. The lack of late-season frost is a bonus for species such as *Magnolia*, where flower blooms avoid being blighted by freezing injury, as might occur more commonly in colder, rural localities.

The threats to city trees can be alleviated significantly by careful planning, design and management of urban green spaces. Integrating green space within the main planning process allows space to be created for appropriate tree species and for these components be linked to other key objectives on, for example, flood mitigation, noise abatement, thermal comfort, biodiversity, etc. Such integrated approaches require dialogue between different disciplines and specialisms, with environmental horticulturists, landscape architects and arboriculturists working alongside planners, engineers, architects, community groups, utility companies and others with an interest in the urban form, with ideally, these dialogues taking place from an early stage in the development process.

Management of the tree stock is depended on professionally trained tree officers and arboriculturists. Their role is to inspect and identify trees for any early indications of pathogens or pest problems, improve aesthetics through appropriate tree choice and structural pruning, and ensure that the urban tree stock remains safe to the public and nearby properties. This will include implementing selected tree limb removal or aiding tree crown stabilisation through bracing techniques. Traditional expertise in these matters is being augmented by newer technologies, such as Geographic Information Systems (GIS) and aerial photography. Imaging systems can aid estimations of urban forestry canopy cover and even species composition, with data being downloaded to centralised data bases. Similarly, evaluations of city trees now employ systems that help integrate information from a variety of sources and enable tree officers to provide an aesthetic or economic value to the trees within their jurisdiction. Such systems vary in their complexity. The Helliwell system is entirely based on expert judgement and focuses on visual amenity value with very little requirement for field data to be collected. The Capital Asset Value for Amenity Trees (CAVAT) process identifies the wider benefits of trees to communities rather than purely visual amenity but does not provide detailed benefit and cost requirements. On the other hand, i-Tree requires data collected from a sample or even a complete inventory of the neighbourhood street trees and interlinks this to information relating to the local community and existing management procedures (e.g. current management costs, local population size and energy use) before providing cost–benefit data outputs. In tandem to developing these highly technical approaches to understanding the tree population, effective community engagement is paramount too. Citizens who identify and care for their neighbourhood trees are much more likely to engage with preservation of existing stocks and support replacement strategies and new additional plantings. Indeed, such community groups have a significant role in aspects such as species choice and planting efforts, and they underpin initiatives such as the Community Forest Schemes in the UK.

Miyawaki and other 'mini' forests

There has been interest in recent years in developing 'mini' forests in urban areas. These comprise a very small area of woodland which is encouraged to grow on vacant plots, parts of parkland or as special features within the urban matrix, e.g. within a shopping precinct or on a traffic roundabout. One special form of 'mini' forest is the Miyawaki Forest system.

The system is named after Akira Miyawaki – the Japanese botanist who specialised in tree seed propagation and natural forest generation/succession dynamics. The Miyawaki Forest method builds on the theory of potential natural vegetation (PNV) and succession (Zhang *et al.,* 2023; Qi *et al.,* 2024). Here, vegetation establishes itself without human intervention, with the species mix and dynamics based on specific climatic, soil and topographic conditions. The claims around the Miyawaki system are that it establishes a woodland more quickly (and naturally!) than conventional planting, and that biodiversity increases rapidly over the initial years of the forest development. Planting of Miyawakis follows the concepts of natural forest establishment (where many seedlings germinate together in close density) by planting small trees (500 to 1000 mm tall) at a very high density (e.g. 200–300 mm apart). A wide range of native woody plants are encouraged within the planting mix. The young trees are given a heavy organic mulch to suppress weed growth (and this possibly acts effectively to retain soil moisture over the early years of establishment and provide nutrients to the young trees as it breaks down). Weeding is also carried out in the initial years of the woodland. The high tree density and weed control means the trees grow rapidly due to natural competition effects (trees etiolating due to competition for light) and readily available water/nutrients. In due time, the trees 'self-thin' as competition for light and other resources results in 'the survival of the fittest' or at least those that can fully exploit the ecological niches created by the semi-natural process. The rapid establishment of a thick grove of trees (Wang *et al.*, 2002) is linked to lower long-term costs, less human management intervention, better (more natural?) plant community structures and reputably higher biodiversity (Lewis, 2022), although some of these claims need to be substantiated in the different contexts Miyawaki Forests are grown in. Certainly, Miyawaki Forests have caught the public imagination in recent years, and small but dense forests are becoming popular in a number of inner city areas (Fig. 6.18). In many ways they have become a metaphor for the re-greening of highly urbanised, densely built urban locations. The total vegetated area of the city is still small (some individual Miyawaki Forests are only the area of a tennis court), but Miyawakis are bringing nature to urban communities and highlighting the value of

Fig. 6.18. Miyawaki Forests do not take up much space and are a useful way of allowing citizens to come together and engage with nature. (Used with permission from H. Qi., 2025)

urban forests with respect to heat and water management, air quality improvements, opportunities for engagement with nature and, indeed, encouraging wider, pro-environmental behaviours.

Conclusions

- Woody plants are divided into the main groups of trees, shrubs and climbers (vines). These groups can be linked to slightly different ecological niches in the natural world.
- Urban form is strongly influenced by the presence of woody plants, with scale and context dictating the types of plants used. Trees provide a strong structural background to many avenues, parks and gardens, whereas shrubs and climbers are effective as screens or to promote visual interest at a smaller scale. These plant types contribute significantly to ecosystem service delivery within the cityscape and can be important in providing resources for wildlife.
- Woody plants are propagated from a variety of techniques, but trees are dominated by seed production, field-budding or grafting. Clonal production of shrubs and climbers tends to be dominated by propagation via cuttings. Woody plants are retailed through wholesale nurseries (for the professional landscape sector) and garden centres, nurseries and online sales for the gardening public.
- The production of ornamental varieties of woody landscape plants supports a large and diverse nursery trade industry, with scheduled production to meet market demands. Large numbers of 'hardy' woody plants will start life within protected or semi-protected facilities and be grown under intensive management regimes before finally being planted in the landscape. This has implications for the management of resources such as water, energy, nutrients, pesticides and substrates such as peat. Transitions are being made in attempts to make production of such plants more sustainable.
- Location within the landscape will determine species choice following the rule 'right plant, right place'. Plant selection should be based on final size, aspect and exposure, temperature range, soil type, structure and pH, moisture availability and predisposition to and prevalence of significant pests and pathogens, as well as aesthetic merit and the overall objectives of the planting design. Increasingly, plant taxa are being selected for the ecosystem services they provide.
- Establishment is a critical phase in the lifespan of woody plants, with the greatest number of plant failures occurring at this post-planting stage. Key factors to account for include providing adequate water and reducing competition from weeds until plants become fully established in the parent soil.
- Urban soils are particularly challenging to plants, as those around built structures are frequently heavily compacted and contain physical and chemical contaminants. They are typified by high bulk densities, limited organic matter and poor structure that impedes movement of water and air. Such soils need amelioration, or the construction of special root zone features, such as spacious tree planting pits to help ensure the longevity of the landscape plants.
- Weed, pest and pathogen control are important in ensuring the survival of woody plants, but in addition their presence/absence is linked to the fundamental component of aesthetic quality, although this can depend on context (being generally less tolerated in highly formalised or designed plantings). Increasingly, mulches are employed to help combat weed pressure with reduced emphasis on chemical herbicides. Biocontrol measures can be effective in dealing with pests. Vigilance is also part of the armoury though, for example, to ensure serious pathogens do not spread from one area to another.
- Urban forest is the term used to denote the tree population and other woody plants found in and around urban conurbations. Urban forestry is tree management within the entire urban area, not only the management of individual trees, and takes account of social and economic, as well as environmental, factors associated with trees. New types of woody landscapes are being considered in the urban environment such as 'mini' forests (Miyawaki) and are gaining some traction with the public and key stakeholders.

References

Abe, T. (2022) Effects of treeshelter on seedling performance: A meta-analysis. *Journal of Forest Research* 27, 171–181.

Álvarez, S. and Sánchez-Blanco, M.J. (2021) Benefits of applying deficit irrigation strategies in ornamental plants. *VIII International Conference on Landscape and Urban Horticulture* 1345, 343–350.

Apostol, K.G., Jacobs, D.F. and Dumroese, R.K. (2009) Root desiccation and drought stress responses of bareroot *Quercus rubra* seedlings treated with a hydrophilic polymer root dip. *Plant and Soil* 315, 229–240.

Armour, T., Job, M. and Canavan, R. (2012) *The Benefits of Large Species Trees in Urban landscapes: A costing, Design and Management Guide* (C712). CIRIA, London. UK, p. 129.

Arnold, M.A., McDonald G.V., Bryan, D.L., Denny G.C., Watson W.T. and Lombardini L. (2007) Below-grade planting adversely affects survival and growth of tree species from five different families. *Arboriculture and Urban Forestry* 33, 64–69.

Bainard, L.D., Klironomos, J.N. and Gordon, A.M. (2011) The mycorrhizal status and colonization of 26 tree species growing in urban and rural environments. *Mycorrhiza* 21, 91–96.

Bergstrand, K.J.I. (2017) Methods for growth regulation of greenhouse produced ornamental pot-and bedding plants - A current review. *Folia Horticulturae*, 29, 63–74.

Bigelow Jr, L.M., Fahey, R.T., Grabosky, J., Hallett, R.A., Henning, J.G., Johnson, M.L. and Roman, L.A. (2024) Predictors of street tree survival in Philadelphia: Tree traits, biophysical environment, and socioeconomic context. *Urban Forestry and Urban Greening* 94, 128284.

Billeaud, L.A. and Zajicek, J.M. (1989) Influence of mulches on weed control, soil pH, soil nitrogen content, and growth of *Ligustrum japonicum*. *Journal of Environmental Horticulture* 7, 155–157.

Cameron R.W.F., Harrison-Murray R.S. Van Campfort K., Kesters K. and Knight L.J. (2001) The influence of branches and leaf area on rooting and development of *Cotinus coggygria* cv. Royal Purple cuttings. *Annals of Applied Biology* 139, 155–164.

Cameron, R.W.F., Harrison-Murray, R.S., Atkinson, C.J. and Judd, H.L. (2006) Regulated deficit irrigation – A means to control growth in woody ornamentals. *The Journal of Horticultural Science and Biotechnology* 81, 435–443.

Cameron, R., Harrison-Murray, R., Fordham, M., Wilkinson, S., Davies, W., Atkinson, C. and Else, M. (2008) Regulated irrigation of woody ornamentals to improve plant quality and precondition against drought stress. *Annals of Applied Biology* 153, 49–61.

Chalker-Scott, L. (2007) Impact of mulches on landscape plants and the environment. A review. *Journal of Environmental Horticulture* 25, 239.

Davies, R.J. (1985) The importance of weed control and the use of tree shelters for establishing broadleaved trees on grass-dominated sites in England. *Forestry* 58, 167–180.

Davies, W.J., Wilkinson, S. and Loveys, B. (2002) Stomatal control by chemical signalling and the exploitation of this mechanism to increase water use efficiency in agriculture. *New Phytologist* 153, 449–460.

Davis, M.A., Wrage, K.J., Reich, P.B., Tjoelker, M.G., Schaeffer, T. and Muermann, C. (1999) Survival, growth, and photosynthesis of tree seedlings competing with herbaceous vegetation along a water-light-nitrogen gradient. *Plant Ecology* 145, 341–350.

Fini, A., Frangi, P., Amoroso, G., Piatti, R., Faoro, M., Bellasio, C. and Ferrini, F (2011) Effect of controlled inoculation with specific mycorrhizal fungi from the urban environment on growth and physiology of containerized shade tree species growing under different water regimes. *Mycorrhiza* 21, 703–719.

Gallo, J., Baláš, M., Linda, R. and Kuneš, I. (2020) The effects of planting stock size and weeding on survival and growth of small-leaved lime under drought-heat stress in the Czech Republic. *Austrian Journal of Forest Science* 137, 43.

Gilman, E.F. (2001) Effect of nursery production method, irrigation, and inoculation with mycorrhizae-forming fungi on establishment of *Quercus virginiana*. *Journal of Arboriculture* 27, 30–39.

Gilman, E.F., Grabosky, J., Stodola, A. and Marshall, M.D. (2003) Irrigation and container type impact red maple (*Acer rubrum* L.) 5 years after landscape planting. *Journal of Arboriculture* 29, 231–236.

Heiskanen, J. (1995) Physical properties of two-component growth media based on *Sphagnum* peat and their implications for plant-available water and aeration. *Plant and Soil* 172, 45–54.

Henrici, A., Andrews, J, Bailey, J. (2006) The Tyntesfield Monster – a spectacular slime-flux. *Field Mycology* 7, 128–131.

Hermann, R.K. (1967) Seasonal variation in sensitivity of Douglas-fir seedlings to exposure of roots. *Forest Science* 13, 140–149.

Jim, C.Y. (1998) Urban soil characteristics and limitations for landscape planting in Hong Kong. *Landscape and Urban Planning* 40, 235–249.

Johnston, M. and Hirons, A. (2014) Urban Trees. In: Dixon, G.R. and Aldous, D. (eds) *Horticulture: Plants for People and Places*, Volume 2. Springer, the Netherlands, pp. 693–711.

King, C.M., Robinson, J.S. and Cameron, R.W. (2012) Flooding tolerance in four 'Garrigue' landscape plants: Implications for their future use in the urban landscapes of north-west Europe? *Landscape and Urban Planning* 107, 100–110.

Klironomos, J.N. and Allen, M.F. (1995) UV-B-mediated changes on below-ground communities associated with the roots of *Acer saccharum*. *Functional Ecology* 9, 923–930.

Kristoffersen, P., Rask, A.M., Grundy, A.C., Franzen, I., Kempenaar, C. *et al.* (2008) A review of pesticide policies and regulations for urban amenity areas in seven European countries. *Weed Research* 48, 201–214.

Lewis, H. (2022) *Mini-Forest Revolution: Using the Miyawaki Method to Rapidly Rewild the World*. Chelsea Green Publishing White River Junction, Vermont, USA.

Liu, M., Zhang, D., Pietzarka, U. and Roloff, A. (2021) Assessing the adaptability of urban tree species to climate change impacts: A case study in Shanghai. *Urban Forestry and Urban Greening* 62, 127186.

Liu, X., Wei, W., Liu, G., Zhu, B., Cui, J. and Yin, T. (2024) Effects of conventional non-biodegradable film-derived microplastics and new biodegradable film-derived microplastics on soil properties and microorganisms after entering sub-surface soil. *Agronomy* 14, 753.

Machado, D.L., Dourado, M.N., de Freitas, M.S., de Souza, L.M., da Silva, E.M. *et al.* (2024) Organic mulching alters the soil microclimate, increases survival and growth of tree seedlings in restoration planting. *Forests* 15, 1777.

Martin, T.P., Harris, J.R., Eaton, G.K. and Miller O.K. (2003) The efficacy of ectomycorrhizal colonization of pin and scarlet oak in nursery production. *Journal of Environmental Horticulture* 21, 45–50.

Metzler, P., Ksiazek-Mikenas, K. and Chaudhary, V.B. (2024) Tracking arbuscular mycorrhizal fungi to their source: Active inoculation and passive dispersal differentially affect community assembly in urban soils. *New Phytologist* 242, 1814–1824.

Montague, T., Kjelgren, R., Allen, R. and Wester, D. (2004) Water loss estimates for five recently transplanted landscape tree species in a semi-arid climate. *Journal of Environmental Horticulture* 22, 189.

Oehl, F. and Koch, B. (2018) Diversity of arbuscular mycorrhizal fungi in no-till and conventionally tilled vineyards. *Journal of Applied Botany and Food Quality*, 91, 56–60.

Pauleit, S., Jones, N., Garcia-Martin, G., Garcia-Valdecantos, J.L., Rivière, L.M., Vidal-Beaudet, L. and Randrup, T.B. (2002) Tree establishment practice in towns and cities. – Results from a European survey. *Urban Forestry and Urban Greening* 1, 83–96.

Quigley, M.F. (2004) Street trees and rural conspecifics: Will long-lived trees reach full size in urban conditions? *Urban Ecosystems* 7, 29–39.

Qi, H., Dempsey, N. and Cameron, R. (2024) Seeing the forest for the trees? An exploration of the Miyawaki forest method in the UK. *Arboricultural Journal* 46, 292–304.

Roman, L.A. and Scatena, F.N. (2011) Street tree survival rates: Meta-analysis of previous studies and application to a field survey in Philadelphia, PA, USA. *Urban Forestry and Urban Greening* 10, 269–274.

Royal Horticultural Society (2023) Top 10 Garden Diseases of 2023: Four New Contenders Hit Charts. Available at: https://www.rhs.org.uk/science/articles/2023-diseases (accessed 18 April 2024).

Schütt, A., Becker, J.N., Gröngröft, A., Schaaf-Titel, S. and Eschenbach, A. (2022) Soil water stress at young urban street-tree sites in response to meteorology and site parameters. *Urban Forestry and Urban Greening* 75, 127692.

Siering, N. and Grüning, H. (2023) Stormwater tree pits for decentralized retention of heavy rainfall. *Water* 15, 2987.

Skroch, W.A., Powel, M.A., Bilderback, T.E. and Henry P.H. (1992) Mulches: Durability, aesthetic value, weed control and temperature. *Journal of Environmental Horticulture* 10, 43–45.

Smart, N., Eisenman, T.S. and Karvonen, A. (2020) Street tree density and distribution: An international analysis of five capital cities. *Frontiers in Ecology and Evolution* 8, 562646.

Smith, K.D., May, P.B. and Moore, G.M. (2001) The influence of compaction and soil strength on the establishment of four Australian landscape trees. *Journal of Arboriculture* 27, 1–7.

Somerville, P.D., Farrell, C., May, P.B. and Livesley, S.J. (2020) Biochar and compost equally improve urban soil physical and biological properties and tree growth, with no added benefit in combination. *Science of the Total Environment* 706, 135736.

Specht, A. and Harvey-Jones, J. (2000) Improving water delivery to the roots of recently transplanted seedling trees: The use of hydrogels to reduce leaf and hasten root establishment. *Forest Research* 1, 117–123.

Stabler, L.B., Martin, C.A. and Stutz, J.C. (2001) Effect of urban expansion on arbuscular mycorrhizal fungal mediation of landscape tree growth. *Journal of Arboriculture* 27, 193–202.

Struve, D.K., Burchfield, L. and Maupin, C. (2000) Survival and growth of transplanted large-and small-caliper red oaks. *Journal of Arboriculture* 26, 162–169.

Thetford, M., Miller, D., Smith, K. and Schneider, M. (2005) Container size and planting zone influence on transplant survival and growth of two coastal plants. *HortTechnology* 15, 554–559.

Trkulja, V., Tomić, A., Iličić, R., Nožinić, M. and Milovanović, T.P. (2022) *Xylella fastidiosa* in Europe: From the introduction to the current status. *The Plant Pathology Journal*, 38, 551.

Uceda-Campos, G., Feitosa-Junior, O.R., Santiago, C.R.N., Pierry, P.M., Zaini, P.A., *et al.* (2022) Comparative genomics of *Xylella fastidiosa* explores candidate host-specificity determinants and expands the known repertoire of mobile genetic elements and immunity systems. *Microorganisms* 10, 914.

Walters, M. and Sinnett, D. (2021) Achieving tree canopy cover targets: A case study of Bristol, UK. *Urban Forestry and Urban Greening* 65, 127296.

Wang, R.Q., Fujiwara, K. and You, H.M. (2002). Theory and practices for forest vegetation restoration: Native forest with native trees-introduction of the Miyawaki's method for reconstruction of. *Chinese Journal of Plant Ecology* 26, 133.

Wang, B., Niu, J., Berndtsson, R., Zhang, L., Chen, X., Li, X. and Zhu, Z. (2021) Efficient organic mulch thickness for soil and water conservation in urban areas. *Scientific Reports* 11, 6259.

Watson, W.T. (2005) Influence of tree size on transplant establishment and growth. *HortTechnology* 15, 118–122.

Watson, G.W. and Himelick, E.B. (1997) *Principles and Practice of Planting Trees and Shrubs*: International Society of Arboriculture. Champaigne, Illinois, USA pp. 200.

Weber, G and Claus, M. (2000) The influence of chemical soil factors on the development of VA mycorrhizas of ash (*Fraxinus excelsior* L.) and sycamore (*Acer pseudoplatanus* L.) in pot experiments. *Journal of Plant Nutrition and Soil Science*, 163, 609–616.

Wells, C., Townsend, K., Caldwell, J., Ham, D., Smiley, E.T. and Sherwood, M. (2006) Effects of planting depth on landscape tree survival and girdling root formation. *Arboriculture and Urban Forestry* 32, 305.

Wirabuana, P.Y.A.P., Sadono, R. and Juniarso, S. (2020) Planting depth management increases early growth, aboveground biomass, and carbon storage of *Eucalyptus pellita* at Ultisols in South Sumatra. *Journal of Degraded and Mining Lands Management* 7, 2253.

Yim, B., Baumann, A., Grunewaldt-Stöcker, G., Liu, B., Beerhues, L. *et al.* (2020) Rhizosphere microbial communities associated to rose replant disease: links to plant growth and root metabolites. *Horticulture Research* 7, 144.

Zemek, R. and Pastirčáková, K. (2023) Pests and pathogens of urban trees. *Forests* 14, 1653.

Zhang, X., Hu, H., Wang, X., Tian, Q., Zhong, X. and Shen, L. (2023) Plant community degradation inquiry and ecological restoration design in south lake scenic area of China. *Forests* 14, 181.

7 Non-woody Landscape Plants

Abstract

The non-woody landscape plants are an extensive and diverse range of taxa. Significant lignification of the tissues is rare, so plant stems are generally softer in texture than perennial woody plants. Many non-woody plants also dieback once a year due to cold winters or hot dry seasons – and survive as a dormant perennial organ for part of the year (herbaceous plants, geophytes and ferns). Plants that do not survive the stressful period within their natural environments, but rather perpetuate the species alone by seeds, are annuals. Others retain some of their stems/leaves and can tolerate the most stressful conditions, including being buried by snow (e.g. alpines). Arguably some of these plant groups provide the greatest colour within the environmental horticultural arena – usually due to flamboyant flower displays. As with woody plant types, the success of these landscape plants depends on good plant community design, management and maintenance.

7.1 Introduction

Managed landscapes rely strongly on non-woody plants. These are plants with little lignin/lignification and thus they lack tall structural stems that persist from one season to the next. The most abundant non-woody plants are the amenity grasses – lawn turf – and these are dealt with in Chapters 8 and 9. Other areas such as woodland may have a ground flora that relies strongly on herbaceous perennials and geophytes (bulbs, tubers etc.), and if heavily shaded, ferns. Key parts of parks, civic areas and gardens particularly will have areas of extraordinary aesthetic value composed of flowering annuals, geophytes, perennials and architectural plants (Fig. 7.1).

Landscapes dominated by non-woody plants range from the natural (e.g. including woodland ground flora [herb layer], salt marshes, prairie and steppe landscapes, tundra and even deserts with their own range of perennials and ephemerals), the semi-natural or nature-'mimicking' landscapes (e.g. meadows, field margins, scree and rock gardens, green roofs and rain gardens) and formal or designed landscapes ('Edwardian' herbaceous flower borders, annual flower borders, island borders, pots, tubs, window boxes and hanging baskets and roof gardens). Both within natural landscapes, such as when desert annual plants (ephemerals) flower en masse after the onset of rain (so called 'desert blooms') or highly formalised bedding plant displays, these non-woody plants arguably provide the greatest colour spectacles in the floral world (Fig. 7.2).

The way non-woody landscape plants are used tends to go in and out of fashion, and what is in vogue can be determined by factors such as convenience of management or the resources available – some plantings require more intensive management than others. Some styles align themselves with garden landscapes (both private and public), e.g. rockeries and the use of alpine plants, whereas others, such as formal, mass bedding plant displays, are perhaps more associated with civic parks. In recent years, there have been moves towards more environmentally related and sustainable designs, with a number of landscapes that mimic nature becoming more popular.

7.2 Definitions, Ecology, Growth Habits and Use in the Landscape

Plant types within the non-woody group include a wide range of growth habits and morphologies, of which the most common or iconic outdoor grown ones are listed below. Annuals (ephemerals) have only one cycle of growth, culminating in seed production to ensure the survival of the next generation of the species. Perennial herbaceous plants live longer with most plants dying-down during the dormant phase (and effectively visually disappearing from the landscape), but not all do – some alpine plants for example retain soft, lignin-free growth

DOI: 10.1079/9781800621763.0007

Fig. 7.1. A spring flower border in a public park, composed of three *Tulipa* cultivars (*T.* cv. Queen of Night, deep purple/black; *T.* cv. National Velvet, deep red; *T.* cv. Grand Perfection, white with red stripe) and the short-lived perennial *Myosotis sylvatica* with pale blue flowers.

Fig. 7.2. A number of annual plants are derived from short-lived desert ephemerals such as *Eschscholzia californica*, which have high-impact floral display in their natural habitat after the infrequent periods of rain.

throughout their lifespan. The herbaceous life form is found in dicots, monocots and pteridophytes (ferns). With herbaceous plants, a number of different life forms have evolved based on the typical duration of longevity, fitting against a number of ecological niches. Herbaceous perennials usually have at least three growth cycles and in most cases many more before they die. There are also 'intermediate' types – biennial herbaceous plants (for example, some *Digitalis* spp. [foxgloves]) – which develop shoots and leaves one year, then a flowering meristem the following, before dying after flowering. There are monocarpic herbaceous plants that share the traits of biennials, but the vegetative growth period prior to the final and fatal flowering is stretched out to 4-7 years. Monocarpism is perhaps most obviously found in herbaceous plants in the Sino-Himalayan region, for example, *Meconopsis* (Himalayan poppy).

7.3 Annuals

These are plants that complete their lifecycle within one growing season. They divide into two broad camps in environmental horticulture, namely (i) directly sown annuals and (ii) 'bedding plants' that are sown and grown on in trays and pots and then planted out in 'flower beds' once they reach a suitable size. In reality, bedding plants can also include some short-lived perennial genotypes – see below.

Annual plants are utilised due to their vibrant flower colours and ability to produce flowers for much of the active growing season (repeat flowering), although some are popular due to foliage characteristics too (e.g. *Plectranthus scutellarioides*, formerly *Coleus blumei,* with amazingly coloured central leaf blotches or brightly coloured 'thread' lines bordering the edge of the leaf). The free-flowering characteristics of annuals relates to key ecological adaptations associated with rapid flower formation and seed setting in their natural environment before adverse conditions ensue. For example, a number of bedding plant annuals are derived from desert ephemerals, which only have a limited period to complete their lifecycles after significant rainfall events have replenished the soil's moisture. Indeed, most annual flowering ornamentals originate from arid biomes where there are distinct 'rainy seasons' either for discrete periods in winter, or alternatively in summer. Species adapted to climates with winter rains include *Eschscholzia, Nemophila, Clarkia* and *Limnanthes* (all from California); *Calceolaria, Salpiglossis* and *Nicotiana* (Chile and other parts of South America); *Gazania, Diascia, Lobelia* and *Dorotheanthus* (southern Africa); and *Calendula, Lathyrus, Antirrhinum* and *Papaver* (Mediterranean basin). Others evolved in biomes with summer rains, namely *Cosmos, Salvia, Zinnia, Tagetes* (Mexico and Central America) and *Impatiens, Primula, Dianthus* (China and central Asia).

Most annual plants excel in full light or semi-shade, whereas heavily shaded areas promote etiolation and limit the number of flowers induced. Other than that, there is some variation for environmental preferences (Table 7.1).

Table 7.1. Common bedding plant species and soil/temperature preferences

Scientific name	Common name	Horticultural traits
Dry or free-draining soil – warm temperatures		
Calendula officinalis	pot marigold	Single or double daisy-like flowers in orange, gold, cream or yellow hues. Long flowering season. Used as a 'companion plant' to help avoid aphid infestation.
Celosia	cockscomb	*Celosia argentea* var *plumosa* and *spicata* have upright 'feathery' flower plumes, whereas *C. cristata* has some vars with crested plume, similar to a cockscomb.
Dorotheanthus bellidiformis syn. *Mesembryanthemum criniflorum*	Livingstone daisy/ice plant	Iridescent, star-shaped, daisy-like blooms in shades of orange, yellow, white, cream, pink and crimson. Low growing with narrow succulent leaves.
Eschscholzia californica	Californian poppy	Single row of petals of yellow or orange in true species, but hybrids now also have reds and pinks.
Erysimum cheiri	wallflower	Adapted to grow out of walls and other dry, stony environments.
Gazania spp.		Hot conditions and well-drained soils.
Iberis umbellata	candytuft	Flat inflorescence in shades of white through to deep pink – easy to cultivate.
Linaria spp.	toadflax	Similar but smaller flowers to *Antirrhinum*, with strains such as 'Northern Lights' and 'Fairy Bouquet' popular.
Nemesia spp.		Adapted to warm, sandy soils, and found growing naturally in sand flats in southern Africa. Wide range of flower colours, including white and red bicolour form called cv. Danish Flag.
Oenothera spp.	evening primrose	Most varieties available are upright 200–800 mm tall with yellow flowers normally, although white and mauve also available.
Portulaca grandiflora	moss rose	A trailing habit with single or double miniature rose-like flowers.
Zinnia spp.		Single, double or multi-petal forms in range of pinks, reds, yellows and orange. Rich, free-draining soil.
Moderately rich, moisture-retentive but free-draining soil – warm temperatures		
Alcea rosea	hollyhock	A biennial. The quintessential cottage garden plant – large spires of single or double flowers in a wide range of colours.
Amaranthus spp.	love-lies-bleeding	Long drooping panicles of flowers often red or yellow. Likes warm conditions, moist but nutrient-poor soils.
Antirrhinum majus	snapdragon	Common name derived from the way the base of the dorsal petal can be squeezed to 'open' the ventral petal, like a dragon (or rabbit's) mouth! Does best on light, but fertile soils.
Centaurea cyanus	cornflower	Wild form has bright, deep blue flowers but cultivated ones include red, pink and white too. Does better in a cool rather than a hot climate.
Clarkia amoena	godetia	Open, fan-shaped flowers of various hues of pink, red, salmon and white.
Consolida spp.	larkspur	Member of *Ranunculaceae,* similar to the perennial *Delphinium.*
Coreopsis spp.	tickseed	Prefers a consistently moist, if not wet, soil. Flowers dominated by red and yellow hues. Good nectar source for pollinating insects.
Cosmos spp.		Open or slightly cupped shaped flowers, often on long stems. Tall plant useful in open, 'airy' spaces.

Continued

Table 7.1. Continued.

Scientific name	Common name	Horticultural traits
Dianthus spp.	pinks, carnations, sweet William	Flowers dominated by pinks, white and red. Strong scent in some cultivars. Most prefer high pH soils with lime.
Diascia spp.		Short-lived perennial now an archetypal patio plant, used in containers and hanging baskets.
Helianthus annuus	sunflower	Popular due to the large size of the plant, large leaves and grand flower heads in yellow, ochre and brown.
Ipomoea tricolor	morning glory	Climbing vine with various colours of white, purple or blue flowers. Cv. Heavenly Blue has luminous sky-blue flowers. Prone to red spider mite, but high humidity and temperatures help keep this pest at bay.
Lathyrus odoratus	sweet pea	Annual climbing plant, used for cutting flowers for interior décor and as colourful screen outside. Nutrient-rich, free-draining soil optimises growth and continued production of flowers
Linum spp.	flax	Rapid growing with flower colours in blue, red or white.
Lobelia erinus		One of the oldest bedding plants, being grown in Europe since 1752. Both compact and trailing forms, with small flowers in blue, red or white. Often used in hanging baskets.
Matthiola incana	stock	Flowers in white, mauves and pink. Dwarf varieties have been bred to stop lodging (flower heads bending over).
Nigella damascena	love-in-a-mist	Habit of self-seeding and returning year on year.
Papaver spp.	poppy	Shades of red, pinks and white. *P. commutatum* is the ladybird poppy with black spots at the base of the petals. *P somniferum* is the opium poppy, with glaucous blue/grey foliage and flowers in red, white, mauve and purple.
Pelargonium x *hortorum*	geranium	Common in red, pink, white and mauve flowering forms. Scarlet varieties very characteristic of window boxes in southern Europe and the Mediterranean region.
Petunia x *hybrida*		Tolerate moderately dry soils but flower damage evident after periods of wind and rain. Highly popular bedding plant but not the easiest to produce via seed due to *Pythium* and *Phythopthora* diseases. Wide range of flower colours including coloured centred, veined and stripped petal forms. Surfinas are a smaller flowered group that are useful in hanging baskets due to their trailing habit.
Tagetes spp.	African and French marigold	Despite the common names for *T. erecta* (African marigold) and *T. patula* (French marigold) both are actually native to Mexico and central America. Typified by bright yellow and orange flower heads.
Tropaeolum	nasturtium	Vigorous trailing plant with flowers in red, yellow and orange. Excessively rich soils result in rampant plants with fewer flowers.
Salvia splendens		Usually a bright scarlet and popular in formal bedding schemes, now other variants available.
Verbena x *hybrida*		Upright forms are usually 200–300 mm high, although trailing forms can have longer stems. Wide range of colours.
Viola x *wittrockiana*	pansy/viola	Pansy flowers tend to be larger than viola, although both forms can be clear colours or with a black blotch at the petal base. Perform best in fertile, moist soil.

Continued

Table 7.1. Continued.

Scientific name	Common name	Horticultural traits
Moist, free-draining soil with some shade		
Begonia semperflorens		Best in part-shade in hot climates. Predominately pink, red and white flowers.
Impatiens x *hybrida*	New Guinea impatiens	Needs shade in hot climates, good for containers and attractive foliage.
Impatiens walleriana	busy lizzie	Tolerates shade and fleshy leaves prone to drought-induced wilting if inadequate moisture supply. Multiple hues of pinks, reds and white, with some stripped petal forms
Limnanthes douglasii	poached egg plant	White flowers with a yellow centre reminiscent of a poached egg.
Lunaria annua	honesty	White or purple flowers with large papery seed heads.
Myosotis sylvatica	forget-me-not	Prefers some shade and moist soil in warm climates, but good spring bedding species in the more temperate regions. Pale blue flowers, but also now white and pink forms.
Primula spp.	primula and polyanthus	Excellent spring flowering bedding plants. Variations in flower colour, size and stem length. Do not perform well in excessively dry soils.
Wet or moist soils – cool temperatures		
Mimulus	monkey musk	*Mimulus lutea* and derived hybrids require damp soil, although other species are xerophytes adapted to arid soils and dry woodlands.

Annuals from 'direct sowing'

Many individual gardeners, local authorities and other landowners produce flowering displays by sowing seed directly into well-prepared soil. These usually rely on so-called 'hardy' annuals, although low temperature tolerance is not the only factor that defines these plants. *Tropaeolum* (nasturtium), for example, is not frost tolerant but is sufficiently vigorous and robust to grow from a seed planted in late spring to become a 2 m long flowering 'vine' within 6 weeks. There are overlaps with many of these species in terms of traits and management approaches to that of extensive displays of arable flowering annuals (cornfield mixes). Nevertheless, such plants have traditionally been used in a different context, particularly with where they have been sown, i.e. in garden settings to fill vacant spots between shrubs and herbaceous plants, as well as in designated flower beds of their own.

Hardy annuals can be sown from seed in autumn (depending on cold tolerance) to give early displays the following summer. Indeed, a first sowing in autumn (e.g. September, when soil temperatures are still conducive to good germination rates) followed by a subsequent sowing in spring, provide an opportunity to maximise the display throughout the summer. Autumn-sown seed often overwinter as seedlings, well able to capitalise on increasing temperatures and photoperiods in spring to maximise photosynthesis. Most sowings, however, are conducted in the spring to coincide with warming soils.

Seed beds are prepared by ensuring the soil is free of weeds, cultivated to provide a fine tilth (good soil crumb structure and even surface without large clods). Direct sowing into this cultivated soil involves either a broadcast technique (scattering seed over the soil surface) or drilling (sowing in rows). Designating areas for the different genotypes to be sown is implemented by marking out with sand or a similar material, and this helps plan for variations in flower colour, height and habit. The aim with broadcasting is to sow the seed evenly and mixing the seed well with sand beforehand improves distribution. A moderate level of soil fertility tends to suit many hardy annuals. Excess nutrients tend to promote aggressive weed growth, but very impoverished soils will not allow these species to attain the rapid growth rates and quickness to flowering that is their characteristic trademark. Heavy, damp 'cold' soils may inhibit germination, whereas a light but fertile free-draining soil is optimum for many genotypes. As annual flowering plants cover many genotypes and even different ecological niches (e.g. cornfield annuals which exploit the light and rapid release of nutrients that corresponds to the annual cultivation of fields, or the aforementioned desert ephemerals and their reliance on temporary periods of rain), some of these points cannot be generalised. Not least with plants such as *Lathyrus odoratus* (sweet pea) which, despite being a nitrogen 'fixer', actually responds well to deep organic, nutrient-rich soils.

Sowing in drills has the advantage of locating where the seedlings will emerge from and hence easing management tasks such as hoeing out competing weeds. It also allows larger seed to be sown at a pre-set distance ensuring the young plants have adequate space to develop, without crowding each other and competing for water and light. Once sown, seed are often covered with a thin layer of soil or sand, e.g. 2–5 mm. Although, some species may be reliant on light to improve germination, shallow covering often still provides enough irradiance, but has the advantage of disguising the presence of the new seed from birds and small mammals. The use of direct sown annuals, compared to intensively managed, glasshouse-produced tray/pot-grown stock, is seen by some as a more sustainable way to provide seasonal colour in the landscape.

Cornfield annuals and 'annual meadows'

The term 'annual meadow' is a bit of a misnomer, as technically meadows are grassland communities, which in practice are more often than not cut for hay (although in reality some meadows such as 'alpine meadows' would not necessarily be cut). The term is used for annual plant communities though, where horticulturally speaking, there is a riot of colour induced by flowering plants and that these plants are arranged in an informal manner. Much of the practice and philosophy around 'annual meadows' relate to the flowering annual plants that were traditionally associated with arable crops, before the advent of chemical herbicides largely removed their presence. The most 'common' colourful annual 'cornfield' plants in western Europe are *Papaver rhoeas* (field poppy), *Glebionis segetum*, formally *Chrysanthemum segetum* (corn marigold), *Anthemis austriaca* (corn chamomile), *Centaurea cyanus* (cornflower) and *Agrostemma githago* (corncockle), but a number of these are anything but common now in the rural landscape. These species are often sown together in ornamental displays to recreate the impression of cornfields of yesteryear. One, the field poppy, is also used on its own within public space plantings, as it has a 'revered' place in western culture as the symbol of death, blood and self-sacrifice associated with World War I: the arable fields of Belgium and northern France 'turning red' with poppies after the soil had been disturbed by munition shells and mortars. These traditional arable field annuals have been augmented by other species to further embolden and extend the season of interest of annual flowering communities (Table 7.2).

Most of these annual species are adapted to germinating and growing quickly, setting flower in as little as 6 weeks after germination. The use of successive sowings or increasing the range of species that flower at different times can help extend the flowering period across the summer. As implied, the flower display is only a single 'annual' event, and although some viable seed may remain in the soil from one year to the next, it is normal practice to re-sow seed again the following winter/spring.

Bedding plants

Bedding plants are annuals, biennials or short-lived perennials (but treated like annuals) that traditionally were 'bedded out' in flower beds in parks and other civil or municipal areas from 1820 onwards. Bedding plants are still used in civic town centre and park displays as well as along roadsides and business parks but are also utilised extensively in the private garden. Indeed, many garden centres in the northern hemisphere rely on the late spring sales of bedding plants to ensure they remain in profit for the rest of the year. Plants tend to be retailed in trays (6, 8 or 12 individual plants per tray) or small pots. Sales are very weather dependent and fair/sunny weather over Easter, Mothering Sunday, May Day bank holiday weekends and other country-specific holidays are often critical in determining if a garden centre or nursery is to have a successful season. The contribution of bedding plants to the horticultural sector is significant and the wholesale value of annual bedding plants in the USA is in the region of US$1.55 billion p.a. (Anon., 2019).

7.4 Herbaceous Perennials

Herbaceous perennials (Fig. 7.3) are characterised by a period of active shoot development, followed

Table 7.2. Range of species with colourful flowers now commonly used in annual flower communities ('annual meadows')

Species	Common name	Predominate colour
Agrostemma githago	corncockle	purple
Ammi majus	bishop's flower	white
Atriplex hortensis	orache	red
Centaurea cyanus	cornflower	blue
Coreopsis tinctoria	tickseed	yellow/red
Cosmos bipinnatus	cosmos	pink/white/red
Delphinium ajacis	larkspur	blue
Dimorphotheca sinuata	African daisy	orange/yellow
Echium vulgare	viper's bugloss	blue
Eschscholzia californica	Californian poppy	orange
Glebionis segetum	corn marigold	yellow
Iberis umbellata	candytuft	pink
Linaria maroccana	fairy toadflax	purple/white/pink
Linum grandiflorum cv. Rubrum	flax	red
Linum usitassimum	flax	blue
Malcomia maritima	Virginia stock	purple/white /pink
Matricaria inodora	scentless mayweed	white
Nemophila insignia	baby blue eyes	blue
Nigella damascena	love-in-a-mist	blue
Papaver rhoeas	field and 'Shirley' poppies	red/pink
Papaver somniferum	opium poppy	purple/red/pink
Rudbeckia hirta	black-eyed Susan	yellow
Salvia horminum	salvia	blue/purple

Fig. 7.3. Herbaceous perennials and ornamental grasses being used at the edge of a pathway to provide colour and form to the landscape.

by the abscission of the above-ground parts as the plants enter a period of dormancy or quiescence. As the shoots and leaves begin to decline and abscise, resources – primarily carbohydrates – are translocated to the roots or ground-level 'crown', where they are stored over the dormant period. These are then used to initiate and power shoot growth again in the following growing season. In western European horticultural practice most of the plants referred to as herbaceous plants are typically dicots. Ecologists refer to dicot herbaceous plants as 'forbs'; this is a convenient term for distinguishing between, for example, non-grassy, and grass-like herbaceous plants, such as sedges and true grasses. The latter two groups are often referred to by ecologists as graminoids, again a useful cover-all term.

Due to the fact that herbaceous plants and geophytes do not make wood, the patterns of growth by which canopy expansion takes place is much more varied than in trees and shrubs. Understanding these patterns is very important when using herbaceous plants, as they determine the outcome of competition between different species in plantings, in particular the capacity of one species to eliminate another. The following growth forms are frequently observed.

Herbaceous perennial growth forms

(i) Gradual shoot expansion but slow canopy expansion

In this growth form, the subterranean or soil surface shoot meristems expand slowly to form a concentric ring. Annual additions to the radius of the shoot ring may be less than a centimetre in very slow growing, small species, for example, *Scabiosa columbaria, Geum coccineum* or *Pulsatilla vulgaris*. As a result, the shoot tissues produced from these buds only expand approximately the same limited distance each year; total canopy extension is very gradual and predictable, with the plants forming neat clumps. Mature plants of *Pulsatilla vulgaris* will, for example, rarely occupy a volume of space wider than 400–500 mm. Where species with this growth habit are very long-lived, as in the case, for example, of some *Hosta* spp., this leads to a very large increase in the space occupied by the foliage between planting and maturity; the canopy radius (but not the ring of subterranean shoots) of *H. sieboldiana* reaches 1.2 m across or more at maturity.

There are also qualitative differences in how canopy expansion precedes in species with this growth strategy. In some species, buds that continue to produce shoots are retained in the centre of the clump, as for example in the case of *Epimedium* or *Omphalodes cappadocica* maintaining a dense clump in perpetuity. In other species, the buds in the centre of the clump are abandoned, resulting in the development of doughnut-like canopies with an empty centre.

(ii) Gradual shoot expansion but seasonally rapid canopy extension

This growth form is common in many *Geranium* species, for example, *Geranium psilostemon, G. wallichianum, Oenothera macrocarpa, Persicaria amplexicaule,* and *Potentilla* hybrids, amongst others. This growth form is relatively common in the herbaceous plants popular in traditional border plantings, where a clump of each species is often separated from its neighbours by a large space that is subsequently filled by the expanding leaf canopy as the plants grow by mid-summer. This growth habit, however, has the disadvantage of eliminating slower, shade-sensitive species in mixed plantings, if sufficient room is not given between different species.

(iii) Endless spread by below-ground or above-ground shoots

This growth pattern varies in terms of the extent of spread possible. At the conservative end of the spectrum, there are sheet-forming plants such as *Ajuga reptans*, *Lysimachia nummularia, Phlox stolonifera* and *Tiarella cordifolia* (above-ground rooting stems) and *Ajuga genevensis* (below-ground rooting stems). These species in time tend to form open, clonal patches rather than a 'smooth' continuous unbroken carpet of foliage, although, because they do root as they grow, they can theoretically colonise large areas. The next step up in scale comes with species such as *Geranium macrorrhizum*, which slowly forms long, permanent, closed carpets. At the extreme end of the sheet-forming spectrum are *G. procurrens* and *Lamiasterum galeobdolen*, which expand their canopies by over 1 m each year. This growth pattern is also found in species with tall leafy stems, for example *Lysimachia clethroides, Euphorbia griffithii*, and *Macleaya cordata*. These species form tall, clonal patches and eliminate everything that is not highly shade tolerant or summer dormant as they spread.

(iv) Slow canopy expansion but with new individuals established by self-sowing

Many of the species that use most of their photosynthates to produce large amounts of biomass, (ii)–(iii), have relatively limited seed production and only occasional self-sown seedlings. Because plants with slow canopy expansion are often associated with unproductive habitats in which there is incomplete vegetation cover, these species tend to have higher levels of seed production and self-sowing because there are opportunities for this strategy. Species that have a higher-than-average capacity to do this include *Berkheya purpurea, Dianthus carthusianorum, Dracocephalum rupestre, Origanum vulgare, Penstemon barbatus, Polemonium reptans, Scabiosa columbaria* and *S. columbaria* subsp. *ochroleuca*.

7.5 Geophytes

The term 'geophyte' is used here rather than the more common horticultural term 'bulb' or 'bulbous' to get around the problem that many of the plants generically referred to as bulbs technically are not. True bulbs have a very distinctive perennation structure with concentric rings of compressed leaves. This growth form is the norm in the Alliaceae (for example *Allium*), and the Amaryllidaceae (*Crinum, Cyrtanthus, Fritillaria, Narcissus* [daffodils] and *Scilla*). Some bulbs have a protective tunic around the leaves, as in *Tulipa* (tulips) whereas others have the fleshy leaf scales open to the air such as is common in the Liliaceae (lilies). Corms, which unlike bulbs have no obvious internal structure, are essentially a block of carbohydrates, with a root meristem at the base and a shoot meristem further up the corm. Corms are the norm in geophytic members of the Iridaceae (*Crocus, Gladiolus, Iris, Ixia* and *Watsonia*). Finally, there are tubers, of which the culinary potato is the most familiar. These can be either a modified stem or modified root that acts as a carbohydrate store with growth points (roots and shoots) distributed across the exterior, although in some tubers (for example *Cyclamen*), the location of the shoot and root meristem is clearly defined as top or bottom. Tubers typically differ from corms in being asymmetric, often fleshy, and sometimes forming on roots and thus extending laterally (creeping) along under the soil surface. Amongst others they are found in *Anemone, Cyclamen, Dahlia* and *Tropaeolum*.

Geophytes have a similar annual growth cycle to that of herbaceous perennials, surviving periods of cold, heat, drought or excessive shade (woodlands in summer when the leaf canopy of trees has closed over) as a dormant resting organ, with (usually) above-ground leaves absent. Although active growth and dormancy in geophytes are controlled by environmental cues, the precise biochemical mechanisms regulating the transition from one to the other remains elusive (Zhao *et al.*, 2022).

The scale of difference in terms of canopy expansion tends to be reduced in geophytes, but it is still possible to identify important differences in growth habit and spread.

Geophyte expansion

(i) Solitary individuals and little or no vegetative expansion

This is the case in species that do not readily produce corm, bulb or tuber offsets. With the exception of some tulips, plants with this habit tend to be uncommon in mainstream environmental horticultural use but potentially are important in amateur horticultural circles. Some examples are *Brunsvigia, Eremerus, Iris latifolia, Iris reticulata* and *Romulea* spp.

(ii) Gradual but slow expansion through vegetative offsets

This clump-forming growth habit occurs because of offset formation and is common in some species within *Narcissus, Eucomis, Gladiolus, Galtonia, Nerine* and *Watsonia*. 'Daughter' bulbs form off the basal plate of the parent bulb but may be too immature to form flowers the year after initiation.

(iii) Solitary individuals with slow offset formation with new individuals established by self-sowing

Where these species are relatively large with abundant foliage, this can be a problem in plantings where there is insufficient maintenance to remove seed heads prior to seed dispersal. *Allium* such as *A.* x *hollandicum* for example eventually produce "Allium forests" where self-seeding is not checked, leading to the elimination of low-growing herbaceous plants. This habit is also highly developed in *Galtonia candicans*, and *Tulipa sprengeri*. It is also found in much smaller-growing species such as many *Chionodoxa, Crocus, Cyclamen*, and *Narcissus*, such as *N. bulbocodium*.

7.6 Ferns

Ferns (Polypodiopsida or Polypodiophyta) are vascular plants (plants with xylem and phloem) again without a permanent strong lignified structure. They reproduce via spores and have neither seeds nor flowers. Most tend to be shade-loving and are used in woodland or, for example, in green walls on a north-facing wall (in the northern hemisphere). They are popular due to their strong structural geometric form. As with herbaceous perennials ferns tend to unfurl their leaves (fronds) after winter – once the risk of frost has passed. Many prefer damp, but not waterlogged, conditions and do not tolerate high winds (Fig. 7.4).

7.7 Alpines

These are short-stature plants, often having a prostrate or hummock-forming habit (Fig. 7.5). They can be herbaceous in terms of the foliage abscising in the dormant period, but they can also have non-lignified stems that are retained all year round, some surviving under snow drifts in their natural habitat. These plants are adapted to mountainous landscapes and so can tolerate low temperatures and wind. Most like a free-draining substrate as most hillsides have stony or gravelly soils and a slope that tends to promote water runoff. Some tolerate high humidity, but other prefer dry or even

Fig. 7.4. Ferns often thrive in shade locations. Here *Matteuccia struthiopteris* (Shuttlecock fern) is used to good effect on a woodland folly.

Fig. 7.5. Alpine plants such as *Primula marginata* often forms clumps or prostrate mats to avoid excessive damage from wind and cold.

arid conditions. Whilst a broad generalisation, most prefer rather nutrient-deficient soils – again typical of alpine regions with low organic matter soils and rapid depletion of nutrients through leaching via rapid rainwater runoff. Alpines are low-growing plants, so have poor weed tolerance – the presence of which shades them out from the light.

7.8 History of Non-woody Ornamentals in the Landscape

It is impossible to know precisely when annuals, herbaceous plants and geophytes were first cultivated in gardens, although wall paintings from the classical civilisations suggest that the more spectacular species, such as *Lilium candidum* have a very long history of cultivation. It seems highly likely that initially the range of flowers used were largely restricted to native and near-native species, as in the case of *Lilium candidum* in the Mediterranean, but some species were clearly being moved substantial distances along trade routes before 1000 CE, for example tulips from central Asia. Geophtyes are easy to move around (when done at the right time of year) because of their desiccation-resistant storage organ. Of the herbaceous plants depicted in tapestries in the Middle Ages, many actually exist in the real world, as opposed to solely in the artists' imagination. Some of these species appear to have been planted in gardens during this period. The range of plants expands rapidly from the Renaissance onwards as world trade increases. The first tulips to arrive in Vienna via the Ottoman Empire in 1512 (Culver-Campbell, 2001) caused a horticultural 'storm'.

There is evidence that many herbaceous plants and geophytes were planted into close-cut lawns; a practice known as enamelling, sometimes in parterre-like compartments; a sort of cultural meadow (Woudstra and Hitchmough, 2000). This style is also represented from 1800 in central European gardens such as Prudhonice in Prague (Czech Republic). Flower gardens are also presumed to have existed at this time; however, given their relatively ephemeral nature evidence is limited. Research by Laird (1999) has shown that even during the height of the English Landscape Movement, flower gardens continued to be in vogue, often at the rear of grand houses, and this tradition continues to develop from *c*.1820, as the herbaceous border, often set within a formal architectural structure. In the latter part of the 19th century, however, there is also a more naturalistic school of use, pioneered by writers and gardeners such as William Robinson (who wrote *The Wild Garden* published in 1870) and his contemporaries in Germany (Woudstra and Hitchmough, 2000). From then on, the 'cottage garden' in which herbaceous plants are used in informal, naturalistic arrangements becomes popular. Very few of these new ideas filter through to public green space, where herbaceous planting is normally restricted to herbaceous borders and geophytes to spring formal bedding or the use of *Crocus* and *Narcissus* (daffodils) 'naturalised' in mown grass. Probably the most dramatic shift in the use of herbaceous perennials in public landscapes occurs in Germany from 1950 onwards, through a more 'ecologically inspired' view of planting herbaceous perennials and geophytes to form naturalistic plant communities (Hansen and Stahl, 1994). Through the institution of garden festivals, these plant communities were developed to a high level of sophistication, using a mixture of native and non-native species, with semi-natural vegetation types such as steppe (a dry grassland extending through Eurasia) and prairie (a summer-growing North American equivalent) as key inspirations.

These ideas are then refashioned through plantsmen/plantswomen designers such as Oudolf (Oudolf and Kingsbury, 2014) and applied to prestige urban landscape projects on a grand scale, as for example in the High Line in New York. Together with the ideas and practice from the University of Sheffield's Department of Landscape (Dunnett and Hitchmough, 2004), this leads to a situation in the second decade of the 21st century where herbaceous planting becomes highly desirable at least as an aspiration in public planting, although in practice it remains relatively uncommon. Nearly all the herbaceous planting now practiced in public landscapes embodies two characteristics that were not much considered in the UK prior to 2000: the application of (i) ecological and (ii) functional ideas.

These ecological ideas often address plant selection (see below) in relation to the environment of the planting site, whilst functional ideas pertain to, for example, how might the plants be used and managed to reduce weed invasion and to fulfil a role beyond the purely aesthetic. Frequent roles are supporting biodiversity and managing rainwater runoff (Dunnett and Clayden, 2007; Nur Hannah Ismail *et al.*, 2023) microclimate modification (Blanusa *et al.*, 2013; Monteiro *et al.*, 2017). Incorporating these ideas into design results in various different forms of planting. The planting styles that arise from this can be classified in different ways, for example in terms of spatial arrangement in the three dimensions of space, and then in terms of the locations and/or conditions to which planting might be used in. With the exception of evergreen geophytes such as, for example, *Dierama*, the extended dormancy period of most geophytes means that within sustainable forms of plant use (as opposed to ephemeral planting displays), geophytes normally have to be planted in conjunction with plants that retain a structure and presence during the period when the *Geophytes* are dormant. Typically, this might involve seasonally unmown grass and meadows; herbaceous planting or underneath deciduous shrubs and trees. *Lilium* (lilies) for example complement woodland planting schemes – giving colour below the tree canopy and between flowering shrubs (Fig. 7.6). These approaches are discussed later in this chapter.

Ferns became popular in Europe in the UK Victorian era – pteridomania was the huge love affair for ferns and all things fern-like in Britain between 1840 and 1890. Ferns and fern motifs appeared everywhere: in homes, gardens, art and literature. Their images adorned fabrics, porcelain,

Fig. 7.6. *Lilium lancifolium* cv. Splendens (tiger lily) and other woodland lily species provide much needed colour in woodland glades in summer.

furniture – even biscuits. Ferns were either held in a fernery – a special glasshouse where ferns were cultivated and displayed – or a shaded outdoor area, such as a fern-gully or grotto. Such locations used boulders and large rocks to create a cool, moist root-run for the ferns, whilst the surrounding trees and shrubs gave shade and protection.

Rocks have been used as decorative and symbolic elements in gardens and can be traced back at least 1500 years in Chinese and Japanese gardens. In Europe, rockeries initially used artificial or exotic minerals (shells, feldspars or lava) but the introduction of alpine plants from expeditions in European, Asiatic and North American mountain ranges encouraged wealthy garden owners to try and replicate the environments the plants were sourced from, from 1830 onwards. Thus, limestone, granite, sandstone and slate became common 'media' for alpine plant collections. Commercial firms and nurseries could supply complete rockeries, at great expense.

Annual bedding plant schemes were at their zenith during the Victorian era and grew in popularity in the UK when the glass tax was repealed in 1845 (glasshouses being required to germinate and grow on the tender 'half-hardy annuals). 'Formal' bedding relates to plants being arranged in specific colour groups and designs, often using straight lines and geometrical patterns, rather than more ad hoc (informal) bedding plant arrangements. Before the tax on glass was relaxed, the usage of 'half-hardy' flowering annuals tended to be restricted to the gardens of large estate houses, the owners of which were the only people wealthy enough to afford to grow on plants under heated glasshouse conditions, before later 'bedding-out' these colourful plants into the garden. After the repeal of the glass tax, the subsequent widespread use of glass and the introduction of newer, larger glasshouses meant that the germination and early development of bedding plants could be more easily justified. Bedding plants (e.g. *Viola*) and short-lived perennials (e.g. *Dianthus*) become popular with working-class people in the Victorian era, who cultivated and bred ornamental plants in pots as an 'antidote' to the hard-working conditions they endured for much of the week.

Although, for today's gardening public, bedding schemes usually connote vibrant flower displays, foliage was important historically in that 'carpet' bedding schemes developed with the aim of providing a close cover of foliage to provide a backdrop for flowering plants, with low-growing species such as *Sedum*, *Echeveria*, *Alternanthera*, *Senecio* and *Sempervivum* being ideal for such situations. Carpet bedding rose in popularity between 1850 and 1900, showing itself to be highly adaptable to commemorative or civic planting incorporating town names, shields and other designs.

7.9 Production of Non-woody Landscape Plants

Seeds

Seed is used for herbaceous perennials, annual bedding, alpines and certain bulb species. The largest numbers of plants produced from seed is with annuals. Seed for spring/summer bedding species are sown from late December into March (northern hemisphere), although batches of the same cultivar may be sown sequentially in a scheduled production to allow these cultivars to be marketed over an extended period. On large-scale production nurseries, seed is sown into cellular plug trays via a mechanised seed planter. Plug trays come in different sizes (number of cells × volume of individual cell), e.g. 50 to 800 individual cells. The use of plug cells, which encourages the development of a robust, coherent rootball, helps the young seedlings transplant into their final trays without excessive root disturbance (transplant shock). The small volume of the plugs, however, requires the growing media to be very fine grade so it flows into the spaces easily. Fine-milled and sieved coir or peat is commonly used for this purpose, as are smaller grades of perlite and vermiculite. As with seed distribution in plug trays, transplanting the young plant from the plug tray to the cell pack or final tray is done mechanically.

Reliable and consistent germination is key to ensuring plants are produced in predictable schedules. Sowing date for bedding plants is calculated based on the marketing period for any given crop, but the length of production schedules can vary depending on time of retail. Early spring season batches, for example, take longer to produce, due to predominately cooler growing temperatures or lower irradiance levels, compared to those seed batches sown later under warmer/brighter conditions. Sowing dates, therefore, are carefully calculated by factoring in not only the time to retail the crop but also the environmental factors that are likely to affect growth rates and flower initiation. Consistency in this respect, within a bedding plant series, is increased by the use of F1 seed.

F1 seed is derived from two parental strains that each have been inbred (self-pollinated) to develop a consistent phenotype (pure line). These parental pure lines may have a specific desirable trait, such as unusual flower colour or good branching habit. As seed comes from crossing two very distinct parental lines, they possess hybrid vigour, i.e. the progeny tend to be stronger, more vigorous plants than either of the parental inbred lines. In this way, it is feasible to express both the outstanding qualities of the parental plants and new desirable characteristics from the hybrid itself. F2 plant hybrids also exist. These result from the self- or cross-pollination of an F1 plant. These retain some of the desirable traits of the original F1 and can be produced more cheaply as no intervention in the pollination is required.

Ferns are propagated from spores. These are collected from the underside of the leaves, when they are ripe, then cleaned and placed on sterile fine growing media and kept moist to let the young fern develop.

Cuttings

Softwood stem cuttings are alternative means to produce larger numbers of young plants in a relatively short time period. As most herbaceous species tend to self-propagate through daughter crowns (off-sets), the capacity for the herbaceous stem to form roots in general is greater than that of woody plant material. Thus, many herbaceous lines 'strike' successfully from cuttings. In commercial nursery production, labelling of the stock-bed rows or the potted-on juvenile material is paramount – as when plants are retailed there is unlikely to be any flower present to identify the cultivar, and sometimes there is not even leaf present to help identify the species!

Stem cuttings are also important for certain types of bedding plants – some of those that are technically perennial but grown every year and disposed of at the end of the growing season. These include plants such as Surfina *Petunia* and *Pelargonium*.

Division

Many herbaceous perennials propagate naturally through divisions – new shoots developing from the crown (interface between shoots and roots) of the plant. Thus, natural propagation in field-grown stock is one way to commercially produce herbaceous material. 'Mother' plants are divided and the divided 'transplants' planted out to grow on and 'bulk-up' in the field and lifted later – or potted in small 90 mm or 1 litre pots and again grown on before sale.

The way geophytes are propagated does depend on the taxa. With *Tulipa* and *Narcissus*, bulbs can be produced as a by-product of the cut-flower industry for tulips and daffodils sold via florists. More specialised propagation of species such as *Lilium* may be through dividing bulbs (new bulbs generate from the base plate of the parent bulb) or peeling away the leaf scales (scaling) of the parent bulb and placing these in moist vermiculite or other free-draining medium. Daughter bulbs regenerate from the base of the leaf scale.

Micropropagation

Micropropagation (tissue culture) is used in the production of certain ornamentals, particularly geophytes (and orchids, see Chapter 11). It is especially important in the 'bulking-up' of a new cultivar, to ensure large numbers of plants become available quickly to the market. Micropropagation involves cutting out a small section of plant, sterilising this with ethanol to restrict pathogen infection and then growing the ex-plant in gel under aseptic conditions. With appropriate exposure to nutrients and growth-regulating compounds (plant hormones) new plants form from the cut ex-plant tissue. This can be through direct organogenesis (e.g. new shoots develop from ex-shoot tissues) or via the formation of undifferentiated cells first (callus – indirect organogenesis). The callus can then be encouraged to form new whole plants depending on the chemical environment it is placed in. There is also a process where an intermediate embryo is formed (similar to that found in the plant seed) – this process being known as somatic embryogenesis (Yasemin and Beruto, 2024). It is thought that the global annual production of ornamental plants through micropropagation has increased from 800 million to 2 billion plants over the last decade, with a value of US$ 2.8 billion by 2030.

Growing on in nurseries and retail

Hardy herbaceous plants are grown on in pots on nursery beds or grown in field soils. Overhead irrigation is often supplied for herbaceous perennials in pots as their water demands can be high in summer. Many plants may be over-wintered in pots, so the growing medium needs to retain good drainage capacity and aeration too. Green waste composts may be mixed with bark or wood fibre products to improve the drainage and aeration properties within the pot. Often these plants are given a top dressing of bark (or grit in the case of alpine plants), to restrict weed growth in the top layer of the growing media. Larger species of herbaceous material (e.g. *Helenium*, *Delphinium*, *Monarda* and *Rudbeckia*) can suffer wind-throw, and holding pots within 'pot-holder' frames tends to reduce the chances of them falling over in the wind. Growers will provide nutrition through the use of controlled-release fertilizer granules incorporated into the media. These granules are coated in a polymer and release nutrients slowly into the medium as the granule weathers, the process being temperature dependent. This form of fertilisation can help growers differentiate between water and nutritional needs, optimize fertilizer use and reduce nutrient leaching. Small granule sizes help ensure each cell receives its due allocation of granules. The granules help extend nutritional feeding through to the retail and even garden stage too (e.g. 2- to 3-month formulations). The drawbacks are that growers can no longer regulate growth via alterations in the nutrients made available to the crop, and that as nutrient release is temperature dependent, mismatches can occur between greatest release periods and crop need. For example, excessively warm periods early on during production release more nutrients that the crop actually requires at that particular stage. Sometimes liquid feed is used to top up any deficiencies or give more precise growth control during the scheduling of specific crop types.

Excessive shoot extension (height) in non-woody ornamentals can be problematic. Herbaceous perennials, many of which are attractive because of their tall, arching flowering stems – e.g. *Alcea rosae* (hollyhock), *Delphinium* spp., and *Helianthus* spp. – do not lend themselves to handling in nurseries, garden centres or squeezing into a small family car after purchase by a garden owner. Garden size has also got smaller over the decades, and many traditional tall, border plants now are out of scale for the small beds and borders that predominate in modern urban gardens. Thus, herbaceous perennials have seen a trend to select and breed shorter-stature cultivars.

Herbaceous ornamentals are sold wholesale to the landscape industry and garden centres or at a retail 'mark-up' to the garden consumer directly. For the latter, most plants are sold in pots with a plant label included, providing an image of the plant in flower and cultivation 'instructions'.

With the exception of evergreen species, the most widely planted geophytes are nearly always retailed as dry bulbs, corms and tubers in the dormant season. In northern regions this means late summer and early autumn for *Narcissus* (daffodils), *Tulipa* (tulips) and *Crocus* but late winter/spring for *Dahlia, Gladiolus, Begonia* and *Lilium* (lilies). Geophytes are usually retailed in net or plastic bags. Top-quality bulb-growing companies specify the grading size on sale; larger bulbs tend to perform more robustly.

7.10 Case Study – Annual Bedding Plant Production

As annual plants need to be grown every year, a whole industry has evolved around the production of these plants with the main sale period being the spring and early summer. Annual bedding plants embrace a large range of species and cultivars, but the most popular are outlined in Table 7.3. In the USA, *Petunia* represents the greatest value of any single crop (Fig. 7.7), with high numbers being used in hanging baskets as well as sold in trays and pots (Table 7.3). Traditionally, bedding plants were defined as hardy or half-hardy, the former being able to be sown from seed out of doors, compared to the latter that required protected environments (glasshouses or polythene tunnels) during the propagation and early growing on phases. These half-hardy species were planted out once the threat from late-spring frosts had receded (Fig. 7.8).

Each year seed houses release new varieties of these 'old favourites' to maintain market interest, improve robustness and flowering attributes or introduce novel strains (new, unusual colours or floral morphologies). Increasingly though, new species are being introduced and novel cultivars raised that fall outwith these old classifications of hardy/half-hardy. These are tender perennials that are propagated each year with the notion of providing summer flower colour but with plants not being retained through until the following year. In the UK and elsewhere, the prime functions of these plants are to fill containers, pots and hanging baskets around the house or on garden patios. Many derive from warm climatic regions such as southern Africa (*Osteospermum, Diascia* and *Gerbera*) or Central (*Zinnia* and *Dahlia*) or South (*Calibrachoa*) America and therefore do well in the hot, summer microclimates of hard-surfaced patios. This has led to the additional term 'patio plants' being used to describe plants with these traits.

Extending the season of interest has been a prime goal for bedding plant producers, and bedding species are not exclusive to summer displays. Indeed, bedding is defined by the season of interest with spring bedding exploiting biennials (sown one year to flower the next), or hardy, short-lived perennials that flower early in the growing season (February–May in northern latitudes). Biennials include *Erysimum* (biennial forms also classified as *Cheiranthus*, wallflowers), *Bellis perennis* (daisy) and *Myosotis* (forget-me-not), and the short-lived perennials typified by *Viola* (pansy) and *Primula* (polyanthus). These plants are frequently used in combination in formal schemes with spring-flowering bulbs including *Tulipa* (tulip), *Hyacinthus* (hyacinth), *Narcissus* (daffodil) or *Muscari*. Bedding schemes were transformed by the introduction of the winter-flowering pansy in 1979. Winter-flowering pansy (and later the smaller-flowered viola version) were bred to produce flowers in the short days of winter, possess enhanced tolerance to cold, rain and wind and to retain a compact habit when weather conditions improved. Summer remains the paramount season for bedding plants, however, with the greatest variety of genotypes being available for flowering from May–September in the northern hemisphere.

Seed companies often produce cultivars of many bedding plant species within a series (or strain). These usually have a similar genetic background, although cultivars within the series will have different flower colours or markings. Selecting a variety of coloured cultivars from the same series allows the grower to provide the market with variety whilst keeping cultural requirements uniform. Much effort is put into developing

Table 7.3. Numbers (millions) of trays, hanging baskets and pots of different bedding plants sold as wholesale products in 2019 in the USA. (Modified with permission from Anon., 2020.)

		Number (million)		
Scientific name	Common name	Trays	Hanging basket	Pots
Petunia x *hybrida*	Petunia	9.21	7.40	36.54
Pelargonium x *hortorum* (cuttings)	Geranium	0.42	4.35	40.18
Pelargonium x *hortorum* (seed)	Geranium	0.42	0.35	16.01
Viola wittrockiana	Pansy and viola	8.51	1.31	37.35
Impatiens walleriana	Busy Lizzie	5.40	2.14	14.29
Begonia semperflorens	Begonia	5.83	2.69	27.02
Impatiens hawkeri	New Guinea impatiens	0.68	2.88	18.11
Tagetes and *Calendula*	Marigold	5.42	0.16	13.99
Other		29.29	18.1	229.87

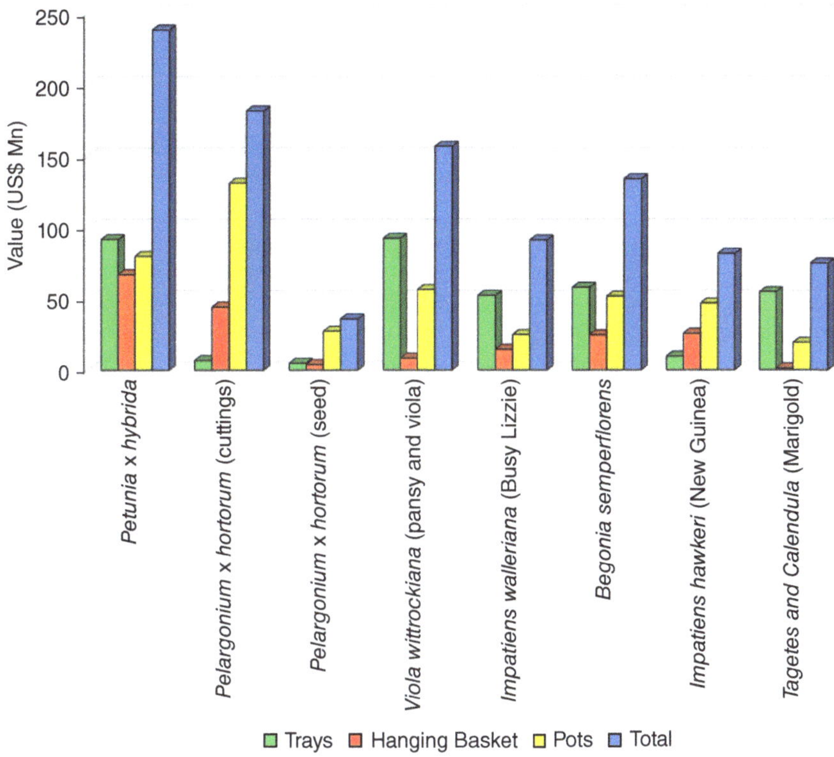

Fig. 7.7. Wholesale value (US$ millions) of trays, hanging baskets and pots of different bedding plants sold in 2019 in the USA. (Modified with permission from Anon., 2020.)

Fig. 7.8. 'Half-hardy' *Begonia* and *Lobelia* cultivars being used in a formal bedding scheme within a UK public park. Plants would have been germinated under glass and grown in modules and pots before being planted (bedded) out in an outdoor location such as this.

new series and cultivars within a series and, due to the demand for novelty, an individual seed house may introduce as many as 80–100 new selections each year.

Much of this breeding effort has been centred on developing more flamboyant strains, with emphasis on double or multiple flower forms, colour 'breaks' (streaks, lines or blotches) in the petals or a fringed edge to the petals (Fig. 7.9). To increase the impact of flower colour, traits have been preferentially selected to increase petal size or alter the hues within the petal. This has often resulted in the elimination, or diminished size, of other parts of the inflorescence, including the nectaries and pollen sacs. To this extent certain genotypes, and bedding plants in general, have been criticised for reducing their potential to supply invertebrates (as well as pollen/nectar, feeding bats and birds) with 'high-energy' food sources (see Chapter 5). Varieties of *Dahlia* for example with 'open flowers' allowing easy access to nectaries/pollen were found to be visited by pollenating insects 20 times more frequently than forms bred for their double or multiple petals (Garbuzov and Ratnieks, 2014). Concerns have also been expressed over the use of genetic modification technology (GM, i.e. where new DNA is inserted into a species to alter its genetic code) for ornamental plant varieties, where the introduction of new genes promises unusual colour 'breaks', e.g. a blue-flowered *Dianthus* (carnation) or increased drought tolerance, but where these modifications may enhance risk due to factors such as pollen-mediated gene flow, 'weediness' characteristics, enhanced vigour and resilience with implications for invasion of natural areas, as well as direct harm to non-target organisms (e.g. invertebrates that may feed off the modified plants). In reality, the introduction of genetically modified ornamental plants has been limited to date, largely by the regulatory costs of introducing the new genotypes.

Fig. 7.9. Flower forms in *Viola wittrockiana* (viola and pansy). Breeding has resulted in the relatively simple viola flower (top) being bred into the larger pansy (bottom), with a wide variety of colours and petal shapes now available.

Although there has been a trend in recent years for bedding plant genotypes to be sold in individual pots, reflecting the marketing of more exclusive cultivars and larger individual plants that can command a greater price, the traditional bedding plant is sold in a tray or 'flat'; these often being composed of 6, 8 or 12 plants within a tray. The original wooden seed trays that seedlings were pricked out into had no compartments, but modern plastic and polystyrene trays are divided into a number of cells. The volume of individual cells dictating the size of plant required at the retail stage. Most bedding plant sales rely on some of the plants in the tray being in flower (in colour) to attract the customer and help guarantee a sale.

Bedding plant propagation is divided into four stages.

1. Radicle emerges from the seed coat.
2. Stem and hypocotyl (seedling leaves) emerge.
3. First true leaves emerge and develop.
4. Seedlings become established in the plug and are ready for transplanting, shipping (moving on to other nurseries or garden centres) or holding.

Plug technology including the requirement for capital investment in mechanised seed sowers, roller benches and carefully controlled environmental germination rooms mean that some nurseries now specialise in plug production alone, selling on plants to 'finishing' nurseries, rather than encompassing the whole production process. Conversely, other nurseries will solely buy in plugs, transplant these on to larger module sizes and become more specialised in the husbandry required to control growth and meet the specific demands for finished plants in the garden centre or landscape markets.

Bedding plant germination

Ornamental plants come from a wide range of ecophysiological backgrounds and have developed evolutionary strategies to germinate under conditions that increase the chances of establishment and survival in those varied locations. As such, some ornamental species require a range of environmental parameters to facilitate germination, whereas others may have a different set of optimal criteria. This is best demonstrated with the requirement for light. Germination is promoted in species such as *Amaranthus*, *Nicotiana*, *Lobelia*, *Cosmos* and *Antirrhinum* when light is provided, whereas light inhibits germination in *Cyclamen*, *Nigella*, *Tropaeolum* and *Phlox*. The wavelength of the irradiance is the fundamental component in many cases, with the ratio of red to far-red light influencing germination through the action of phytochrome.

Plant origin also impacts on temperature requirements. Species from tropical and subtropical regions require germination temperatures >10°C (below which physiological chilling injury of the seed/seedling may occur). Other species can tolerate temperatures as low as 5°C. Some 'cool-requiring' species require temperatures <25°C, with germination being inhibited at higher temperatures (e.g. *Delphinium*, *Primula* and *Viola*). During germination, media moisture content should be sufficient to allow the seed to absorb water, whilst retaining oxygen within the medium. The media needs to be consistently moist as temporary drying out shortly after germination is a common source of seedling failure.

Growing media for bedding plants

The growing media (substrate) for ornamental plants does vary with what is available in different countries. Peat is still used extensively, although its harvesting causes damage to peat bogs and releases CO_2 to the atmosphere. Coir (derived from coconut husks), tree bark, wood fibres, anaerobic digestate and green compost (composted garden and landscape vegetation material) are all used too, with a number of media manufacturers trying to replace peat with these, or with blends of these materials. For bedding plants, the relatively small volumes of growing media accessible to individual plants in plug or cell trays require that the medium must have high performance criteria. The medium needs to hold water between consecutive irrigations, maintain aeration through an effective matrix of pores, retain sufficient nutrients and make these available to developing roots, whilst buffering against excess ion release by providing a high cation exchange capacity (CEC). CEC regulates the form and proportion of chemical ions that bind to the particles of the medium, compared to those found 'freely available' in the substrate solution.

Nutrition for bedding plants

Fertility is not an essential factor during germination as most seeds have enough stored nutrients. Fertilisation in plug production normally starts as liquid feeding at production stage 2 with dilute concentrations (e.g. 25–50 ppm nitrogen, N) being applied at first, followed by increasing concentrations (100–330 ppm N) from stage 3. Actual rates of fertilising will vary dependent on species, growth stage, frequency of application and leaching rates, as well as environmental conditions such as temperature and irradiance. The ratio of the macronutrients $N:P_2O_5:K_2O$ and proportion of micronutrients applied alters with growth stage or time of year. [Note the ratio $N:P_2O_5:K_2O$, sometimes simplified to N,P and K, is the convention to determine relative proportions of nitrogen, phosphate and potassium fertilisers, but it does not mean that the applications are necessarily in these precise chemical forms, for example the compound ammonium phosphate $(NH_4)_3 PO_4$ is often used to provide both the 'N' and the 'P' proportions.] During production stages 3 and 4, it is not uncommon for growers to fertilise the crop with liquid feeds based on 15-15-15 or 20-10-20, $N:P_2O_5:K_2O$, perhaps with supplemented additions of key formulas that include calcium or magnesium.

Temperature

Temperature control is provided within ornamental plant production to encourage growth early in the season or sometimes to protect plants from frost. Additional heat traditionally used oil or gas boilers, but green energy sources are becoming more common and waste heat streams (from power stations, distilleries, etc.) have been exploited in appropriate locations. Cooling tends to be provided through ventilation, e.g. ridge or side vents in a glasshouse, or roll-up sides in a polytunnel. In general, crops respond to increasing temperature by increasing vegetative growth and rate of flowering. For bedding plants, temperatures are somewhat higher for stage 1, lowered for stages 2 and 3, and reduced further during stage 4. Prior to shipping and retail stages, temperatures may be dropped even further. Indeed, many crops are 'finished' outdoors during the warmer spring/summer months where, although there is less control over temperature, the generally cooler temperatures and wind movement (thigmomorphogenesis) help plants harden off (acclimate) before retail.

Optimum temperature ranges within each stage vary between species, but a number of principles are evident. Temperature affects the plants' vegetative habit, with high temperature (>25°C) encouraging internode extension, whereas lower temperatures (10–14°C) promote more compact plants. Day temperatures have a stronger role in inducing these characteristics than night temperatures. Research with *Petunia* showed that plant height increased by 25 mm when night temperatures were increased from 10 to 30°C, whereas the same temperature changes when applied during the day enhanced height by 114 mm (Kaczperski *et al.*, 1991). Flower development is also affected. In two cultivars of *Petunia*, cv. Easy Wave Coral Reef and cv. Wave Purple flower development was faster and date of first flower emergence advanced when the daily mean temperature rose from 14 to 26°C (Blanchard *et al.*, 2011). Time of first flowering decreased from 51 to 22 days for *P.* cv. Easy Wave Coral Reef and from 62 to 30 days for *P.* cv. Wave Purple. There can be a trade-off though, with warmer production temperatures reducing the overall flower number.

Irradiance

Annual bedding plant development is regulated by three components of light – irradiance (intensity of light), photoperiod (daylength) and wavelength or spectrum (light quality). For those growing in high latitudes, natural irradiance levels early on in the production schedule can be insufficient, e.g. winter light levels in northern European countries being only 20% of those of mid-summer. Under such conditions, supplementary 'growth' lighting is provided to increase the photosynthetic capacity of the crop by augmenting natural light. Electric lights designed to emit light between 400 and 700 nm wavelength (photosynthetically active radiation, PAR) are used and increasingly rely on low-energy light-emitting diodes (LEDs) sources. Despite the additional costs, supplementary lighting is vital for some crops due to the positive effects on plant development. Providing supplementary lighting during early phases correlates with shorter production schedules, better plant quality (more compact, better branched specimens) and accelerated flower development. In *Cyclamen persicum* cv. Metis Scarlet Red, for example, exposing plants to increasing daily light integrals reduced time to flowering to 75 days (17.3 mol m^{-2} day^{-1} irradiance) compared to 133 days (1.4 mol m^{-2} day^{-1}). Increasing daily light integrals from 1.4 to 11.5 mol m^{-2} day^{-1} increased flower numbers from 0 to 15 per plant, and average leaf number from 9 to 28 (Oh *et al.*, 2009).

Irrigation

Uniform irrigation across a crop is important to ensure each tray of plants is identical to the next in terms of stature and coverage (Fig. 7.10). Growth is promoted by maintaining a media moisture content that avoids any check in growth via localised drying whilst still ensuring good aeration to the roots. Small plug/pot volumes mean that irrigation tends to be regular (two or more times a day in warm periods) and consistent. Irrigation is not necessarily applied to optimise growth though, but rather to promote a consistent uniform crop. Indeed, reducing the amount of irrigation can improve crop quality by fostering a more compact plant habit, and potentially improve crop shelf life by pre-acclimating the plants to subsequent drought events. Growers often 'dry down' their crop during the final stages of production to both control excessive shoot growth and as a preconditioning treatment (hardening) prior to shipping the plants off the nursery. This should not be confused with sending the plants out with dry growing media, however; final irrigations on the nursery should be thorough to ensure water availability is kept at a maximum during the shipping period.

Fig. 7.10. *Dianthus chinensis* x *barbatus* (hybrid pinks) growers are trying to achieve a quality product at the point of sale. Plant size and shape should be uniform across the tray, and one or two plants should be in flower to allow the consumer to evaluate the colour. Foliage should be a uniform green with little signs of chlorosis or necrosis.

The value of regulated irrigation though can vary with taxa. Reducing substrate moisture from 54% (container capacity) to a drier regime of 20% improved post-production quality in *Solenostemon scutellarioides* cv. French Quarter (coleus), *Petunia* x *hybrida* cv. Colorworks Pink Radiance (petunia), *Impatiens* x *hybrida* cv. Sunpatiens Spreading Lavender (busy lizzie) and *Salvia splendens* cv. Red Hot Sally II (salvia) (Guo *et al.*, 2019). In contrast, quality was not enhanced or was reduced in *Lantana camara* cv. Lucky Flame (lantana) and *Impatiens* x *hybrida* cv. Sunpatiens Compact Hot Coral (busy lizzie). Lower irrigation regimes also resulted in less labour cost, lower levels of water consumption and less glasshouse space being used. The loss of quality included fewer flowers in *Lantana* and *Impatiens* cv. Sunpatiens Compact Hot Coral. 'Drying down' of crops is linked not only with a lower tissue water content, but with physiological adaptations including reductions in leaf water potential, osmotic potential and tolerating some loss of turgor pressure (van Iersel *et al.*, 2010).

7.11 Sustainable Production of Non-woody Ornamentals

The predominant use of polypropylene and polystyrene trays and pots to hold bedding plants and other ornamentals has led to criticism relating to excessive waste associated with production. Recycling of these materials takes place in some countries, but large amounts still end up in landfill sites or other waste streams. This has acted as a catalyst for research into biodegradable containers that intrinsically have a lower carbon footprint and degrade rapidly after use (although retaining their integrity during production remains an important consideration). Some products are now composed of biodegradable materials such as paper, cardboard, wood, plant-based polymers (bio-plastics) or indeed growing media held in place via gels and polymers – for example, pots comprising coir fibre.

A number of these pots allow the plant to be placed directly in the ground without removing it from the container; the roots break through the organic material to reach the parent soil. Although such approaches are considered more sustainable, concerns arise as to how they impact on production (growers citing problems due to higher costs, problems with storing and handling, premature degradation, lack of consistency/reliability in the product and greater volumes of irrigation water required, as well as potential impacts on crop quality). Nevertheless, some materials have proven useful. Some of the bio-polymer pots can be reinforced with wood for example (McCabe *et al.*, 2014; Tomadoni *et al.*, 2020). Kuehny *et al.* (2011) evaluated a range of biodegradable pot types with crops of *Pelargonium, Catharanthus* and *Impatiens*, and they concluded that marketable plants for both the retail and landscape markets were achievable. Specimens held within bio-containers and directly planted in the landscape generally performed well, matching the performance of plants grown in plastic containers and planted out in the conventional manner. Schrader *et al.* (2018) added fertilizer to the bio-polymer pot and found it improved glasshouse production, and then performance in the garden after planting out. Pots and trays of such material should conform to conventional sizes and shapes, so as to fit with existing potting machines.

7.12 Design with Herbaceous Perennials and Geophytes in the Landscape

Unlike annual bedding plants and meadows, the use of perennials in public landscapes takes more planning and preparation to ensure the plant community is appropriate and meets its long-term goals. Environmental horticulturists should 'know' their plants, adapt the planting to the local conditions and often need to work to a brief from a client. Knowledge in landscape design and landscape management helps ensure successful plant displays!

Specifications and planting

Using pot-grown specimens of herbaceous perennials allows planting to occur throughout the year (with irrigation post-planting during summer being a requirement). Container plants can be purchased as 200 cm^3 (root volume) 'plugs' in trays, or in pots of increasing size and cost – i.e. 90 mm diameter, 1-, 2-, and 3-litre. As a general principle, plugs are best restricted to sites with reasonable control of planting conditions and reliable post-planting maintenance. The small media volume in a plug is a disadvantage elsewhere, as the plug dries out quickly and small plugs can be buried in large planting holes leads to a high level of mortality under average site conditions. Plants grown in 90 mm and 1-litre pots are preferred on most sites by the landscape industry because they are large enough to handle and plant easily, large enough to withstand suboptimal planting practice and the cost per unit is inexpensive enough to justify high planting densities. In contrast, pots of 2–3 litres are expensive and problematic for large-scale planting/contracts in terms of the size of the holes that need to be dug and the level changes that result from the spare excavated backfill. These pot sizes though are often preferred by the amateur gardener.

An alternative to container-grown stock is bare-root transplants from field-production specialist nurseries. These restrict plantings from autumn to spring, but when correctly handled often produce very good results. The standard bare-root herbaceous plant product is typically the same price as a 90 mm pot specimen, but the size of the plant and the number of buds, etc. is much greater in the bare ground product.

Procuring and planting geophytes should be done as early as possible (usually in late summer to early autumn for mainstream species like *Tulipa* and *Narcissus*). Delays in planting have a very significant effect on performance; for example, autumn bulbs planted at the tail end of the season in say late October in the northern hemisphere are likely to be desiccated and often with roots or shoots protruding that will be broken on planting.

For the landscape professional, procuring herbaceous plants during the dormant season is more problematic than with woody plants, both in terms of judging the quality of plants that are dormant and confirming precise identification. On projects in which the plant material is procured via tenders issued by the landscape contractor, there is often a high level of undisclosed substitution either with or without the knowledge of the contractor. The net result of this process is that by the time identification can be confirmed it is often mid-summer and there is a general reluctance amongst all parties to reject and re-plant plantings. This leads to plantings that at best may look completely different from what was designed and at worst fail to work as intended. For the environmental horticulturist, procuring through contractors or nurseries with whom a personal relationship exists is a valuable means to minimise this malpractice. Another approach is to commission nurseries to 'contract grow' material where timescales between production and planting in the landscape make this feasible. Procuring plant material during the growing season, where this is possible, has the advantage in that it is much easier to confirm identification/quality.

In landscape practice, planting is often carried out at times that fit a work schedule, rather than those that align best with the biological needs of a plant. In the UK, for example, plant performance is often optimised by planting taking place in September (early autumn) – when soils are warm enough to encourage root development, but not too wet to cause compaction of the soil structure during the planting process. Plants planted in autumn are typically much larger in the following year than those planted 5 months later in spring. Exceptions to the autumn planting rule of thumb are species that may suffer damage in their first winter, for example *Begonia* species, *Dierama, Hedychium, Moraea, Roscoea* and *Watsonia*.

Planting protocols

When planting between spring and autumn, it is important that rootballs are soaked prior to planting as growing media tend to be hydrophobic in nature and thus difficult to wet up again if they have dried out, especially so after planting *in situ*. Many nursery-grown herbaceous plants often come with a weed flora growing on the top of the pots that then seed post-planting and contaminate the planting mulch. Better-quality herbaceous plants will have less of this, but it is useful to specify in all cases that the upper 20–30 mm of the pot medium is manually removed prior to planting. For geophytes, follow the instructions supplied with different species. *Iris* x *germanica*, for example, requires the rhizome to only be partially buried in the soil, where some bulb species may be planted two to three times deeper than the longitudinal size of the bulb. Some bulb species possess contractile roots and can 'pull themselves' to a lower position within the soil should it be perceived as too dry.

7.13 Spatial Arrangements for Herbaceous Planting

Block-based planting

This is the standard 19th-century style of planting that forms the basis of most planting (herbaceous and shrub) throughout the subsequent 20th century. Plants are placed in groups (blocks) of the same species or cultivar to form island monocultures which are then juxtaposed with other species/cultivars in adjacent blocks (Fig. 7.11). The size of block can vary from the small, say five to ten plants, up to hundreds or thousands depending on the scale of the landscape. This approach reached its apotheosis in the USA in the work of James Van Sweden and Wolfgang Oehme in the later part of the 20th century (Oehme and Van Sweden, 1991). On a much smaller scale it was also the style adapted in the early work of Piet Oudolf (Oudolf and Gerritsen, 2013). The advantage of this approach is that it is simple to conceptualise, making it easy to plant; each species can be set at the spacing that best suits, i.e. perhaps highly vigorous species at 600 mm centres, very slow species at 250 mm and so on. It also looks highly designed and intentional, so it provides a strong sense of 'cues to care'. It is relatively easy to maintain, as anything that invades the monoculture is clearly either a weed or a volunteer from elsewhere in the planting and can be readily identified and removed. Research in the Netherlands has investigated plant spacing and the effect this has on establishment and subsequent resistance to weed invasion. Planting small plants (plugs or 90 mm pots) at high densities, for example 200 mm apart, produces a very dense sward that is potentially very resistant to weed invasion. Using the right plant selection, and implementing sensitive maintenance, such plantings can look good for 10–15 years or even longer; for example, Oudolf designs at Pensthorpe Natural Park in Norfolk, UK.

Fig. 7.11. Block planting creates large swathes of the one taxa within the overall plant composition.

The negatives of this planting style are largely aesthetic; unless the blocks repeat, they potentially look uncomfortable because they have a clear sense of grain or direction that may be at odds with the surrounding spatial form. The design looks decidedly 'blocky' and not necessarily very 'natural'. Each block can only look really attractive when in flower, so only one flowering event can be generated from each square metre of planting because of the monocultural basis. In structural terms, this form of planting is simple; there is only ever one layer per unit area. The scientific evidence suggests that this style of planting may not be as useful to invertebrates as habitat with a more complex structural form. This form of planting also suffers on sites when there is a gradation of soil and other environmental factors across the blocks, resulting in some plants or patches of plants within the block performing poorly or even dying. Because there is only one species present, block plantings have the lowest capacity to respond positively to changing site conditions.

Obviously smaller blocks, with fewer plants per block (e.g. one to five) give more opportunities for repeat plantings of the same species, and thus as block size gets smaller, the design can look more fluid and natural. This allows the creation of vegetation that is more complex per unit area than traditional block planting.

Matrix planting

When the block is typically composed of only one plant with potentially a different species or cultivar at each planting location, this has often been referred to as matrix planting, although there is little consistency in the application of this term. Because it is essentially impossible to show individual plants on most planting plans, matrix plantings are typically designed as a planting mix with only the key visual dominants (the biggest plants typically) shown on a plan, and then often only to inform the judgement of the planters about intended spatial distribution. These mixes are planted at a uniform spacing typically around nine plants per square metre, although this varies depending on soil productivity, with density being decreased as soil productivity increases and vice versa. The key design decisions in creating these mixes are how many species, which species are to be used to create the plant community, and how many of the typical nine planting positions in each square metre should be allocated to each species? The judgements in making these decisions are essentially exactly the same as for any other type of planting. Key factors are the visual and physical degree of plant diversity required, the surface topography of the resulting planting (i.e. how tall, how low, how bumpy) and the duration of the flowering season required.

The number of layer structures to build in

Block planting is by definition a monolayer within each block, and the larger the block the more this is the case. In semi-natural 'wild' vegetation, monolayers are uncommon except on very productive soils where one or perhaps two large fast-growing species dominate and inhibit the establishment of additional species. On less productive sites more mixed communities are the norm because there is always a diversity of species that can invade; some are short ones which will be shade tolerant, whereas many of the tall species will require full sun. What are the advantages and disadvantages of building in more than one layer above each square metre of ground? Most plantings using the matrix approach effectively have two or three layers. The lower or ground layer in a two-layer system will generally consist of a spring, vernal growing layer of species that are associated with woodlands and woodland edges. These plants start to grow very early in the year and tolerate being shaded from late spring to early summer onwards by the leaf canopies of taller, summer- and autumn- flowering herbaceous species. Typical examples would be *Primula vulgaris* (primrose) and *P. elatior* (oxslip), *Ajuga reptans* (bugle), *Tiarella cordifolia* and *T. wherryi*, *Omphalodes verna* and *Saxifraga stolonifera*. By having this structure, the ground is covered by foliage (many of these species are evergreen or at least semi-evergreen) in winter, hence reducing winter weed colonisation, and they also look more attractive in winter than the frequently winter-dormant taller species and produce a dramatic spring to early summer flowering display in their own right. They then disappear under the emerging shoots of the later flowering species, the first of which flower in mid-summer and continue through to autumn, giving a 6- to 8-month period of flowering. The key factor for the summer- and autumn-flowering species is that they must be unattractive to slug and snail grazing as the shoots push through, as these herbivores will be more abundant where a vernal layer is present rather than absent. Species that most tolerate this elevated spring grazing pressure include *Symphyotrichum novae-angliae, Doellingerie umbellatus, Aconitum* spp., *Euphorbia* spp. most grasses, *Leucanthemum maximum, Leucanthenella serotina, Lythrum salicaria, Sanguisorba* spp., *Veronica longifolia* and *Veronicastrum*. Because the understorey species are deeply shade tolerant and often have some capacity to become summer dormant, the presence of complementary species that subsequently form the upper canopy layer, and thus intercept much of the light, do not pose a problem for this system.

Due to their complexity, designing and creating matrix-type plantings is more challenging than block monocultures, not least in attempting to identify what is a cultivated plant and what is a weed. What are the advantages of this planting style that might make these additional challenges worth considering? In aesthetic terms, matrix-type plantings provide the capacity to produce a succession of flowering events from each single square metre of planting, for perhaps 5–6 months rather than 2 weeks. They also allow spring-flowering species to be used, many of which are very unattractive in summer and autumn, due to dying or untidy foliage with the taller emergent species visually masking their decline. Because of the repeating pattern, when a species, present at a density of say 1–2 plants m^{-2}, comes into flower, the effect generated is of the whole area of the planting turning this colour. Matrix-like plantings potentially generate much greater levels of drama for longer than with block planting in which only a small area is typically flowering at any point in time. On the functional level, these matrix-like plantings have much greater capacity to fix themselves when soil or other environmental gradients are inherent (as they always are) in a planting. Whilst the aim of planting design is to try to get the placement of plants 'right' so, for example, the species in question thrive, this is often much more difficult in practice due to the difficulty of knowing just how tolerant a species is to a given set of environmental conditions. Matrix plantings are more forgiving in terms of getting these decisions 'wrong'. For example, where plantings run from exposure to full sun to under newly planted trees, as the shade generated by the trees increases, the most shade-tolerant species/cultivars in the herbaceous layer will be preferentially advantaged and the more shade-intolerant species disadvantaged. Thus, providing both shade- and sun-tolerant species within the planting design allows the planting mix to respond to changing conditions even in the absence of intelligent maintenance.

The ratio of plants of later-flowering species (often the tallest) and shorter summer-flowering species needs to be monitored to assure that the tall, later-flowering species do not eliminate the earlier summer-flowering species though shading. For plantings involving nine planting sites per square metre, typically two would be a tall autumn species (but not the same species in each square metre), three would be summer flowering (but not the same species in each square metre), and four would be spring flowering.

Three-layer systems involve the same principles. If the upper two emergent layers are kept relatively sparse, sufficient light will be available to the ground-level species that are relatively shade intolerant. For this to work the ratio of low:medium:tall plants, needs to be in the order of 10:2:0.5. In addition to the species listed previously, species such as *Ajuga genevensis, Geum rivale/coccinea* cultivars, *Heuchera* cvs and species, *Omphalodes cappodocica,* and *Pulsatilla vulgaris* can be used, with the final choice dependent on soil moisture levels and productivity. Emerging species that have basal foliage, but naked flowering stems are preferred for the two upper layers as this plant form prevents casting too much shade and hence eliminating shade-intolerant species in the lower layers. Species that form clumps are preferred too, as these reduce the likelihood of growing over the crowns of the emergent species in spring, casting shade and leading to their extirpation.

When complementary species are successfully included in a design the impacts can be both colourful and help reduce maintenance. Ecological dynamics do come in to play though, and the community composition can change with time. Hitchmough *et al.* (2017) studied the dynamics in a two-layer planting over 5 years. The original planting involved a plant community with a tall over-canopy layer of 18 species of North American prairie or woodland edge forbs, and a shade-tolerant under-canopy of 8 European or North American woodland forbs. After 5 years, the community was dominated by four over-canopy (*Silphium integrifolium, Symphyotrichum novae-angliae, Andropogon gerardii* and *Helianthus mollis*) and two wintergreen under-canopy species (*Primula elatior* and *P. vulgaris*). Of the original 26 taxa used, 22 were still present after 5 years, albeit in some case in very small numbers (Fig. 7.12). Notably, the interspecific

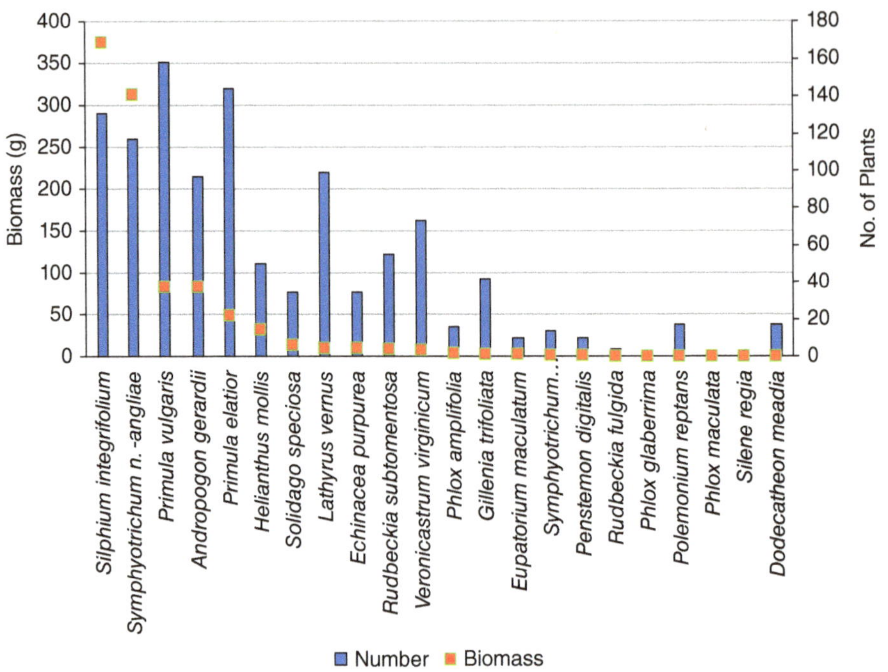

Fig. 7.12. Ecological succession in matrix plantings. Biomass and number of individual plants of different taxa present 5 years after the original community of 26 taxa was planted. (Modified with permission from Hitchmough *et al.*, 2017.)

competition generated by the sown biomass restricted weed colonisation to very low levels, despite the experiment being surrounded by a weedy brownfield. Careful selection of the community can result in weed suppression, a close winter canopy that maintains interest levels and results in good spring flower colour, and finally a few dominant summer species to create impact.

Plant robustness

Annual plants previously grown in trays/pots (bedding plants) are relatively easy to look after in the landscape – due to the short-term nature of 'their stay'. The main criteria are irrigation after planting and perhaps some weed control during the summer months. Annuals grown directly from seed are more vulnerable at the seedling stage to slug/snail pressure or being crowded out by weeds.

Herbaceous perennial and geophyte choice depends more on what the purpose/design of the planting is (see above). Highly formalised displays will be managed differently to those more naturalistic ones. For the latter, plant robustness is a factor. In practice the criteria on which as assessment of plant robustness might be made are typically the capacity (i) to do what is expected of them in terms of growth and flowering and (ii) to persist on a site at relatively low levels of care (Fig. 7.13). But why are some plants demonstrably much more robust than others? Whilst it would be elegant to assume that all plants are of equivalent robustness when grown under the conditions they have evolved for, when cultivated in designed settings in gardens and green space, some species are clearly unfit, whilst others seem very fit. Some species seem to have inherent resilience and just refuse to give up, no matter how badly they are treated, and this seems to occur across a range of conditions. In many cases, this robustness is derived from self-evident aspects of their biology: large, fast-growing but long-lived plants with multiple growth points and/or colonising rhizome, leaves and growth points that are unpalatable to slugs and snails. In stark contrast, very small, slow-growing plants with few growth points and high palatability to herbivores are unlikely to be performing well in 10 years' time!

Partially to improve success levels, the use of local flora in highly urbanised environments is gaining popularity. Matyashuk *et al.* (2023) evaluated 147

Fig. 7.13. Fitting in! Taxa such has *Rudbeckia* are relatively robust as landscape plants as long as moisture levels are sufficient and they can survive mollusc browsing pressure during the establishment phase.

florally attractive/ecologically important herbaceous perennials from the forest steppe region of Ukraine for use in the country's cities. Plants were well adapted to the local conditions, as well as providing unusual flower types across a wide range of colours. Demand for using native plants along urban highways, on green roofs and in private gardens is resulting in a wider range of natives being incorporated in commercial production and the development of new specialist nurseries (Rihn *et al.*, 2022).

In many cases, there is a relationship between soil productivity and longevity in herbaceous plants, with longevity, particularly in short-lived species, declining as soil productivity *increases*. Species such as *Achillea*, for example, are relatively long-lived in their habitats without recourse to division and other cultural practices. In cultivated soils with much higher nutrient and water levels, these species decline relatively quickly. This is presumably because their preprogrammed cycles involved in longevity and 'programmed' necrosis are completed far more quickly.

Planting species in the 'wrong' habitat often has a very negative effect on robustness, and 'plant fitness' for a particular environment is extremely important, if sometimes a less obvious factor, in determining robustness. Key elements in this fitness are the ecological range and amplitude of a species in its habitat: how widespread is the population and how many different environmental variables are encountered in nature. This fitness aspect includes many factors, for example, tolerance of stress in terms of moisture, light and nutrients. The amplitude of a species in response to these ecological factors depends on a mixture of evolutionary history and chance. Some species demonstrate broad amplitude to these sorts of ecological factors, i.e. they might grow perfectly well in sun or shade or from wet soil to dry than would otherwise be expected from their actual habitats in the wild. Other species are robust but only within a much narrower range of possibilities, for example *Primula florindae* (Tibetan cowslip) – one of the so-called bog primula – is extremely robust in cool summer climates when grown in water-saturated soils to very wet soils. (A good example of this narrow robustness is that this species has become naturalised in the peat bogs of Scotland's 'Flow Country' – approximately 7000 km from its own natural range.) Yet this species lacks resilience elsewhere and is short-lived in dry soil. Sometimes these interactions are complex; *Salvia nemorosa* is a continental, dry-summer species that can be robust in maritime wet summer climates but only when the grazing pressure from slugs in spring and summer is controlled. Where this is not the case, this species disappears due to slug herbivory where rainfall exceeds a threshold value of frequency. In geophytes, an important factor in robustness is the degree of soil wetness during the dormant periods. Many geophytes evolved in parts of the world with dry summers, and the resting organs are often subject to damage from fungal pathogens when soils remain moist in summer. *Tulipa* is an important example of this, although species and cultivars are very variable in their response – see, for example, Wilford (2006), on persistence in the different wild-occurring *Tulipa* species. Relationships between plant robustness and fitness are shown diagrammatically in Fig. 7.14.

Despite the implication in the horticultural literature that fitness operates at the species level, this is rarely the case. Within a given true species, each plant is an individual, with genetics that predispose it, or alternatively give it resistance, to given ecological stress factors. When herbaceous plants are derived from seed (not cuttings, which are clonal), there will be some individuals within the population that have greater fitness for the conditions found in a given site than others. Hitchmough (2016, personal communication) evaluated South African species for landscape use, finding that even when a species is not particularly well adapted to a site there is often a small percentage of individuals (typically $<5\%$) that are demonstrably much better fitted to the conditions imposed than others.

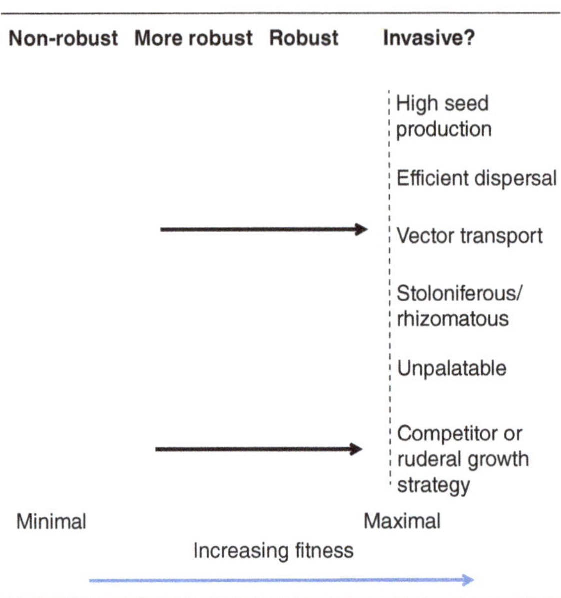

Fig. 7.14. The relationship between increasing plant fitness for a planting site and its behaviour in terms of robustness or ease of cultivation. To become invasive, plants need more than just high fitness and to possess at least some of the traits in the right-hand column.

Across the entire natural distribution of a species the very different climatic and edaphic conditions inevitably result in hugely varying levels of intrinsic fitness to a given set of conditions. Currently cultivated commercial populations of *Gladiolus cardinalis* are derived from plants originally collected from 1000–1400 m altitudes in the mountains of western South Africa; however, populations extend to 2000 m, with these being much more tolerant of severe winter cold and increasingly less tolerant of summer heat. For example, this particular species is almost unable to be grown in the Mediterranean climate of Cape Town, South Africa even though it is just 50 km from its native mountain habitats. Outside of the world of alpine plants (where scholarly appreciation of the habitat of the plants in relation to their appearance and behaviour is very much understood), for most horticultural practitioners there is very limited understanding of how fitness might differ depending on the highly localised geographical, altitudinal and habitat origin of the particular plant population.

The nursery production industry has rarely consciously utilised these processes. It has, however, simply selected the seedlings that grew best, or even just survived, thereby typically increasing 'fitness' in the process. Cultivation typically increases fitness (at least for the urban or garden environment) from one generation to the next from the point of introduction from the wild through these selection processes. This is assuming new, much less fit genes are not incorporated by intentional or unintentional plant breeding. In some cases, fitness is also increased by a change in ploidy levels in seedlings.

The practical outcome of these discussions is that plant selection is vital with respect to species, cultivars and the other sub-specific variants of a species in actually achieving the best long-term fitness possible. Plant selection protocols for landscape planting should deal with two key ideas in addition to the standard morphological and phenological questions about flower colour and timing:

1. How close is the habitat type of the planting site in relation to the wild growing site.
2. How close climatically is the planting site environment to the environment from which the plant is derived in the wild.

Habitat types are typically shaped by interactions between two key factors: the degree of soil moisture and shade as shown in Table 7.4. Steppe plants, for example, are typically best adapted to soils that are relatively dry in summer and often in

Table 7.4. Fitness is relative rather than absolute and changes according to interactions between overall climate and planting locations within these environments. The table shows how groups of species become more or less fitted when placed in different planting combinations of soil moisture and light. (Used with permission from J. Hitchmough.)

Condition	Highly fitted	Moderately fitted	Poorly fitted
Wet shady	*Astilboides tabularis*	*Aconitum carmichaelii*	*Allium christophii*
	Carex pendula	*Aruncus dioicus*	*Euphorbia rigida*
	Matteuchia struthiopteris	*Cimicifuga simplex*	*Origanum vulgare*
	Petasites x *hybridus*	*Geranium sylvaticum*	*Penstemon strictus*
	Primula sieboldii	*Geum rivale*	*Pulsatilla vulgaris*
	Primula vulgaris	*Primula pulverulenta*	*Salvia nemerosa*
Wet intermediate	*Aconitum carmichaelii*	*Aster divaricartus*	*Anemone sylvaticum*
	Aruncus dioicus	*Cyclamen hederifolilum*	*Bupthalmum salicifoliurum*
	Cimicifuga simplex	*Geranium macrorrhizum*	*Centaurea scabiosa*
	Geranium sylvaticum	*Lathyrus vernus*	*Geranium pratense*
	Geum rivale	*Tellima grandiflora*	*Hemerocallis flava*
	Primula pulverulenta	*Trachystemon orientalis*	*Knautia arvensis*
Wet sunny	*Camassia leichtlinii*	*Astilboides tabularis*	*Aster divaricartus*
	Hesperantha coccinea	*Carex pendula*	*Cyclamen hederifolilum*
	Persicaria bistorta	*Matteuccia struthiopteris*	*Geranium macrorrhizun*
	Primula pulverulenta	*Petasites* x *hybridus*	*Lathyrus vernus*
	Succisa pratensis	*Primula sieboldii*	*Tellima grandiflora*
	Trollius europaeus	*Primula vulgaris*	*Trachystemon orientale*
Dry shady	*Aster divaricartus*	*Allium christophii*	*Camassia leichtlinii*
	Cyclamen hederifolilum	*Euphorbia rigida*	*Hesperantha coccinea*
	Geranium macrorrhizum	*Origanum vulgare*	*Persicaria bistorta*
	Lathyrus vernus	*Penstemon strictus*	*Primula pulverulenta*
	Tellima grandiflora	*Pulsatilla vulgaris*	*Succisa pratensis*
	Trachystemon orientale	*Salvia nemorosa*	*Trollius europaeus*
Dry intermediate	*Anemone sylvaticum*	*Camassia leichtlinii*	*Aconitum carmichaelii*
	Bupthalmum salicifoliurum	*Hesperantha coccinea*	*Aruncus dioicus*
	Centaurea scabiosa	*Persicaria bistorta*	*Cimicifuga simplex*
	Geranium pratense	*Primula pulverulenta*	*Geranium sylvaticum*
	Hemerocallis flava	*Succisa pratensis*	*Geum rivale*
	Knautia arvensis	*Trollius europaeus*	*Primula pulverulenta*
Dry sunny	*Allium christophii*	*Anemone sylvaticum*	*Astilboides tabularis*
	Euphorbia rigida	*Bupthalmum salicifolium*	*Carex pendula*
	Origanum vulgare	*Centaurea scabiosa*	*Matteuccia struthiopteris*
	Penstemon strictus	*Geranium pratense*	*Petasites* x *hybridus*
	Pulsatilla vulgaris	*Hemerocallis flava*	*Primula sieboldii*
	Salvia nemerosa	*Knautia arvensis*	*Primula vulgaris*

winter too but also linked with plants that have a requirement for full sunlight. Prairie plant communities are associated with sunny, open sites with often substantial summer rainfall. These plants are not necessarily tolerant of summer drought. Woodland edge communities are composed of species that tolerate sun and moderate shade and often low soil moisture levels. Plant selectors need to undertake research as to the conditions a plant is known to grow in the wild to answer this question. Climatic similarity/dissimilarity interacts with habitat types to provide either further restrictions or alternatively more freedom of choice.

Climatic similarity is driven by three key factors: (i) latitude, (ii) altitude and (iii) overall patterns of precipitation/evaporation. However, plants adapted to grow in soils that are only wet in winter and spring, for example, typically cope poorly in soils that are wet all year round and so are not ideally 'fitted'. Many locations, for example South Africa and North America, have winter rainfall on western sides and summer rainfall on eastern sides, with a mix of both in the middle. European horticulture literature on plants from South Africa, for example, typically ignores this and sees South Africa as a climatically 'homogenous' place with potential catastrophic consequences (Fig. 7.15). Altitude interacts with precipitation patterns; typically, as altitude increases precipitation increases and even in Mediterranean climates the rain-free season become much less severe or even non-existent, the higher one moves in the montane landscape. Plants from really high altitudes are often much less tolerant of high summer temperatures but far more tolerant of severe winter minimum temperatures. Latitude interacts with altitude. The most common origin of plants in UK gardens is temperate Asia often at latitudes much more southerly than the UK, for example Western Sichuan and Yunnan in China are at approximately 38–40° N but plants are naturally occurring at approximately 3000–5000 m altitude. Notions of plant fitness currently accepted will no longer be valid in future as the impacts of climate change take hold. In 50 years' time, the currently slightly winter-cold-limited *Gladiolus cardinalis* will be a ubiquitous, widespread species in the UK, assuming it can cope with the other factors climate change will impinge on it.

Understanding plant responses is important in determining the success of a designed plant community. Only limited numbers of studies though have investigated tolerance to abiotic stress outwith the observations of previously designed landscapes and investigated physiological adaptations. Hillová *et al.* (2016) showed that drought stress led to considerable declines in stomatal conductance (75% in *Stachys macrantha*, 60% in *Geranium macrorrhizum* and 42% in *Brunnera macrophylla*) and in canopy leaf areas (*Stachys* 59%, *Geranium* 53% and *Brunnera* 45%). Nevertheless, both *Brunnera* and *Geranium* remained well adapted to dry conditions unlike the *Stachys* that lost quality/performance criteria; they concluded the former two taxa would be suitable for use in landscapes prone to drying soils. Over a 2-year study in the USA, where plants were established in the first year, then irrigated at different frequencies in the second year, *Penstemon barbatus* var. *praecox nanus rondo* showed the greatest tolerance to all levels of drought, avoiding desiccation by increasing root-to-shoot ratios and decreasing stomatal conductance as water became limiting (Zollinger *et al.*, 2006). *Penstemon* x *mexicali* cv. Red Rocks and *Lavandula angustifolia* (lavender, a Mediterranean sub-shrub) both showed tolerance to moderate drought (2-week watering period) but not the more severe drought pressure (4 weeks without water). Two species demonstrated drought avoidance mechanisms via dying back when water was limited and showed new growth after they were re-watered, i.e.

Fig. 7.15. Knowledge of South African flora and its tolerances, provenances and interrelationships was important for a successful plant community designed by James Hitchmough for the Olympic Park in London in 2012 (Used with permission from J. Hitchmough.)

Gaillardia aristata and *Leucanthemum* x *superbum* cv. Alaska (shasta daisy). In contrast, *Echinacea purpurea* (purple coneflower) exhibited poor visual quality at all irrigation intervals, in particular wilting severely in both drought treatments; although it regained turgor when watered again, overall this taxon was not recommended for landscapes that experienced drought.

Structural form in plant selection

The orientation of leaves and stems of non-woody plants in space and time has a major effect on their performance and how they can be used in plantings, and is a major factor to consider in plant selection. It is possible to think about these plant form issues from both a functional and aesthetic perspective.

From a functional perspective one of the most important foliage factors is the duration of leaf retention. Herbaceous plants with evergreen leaves typically inhibit weed invasion during the winter months more than deciduous species, although herbaceous plants with very slow-to-decompose dead foliage can also be very effective; for example, *Iris sibirica*. In multilayer plantings it is highly desirable to have as many of the ground-layer species having evergreen or semi-evergreen foliage to help suppress weeds; this trait is relatively common in many slow-growing, highly shade-tolerant woodland species, for example *Tiarella wherryii, Tellima grandiflora Asarum* and *Epimedium* spp.

The distribution of the foliage mass in space has a major effect on the capacity of a species to outcompete other herbaceous plants or be outcompeted. Leaf cover blocks out light on neighbouring plants, but it can also be a refuge for snails and slugs. As herbaceous plants grow, their leaves and stems cast increasing amounts of shade, which potentially has a deleterious effect on lower-growing neighbours. The species that are most sensitive to competition for light when grown alongside other herbaceous plants are those that are well adapted to open, sparsely spaced plant communities – such as occur in drier climates or poorer sandy soils with either mat-like (e.g. *Thymus*, *Oenothera macrocarpa*) or rosette-like foliage (e.g. *Verbascum phoeniceum* and *Salvia argentea*). In mixed communities, low-canopied species need to have only the occasional tall species adjacent to them if their longer-term survival is to be ensured.

Ironically, the most problematic herbaceous plants for mixing with shade-intolerant smaller species are generally those that are historically the most popular, namely those with tall leafy stems. In these the architecture normally involves erect stems furnished with leaves from ground level to the inflorescence. A classic example would be the 'leafy, well-furnished' *Helianthus* cv. Lemon Queen. Other examples include *Phlox paniculata, Symphyotrichum novae-angliae* cv. Helen Picton and *Eutrochium maculatum.* Leafy stem morphologies are least problematic when the vertical leafy stems form a narrow, tightly packed bundle, as in *Veronicastrum virginicum* or *Andropogon gerardii.* An alternative morphology are those plants with essentially 'naked stems'. Here the foliage forms a low clump or mound with tall, leafless flowering stems that do not significantly restrict irradiance from reaching the lower canopies. Species with these growth forms include *Echinacea pallida, Ferula communis, Geum* spp., *Meconopsis* spp. (Himalayan poppy), *Silphium terebinthaceum,* and *Stipa gigantean* (Fig. 7.16). Many geophytic genera have this architecture too, for example *Agapanthus, Allium, Dierama, Bulbinella, Eremurus, Iris, Galtonia* and *Watsonia.*

The naked stem architectures also have a particularly useful role aesthetically in that they do not block views through plantings allowing the designer to create far more contrasts of colour and form as a result.

Scale and size of plant are important from an aesthetic perspective too. Large, open areas for planting can accommodate larger and taller taxa, but more confined areas call for plants that are better fitted to a smaller scale. A good example of this would be with ornamental grasses. Tomaskin and Tomaskinova (2020) linked grass taxa to their appropriateness, performance and scale in the landscape. For example, for (i) open spaces, parks and larger gardens – *Miscanthus* spp., *Pennisetum* spp., *Deschampsia cespitosa*, *Chasmanthium latifolium* and *Chionochloa conspicua*; (ii) rocky areas, screes dry steppe gardens - *Festuca gautieri, F. pallens*, *Stipa* spp., *Bouteloua gracilis*, *Koeleria glauca*, *Melica ciliate*, *Helictotrichon sempervirens* and *Panicum* spp. (the smaller species will also do on green roofs); (iii) riparian areas associated with streams, ponds and lakes – *Glyceria maxima*,

Phalaris arundinacea and *Phragmites australis*; and (iv) aquatic environments – *Arundo* spp.

Fig. 7.16. Plants with tall, non-foliated flowering stems such as *Meconopsis* spp. (top) and *Iris* spp. (bottom) are useful for ensuring light passes through to other ornamental species growing below.

Phenology

To plan the colour sequence within the floral border or meadow, it is important in plant selection to know when a particular species or cultivar commences flowering and when it typically ends. This information is often poorly documented in the horticultural literature, but through experience the environmental horticulturist can design plantings where the harmonies and transitions in flowering can be carefully choreographed.

Phenology is also very important in relation to when foliage is produced and then declines, both in herbaceous perennials and geophytes. Herbaceous plants that come into growth late in the growing season are often associated with warmer habitats in which a relatively high-temperature soil/air threshold must be exceeded to initiate growth; some examples include *Incarvillea, Oenothera macrocarpa, Gladiolus* (South African summer rainfall species) *Roscoea, Ruellia* spp., *Schizachryium scoparium, Zantedeschia albomaculata* and *Zinnia grandiflora*. There are, however, some species native to western Europe that also respond in this way, for example *Eryngium maritimum*. These plants are often problematic in terms of weed colonisation in that they do not occupy space effectively until June or July. They are also prone to heavy shading and therefore extinction when planted on the shady side of taller, earlier-to-emerge herbaceous plants. On the other hand, they can often be combined in interesting ways with species that emerge or flower very early in the year to create plantings in which there is little competition for space in the first part of the growing season.

Mulching

In most designed landscapes plantings are only maintainable when some form of mulch is employed on a regular basis to blanket the weed seed bank in the soil and present a 'quick to dry' surface that is hostile to in-blown weed seed. The critical input in year one of herbaceous planting is to prevent blown-in weeds establishing and producing seed whilst the planting is relatively open. If plantings can be kept largely weed-free during this period, year two and subsequent maintenance is greatly reduced, as the canopy of the planted material provides much more competition for light. Horticulturally trained staff are sometimes indifferent to these principles; they see weeding as about doing maintenance to a timetable rather than breaking the lifecycle of weed development. Subsequent maintenance to manage weeds is much greater where weeds are firstly allowed to self-seed before removal.

The least effective materials for mulching are colloidal organic materials such as composted green waste, as whilst these are effective for blanketing weed seed banks and even shading out newly germinated weed seedlings on the layer below, they are an excellent germination surface for seed blown in by the wind. Garden management organisations such as the National Trust in the UK, which generally have masses of composted arisings to use, often mulch their herbaceous plantings twice a year to get around this problem (Fig. 7.17). Coarser, granular materials such as composted bark (particles larger than 15 mm), pea gravel and grit/very coarse sand are highly effective mulches. The advantages of the mineral materials are that they are more persistent, while their disadvantages are that they need to be applied in deep layers to be effective; a minimum of 100 mm, but 150 mm plus is better as it makes it more difficult for seedling weed roots to reach the moist underlying soil prior to dying of moisture stress. They also are relatively hostile environments for slugs, compared to composted organic material that may actually increase local slug densities, with significant ill effects on sensitive planted species. Surface scarification of gravel-type mulches on a warm, dry day is very effective in terms of eliminating cohorts of blown-in annual and biennial weeds. Planting in deep mulches of sands as a deliberate high-stress strategy to minimise weed colonisation in naturalistic plantings is covered in Chapter 8.

Fig. 7.17. A thick mulch of composted green waste/straw mix smothers annual weed seedlings in the soil, whilst being loose enough to let the growing tips of geophytes push through. The material itself though can support the germination of blown-in weed seed, so where this may be a problem two to three applications a year may help as a process to keep burying the weeds.

7.14 Management Factors

Irrigation in the landscape

Bedding plants have a high requirement for water until they become established, and irrigation should take place at planting and on a frequent basis thereafter. Hanging baskets, pots and small containers may need irrigation daily, in some cases twice daily, throughout the growing season; automated drip systems with timer control prove useful under such circumstances. One downside of automated irrigation systems is that, once set up, they are infrequently recalibrated and can result in overwatering of bedding plant schemes.

The reputation bedding plants have for high water use in landscape settings has meant many local authorities have questioned their future use due to climate change and potential restrictions on irrigation water. This has spurred research to assess more closely what are the water requirements of bedding plant species, and how would performance be compromised if more restrictive irrigation regimes were employed. Blanusa *et al.* (2009) investigated the competitive effects for water in a hanging basket scenario by varying numbers of individual plants of two bedding species (*Petunia* cv. Hurrah White and *Impatiens* cv. Cajun Violet) within the basket and recorded responses to reduced irrigation volumes. To optimise performance *Petunia* required approximately 30% more water than *Impatiens*. Under a reduced irrigation regime in which plants only received 25% of the full watering, flower number, plant height and flower size in *Petunia* were reduced by 50%, 33% and 13%, respectively. Despite there being significant reductions in key ornamental plant parameters, these were not proportional to the reduced irrigation. Indeed, visually, there was often little loss of ornamental properties in *Petunia* under the lower irrigation regime; the plants appeared compact and retained relatively good flower cover in proportion to their biomass. For *Impatiens*, however, the growing of single plants at 25% watering was plausible, but the addition of a *Petunia* plant at this regime (i.e. low watering plus competition) was detrimental to plant quality in the *Impatiens* (i.e. flower numbers were reduced by 75% compared with a well-watered single specimen). Therefore, for *Impatiens* to have acceptable quality in a mixed container with *Petunia*, near-optimal irrigation was required. Rather than eliminating bedding plant displays outright, however, the data presented suggested that acceptable bedding displays could be still achieved even when water consumption was reduced by 67% below optimum.

The use of recycled or grey water may become more prominent in future to retain civic bedding plant displays. The question arises, however, as to how much can the quality of water be compromised before affecting plant development or soil ecology? When comparing water from various sources, Arnold *et al.* (2003) demonstrated that *Viola* and *Antirrhinum* grown in the landscape performed well when irrigated with recycled water or water runoff from nursery beds. *Viola* growth during the cool season in Texas, USA increased steadily during the winter months before increasing rapidly in April prior to suffering dieback and then declining under hotter conditions. Differences in growth and flowering were small between recycled, runoff and tap water treatments, but the extra addition of salt from recycled or runoff water reduced plant growth increments, particularly in later winter and early spring when more frequent irrigation was required. Growth in *Antirrhinum* tended to be greatest early on with the use of tap water, with those irrigated with nursery runoff water performing better than with either wetland recycled or elevated salt water. Differences between runoff- and tap water-irrigated plants, however, were not evident in later spring measurements.

Herbaceous perennials usually are only irrigated during the establishment phase, and certainly on larger-scale plantings appropriate species choice is the best tool to try and match landscape performance to the soil type and water available within. Understanding whether the planting links to wet-adapted or dry-adapted perennials is thus important.

Nutrition

On nutrient-poor soils, fertilizer addition helps plant performance in the landscape, aiding establishment and promoting longer periods of growth and flowering. Although often recommended, the precise volume of fertilizer to add is based on limited research. Comparisons using different rates of nitrogen applied to nutrient-poor subsoil (Shurberg *et al.*, 2012) showed that less nitrogen is needed than may commonly be assumed. These studies involving *Zinnia, Catharanthus, Melampodium, Dianthus, Viola* and *Antirrhinum* used N fertilizer applied at five rates in the landscape: 0, 96, 192, 288 and 576 kg ha^{-1}. Regression analysis indicated that all species required N inputs at annual rates exceeding 400 kg ha^{-1} to achieve maximum size, shoot biomass and photosynthetic activity. Plants of acceptable quality, however, were produced at lower rates of nitrogen, 200–300 kg ha^{-1} N. Shurberg *et al.* (2012) believe that adopting these lower rates will provide acceptable aesthetic performance whilst reducing fertilizer costs and the potential for nutrient losses in runoff or leachate. Where the addition of inorganic fertilizers is not appropriate due to concerns linked to environmental sustainability, then organic alternatives can be sought or the environmental horticulturist may consider species better suited to nutrient poor soils, e.g. leguminous plants that can fix their own nitrogen such as *Trifolium* spp. (clover) or *Lathyrus* and *Pisum* (peas) or those that can still flower relatively well under such conditions (e.g. *Calendula, Dianthus* and *Tropaeolum*).

Concerns over peat extraction and limited supplies of bark-based growing media have led to research investigating the potential of wood-based growing media. These notoriously immobilise nitrogen through the activation of microbial activity, resulting in potential nitrogen deficiencies in the plants. These concerns have extended to the exterior landscape where wood amendments may interfere with nitrogen availability. To some extent these issues can be surmounted by the addition of higher-than-normal rates of nitrogen fertilizer, thus providing sufficient nitrogen for microorganism activity and plant development. Wright *et al.* (2009) compared how bedding plants (*Begonia, Solenostemon, Impatiens, Tagetes, Petunia, Plectranthus, Salvia* and *Catharanthus*) grown in either composted pine bark media or composted pine wood media developed once planted out into a landscape setting. As long as some fertilizer (48.6 kg ha^{-1} N) was added to the landscape, soil plant development was comparable irrespective of the initial growing media used in the growing container.

For perennials, biennials, geophytes and direct-sown annuals, nutrition, as with water demand, often links to an understanding of the ecological conditions the species is derived from. Some taxa prefer open, highly organic soil, whereas other may be best adapted to relatively nutrient-poor soils (Bretzel et al., 2009) (Fig. 7.18).

Weed control

Prior to planting, soil should be cultivated to provide a good 'crumb structure' with any resident weeds being removed. The cultivation of the soil will encourage weed seed germination (e.g. in northern Europe species such as *Senecio vulgaris* and *Stellaria media*), and cultivation sometime prior to the planting of the ornamental plants can be used positively to promote early weed seed germination, at which point these can be hoed off. Even so, some further hoeing of the ground may be required to keep weeds in check with annual bedding until these themselves cover the soil surface.

The essential quality, however, of many herbaceous plantings and summer-growing geophytes is the loss of canopy cover between late autumn and spring that facilitates weed invasion. As has previously been discussed, the use of mulches is key to making herbaceous plantings manageable when maintenance resources are limited. These are most cost-effective when applied in winter (e.g. January–February in the UK) and spread over the entire surface of the planting. This is required very occasionally with mineral mulches, but with organic debris mulches such as bark, this is probably required every other year. For any perennials with leaf present, mulch applications need to be done carefully so as to not damage the basal rosette of leaves of some species.

The weeds species that invade herbaceous planted landscapes tend to reflect the nature of those plantings and the forms of management used. In the first year after planting the weed species are mainly those with small seeds that are well distributed by wind, etc. Many of these species are ruderals, such as *Poa annua,* (annual meadow grass) *Cardamine hirsuta* (hairy bittercress) and *Epilobium* (willow herb) hybrids. The 'invasion' rapidly gains momentum from a few weeds to many and is often greatest between autumn and winter when the site is most open and the surface of mulches are permanently moist. These invasion events are predictable and can be managed by either burning the surface over in March (a process being pioneered by Hitchmough and others) or applying a contact herbicide that defoliates but does not kill established planted perennials.

Even when managed 'disturbances' such as spraying and burning are employed there will always be some establishment of perennial weed species that tolerate these regimes. The most common examples of these species in the UK are grasses, and in particular *Elymus repens* (couch grass), *Agrostis stolonifera* (creeping bent), *Holcus lanatus* (Yorkshire fog), *Ranunculus repens* (creeping buttercup), *Taraxacum officinale* (dandelion), *Urtica dioica* (nettle) and *Calystegia sepium* (bindweed). In the early stages of colonisation all of these plants can be managed satisfactorily by careful physical removal. However, with deep gravel mulches shoots may snap off leaving residual roots in the gravel. Herbicides are sometimes resorted to in these situations, or deeper physical excavations are required.

Plant combinations have been trialled that are designed to reduce weed pressure, when there are few opportunities for conventional maintenance. A plant community based solely on *Phlox subulata* cv. Trot Pink and *Potentilla neumanniana* was shown to suppress weeds in research carried out in Italy (Pomatto *et al.*, 2023), with only 0.5 g m^{-2} of dried weed biomass harvested after 3 years. In contrast, a species-rich planting was least effective at weed suppression, with 1115 g m^{-2} of weed biomass being recorded in this treatment.

7.15 Pests and Pathogens

Herbaceous plants and geophytes are particularly sensitive to slugs and snails (Table 7.6) because, in most species, they have to push their new growth through the soil, creating feeding opportunities for slugs and snails that are not offered by shrubs. Because slugs and to a lesser degree snails feed nocturnally, the impact of these herbivores is completely underestimated in the horticultural literature, with the exception of the most highly palatable genera such as *Delphinium*. Even in *Delphinium*, palatability varies considerably at the species level. Many Asian *Delphinium* such as *D. bulleyanum* are remarkably unpalatable, and the same is also true of the North American *D. glaucum*. The patterns in herbaceous plants as a whole do not mirror the situation in *Delphinium*. In species derived from the northern hemisphere, palatability tends to be

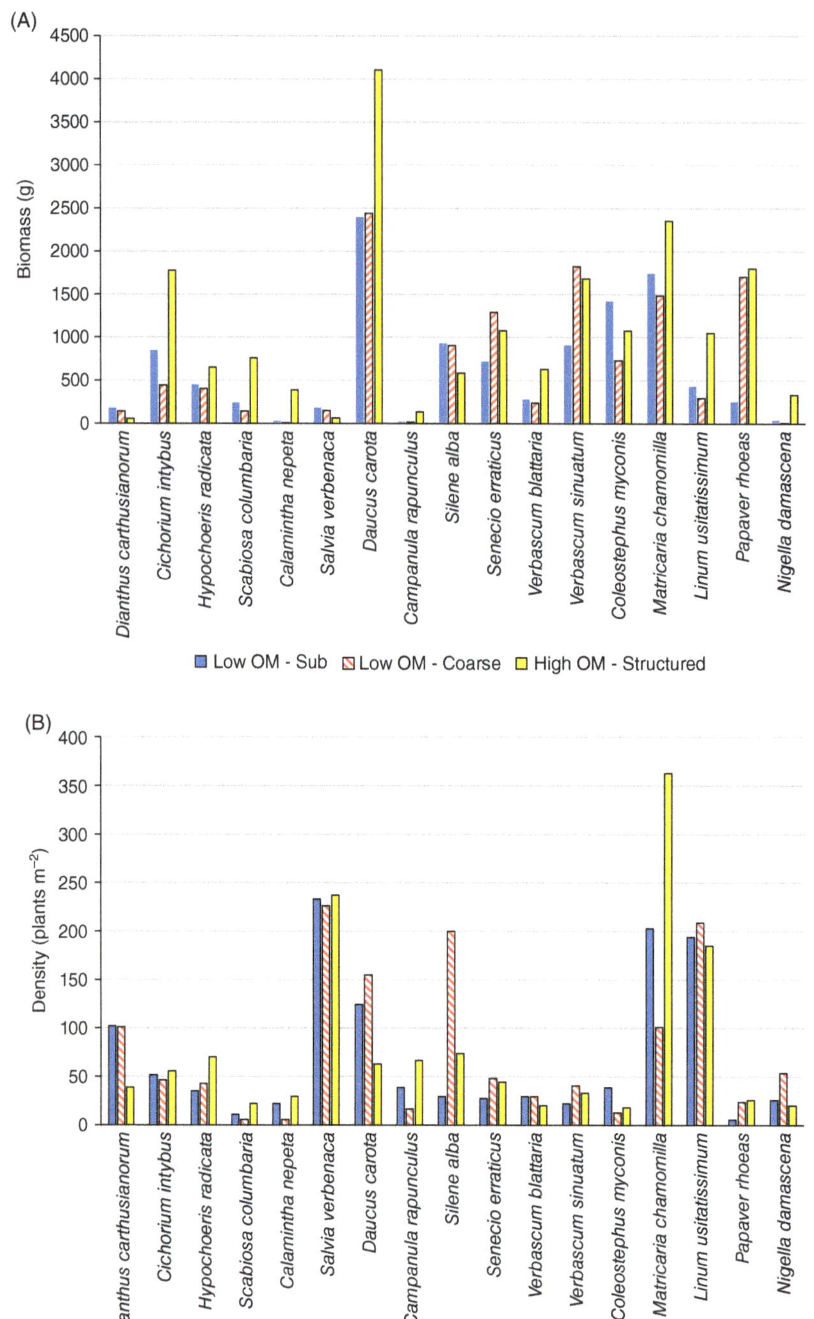

Fig. 7.18. The effect of soil character on plant development across a range of ornamental taxa. Plant biomass (A) and the number of plant specimens per square metre (B) after seedlings were established on three different soil types: sub-surface-type soil with low organic matter content (Low OM - Sub); coarse gravel-like soil with low organic matter (Low OM - Coarse); or 'garden-like', well-structured soil with loam and high organic matter (High OM - Structured). (Modified with permission from Bretzel *et al.,* 2009.)

Table 7.6. Herbaceous and geophyte species and their relative palatability to molluscs

Highly palatable	Intermediate	Unpalatable
Arnica	*Allium*	*Aconitum*
Asclepias	*Aster*	*Agapanthus*
Cacalia	*Bulbinella*	*Astilbe*
Delphinium	*Camassia*	*Crocosmia*
Dracocephalum grandiflorum	*Crinum*	*Crocus*
Echinacea	*Dianthus*	*Euphorbia*
Monarda	*Dodecatheon*	*Filpendula*
Ratibida pinnata	*Eremurus*	*Geranium*
Rudbeckia	*Eucomis*	*Helleborus*
Salvia	*Galtonia*	*Iris*
Trillium	*Liatris*	*Kniphofia*
	Narcissus	*Limonium*
	Primula	*Linum*
	Ranunculus	*Penstemon*
	Solidago	*Potentilla*
		Sanguisorba
		Succisa
		Trollius
		Tulipa

very high in genotypes from the eastern and central USA, where molluscs are not an important part of the herbivore fauna. Therefore, many of these species have not expended resources on developing the physical and chemical weaponry (hairs on leaves or silicon crystals) to deter grazing. Hitchmough and Wagner (2011) investigated the relative palatability of 26 North American prairie plant species to slugs and snails, with most (21) being susceptible to mollusc damage at the seedling stage. By contrast, most European and Eurasian species (*Delphinium elatum* aside!) are much less palatable, with temperate Asian species intermediate, although within each of these very broad groupings there is still distinct, contrasting palatability between species. Other than the spin of the evolutionary dice, many meadow species of high altitude are often very palatable because the climate precludes high mollusc densities (as do cold continental regions such as Eurasia–Asia) and hence effective defences are often absent. In the European Alps, the distribution of palatable species such as *Arnica montana* are known to be determined by slug densities (Bruelheide and Scheidel, 1999).

Certain bedding plant species are also very susceptible to damage by molluscs (slugs and snails) and protection may involve either chemical or physical means (manual removal, use of mulches, physical barriers such as copper bands or biological control via nematodes). Susceptibility often depends on growth stage (seedlings being most prone to 'fatal' browsing) or plant organs. Species used as bedding specimens where there is some evidence of tolerance or unpalatability with slugs and snails include *Ageratum, Alyssum, Anchusa, Antirrhinum* (some), *Aquilegia, Arabis, Argyranthemums* (some), *Aubrietia, Begonia, Bidens, Calendula, Cheiranthus, Coleus, Cosmos* (some), *Dianthus, Diascia, Eschscholtzia, Fuchsia, Gaillardia, Gazania, Geranium, Gypsophila, Iberis, Impatiens, Lantana, Leucanthemum, Myosotis, Nasturtium, Nemesia* (some), *Oenothera, Origanum, Oxalis, Papaver, Pelargonium, Phlox, Polygonum, Portulaca, Pulmonaria, Rudbeckia* (some), *Saxifraga, Scabiosa, Sedum, Sempervivum, Verbascum, Verbena* (some) and *Veronica.*

Many bedding plants are (partially) protected from molluscs using aluminium sulphate and copper sulphate; these work by direct contact with the mollusc and efficacy may be limited to the smaller slug and snail species. Various forms of traps and chemically treated bands are also used to deter slugs from moving towards the susceptible plants in garden settings. Biocontrol products rely strongly on nematodes, e.g. *Phasmarhabditis hermaphrodita*, which are applied to the planting area to increase the parasitic nematode population and increase likelihood of slugs and snails being infected.

Molluscs are not the only problem facing landscape ornamentals. As non-woody landscape plants represent a wide range of genotypes, this is reflected in an extensive range of pests and pathogens. Nevertheless, certain pests and pathogens are encountered on a frequent basis when growing landscape ornamentals.

During early phases of production of bedding plants and other ornamentals, it is important the young seedlings are not stressed. Inappropriate (or irregular) watering, temperature and humidity control leads to fungal diseases commonly described as damping off (*Pythium, Phytophthora* and *Rhizoctonia* spp.), grey mould (*Botrytis cinerea*), root rot (*Pythium, Phytophthora* and *Thielaviopsis* spp.) and crown rot (*Sclerotinia* spp.), with some crops also susceptible to rusts (e.g. *Puccinia* spp.) and powdery mildews (*Oidium, Plasmopara* and *Peronospora* spp.). Environmental conditions that stress the young plants through excessively high or low temperature, overwatering or poor drainage, oscillations between too wet and too dry a growing media, or prolonged periods when moisture droplets are present on the leaf surfaces predispose plants to these pathogens. Similarly, excessive nutrient levels or pH that cause damage to root tissues through high-ion concentrations cause surface wounding and weakening of the plant, thus allowing pathogens to infect more readily.

Common pests encountered across a range of bedding plants and other ornamentals grown under protection during the production phases include aphids (e.g. *Aphis gossypii* and *Myzus persicae*), thrips (e.g. *Frankliniella occidentalis* and *Heliothrips haemorrhoidalis*), whitefly (e.g. *Trialeurodes vaporariorum*), fungus gnat (*Bradysia* spp.) and red spider mite (*Tetranychus urticae*). The use of integrated pest management (IPM) and biocontrol agents is now mainstream in controlling such pests in protected environments. IPM is a change in philosophy in how pests (and pathogens) are dealt with. Rather than waiting for a large infestation of a pest to occur and reacting by applying a heavy dose of pesticides, the approach is one of proactive management. For example, growers develop a site-specific management strategy that includes careful assessment of pest problems through vigilance and regular monitoring of the crop. In this way, potential problems can be identified and dealt with quickly before they become a significant threat to the crop. Outdoors, in the final settings landscape plants are used in, biocontrol tends to rely on natural predators and seeing 'problem' species as part of the food chain. Increasingly, selecting genotypes that have less prevalence of a particular pathogen is the way forward to reduce the incidence of pathogen outbreaks (e.g. within *Alcea rosea* [hollyhock], the Happy Lights series has a reputation for greater resistance against *Puccinia malvacearum* and *P. heterospora* – hollyhock rust).

Conclusions

- Non-woody landscape plants can be categorised into a number of discrete groupings including the following:
  - Annual plants (direct-sown *in situ*) – annual 'meadow' plants.
  - Intensively produced 'bedding plants'.
  - Herbaceous perennials (where the foliage usually dies back during winter). They tend to divide into forbs – dicotyledons (broad-leaves) – and ornamental grasses.
  - Geophytes – plants with storage organs to facilitate a dormant phase and thus help survive periods of stress, such as cold, drought and lack of irradiance. These include the bulbs, tubers and corms.
  - Ferns – non-flowering plants with aesthetic foliage that reproduce via spores.
  - Alpines – low-growing perennials or short-lived plants that are adapted to cold montane landscapes.
- These plants provide the greatest colour in the urban landscape and are frequently utilised in urban parks, in roadside plantings as ornamental components of street furniture (hanging baskets from lamp posts, containers, window boxes, etc.), as part of rain gardens or green roof plantings and within private and public gardens.
- There have been moves in recent years to use some in more naturalistic planting styles. The success of such naturalistic planting communities depends on careful choice in matrix plantings to ensure the community effectively resists weed pressure, and that the plantings are self-sustaining. Careful selection can also help provide resources for local wildlife and help with a range of ecosystem services.
- Commercial production of these non-woody plants entails field production, lifting and sale of bare-root plants at one end of the spectrum, and highly scheduled, intensely produced 'pack-bedding' plants at the other end.
- There is an ambition to produce and use such plants more sustainably, with some progress in the use of non-peat growing media and less chemical use to control pests and pathogens, and to regulate plant form. Alternative sources of irrigation water and pot/packing materials are being sought, although practical progress is still slow on some of these agendas.

References

Anon. (2019) United States Department of Agriculture, National Agricultural Statistics Service. Floriculture Data 2017. Available at: chrome-extension://hbgjioklmpbdmemlmbkfckopochbgjpl/https://www.nass.usda.gov/Publications/AgCensus/2017/Online_Resources/Census_of_Horticulture_Specialties/hortic_1_0004_0004.pdf (accessed 14 January 2025).

Anon. (2020) United States Department of Agriculture/National Agricultural Statistics Service Floriculture – Production and Propagation Materials (2019). Available at: https://www.nass.usda.gov/Publications/AgCensus/2017/Online_Resources/Census_of_Horticulture_Specialties/HORTIC.txt (accessed 14 January 2025).

Arnold, M.A., Lesikar, B.J., McDonald, G.V., Bryan, D.L. and Gross, A. (2003) Irrigating landscape bedding plants and cut flowers with recycled nursery runoff and constructed wetland treated water. *Journal of Environmental Horticulture* 21, 89–98.

Blanchard, M.G., Runkle, E.S. and Fisher, P.R. (2011) Modeling plant morphology and development of *Petunia* in response to temperature and photosynthetic daily light integral. *Scientia Horticulturae* 129, 313–320.

Blanusa, T., Vysini, E. and Cameron, R.W. (2009) Growth and flowering of *Petunia* and *Impatiens*: Effects of competition and reduced water content within a container. *HortScience* 44, 1302–1307.

Blanusa, T., Monteiro, M.M.V., Fantozzi, F., Vysini, E., Li, Y. and Cameron, R.W. (2013) Alternatives to *Sedum* on green roofs: Can broad leaf perennial plants offer better 'cooling service'? *Building and Environment* 59, 99–106.

Bretzel, F., Pezzarossa, B., Benvenuti, S., Bravi, A. and Malorgio, F. (2009) Soil influence on the performance of 26 native herbaceous plants suitable for sustainable Mediterranean landscaping. *Acta Oecologica* 35, 657–663.

Bruelheide, H. and Scheidel, U. (1999) Slug herbivory as a limiting factor for the geographical range of *Arnica montana*. *Journal of Ecology* 87, 839–848.

Culver-Campbell, M. (2001) *The Origin of Plants*. Transworld Publishers, London, UK.

Dunnett, N. and Clayden, A. (2007) *Rain Gardens, Managing Water Sustainably in the Garden and Designed Landscape*. Timber Press, Portland, Oregon.

Dunnett, N. and Hitchmough, J.D. (2004) *The Dynamic Landscape, Design, Ecology and Management of Naturalistic Urban Vegetation*. Taylor and Francis, London, UK.

Garbuzov, M. and Ratnieks, F.L. (2014) Quantifying variation among garden plants in attractiveness to bees and other flower-visiting insects. *Functional Ecology* 28, 364–374.

Guo, Y, Starman, T. and Hall, C. (2019) Growth, quality, and economic value responses of bedding plants to reduced water usage. *HortScience* 54, 856–864.

Hansen R. and Stahl F. (1994) *Perennials and their Garden Habitats*. Oxford University Press, Oxford, UK.

Hillová, D., Lichtnerová, H., Mitošinková, V., Brtáňová, M. Raček, M. and Kubus M. (2016) Effects of drought treatment on three matrix planting perennials. *Acta Scientiarum Polonorum – Hortorum Cultus* 15, 133–144.

Hitchmough, J.D. and Wagner, M. (2011) Slug grazing effects on seedling and adult life stages of North American Prairie plants used in designed urban landscapes. *Urban Ecosystems* 14, 279–302.

Hitchmough, J., Wagner, M. and Ahmad, H. (2017) Extended flowering and high weed resistance within two layer designed perennial "prairie-meadow" vegetation. *Urban Forestry and Urban Greening* 27, 117–126.

Kaczperski, M.P., Carlson, W.H. and Karlsson M.G. (1991) Growth and development of *Petunia × hybrida* as a function of temperature and irradiance. *Journal of the American Society for Horticultural Science* 116, 232–237.

Kuehny, J.S., Taylor, M. and Evans, M.R. (2011) Greenhouse and landscape performance of bedding plants in biocontainers. *HortTechnology* 21, 155–161.

Laird, M. (1999) *The Flowering of the Landscape Garden, English Pleasure Grounds, 1720–1800*. University of Pennsylvania Press, Philadelphia.

Matyashuk, R.K., Pirko, I.F., Gubar, L.M. and Tkachenko, I.V. (2023) Decorative perennials of the regional flora for recreation landscapes in the forest-steppe zone. *Biosystems Diversity* 31, 319–326.

McCabe, K.G., Schrader, J.A., Madbouly, S., Grewell, D. and Graves, W.R. (2014) Evaluation of biopolymer-coated fiber containers for container-grown plants. *HortTechnology* 24, 439–448.

Monteiro, M.V., Blanuša, T., Verhoef, A., Richardson, M., Hadley, P. and Cameron, R.W.F. (2017) Functional green roofs: Importance of plant choice in maximising summertime environmental cooling and substrate insulation potential. *Energy and Buildings* 141, 56–68.

Nur Hannah Ismail, S., Stovin, V. and Cameron, R.W. (2023) Functional urban ground-cover plants: identifying traits that promote rainwater retention and dissipation. *Urban Ecosystems* 26, 1709–1724.

Oehme, W. and Van Sweden, J. (1991) *Bold Romantic Gardens: The New World Landscape of Oehme and Van Sweden*. Acropolis Books, New York.

Oh, W., Cheon, I.H., Kim, K.S. and Runkle, E.S. (2009) Photosynthetic daily light integral influences flowering time and crop characteristics of *Cyclamen persicum*. *HortScience* 44, 341–344.

Oudolf, P. and Gerritsen, H. (2013) *Dream Plants for the Natural Garden*. Francis Lincoln Press, London, UK.

Oudolf, P. and Kingsbury, N. (2014) *Planting a New Perspective*. Timber Press, Portland, Oregon.

Pomatto, E., Larcher, F., Caser, M., Gaino, W. and Devecchi M. (2023) Evaluation of different combinations of ornamental perennials for sustainable management in urban greening. *Plants* 12, 3293.

Rihn, A.L., Knuth, M.J., Peterson, B.J., Torres, A.P., Campbell, J.H. *et al.* (2022) Investigating drivers of native plant production in the United States green industry. *Sustainability* 14, 6774.

Schrader, J.A., Currey, C.J., Flax, N.J., Grewell, D. and Graves, W.R. (2018) Effectiveness of biopolymer horticultural products for production and postproduction nutrient provision of garden and bedding crops and container ornamentals. *HortTechnology* 28, 257–266.

Shurberg, G., Shober, A.L., Wiese, C., Denny, G., Knox, G.W., Moore, K.A. and Giurcanu, M.C. (2012) Response of landscape-grown warm-and cool-season annuals to nitrogen fertilization at five rates. *HortTechnology* 22, 368–375.

Tomadoni, B., Merino, D. and Alvarez, V.A. (2020) Biodegradable materials for planting pots. *Advanced Applications of Bio-degradable Green Composites* 68, 85.

Tomaskin, J. and Tomaskinova, J. (2020) Evaluation of assortment of ornamental grasses and their environmental importance in the urban landscape. *Journal of Environmental Protection and Ecology* 21, 1673–1682.

van Iersel, M.W., Dove, S., Kang, J.G. and Burnett, S.E. (2010) Growth and water use of *Petunia* as affected by substrate water content and daily light integral. *HortScience* 45, 277–282.

Yasemin, S. and Beruto, M.I. (2024) A review on flower bulb micropropagation: challenges and opportunities. *Horticulturae* 10, 284.

Wilford, R. (2006) Tulips: *Species and Hybrids for the Gardener*. Timber Press, Portland, Oregon.

Woudstra, J. and Hitchmough, J.D. (2000) The enamelled mead: History and practice of exotic perennials grown in grassy swards. *Landscape Research* 25, 29–47.

Wright, R.D., Jackson, B.E., Barnes, M.C. and Browder, J.F. (2009) The landscape performance of annual bedding plants grown in pine tree substrate. *HortTechnology* 19, 78–82.

Zhao, Y., Liu, C., Sui, J., Liang, J., Ge, J. *et al.* (2022) A wake-up call: Signaling in regulating ornamental geophytes dormancy. *Ornamental Plant Research* 2, 1–10.

Zollinger, N., Kjelgren, R., Cerny-Koenig, T., Kopp, K. and Koenig, R. (2006) Drought responses of six ornamental herbaceous perennials. *Scientia Horticulturae* 109, 267–274.

8 Semi-natural Grasslands and Meadows

Abstract

This chapter deals with grasses when grown high for aesthetic or ecological purposes. In the natural world, grass ecosystems prevail where it is too dry/cold for trees or where grasses and allied herbaceous plant communities are promoted through fire, herbivory or human management regimes. A 'grassy location' with regular cutting down of the vegetation for hay or grazing by livestock is conventionally known as a meadow. Meadow plant communities, however, are not just composed of grasses. Many flowering herbaceous plants (forbs) are important constituent components of meadows and indeed provide most of the colour. Semi-natural grasslands and meadows have become popular in urban environments, due to their capacity to provide plant diversity, colour and habitat for wildlife. The success of achieving biologically rich semi-natural grasslands, however, depends on soil nutrition, water status, weed pressure, frequency of management interventions (e.g. hay cutting) and the capacity of desirable species to perpetuate themselves.

8.1 Introduction

In terms of urban space, grasslands fall in two main categories with various sub-categories in each. There are (i) the short, regularly mown grasses that constitute turf landscapes and lawns and (ii) areas where the grass is allowed to grow longer to mimic natural grasslands or create meadow landscapes. The short-grown grass swards are dealt with in Chapter 9, and it is the longer sward types that are covered here. It should be noted though that ornamental grasses are also used as 'stand-out' features in the aesthetic landscape and gardens and these are included in Chapter 7.

Semi-natural grasslands and meadows can be found in a number of different urban green spaces (Robbins and Birkenholtz, 2003; Ignatieva *et al.*, 2015; Bretzel *et al.*, 2016). These include rough grassland in parks, alongside roads, motorways verges and on roundabouts, within cemeteries, golf-course 'roughs', school grounds, private gardens, on ex-industrial brownfield sites, within urban nature reserves, within swales and on green roofs. Such grassland can be completely dominated by grass species (rough or rank grassland) or have a very high component of flowering dicotyledonous species (broad-leaved flowering plants of forbs), as in highly floristic meadows or perennial meadows.

Semi-natural grasslands and meadows are becoming more popular as local authorities explore alternatives to traditional lawns. The drivers for this are more sustainable management practices (less mowing and less petrochemical use/emissions), to help reduce labour costs and to promote biodiversity, through more diverse plant communities (Ignatieva and Ahrné, 2013).

8.2 Grasslands Around the World

Climate, soil conditions and management regimes will dictate what sort of grassland plant communities establish/and or evolve in urban green space. Planted urban grassland landscapes can mimic natural biome types. It is worth briefly exploring the 'natural' grassland types to better understand how environmental conditions affect growth and ecology in the city grasslands. There are two main types of grassland biomes found across the globe: tropical grasslands and temperate grasslands.

Tropical grasslands are often referred to as savannah grasslands. These are not solely grasslands because there are often plenty of intermittent trees in these landscapes. These include the llanos in Venezuela and the savannahs of Africa, South America, India and Australia, and the campos of southern Brazil and Uruguay. Tropical grasslands are located in the periphery of the equatorial forest region towards the eastern coasts of continents. It is a transition zone between equatorial rainforests

DOI: 10.1079/9781800621763.0008

and the completely dry deserts. Often tropical grasslands are influenced by monsoon climatic zones and experience seasonal rainfall in summer and remain mostly relatively cool and dry during winter. They often have distinct wet and dry seasons. During wet periods the grass growth can be vigorous, and the grasslands can easily conceal even large mammals such as *Loxodonta africana* (African elephant), *Elephas maximus* (Asian elephant) *Ceratotherium simum* (white rhinoceros), *Rhinoceros unicornis* (Indian rhinoceros) and *Giraffa* spp. (giraffe). Grass species can be up to 5 m tall and key genera are *Hyperrhenia* (elephant grass), *Panicum, Andropogon* and *Pennisetum* spp.

The temperate grasslands include the steppes in Eurasia, prairies in North America, pampas in Argentina, velds in South Africa, downs in Australia, Manchuria grasslands in China, pustaz in Hungary and Canterbury grasslands of New Zealand. An additional type – cold grasslands – includes the tundra of the Arctic. The temperate grasslands are primarily located between 30 to 60° latitudes in interior parts of the continents in the northern and southern hemispheres. Temperatures range between -40°C to +30°C, with mean rainfall ranges of 750–1000 mm per year. Organic matter can build up in temperate grasslands, resulting in dark chernozems or black soils. Important grass genera include *Festuca, Stipa, Bouteloua* (buffalo grass) and *Nassella* spp. Grassland forbs include floriferous genera such as *Rudbeckia, Echinacea* and *Solidago* within the North American prairies.

As implied from the information above, grassland communities occur where some form of abiotic stress inhibits vegetation succession to trees and shrubs. This may be cold, fire, drought or grazing pressure. Varied and special types of grassland communities though can occur even with similar/the same climatic regions. Many meadows are semi-natural landscapes in that it is through the activities of humans that the florally rich plant communities are encouraged, for example grazing of domestic livestock, cutting for hay or even flooding to improve the quality of pastureland (traditional wet-meadows) (Fig. 8.1). These are augmented by other grassland communities, such as in the UK, machair, heaths, moorland, freshwater marshland, reedbeds, saltmarsh, chalk downland and pasture.

Natural grasslands are important ecosystems that face threats from human activities including conversion to arable agriculture, overgrazing and climate change. The grasslands cover approximately

Fig. 8.1. Different ranges of herbaceous flowering plants (forbs) are found in different meadow types. Top: *Campanula rotundifolia* (Scottish bluebell/harebell) is evident in free-draining, nutrient-poor grasslands, heaths and at the edge of sand dunes. Bottom: In contrast, *Fritillaria meleagris* (snake's-head fritillary) prefers damp meadows, where river water inundates the land in winter.

20–40% of the Earth's land surface, making them one of the largest and most agriculturally useful biomes. Partially, it is the large loss of habitat such as the North American prairie ecosystem (where only <5% of the original area is thought to remain, with perhaps <1% of the tall-grass prairies left) that has spurred the popularity and conservation initiatives around urban prairies and meadows. Similarly, in countries such as the UK, changes in agricultural practices have seen the demise of colourful flowering meadows to be replaced by near-monocultures of *Lolium perenne* (perennial ryegreass) as more productive pasture, or the ploughing up of old meadows. These activities, however, have acted as a catalyst to preserve what meadow systems still remain in rural locations and also for the ambition to try and replicate typical traditional meadow communities within urban settings. Meadow-making, particularly in urban situations, is becoming popular as an alternative to more intensive horticultural lawn landscapes, for which the resources to manage are increasingly not available. Meadow-like vegetation has become seen as a sort of 'magic bullet', colourful and attractive to people, good for supporting biodiversity and yet also relatively inexpensive to create and manage.

In the UK, in those locations where dense trees are absent, grass-dominated landscapes tend to be the mainstream, much more so, for example, than in continental Europe. This is due to lower levels of summer moisture stress, mild winters that facilitate grass dominance during this period and high levels of slugs and snails that reduce the competitiveness of many grassland broadleaves. Because of the presence of silicon in the leaves, grasses are generally less palatable to slugs than are forbs (O'Reagain, 1989). Urban semi-natural grasslands gain their inspiration from hay meadows, prairies and sometimes steppe landscapes.

8.3 Florally Diverse Grasslands

Meadows

A meadow is a type of open habitat characterised by grasses, herbaceous plants (forbs) and other non-woody plants, with sparse or no tree cover. They can occur naturally under favourable conditions or be artificially created from cleared woodland or shrubland. Meadows support diverse flora and fauna, providing habitats for pollinating insects, nesting areas for birds and food sources for various wildlife. Depending on the depth of the water table and the likelihood of seasonal flooding, meadows can be referred to as dry-meadow and wet-meadow landscapes.

The modern trend to create urban meadows, referred to as 'meadowscaping' (Cerra, 2017), is linked to either completely replacing lawns with native meadows, managing lawns less intensively (i.e., reduction of mowing and no fertilizer and pesticide inputs) or creating new meadows from bare soil (Smetana and Crittenden, 2014; Norton *et al.*, 2019).

Traditional meadows are linked to grazing or the harvesting of hay and would have evolved their own community of native grasses and forbs. The term 'meadow' is most specifically applied to vegetation that is regularly cut for hay every year, normally between summer and autumn. After the clearing of forest, generally well underway in many temperate parts of the world by the Bronze Age, grazing of the resulting herbaceous understorey alternated with cutting for hay for winter feed. This resulted in the classic lowland hay meadow model. Lowland (and many upland) meadows are therefore a semi-natural agricultural vegetation, based on a reordering of native plant species through human management.

In addition to these, there are another type of meadow found at high altitudes, often above the tree line. These 'alpine' meadows are natural and exist without cutting as the summer growing season is too cool for woody plants to invade; unlike agricultural meadows they cannot become anything else without a major shift in the climate. Typically, alpine meadows occur on unproductive soils, in high rainfall climates, throughout the mountains of the world. Classic examples are found in the Rocky Mountains of North America, the Alps of Europe and Caucasus, Himalaya and other mountains chains in Asia. They are often grazed by wild (or sometimes domesticated) ungulates but are otherwise stable.

Modern urban 'perennial' meadows, however, are designed landscapes and may mimic a range of natural grasslands; often, the rationale is to create a strong visual impact on the landscape – with flower colour and plant form being important to the visual aesthetics. They may use native, but also non-native, herbaceous perennials. Thus, the term 'meadow' is used in this text to cover open habitats or fields, created and managed by humans within urban and suburban landscapes, vegetated by perennial native grasses and flowering perennial broadleaf forbs which can mimic natural and semi-natural vegetation such as North American prairies or Eurasian steppes (Hitchmough and De la Fleur, 2006; Köppler *et al.*, 2014). Meadow plant communities develop diverse, self-sustaining vegetation through self-propagation via seed or offshoots (daughter plants).

Climate determines the plant community composition and in some drier climatic locations the term xeric meadow or xeriscapes are used. These xeriscapes, meadow-like plant communities, consists of drought-tolerant native grasses, succulents, flowering forbs and small bushes in a xeric (dry) environment. The creation of xeriscape meadows is a popular practice for sustainable management of urban green spaces in dry and drought-prone environments as an alternative to lawns (Smith and Hilaire, 1999).

Steppe

Steppe was derived from the Russian word 'step' before the 17th century but the origins of this word are obscure. Unlike meadows, steppe are not regularly cut for hay although often grazed by domestic livestock. The general conception of steppe is a dry grassland associated with continental climates, i.e. with alternating hot, relatively dry summers and very cold winters, with often large diurnal temperature fluctuations even during the summer. The most characteristic genus of steppe, certainly in Eurasian and in some southern hemisphere steppe, is the spear or feather grass. Although associated with large tracts of central Asia, naturally occurring steppe is also found in eastern Europe, parts of central Spain, and southern central France. Some steppe are historically burnt every spring to remove dead overwintering plant material and produce a new, more palatable growth for livestock.

Prairie

The third major meadow-like grassland type is prairie, typically considered to be restricted to North America but which has similarities with tall grassland and forb communities in other parts of the world, such as eastern South Africa and parts of China and Asia. Prairie appears to be a semi-natural vegetation, in which the species associated with clearings and edges of woodland have been reordered into a distinctive plant community through the combination of grazing by bison, deer and antelope in combination with regular aboriginal spring burning. Some plant geographers see prairie as a component of steppe, which makes sense if one considers the very dry end of the prairie continuum, but not at all when one looks at the moist (mesic) to wet end of the spectrum, where prairie may be 2 m tall, and essentially closed in with a huge biomass.

Cultural links

Meadows have a strong cultural identify in many counties – and indeed to many individuals, the idealistic landscape of choice is one of swaying grasses intermingled with an array of colourful and beautiful flowers. They are the epitome of a pastoral, 'golden' past era, when life was simple and carefree (Fig. 8.2), and of course, in reality was probably anything but. Alpine floral meadows of the Alps in Europe, the Himalayas of Asia or the Rockies in North America and many other mountain ranges are linked with diverse, colourful, interesting natural flora (Fig. 8.3), as well as pictorially dramatic landscapes. Water meadows and hay meadows feature commonly in art and literature, and even children's nursery rhymes are frequently themed around a meadow flower. Haymaking was central to agricultural communities for millennia – a source of fodder that allowed livestock to survive winter – and thus provide food on the table for village people during the very bleakest part of the year. Haymaking was central to the social calendar of the village – when the community would come

Fig. 8.2. Meadows are often associated with a romantic, rustic, simpler era and a slower pace of life.

Fig. 8.3. Many diverse meadow/grassland types are associated with strong colour themes, e.g. wet meadows in Asia where *Primula pulverulenta* finds a niche.

together for a common purpose. This, combined with the dramatic loss of meadow landscapes since World War II through agricultural intensification and efficiencies, has left a strong 'collective cultural memory' around meadows. Arguably, they are 'missed and yearned for', even by citizens that never actually experienced them in reality. This may be one driver for why 'new' urban colourful meadows are gaining popularity.

Similarly, the desire to recreate prairie landscapes (albeit in a much smaller scale) in the gardens and parks of suburban North America has been linked to cultural aspects as well as simply ecological/environmental reasons. Prairie landscapes are linked to Indigenous peoples' heritage, with traditionally strong bonds between the natural world and people's spirituality. The prairie is a unique landscape and one that North Americans in general are embracing and identifying with because of that. Even for those who claim their heritage as European, there is now a desire to celebrate native plant heritage. In addition, growing 'grass long' in prairie-style gardens has been seen as a notion of independence and freedom, breaking away from the constraints and conventions of the well-manicured lawn. There are also those connections to an open, 'unbounded' expansive landscape, which evokes feelings of freedom, simplicity and wilderness, thus acting as a counterpoint to the pressures of modern life. The beauty of the prairie has long been a source of inspiration for North American artists, writers, and photographers, contributing to a distinct cultural aesthetic and sense of place.

8.4 Why Let the Grass Grow? Ecosystem Services and Disservices

Naturalist grasslands and meadows provide a range of ecosystem services and some drawbacks (disservices) to the urban environment. Some of the services overlap with that of more formal mown lawns (see Chapter 9), including their capacity to help rainwater infiltration (Paudel and States, 2023), act as pollutant traps (e.g. oil spills off roadways, Yang *et al.*, 2020), provide urban cooling through evapotranspiration (Francoeur *et al.*, 2021) and stop erosion and stabilise soils (Monteiro, 2017).

Some services may be lost though as lawns morph into meadows, for example, their capacity to act as a sports surface and facilitate other recreational activities and pastimes, such as picnicking. The drivers for more urban meadows and their expansion (Fischer *et al.*, 2020) are often cited as follows:

- Being better for biodiversity than close mown grasslands/lawns (Tilman *et al.*, 2006; Southon *et al.*, 2017, Chollet *et al.*, 2018; Baldock *et al.*, 2019).
- Creating a different form of aesthetic, e.g. opportunities for more flower colour and plant forms or seeing the 'grass blow in the wind'.
- Based on the points above, potentially greater capacity to improve human psychological well-being.
- Less maintenance and labour cost due to much-reduced mowing regimes.
- Better 'for the environment' due to less reliance on artificial fertilizers and irrigation water and the absence of pesticide use (Ignatieva *et al.*, 2015; Smith *et al.*, 2015; Norton *et al.*, 2019).

Other advantages over formal lawns include:

- Better capacity to abate poor air quality – as the longer grasses increase the surface area that can help trap gaseous and particulate pollution (Przybysz *et al.*, 2021; Nawrocki *et al.*, 2023).
- Catch more rainwater than lawns and dissipate more effectively through greater evapotranspiration.
- Act as more effective ecological corridors – invertebrates, small mammals, amphibians and reptiles have more cover when moving through.

Disadvantages compared to formal lawns, however, include:

- More likely to produce allergens from grass and other plant pollens (the tall grasses in meadows

are usually allowed to flower, i.e. produce pollen, compared to close-mown swards where the grasses may never get a chance to form a flowering panicle).
- Greater capacity to trap litter and dog faeces (the latter more difficult to clean up in long grass too).
- Grass and tall forbs can lodge (stems fall over in wind and rain) resulting in the meadow looking 'untidy' to some.
- The volume of off-cuts (arisings) can be greater at any one time, with less frequent cutting, and this can be difficult or more expensive to dispose of.

8.5 Meadows Through Management – Altering the Cutting (or Grazing) Regimes

Many local authorities and other stakeholders are attracted to transforming their close-mown turf to a longer (and hopefully a more floristically diverse) sward to (i) reduce mowing costs, (ii) meet targets on biodiversity and (iii) potentially increase the aesthetic value of a location, although this latter point largely depends on how successful the arrangement is in terms of increasing the number of plant species present and their capacity to increase public interest through great flower colour.

Conventional regular gang mowing

Most urban grass in the UK is managed by gang mowing – a gang being a number of individual cylinder (reel) mowing units working in combination behind a tractor. The dominance of mown grass in the UK and other maritime climates is largely due to its multifunctional characteristics, but also because it is generally the cheapest vegetation surface in terms of maintenance (Cobham, 1990). Mown grass is cheapest to manage in large spaces, without obstacles interfering with the action of the multiple gangs. Areas of gang-mown parkland may receive in excess of 30 cuts per year (about weekly during spring and summer). In the UK, turf is ubiquitous and for most people it maintains a sort of neutral constant presence. Because it is almost never irrigated or fertilized or treated with pesticides (in general public space) it attracts little of the environmental opprobrium that it does, for example, in much of the USA (Herbert-Bormann *et al.*, 2001).

On the negative side, mown grass, because it is used so extensively often to the exclusion of all other urban vegetation in UK green space, is both relatively limited in terms of the experiences it can offer users (it is often seen as rather boring) and quite limited in its biodiversity value (Smith *et al.*, 2015). Debates have raged about the capacity of mown grass to store carbon (Townsend-Small and Czimczik, 2010; Wang *et al.*, 2022; Gillman *et al.*, 2023), and whilst it does incorporate carbon into the soil from grass clippings post-mowing, these levels are similar to the carbon stored from alternative management options such as managing as meadow, etc. All in all, from a sustainability perspective, the UK probably has much more gang-mown grass than can rationally be justified; hence, there is interest in how the mown-grass estate could be managed differently.

Rough grassland

Rough grassland is where the grass is allowed to grow long but does not necessarily increase the floral diversity of the sward. It can be carried out where the work rate of gang mowing is slowed by topography or obstacles such as trees and edges. It can also be the preferred management regime in locations such as golf course roughs. There is a trend in some public open spaces to switch from gang mowing to cutting once or twice a year using a flail mower, often in winter to form rough grassland. This treatment has similar, or in some cases, slightly reduced costs to gang mown turf (Cobham, 1990).

With rough grass, flails macerate the cut grass into lumps which are left on the cut surface. As a result, as the cut grass decomposes the nutrients present in it are recycled back into the soil, and hence on highly fertile clay or loam-based soils there is no reduction of soil productivity. Apart from situations where there is exceptionally infertile sandy soil, this tends to result in a plant community that shifts from rhizomatous–stoloniferous grasses and rosette-forming forbs such as *Bellis perennis* (daisy) and *Taraxacum officinale* (dandelion) to tall, coarse, tussock-forming grasses such as *Dactylis glomerata* (cocksfoot) and *Arrhenatherum elatius* (false oat grass) The shading and competition provided by the latter grasses leads to the loss of the broad-leaved forbs and a subsequent drop in plant diversity. This management practice is often 'sold' to residents on the basis that cessation of regular cutting will encourage wildflowers to develop, but this does not normally occur on productive soils because the dense growth of tussock grasses is extremely hostile to the establishment of meadow forbs. Even when seedling forbs do establish after cutting of the grassland in winter, the subsequent vigorous growth and large leaf canopies of the tussock grasses typically cause the elimination of most of the established seedlings by mid-spring.

The pejorative term 'rough grassland' is used for these types of grassland and that partially reflects their relative unpopularity in urban settings (Junge *et al.*, 2009). In reality, these grasslands can be very supportive of animal biodiversity (if not plant biodiversity). Species richness and abundance of small mammals in particular is good (e.g. in the UK, *Microtus agrestis*, field vole; *Myodes glareolus*, bank vole; *Sorex araneus*, common shrew; *Sorex minutus*, pygmy shrew; *Apodemus sylvaticus*, wood mouse; and *Apodemus flavicollis*, yellow-necked mouse) with the dense foliage providing cover from aerial (e.g. *Tyto alba*, barn owl; *Strix aluco*, tawny owl; and *Falco tinnunculus*, kestrel) and larger mammal predators (*Vulpes vulpes*, red fox; *Meles meles*, Eurasian badger; *Mustela erminea*, stoat; and *Mustela nivalis*, weasel) (Humbert *et al.*, 2012). Rough grass is also an excellent habitat for sap-sucking insects, beetles, spiders, slugs and snails.

In other parts of the world mown grass is more difficult to maintain and hence often retains a cachet of exclusivity, particularly where water is short as in southern Europe, California or the Gulf States of the USA.

Florally diverse lawns

Relaxing the mowing regime of a formal lawn can encourage a surprising range of flowering forbs that have adapted to growing low, yet retain their capacity to produce flowers. Indeed, most non-intensively managed lawns will have persistent forb species present. Traditionally such species would have been seen as weeds and cultivation techniques and herbicides applied to get rid of them. Now schemes such as 'No Mow May' (Plantlife, 2019) in the UK encourage the grass cutting to be delayed in the spring/summer, allowing these forbs to produce flowers that help support pollinating insects and focusing attention on more sustainable lawn management. Indeed, the No Mow May initiative has gained significant support, with over 40 local councils in the UK participating by leaving some public green spaces unmown and thus complementing the effort of private garden owners. Florally diverse lawns encourage flowers to form, create aesthetic interest and boost nectar and pollen supply for insects (Del Toro and Ribbons, 2020), but the longer grass is not encouraged in perpetuity –usually the traditional lawn regime is restored by a close cutting after flowering finishes. However, the process can be repeated even within the one growing season, allowing flowers to again form late in the summer. Species well adapted to these temporary cessations in mowing include *Bellis perennis* (daisy) *Lotus corniculatus* (bird's foot trefoil), *Myosotis arvensis* (forget-me-not), *Ranunculus repens* (creeping buttercup), *Trifolium repens* (white clover), *Cerastium fontanum* (common mouse ear) and *Taraxacum officinale* (dandelion).

Florally diverse lawns tend to be promoted by those homeowners with an interest in wildlife or a more relaxed attitude to gardening (Fuentes, 2021). Floral diversity decreases as more emphasis is placed on traditional lawn aesthetics, and for those that want to manage their lawns as locations for children to play on. Enhancing the species of plants that are allowed to flower and produce nectar/pollen increases the invertebrate abundance and diversity. Wolfin *et al.* (2023) found that more florally enriched lawns supported more diverse and different bee communities compared to lawns where just *Trifolium repens* was the sole forb present. Smith and others (Smith and Fellowes, 2015; Smith *et al.*, 2015) promote even the notion of grass-free (but forb-rich) lawns to support urban insect populations. Despite these activities being driven largely to help ailing wildlife populations, the concept of flowering lawns is not new. 'Flowery meads' were medieval 'lawns', traditionally found in the cloisters of monasteries. These grassy areas were adorned with a variety of small, colourful flowers, creating a picturesque and naturalistic landscape. The concept was an essential feature of medieval gardens and is often depicted in medieval tapestries known as 'mille fleurs' (a thousand flowers). Flowery meads are associated with the idyllic and romanticized landscapes of medieval Europe, appearing in works by Shakespeare, as well as in Pre-Raphaelite paintings.

Spring meadows

A number of forb species and geophytes are programmed to flower early in the growing season, before the canopies of competing plants close over and exclude light. These plants can be exploited to provide colourful spring meadows. Here the grass is mown as normal throughout the year, but with a break imposed between February and May to allow plants that can flower during this time to do so. In Europe, dominant forb species include *Taraxacum* spp. (dandelion), *Bellis perennis* (lawn daisy), *Ranunculus repens* (creeping buttercup) and *Plantago lanceolata* (lawn plantain). Colour can be augmented by the inclusion of geophytes (bulbs, see Chapter 7). Colourful species include *Galanthus* (snowdrop), *Crocus*, *Narcissus* (daffodil, Fig. 8.4), *Tulipa* (tulip, e.g. the more vigorous Darwin hybrids such as Apeldoorn can survive a number of years in meadow culture), *Hyacinthoides* (bluebell, Fig. 8.5) and if a slightly later cut is employed then *Camassia* (quamash). Flailing grass in late autumn to early winter inhibits the formation of tussock structures, thus allowing even small geophtyes to project their foliage into the light and photosynthesise adequately to increase and spread, both vegetatively and sometimes also from seeds.

These species are relatively easy to establish because they compete for resources such as light and nutrients when the grasses are at their least competitive, i.e. they utilise a window or reduced competition within the grassland. Despite this, the higher the standing biomass of the flailed grass, the more attention needs to be given to selecting more vigorous geophytes with taller leaves, such as, for example, taller-growing *Narcissus* and *Camassia*. Small geophytes such as *Crocus* are more successful in less productive grasslands. Planting needs to be undertaken in autumn for best establishment; hence, cutting the grassland in September in the year of planting is recommended, as is mowing it in November to ensure competition for light is limited in this first year. Commercial bulb mixes for mass planting into mown grass have been developed in the Netherlands and are readily commercially available. As with other meadow/grassland systems, moisture levels can determine the flora present. In Europe, iconic species such as *Fritillaria meleagris* (snake's-head fritillary) are associated with traditionally managed damp river meadows, with the fields being full of flowers for a few weeks in April/May.

Fig. 8.4. A range of *Narcissus* (daffodils) species and cultivars dominate this grassy slope in spring.

Summer meadows

These follow the principles of traditional hay meadows, with the vegetation left to grow throughout the spring and early summer and then the meadow being cut once or more from mid-summer onwards. In urban situations, the need to cut and remove the herbage from the site can increase the cost over that of flail cutting. The plus side is that a more diverse plant community can arise over time (sometimes taking as long as 10 years). The annual removal of the peak standing biomass in summer

Fig. 8.5. *Hyacinthoides non-scripta* (bluebell) is a woodland plant but can establish on damp, free-draining meadows.

gradually reduces the productivity of the grass sward, creating opportunities for natural colonisation of native forbs, plus more reliable establishment of forbs through planting or over-seeding (Wells, 1980; Marrs, 1985). Other forms of biodiversity increase too (Sehrt *et al.*, 2020). One needs to be realistic, however, about what can be achieved through meadow cutting alone. It will not lead to highly floristically diverse meadows in the short term where soil productivity is high and where there are few if any meadow forbs growing adjacent to the site. In the absence of the latter, a situation known as seed limitation prevails: forbs do not establish simply due to the fact that there is no seed of these species present. Meadows can therefore remain forb species poor for decades or longer under these circumstances. Most urban sites are essentially islands separated from more biodiverse patches of land from which 'seed rain' might derive. Nor can much, if any, faith be placed in the idea that desirable forbs might recruit following a switch to a meadow regime. Most of these species do not persist or do not persist for long within the 'seed bank' of the soil. To overcome a lack of forbs, sowing or direct planting of forbs may be required to help boost diversity (see below). One exception to this 'rule' are the terrestrial orchids; their seed is microscopic and so light that it can be blown by the wind many kilometres from the parent plant. The soil does need to have the recipient mycorrhizal fungi species present, however, to help the orchid seed geminate and develop. Where this is the case, orchid species can seem to 'appear out of nowhere'! Three orchid species migrated into the author's garden, 4 years after the implementation of a 'meadow management' regime to the garden lawn.

The costs associated with cutting the sward in summer as hay vary substantially depending on the scale of the operation and the machinery available. Much of the green space network of Stockholm, Sweden is dominated by meadow-like grasslands (floristically relatively poor) and are managed by cutting in August followed by baling using large-scale agricultural equipment. The actual management costs of this operation per unit area are unknown but are likely to be substantially lower than the 4 h/100 m^2 typically required for hand cutting with a strimmer and brushcutter followed by manual raking and removal. There is an extensive range of cut and then independently bale equipment on the market. Some of the machinery is designed for small-scale meadow cutting in the mountains of Europe and often involves self-propelled reciprocating blade mowers and small-scale baling machines.

Timing of cutting strongly determines which plant species survive and thrive and those which do not. As the timing of cutting is interlinked with plant lifecycle, it is a useful rule of thumb to cut meadows at a consistent time each year (or at least at a consistent place in the plant lifecycle, i.e. seasonal shifts mean calendar days cannot be totally relied on). Cutting time and frequency have a major impact on competitive relationships between species in the meadow and ultimately on which species are dominant and which subordinate. When dealing with grass-dominated systems (gang-mown grass allowed to grow long) in which the forbs present are initially mostly low-growing creeping or rosette-forming species, cutting early in the summer, typically late June or early July (northern hemisphere) is likely to be the most appropriate strategy to maximise forb dominance. Cutting at this time checks the physical dominance of the tall grasses such as *Arrhenatherum* and *Dactylis* that are likely to be shading the community, allowing light back into the system soon enough to limit the extirpation of light-demanding low forbs. Implementing this regime during the first 5–10 years is going to be beneficial for moving towards a more diverse rather than tall and rank meadow. On really fertile soil a second autumn cut and removal of the cuttings (arisings), in addition to an early spring mow to 'tidy up', is advantageous.

Whilst this regime is likely to be effective in removing the shading from tall grasses, one needs to be aware it may also disadvantage those forb species with taller leafy stems (for example, *Malva moschata*, or *Knautia arvensis*). To favour these species a delayed cut is required, for example in August or even September.

Such cutting regimes are a cultural 'hangover' from traditional hay meadow management, where the primary objective, of course, was to maximise the yield of hay to feed to animals – little conscious regard was given to optimising plant biodiversity. Thus, alternative mowing times/frequencies can be considered. Although it involves two hay cuts per annum, a regime in which the meadow is cut in late May and then again in October is worth considering for high-profile sites as it pushes the main flowering peak back into summer and reduces meadows looking very 'dry and strawy' and unattractive during this period. Again, this system tends to check the dominance of tall grasses and removes the flowers and developing seeds of the grasses. In practice, environmental horticulturists need to experiment with individual sites in question, as local conditions can influence species competitiveness. Effective managers will observe the changes and judge whether the cutting regime favours the desired species and is preferred by those people using the landscape.

8.6 Encouraging the Establishment of Forbs in Summer Meadows

When conditions are right (i.e. soil nutrient levels decrease, thus increasing the competitiveness for forbs against the grasses), certain forb species do become established in the sward and have been linked to traditional meadow plant communities (Table 8.1).

In general though, in new urban meadows, where grass growth is still strong, forbs are more difficult to establish because their growth period generally coincides with that of the grasses, and hence there is no separation of competition in time as is the case with the geophtyes. The critical factor in establishment in these highly competitive grass swards is whether the proposed forbs are potentially 'ecologically fit' for the grassland in question, for example, as determined by their palatability to slugs and snails and the height and shade tolerance of their leaves.

Fitness often involves a collection of factors: how similar is the climate the species in question evolved in to that of the planting site and how similar are the soil conditions in terms of wetness–dryness and fertility. In the UK for example, this does not mean that only species native to Britain will grow well in grasslands in the UK, but rather other species can be drawn in from habitats around the world, where climatic conditions are similar to the UK, including higher altitudinal locations within warmer regions. An example of this is the mountains of Europe or the Caucasus, geographically distant but supporting many species that are well fitted to the UK. *Lychnis chalcedonica*, for example, is found in Russia in wet grasslands along streams but is very tolerant of growing in rough grass in the UK, as is *Geranium psilostemon* from northern Turkey and *Euphorbia palustris* (Hitchmough, 2009) from fens and boggy grassland in Europe. Most North American species are not very tolerant of these grasslands because as a rule they are prone to be highly palatable to slugs and snails (Hitchmough and Wagner, 2011).

Species with taller leaves that project into the light are in general much easier to establish in rough grass than are species with basal leaves which are heavily shaded by surrounding tussocky grasses. Exceptions to this are rosette-forming species with high levels of shade tolerance that grow actively during the winter months, for example *Primula veris* and *P. vulgaris*. *Geranium* species are particularly tolerant to growing in rough grass

Table 8.1. Native UK/western European forb species well suited to summer–early autumn cut hay meadows

Species	Common name	Soil preference/tolerance
Centaurea nigra	common knapweed	
Centaurea scabiosa	greater knapweed	drier
Galium verum	lady's bedstraw	
Geranium pratense	meadow cranesbill	
Hypochaeris radicata	cat's ear	
Leontodon autumnalis	autumn hawkbit	
Leucanthemum vulgare	ox-eye daisy	
Lotus corniculatus	bird's foot trefoil	
Malva moschata	musk mallow	
Primula veris	cowslip	
Prunella vulgaris	primrose	
Ranunculus acris	meadow buttercup	wetter
Sanguisorba officinalis	great burnet	wetter
Scabiosa columbaria	small scabious	
Stachys officinalis	betony	drier
Succisa pratensis	devil's-bit scabious	wetter

because of their capacity to continue to elongate their leaf petioles so that their leaves are always in the sun, even in very rank grassland.

Species that have demonstrated the capacity to compete in grasslands managed on a flail cutting (either in practice or in research) are shown in Table 8.2, with examples in Fig. 8.6.

8.7 Forb Establishment in Existing Grasslands

The gap between the competitive capacity of established grasses and young forbs (technically referred to as competitive asymmetry) can theoretically be narrowed by trying to establish forbs as container-grown plants. Historically these have often been grown in small, plug cells for this purpose, but the use of large pot-grown material, such as those in 90 mm pots, results in larger plants, which take longer to be defoliated by slugs. Larger plants too have more carbohydrate reserves to push their leaves up into the light and are to be preferred over plug-grown plants, when budgets allow. Plants that do establish from the smaller plugs tends to be those with a high degree of shade tolerance and are not particularly palatable to slugs (Davies *et al*., 1999). *Primula veris* (cowslip) is one such species, evidenced by its widespread establishment on motorway verges over the

Table 8.2. Forb species capable of growing in rough grass swards cut in late autumn or winter

Wet to moist soil types		Moist to dry soil types	
Species	Common name	Species	Common name
Euphorbia palustris	marsh spurge	*Geranium pratense*	cranesbill
Iris sibirica	Siberian iris	*Geranium psilostemon*	
Lychnis chalcedonica	Maltese cross	*Geranium* x *magnificum*	
Lythrum salicaria	purple loosestrife	*Hemerocallis* (vigorous cvs)	day lily
Persicaria bistota	bistort	*Malva moschata*	musk mallow
Primula vulgaris	primrose	*Malva alcea*	greater musk mallow
Veronicastrum virginicum	Culver's root	*Papaver orientale*	oriental poppy
		Papaver bracteatum	Iranian poppy
		Primula veris	cowslip
		Sanguisorba officinalis	great burnet

Fig. 8.6. *Iris sibirica* (Siberian iris, blue) and *Persicaria bistota* (bistort, pale pink) have gained a foothold in this damp meadow and are beginning to thrive.

past 25 years. Another species that is probably effectively established in this way is *Succisa pratensis* but there is relatively little research on this. Planting is likely to be best done in the autumn (e.g. October in the northern hemisphere) after the final hay cut or winter tidy-up mow. Soils are still warm enough to get some root establishment into the parent soil profile. Planting in spring although feasible runs the risk of the soils getting drier and the organic medium within the pot shrinking away from soil, leaving (a difficult) air gap for the young roots to bridge and ultimately resulting in plant failure.

One of the problems with planting is that it inevitably involves relatively few plants being added per unit area of grassland. Even with locally native species, existing grass swards are very heterogeneous both below and above ground in terms of the degree of competition: some plants will fail in one planting location that would succeed 50 mm away. Serendipity or to use the ecological term 'stochasticity' is important in all ecological processes but is a major issue in establishment within meadows. This competitive heterogeneity is only rarely visually evident and is a challenging idea to environmental horticulturists, who generally manipulate environments and competition to avoid the unpredictability that comes with heterogeneity. The more plants that are planted, the greater the chance some will succeed, due to the competition pressure being relaxed and hence permitting establishment in at least some locations. To achieve establishment of a wide range of species it is generally necessary to plant every year over a long period of time and to see this process as an ongoing management one rather than a one-off. This will statistically greatly increase the likelihood of positive outcomes in a process that is largely uncontrollable. As a result, this process is ideally undertaken by community group labour, supported logistically by the managing authority.

There has been some research on whether it is possible to increase successful establishment of forbs in grassy swards by providing a competition-free space (normally referred to in ecological terms as a 'gap') (Davies *et al.*, 1999) around each planted specimen. These gaps can either be made mechanically by stripping off small patches of turf or more economically by spraying out a circle or square of grass with a herbicide 8 weeks before planting. Work on trees shows that the creation of competition-free gaps is immensely effective, but this is not necessarily the pattern that is generally observed in planting forbs in grass-dominated swards. In these, providing a gap significantly improves the amount of forb biomass made in the first year, but then in the second year, when the gap generally closes due to grass reseeding or vegetative encroachment from the edge, forb biomass is either static or declines in the second year (Hitchmough, 2000, 2009). In the third year, if the forbs are (i) well fitted to the environment they are planted into and (ii) placed in a microsite which allows them to establish, they may then begin to increase their biomass. Where this is not the case, most of the planted individuals either remain as very stressed, dwarfed individuals or simply die (Davies *et al.*, 1999). This fitness for the site can often be thought of in terms of soil moisture regimes, for example, *Ranunculus acris* (meadow buttercup), *Succisa pratensis* (devil's-bit scabious) and *Lychnis flos-cuculi* (ragged robin) are species designed to compete in moist to wet soils, whereas *Salvia pratensis* (meadow clary), *Centaurea scabiosa* (greater knapweed) and *Scabiosa columbaria* (small scabious) are plants that are most competitive in dry soils. In horticultural terms, all of these species will grow satisfactorily under low levels of competition in an average garden soil, but when subjected to competitive pressure as in a meadow sward, the range of tolerance shrinks back to that defined by the evolutionary niche of the species, i.e. in the case of the examples given, wet soils or drier soils.

Fitness is also much affected by growing season air temperatures, sometimes directly, but also indirectly. *Salvia pratensis* is a continental European species adapted to warm summers, which has outlier populations in the most continental (climatically speaking) parts of the southern UK. In northern parts of the UK, it will grow seemingly satisfactorily but is often eliminated by the greater slug densities associated with cooler, moister sites. On very sandy or gravelly soils outside its natural range it may perform well due to reduced densities of molluscs, but in competitive grassy swards this is unlikely to be the case.

The statistical likelihood of overcoming the problem of serendipity, i.e. a plant finding itself in the right microsite, clearly increases with the number of individuals planted providing the species used are sufficiently fit for the meadow environment under consideration. Establishing forbs from sowing seed *in situ* potentially involves very large numbers of seedlings and is therefore at least notionally an appealing way to diversify grass-dominated swards.

To maximize the chances of success, sowing needs to be undertaken in autumn within the northern hemisphere, immediately after cutting the grass as close to the soil level as possible, and then heavily scarifying the surface to ensure seeds are in physical contact with mineral soil. Some UK meadow forbs (for example *Ranunculus acris*, meadow buttercup) are typically autumn germinators (Grime *et al.*, 1996) and will potentially germinate in the months post-sowing. Many species, however, have some type of dormancy, either mechanical (for example, hard seed coats) or physiological, which inhibits germination. In many species this dormancy is effectively overcome by sitting in cold, wet soil over winter and experiencing thawing and freezing cycles, with germination occurring mainly in March to May.

Once these seedlings germinate and emerge, however, the challenge is to reduce their loss through competition for light and water with the surrounding established grasses. The more fertile the site, the more productive will be the grasses, and the greater the mortality of the seedlings. Practices such as mowing in spring to defoliate grass and maintain light access to the soil level seem sensible practices, but because cutting may cause lateral grass shoot growth, this may actually negate many of the expected benefits (Hitchmough *et al.*, 2008). As a result, high levels of mortality are to be expected and in productive grasslands this over-sowing practice will only very slowly lead to species establishment and diversification. As with planting, over-sowing will be most successful when carried out every year, until good-sized populations of the desired species have been established. At that point, seed rain from within the swards will become much more intense, and the system becomes self-perpetuating, with no further requirement for over-sowing.

Hemiparasites

One strategy useful to establishment is to reduce the productivity of the sward by establishing the hemiparasite *Rhinanthus minor* (yellow rattle). This is an annual species which after germination attaches via haustoria onto the roots of the most common species in a grassland. In a grass-dominated meadow these will mainly be grasses; however, *Rhinanthus* does not parasitise grass alone and will attach onto forbs where these are also abundant (actually controlling overly vigorous forbs too). This is a process of statistical chance; species that are common get parasitised more because there is a greater chance of one being next to each germinating *Rhinanthus*.

Legumes like clover are often preferred as hosts (Jiang, *et al.*, 2008). *Rhinanthus* then taps into the supply of carbohydrates and water from the host, thereby reducing the growth of the host, although the amount of biomass that the *Rhinanthus* then produce is less than would have been the case if the hosts had kept their own water and carbohydrates. The net result is therefore a reduction in the standing biomass of the meadow – in effect, a reduction in total productivity as might occur with meadows on much less fertile soils. In one experiment (Hitchmough, 2014, unpublished data) *Rhinanthus* reduced dry biomass from approximately 800 g m^{-2} to around 375 g m^{-2}). So *Rhinanthus* can potentially be very helpful (Fig. 8.7). Once *Rhinanthus* is established in one part of a field, its action can be aided by manually harvesting the seed (just as the pods are about to open and the wind causes these to 'rattle') and distributing these 'fresh seeds' on recently mown areas, ensuring some seed falls far enough to make contact with the bare soil. This then allows the seed to experience a long period of chilling during the autumn (typically, in excess of 190 days is required for successful germination). There are many other hemiparasites that could be used in meadows in other parts of the world, although the evidence is that they do not reduce productivity in the same way as *Rhinanthus*. *Pedicularis* spp. have their main distribution in Western China but are found everywhere in the northern hemisphere and include many highly attractive species. In western North America, *Castilleja* is another major hemiparasite genera, again with spectacular floral structures.

8.8 Creating Meadow-like Communities from Scratch

Creating meadows from scratch has a number of advantages and disadvantages over working with existing grasslands. In contrast to working with a starting point that is almost 100% grasses, this approach has the potential to create heavily forb-dominated, even grass-free, 'grassland' within 2 years that is florally dramatic (Hitchmough, 2017; Bretzel *et al.*, 2024). Research into public attitudes to wildflower meadows confirm that the more flowers there are, and the more structurally

Fig. 8.7. *Rhinanthus minor* (yellow rattle) (left) is a hemiparasite of other plants. In the image on the right, note the brown seed pods of *Rhinanthus* in the middle, standing proud of the other plants and the more vigorous vegetation behind, where it has yet to colonise.

diverse the planting is, the more attractive the meadow (Lindemann-Matthies and Bose, 2007; Southon *et al.*, 2017). In the urban political arena, this can be used to get a broad section of the community as supporters, in contrast to the 'glacially' slow approach of working with the existing grasslands and their slow transformation. Creation from scratch allows a much more bespoke approach. This minimises the input/influence of grass (at least at the early stages), creates greater floral colour impact and allows a much greater range of species, both native and non-native, to be sown and established, successfully. This allows the meadow to be designed to flower for longer, or at times of the year when grass-dominated meadows are potentially very unattractive. This approach can also be used to better target resources such as nectar and pollen to invertebrates at times when these are in short supply.

On the negative side, creating meadows from scratch involves significant capital costs for site preparation, seed/plants, sowing and initial maintenance that is largely additional to working with an existing meadow. In addition, to be successful requires a workforce with the capacity to develop a good understanding of the process and how to deliver on this in practice. When dealing with new projects in conjunction with construction work, many of these costs will be expended in any case, whatever the landscape design envisaged, and indeed even the most ambitious sown meadows will appear to be relatively inexpensive compared to most of the alternatives, such as conventional decorative planting, roll-out turf or hard surfaces.

Choice of plant community

When we talk of meadows most people assume this refers to romantic images of purely native herbaceous plant communities in the countryside (see above). By the early 21st century, outside of a few national parks and areas of outstanding natural beauty, this is something of a bygone world in the intensive agricultural landscape of many developed countries. Designed, sown meadows offer a far more extensive range of possibilities based on either semi-natural stereotype meadows drawn from the native country or elsewhere in the world, through to completely synthetic, designed communities of native and exotic, or indeed entirely exotic, species.

Probably the main issues to bear in mind when making decisions on what type of floral grassland community to create is the environmental conditions on the site and the resources available for management. To be successful there is a fundamental need to match the ecological requirements of the proposed species for use with site conditions. If, for example, the site presented is a post-industrial site around a new building where the soils are composed of crushed concrete and rubble, then this would be an excellent site for a very dry meadow, steppe vegetation or a very dry prairie. Conversely, if the site is on very good, highly productive soil, species characterised by large biomass, such as tall leafy stem meadow species, or prairie, should be preferred. In contrast, smaller-growing, stress-tolerating species associated with dry unproductive sites will quickly be eliminated by the most vigorous species sown or planted as part of the community, and in the longer term by invading species from the outside. Responding thoughtfully to productivity and moisture gradients is the bedrock for creating sustainable meadow-like vegetation.

All meadow-like plant communities in the UK ultimately are invaded by tall, weedy, native grasses such as *Dactylis* and *Arrhenatherum*, unless active management is put in place to prevent this. The seed of these species is dispersed by small mammals and is widely distributed sooner or later. If it is anticipated that there are not the resources to prevent this process taking place, then it is best advised to use species and communities that are more tolerant of invasion by these and other native grasses. Tolerance of these grasses is best developed in forb species that have co-evolved with what are cool season C3 (spring growing) grasses, i.e. species native to maritime western Europe (Table 8.3 and Fig. 8.8). These species have deployed chemical defences to be relatively unpalatable to slugs and snails, and whilst their growth is reduced by competition for light with grasses, unlike, for example many North America prairie forbs, they are not typically eaten out by slugs as shading and humidity in the meadow increase due to greater cover of grasses.

As noted, floral meadows are introduced into urban green space as a mechanism for aesthetic enhancement and support biodiversity. More complex vegetation communities (more varied species and differences in height/plant form) are associated with increasing invertebrate biodiversity (Norton *et al.*, 2019; Fernandes *et al.*, 2023; Marshall *et al.*, 2023). However, Rust *et al.* (2024) suggested that sowing flowering species to establish new urban meadows had limited added benefits for soil, plant and pollinator diversity compared to reduced mowing that allowed resident wildflower species to establish. This may partially relate to the longevity of these plant communities before studies were conducted.

Designing a seed mix

One option is to buy 'off-the-peg' seed mixes that are readily available particularly for native species and, to a lesser degree, native and non-native species. If using the conventional native wildflower meadow products, it is strongly advised that the grass component of the seed mix is omitted (or at least reduced to below 10% of the total). Too much grass seed will result in a grass-dominated community from the outset, rather than the forb-dominated one most people want to see. If environmental horticulturists wish to produce a bespoke sowing mix, they need to develop a list of the species to include and then try to imagine how the various

Table 8.3. Species that tend to hold their own in grass-dominated plant communities in the UK

Species native to western Europe		Exotic species	
Centaurea nigra	common knapweed	*Buphthalmum salicifolium*	ox-eye
Galium mollugo	hedge bedstraw	*Euphorbia palustris*	marsh spurge
Geranium pratense	meadow cranesbill	*Malva alcea*	greater musk mallow
Knautia arvensis	field scabious	*Papaver orientale*	oriental poppy
Lythrum salicaria	purple loosestrife	*Veronicastrum virginicum*	Culver's root
Malva moschata	musk mallow	*Silene chalcedonica*	Maltese cross
Primula veris	cowslip		
Primula elatior	oxlip		
Primula vulgaris	primrose		
Sanguisorba officinalis	great burnet		

Fig. 8.8. *Primula vulgaris* (primrose) is another species that is associated with woodland but can spread into meadow communities, partially because it can tolerate the shade imposed by the taller grasses.

elements will compete with one another. Based on this, they will need to calculate approximately how many seedlings of each species they would wish to establish in each square metre. In most commercial seed mixes this is little more than a very crude guess! Hitchmough (2004) claims a more precise way is to use a spreadsheet in which the numbers of seed found in a gram weight is ascertained and correlated to estimates of likely percentage field emergence. Based on a formula integrating seed number per gram weight, typical field emergence rate and desired plant density per species, this then calculates the number of grams of seed for each square metre to be sown within the mix.

Examples of seed sowing rates with respect to the desired number of plants per square metre are given for a number of different communities/environmental scenarios (Tables 8.4–8.8).

Table 8.4. Native meadow seed mixes used for species-rich lawn/spring meadow community at the London Olympic Park (2012) (Hitchmough, 2016, personal communication.)

	Target plants (number m^{-2})	Weight of seed required (g m^{-2})
Grasses		
Agrostis vinealis	50	0.011
Agrostis capillaris	50	0.011
Anthoxanthum odoratum	30	0.067
Carex flacca	20	0.111
Cynosurus cristatus	50	0.104
Festuca rubra	50	0.167
Lolium perenne	50	0.333
Trisetum flavescens	50	0.067
Total grasses	350	0.871
Forbs		
Achillea millefolium	50	0.042
Bellis perennis	50	0.031
Cardamine pratensis	50	0.123
Galium verum	50	0.132
Hypochaeris radicata	50	0.208
Leontodon autumnalis	50	0.179
Leontodon hispidus	50	0.278
Lotus corniculatus	50	0.500
Plantago lanceolata	50	0.125
Primula veris	50	0.263
Prunella vulgaris	50	0.250
Ranunculus bulbosus	50	0.952
Thymus polytrichus	50	0.125
Trifolium pratensis	50	0.333
Total forbs	700	3.542

Site preparation for sowing

A critical factor for a successful flower meadow is the elimination of weed pressure from the outset. Sowing seed onto land where there are viable fragments of vegetative weeds and/or a numerically huge bank of viable weed seeds will lead to failure and disappointment. Thus, control of perennial standing weeds should be undertaken at least 4 months before the proposed autumn sowing. The traditional way to deal with a potential weed pressure would be to spray the weeds or grasses currently occupying the site using two applications of glyphosate with any regrowth of weed seeds being eliminated by the second application. Where concerns over glyphosate are voiced, other chemical means are now being considered, including the use of steam, foams, organic oils and acids, as well as other herbicide formulations (Fogliatto *et al.*, 2020; Hudek *et al.*, 2021). Where glyphosate is used then the objective would be to minimise soil cultivation afterwards – as this reactivates the dormant weed seed bank. To gain more reliable establishment of species, it is necessary to switch off the germination of these weed seeds. Interestingly and in contrast, actual frequent repeat cultivation of the soil, when glyphosate is not used, is a conventional, non-chemical way of exhausting this seed bank. The repeat use of tractor-led cultivators and tines, based on petrochemical energy is not particularly sustainable and environmentally friendly though!

'Switching off' the seed bank is most readily achieved by spreading a 75–100 mm layer of sowing mulch, some material that does not itself

Table 8.5. Seed mixes used for hay meadow for *moister* slopes at the London Olympic Park (2012) (Hitchmough, 2016, personal communication.)

	Target plants (number m^{-2})	Weight of seed required ($g\ m^{-2}$)
Forbs		
Achillea millefolium	5	0.003
Agrimonia eupatoria	1	0.200
Betonica officinalis	10	0.250
Centaurea scabiosa	3	0.133
Deschampsia cespitosa	5	0.005
Festuca ovina	20	0.077
Galium mollugo	5	0.016
Galium verum	15	0.036
Geranium pratense	5	0.303
Geranium sanguineum	3	0.200
Knautia arvense	5	0.083
Leucanthemum vulgare	10	0.017
Linaria vulgaris	10	0.006
Malva moschata	5	0.125
Origanum vulgare	15	0.100
Primula veris	15	0.079
Prunella vulgaris	10	0.042
Ranunculus acris	10	0.250
Sanguisorba officinalis	5	0.179
Succisa pratense	5	0.238
Trifolium pratense	1	0.006
Total forbs	163	2.348

Table 8.6. Seed mixes used for hay meadow for *drier* slopes at the London Olympic Park (2012) (Hitchmough, 2016, personal communication.)

	Target plants (number m^{-2})	Weight of seed required ($g\ m^{-2}$)
Forbs		
Calamintha nepeta	10	0.01
Campanula glomerata	10	0.01
Centaurea scabiosa	10	0.44
Daucus carota	10	0.05
Echium vulgare	5	0.08
Festuca ovina	10	0.04
Galium verum	20	0.05
Leontodon hispidus	10	0.11
Leucanthemum vulgare	10	0.02
Linaria vulgaris	5	0.00
Lotus corniculatus	5	0.03
Malva moschata	5	0.13
Origanum vulgare	20	0.01
Primula veris	20	0.14
Prunella vulgaris	10	0.04
Salvia pratense	5	0.05
Scabiosa columbaria	20	0.29
Thymus polytrichus	20	0.04
Total forbs	205	1.56

contain weed seeds, on top of the existing soil surface. This layer inhibits germination from the soil seed bank below by reducing diurnal temperature fluctuation and increasing carbon dioxide (CO_2) levels, but also by causing the death of many small-seeded weeds as they try to emerge through the layer to reach the light – they run out of carbohydrates and other resources. Any mulch material that is relatively low in nutrients and which can store some water and supply this to germinating seeds can be used. Choice comes down to cost, availability and its capacity to supply moisture to the germinating seeds of the flowering species. Two materials that are particularly widely available in urban areas are sand and composted green waste. Sand is heavy to move around but it creates a very well-drained surface which on a large scale is hostile to slugs, and it also reduces productivity to some degree. Green compost (composted green waste) is lighter but depending on its carbon-to-nitrogen (C:N) ratio may increase productivity. With sand, irrigation is required to ensure good germination in the absence of substantial and regular spring rain, whereas germination on compost is normally more reliable without irrigation. The flip side to compost is that it is a great receptor site for the germination of blown-in weeds during the first year. Where it is possible to do so, and particularly on sites where visual expectations of the meadow vegetation will be very high, temporary irrigation between March and June is essential for success. It is the combination of severe spring moisture stress plus competition with weeds from the soil seed banks that causes poor performance in many sown perennial meadows.

Sowing practice

Distribution of forb seed on the soil should be as uniform as possible. Where bare patches occur, these are likely to be colonised by weeds very rapidly. Conversely, areas that are over-sown, with high seedling mergence, lead to competitive elimination of sown seedlings by a process known as self-thinning (Hitchmough *et al.*, 2004).

Given the substantial difference in the physical size of seeds in sowing mixes (from 50 to 20,000 seeds/g), in most cases the most successful way to establish meadows is by hand sowing. This sounds

Table 8.7. USA Prairie seed mix used at Oxford Botanic Gardens (Hitchmough, 2016, personal communication.)

	Target plants (number m^{-2})	Weight of seed required (g m^{-2})
Forbs		
Agastache rupestris	1	0.0056
Amorpha canescens	1	0.0101
Asclepias tuberosa interior form	5	0.1563
Aster oblongifolius	1	0.0023
Echinacea pallida	2	0.0400
Echinacea paradoxa	5	0.1250
Echinacea purpurea cv. Prairie Splendor	4	0.0533
Erigeron glaucus cv. Albus	2	0.0027
Eryngium yuccifolium	2	0.0294
Geum triflorum	5	0.0196
Helianthella quinquenervis	1	0.0011
Liatris aspera	5	0.0455
Liatris scariosa cv. Album	5	0.0549
Mirablis multiflora	1	0.0114
Oenothera macrocarpa var. *incana*	5	0.0980
Penstemon barbatus cv. Coccineus	4	0.0211
Penstemon cobaea	3	0.0364
Phlox pilosa	5	0.0490
Rudbeckia maxima	0.25	0.0091
Rudbeckia missouriensis	3	0.0100
Ruellia humilis	1.5	0.0556
Silphium laciniatum	0.33	0.0314
Silphium terebinthinaceum	0.33	0.0413
Solidago speciosa	2	0.0016
*Stokesia laevis*cv. Omega Skyrocket	1.5	0.0682
Total forbs	65.91	0.9789

Table 8.8. Steppe-like mix used at Oxford Botanic Gardens (Hitchmough, 2016, personal communication.)

	Target plants (number m^{-2})	Weight of seed required (g m^{-2})
Allium senescens	8	0.1778
Aster oblongifolius	0.2	0.0005
Astragulus centralpinus	0.5	0.0431
Campanula persicifolia cv. Grandiflora	2	0.0013
Dianthus carthusianorum	2	0.0080
Dianthus carthusianorum cv. Ruperts Pink	5	0.0227
Dracocephalum argunense cv. Fuji Blue	2	0.0400
Echinops ritro	1	0.1111
Eryngium maritimum	1	0.1667
Eryngium planum cv. *Blaukappe*	1	0.0147
Euphorbia polychroma	0.5	0.0172
Euphorbia nicaensis	3	0.0606
Galium verum	3	0.0048
Hyssopus officinalis var. *aristatus*	3	0.0141
Incarvillea delavayi cv. Bees Pink	1	0.0333
Incarvillea zhongdianensis	1	0.0250
Inula ensifolia	3	0.0125
Laserpitum siler	0.2	0.0222
Limonium latifolium	3	0.0150
Linum narbonense	3	0.0791
Malva alcea cv. Fastigiata	0.33	0.0047
Marrubium supinum	1	0.0070
Papaver orientale cv. Brilliant	1	0.0011
Paradisea lusitanica	1	0.0500
Perovskia atriplicifolia	1	0.0061
Pulsatilla vulgaris	2	0.0333
Salvia nemorosa cv. Blaukonigen	1	0.0056
Scabiosa lachnophyllya	2	0.0152
Scabiosa ochroleuca cv. Moon Dance	3	0.0182
Scutellaria baicalensis	3	0.0174
Sedum telephium cv. Emperor's Waves	5	0.0020
Silene schafta cv. Persian carpet	5	0.0095
Teucrium chamaedrys	2	0.0182
Veronica incana	5	0.0018
Total forbs	75.73	1.0598

very slow but a team of one sower and two seed mixers can typically sow 2–3000 m^2 per day. Given that sowing rates are generally (forbs only) about 1–1.5 g m^{-2}, it is necessary to sow with a bulking carrier. The best material for this (because it is light and very visible on a sand surface) is sawdust, although, unless there is a sawmill close by, this is sometimes challenging to procure in volume. Sowing in two walking passes tends to produce the required uniformity. Thus here, the seed calculated to be required for the area to be sown is divided in half. For inexperienced sowers, three string lines at metre intervals are laid across the long axis of the site and the sower carrying a bucket of seed and sawdust walks down the lanes created sowing the seed with a swinging arm action. The entire area is sown in this way, moving the string lines as required across the site. The process is then repeated as a second pass with the remaining half of the seed. The seed is then raked into the sand with a landscape rake. Where disturbance by local wildlife or domestic animals is anticipated (e.g. foxes or cats) and no irrigation is possible, covering the sown area with an open-weave jute erosion matting appears to improve germination but adds about £1.00 m^{-2} (US$1.20 m^{-2}) to the cost of the process, so it is most likely to be used on prestige, relatively small-scale sites.

Post-sowing maintenance

Species have different temperature requirements for germination, so germination is staggered across quite a long period of time. Woodland species often germinate at very low temperatures; *Primula elatior* (oxlip) and *P. vulgaris* (primrose), for example, may germinate in February under UK conditions, whereas heat-demanding prairie species such as *Asclepias tuberosa* (butterfly weed) do not emerge until June. In the main, however, seedlings of most species emerge in April and May. The main factor that maximises field emergence (the percentage of seed sown that results in a seedling) is freedom from soil moisture stress (Rózová *et al.*, 2023); hence, irrigation during April and May has a very large effect on 'success'. Irrigation should only involve relatively small amounts at each irrigation event, enough to saturate the sowing mulch but not much more, i.e. to avoid encouraging the weed seed below the mulch. Even with sand or compost sowing mulches, some weeds will emerge, particularly where the spread layers are too thin (layers shallower than 50 mm are not very effective at preventing weed emergence) or where sand/compost has been stored in builders' yards surrounded by weedy vegetation that distributes its seeds onto the heaps prior to sale. Also, if irrigation is excessive in terms of applying large amounts frequently, this will de-oxygenate the soil beneath the mulch and this leads to massive worm casting and the deposition of weed seeds from the soil beneath on the surface of the sowing mulch.

Roguing out weeds is very valuable, especially if this is combined with the establishment of the sown vegetation by the end of the first growing season and where this canopy has just about closed over. If at this point the landscape is almost free of major weeds, then typically, the future is very promising as the balance in the community is very much in favour of sown as opposed to invading species. Where weeds dominate, the long-term prognosis is generally poor, although in meadow vegetation types that are cut as hay in summer, this process is very helpful in checking many common weeds. Roguing out weeds is generally best left to about June–July, as by then the difference in the vigour of the weeds and sown species is obvious in most cases. Where staff have limited ability to differentiate between weed and ornamental plant seedlings, then a useful process is to 'arm' these staff with 'thumbnail' photographic images (laminated on an A4 card which the weeders hang around their necks on a lanyard) that can be used as a quick identification guide (Hitchmough, 2016, personal communication). This greatly helps identification and reduces the risk of weeding out the sown species. Sowing some of the seeds in small, labelled pots helps maintenance staff become familiar with the seedlings too, a vital part of the training process.

With hay meadow-type vegetation, mowing off at 75 mm can be used to reduce weed competition in the first year and is particularly useful with annual weeds. With prairie vegetation types that are less tolerant of cutting during the growing season, this process is potentially less beneficial as regrowth of the prairie species post-cutting is slower. Prairie-like vegetation typically requires more input during the first year to establish successfully. Over the longer term, leaving some areas of the meadow uncut acts as a haven for some invertebrates and allows a carry-over of populations from one year to the next. For example, studies have investigated how delaying mowing and leaving uncut refuges benefit certain insect populations in extensively managed meadows (Révész *et al.*, 2024) (Fig. 8.9).

Fig. 8.9. Leaving patches of grassland uncut allows invertebrate eggs, larvae or pupae to survive from one year to the next without disturbance – thus, allowing the populations to maintain their presence in the grassland. *Zygaena filipendulae* (six-spot burnet moth) on *Centaurea scabiosa* (greater knapweed).

Conclusions

- Grasslands and forb-rich meadow communities generally predominate where abiotic or biotic factors inhibit the establishment of taller, woody vegetation. Such factors include temperature and moisture extremes but also fire, grazing pressure or mowing (haymaking).
- These types of communities are found as native plant communities in many parts of the world, and they often contain a high proportion of highly attractive flowering species that can be used to support biodiversity and provide exciting visual experiences for people living in urban areas.
- In urban areas, a move away from close-mown turf grass to 'meadow-like' communities has a number of advantages in terms of energy use, labour cost and urban biodiversity. Developing an iconic, species-rich, colourful 'traditional meadow' with consistent performance from one year to the next can be challenging, however. A useful compromise is to introduce spring-flowering geophytes or limit the floral contribution to the more vigorous forbs such as *Taraxacum* (dandelion) and *Ranunculus* (buttercup) spp. through altered management of the sward, e.g. timing and frequency of cutting.
- Due to eutrophication increasing soil productivity, in most urban grasslands, forb density and diversity is low and very slow to develop even with sensitive management. Most desirable species are often locally extinct and hence there is no seed of these species being transported to these potential grassy habitats. Similarly, the chances of highly floral, meadow dicot species regenerating from the seed bank are low.
- Flowering forbs can be introduced to grass-dominated swards, but the chances of successful establishment are strongly determined by the vigour of the grass species present (e.g. as determined by nutrient/moisture levels), the size of transplant (larger plants being more resilient than plugs or seed for example), shade and mollusc tolerance of the introduced species and the persistence/patience of the management team to introduce plants over a number of years to try and build up a viable population.
- In many cases, a more realistic strategy is to create meadows by sowing from scratch, as a means of speeding up the developmental process. Although this requires a larger input of capital resources at the outset, it allows low-productivity substrates such as crushed building materials or subsoil to be used and thus change the site to allow a much wider range of species to persist and thrive.
- To develop a community of colourful flowering forbs, the key is choosing species 'ecologically fit' for the site. This includes using species adapted to similar climates, soil fertility and moisture regimes, and in many situations, using species that also possess some tolerance to mollusc grazing pressure and the ability to compete with grasses and other taller-growing forbs for light.
- A combination of severe spring moisture stress plus competition with weeds from the soil seed banks often causes poor performance in many sown perennial meadows.

References

Baldock, K.C., Goddard, M.A., Hicks, D.M., Kunin, W.E., Mitschunas, N. *et al.* (2019) A systems approach reveals urban pollinator hotspots and conservation opportunities. *Nature Ecology and Evolution* 3, 363–373.

Bretzel, F., Vannucchi, F., Romano, D., Malorgio, F., Benvenuti, S. and Pezzarossa, B. (2016) Wildflowers: From conserving biodiversity to urban greening - A review. *Urban Forestry and Urban Greening* 20, 428–436.

Bretzel, F., Vannucchi, F., Pezzarossa, B., Paraskevopoulou, A. and Romano, D. (2024) Establishing wildflower meadows in anthropogenic landscapes. *Frontiers in Horticulture* 2, 1248785.

Cerra, J.F. (2017) Emerging strategies for voluntary urban ecological stewardship on private property. *Landscape and Urban Planning* 157, 586–597.

Chollet, S., Brabant, C., Tessier, S. and Jung, V. (2018) From urban lawns to urban meadows: Reduction of mowing frequency increases plant taxonomic, functional and phylogenetic diversity. *Landscape and Urban Planning* 180, 121–124.

Cobham, R. (1990) *Amenity Landscape Management, A Resources Handbook*. E & F.N. Spon Ltd., London, UK.

Davies, A., Dunnett, N.P. and Kendle, T. (1999) The importance of transplant size and gap width in the botanical enrichment of species-poor grassland in Britain. *Restoration Ecology* 7, 271–280.

Del Toro, I. and Ribbons, R.R. (2020) No Mow May lawns have higher pollinator richness and abundances: An engaged community provides floral resources for pollinators. *PeerJ* 8, e10021.

Fernandes, M.P., Matono, P., Almeida, E., Pinto-Cruz, C. and Belo, A.D. (2023) Sowing wildflower meadows in Mediterranean peri-urban green areas to promote grassland diversity. *Frontiers in Ecology and Evolution* 11, 1112596.

Fischer, L.K., Neuenkamp, L., Lampinen, J., Tuomi, M., Alday, J.G., Bucharova, A., Cancellieri, L., Casado-Arzuaga, I., Čeplová, N., Cerveró, L. and Deák, B. (2020) Public attitudes toward biodiversity-friendly greenspace management in Europe. *Conservation Letters* 13, e12718.

Fogliatto, S., Ferrero, A. and Vidotto, F. (2020) Current and future scenarios of glyphosate use in Europe: Are there alternatives? *Advances in Agronomy* 163, 219–278.

Francoeur, X.W., Dagenais, D., Paquette, A., Dupras, J. and Messier, C. (2021) Complexifying the urban lawn improves heat mitigation and arthropod biodiversity. *Urban Forestry and Urban Greening* 60, 127007.

Fuentes, T.L. (2021) Homeowner preferences drive lawn care practices and species diversity patterns in new lawn floras. *Journal of Urban Ecology* 7, juab015.

Gillman, L.N., Bollard, B. and Leuzinger, S. (2023) Calling time on the imperial lawn and the imperative for greenhouse gas mitigation. *Global Sustainability* 6, e3

Grime, J.P., Hodgson, J. and Hunt, R. (1996) *Comparative Plant Ecology, A Functional Approach to Common British Species*. Chapman and Hall, London. UK.

Herbert-Bormann, F., Balmori, D. and Gebale, G.T. (2001) *Redesigning the American Lawn: A Search for Environmental Harmony*, Yale University Press, Yale, Connecticut, USA.

Hitchmough, J.D. (2000) Establishment of cultivated herbaceous perennials in purpose sown native wildflower meadows in south west Scotland. *Landscape and Urban Planning* 51, 37–51.

Hitchmough, J.D. (2004) Naturalistic herbaceous vegetation for urban landscapes. In: Dunnet, N. and Hitchmough, J.D. (eds). *The Dynamic Landscape, Design, Ecology and Management of Naturalistic Urban Planting*. Taylor and Francis, London. UK.

Hitchmough, J.D. (2009) Diversification of grassland in urban greenspace with planted, nursery-grown forbs. *Journal of Landscape Architecture* 4, 16–27.

Hitchmough, J. (2017) *Sowing Beauty: Designing Flowering Meadows from Seed*. Timber Press, Portland, Oregon, USA.

Hitchmough, J. and De la Fleur, M. (2006) Establishing North American prairie vegetation in urban parks in northern England: Effect of management and soil type on long-term community development. *Landscape and Urban Planning* 78, 386–397.

Hitchmough, J.D. and Wagner, M. (2011) Slug grazing effects on seedling and adult life stages of North American prairie plants used in designed urban landscapes. *Urban Ecosystems* 14, 279–302.

Hitchmough, J.D., De La Fleur, M. and Findlay, C. (2004) Establishing North American prairie vegetation in urban parks in northern England: 1. Effect of sowing season, sowing rate and soil type. *Landscape and Urban Planning* 66, 75–90.

Hitchmough, J.D., Paraskevopoulou, A. and Dunnett, N. (2008) Influence of grass suppression and sowing rate on the establishment and persistence of forb dominated urban meadows. *Urban Ecosystems* 11, 33–44.

Hudek, L., Enez, A. and Bräu, L. (2021) Comparative analyses of glyphosate alternative weed management strategies on plant coverage, soil and soil biota. *Sustainability* 13, 11454.

Humbert, J.Y., Ghazoul, J., Richner, N. and Walter, T. (2012) Uncut grass refuges mitigate the impact of mechanical meadow harvesting on orthopterans. *Biological Conservation* 152, 96–101.

Ignatieva, M. and Ahrné, K. (2013) Biodiverse green infrastructure for the 21st century: from "green desert" of lawns to biophilic cities. *Journal of Architecture and Urbanism* 37, 1–9.

Ignatieva, M., Ahrné, K., Wissman, J., Eriksson, T., Tidåker, P. *et al.* (2015) Lawn as a cultural and ecological phenomenon: A conceptual framework for transdisciplinary research. *Urban Forestry and Urban Greening* 14, 383–387.

Jiang, F.W., Dieter Jeschke, W.D., Hartung, W. and Cameron, D.C. (2008) Does legume nitrogen fixation underpin host quality for the hemiparasitic plant *Rhinanthus minor*? *Journal of Experimental Botany* 59, 917–925.

Junge, X., Jacot K.A., Bosshard, A. and Lindemann-Matthies, P. (2009) Swiss people's attitudes towards field margins for biodiversity conservation. *Journal of Nature Conservation* 17, 150–159.

Köppler, M.R., Kowarik, I., Kühn, N. and von der Lippe, M. (2014) Enhancing wasteland vegetation by adding ornamentals: Opportunities and constraints for establishing steppe and prairie species on urban demolition sites. *Landscape and Urban Planning* 126, 1–9.

Lindemann-Matthies, P. and Bose, E. (2007) Species richness, structural diversity and species composition in meadows created by visitors of a botanical garden in Switzerland. *Landscape and Urban Planning* 79, 298–307.

Marrs, R.H. (1985) Techniques for reducing soil fertility for nature conservation purposes: a review in relation to research at Roper's Heath, Suffolk, England. *Biological Conservation* 34, 307–332.

Marshall, C.A., Wilkinson, M.T., Hadfield, P.M., Rogers, S.M., Shanklin, J.D. *et al.* (2023) Urban wildflower meadow planting for biodiversity, climate and society: An evaluation at King's College, Cambridge. *Ecological Solutions and Evidence* 4, e12243.

Monteiro, J.A. (2017) Ecosystem services from turfgrass landscapes. *Urban Forestry and Urban Greening* 26, 151–157.

Nawrocki, A., Popek, R., Sikorski, P., Wińska-Krysiak, M., Zhu, C.Y. and Przybysz, A. (2023) Air phyto-cleaning by an urban meadow – Filling the winter gap. *Ecological Indicators* 151, 110259.

Norton, B.A., Bending, G.D., Clark, R., Corstanje, R., Dunnett, N. *et al.* (2019) Urban meadows as an alternative to short mown grassland: Effects of composition and height on biodiversity. *Ecological Applications* 29, e01946.

O'Reagain P.J. (1989) Leaf silification in grasses; A review. *Journal of the Grassland Society of South Africa* 6, 37–42.

Paudel, S. and States, S.L. (2023) Urban green spaces and sustainability: Exploring the ecosystem services and disservices of grassy lawns versus floral meadows. *Urban Forestry and Urban Greening* 84, 127932.

Plantlife (2025) No Mow May. Available at: https://www.plantlife.org.uk/campaigns/nomowmay/ (accessed 23 January 2025).

Przybysz, A., Popek, R., Stankiewicz-Kosyl, M., Zhu, C.Y., Małecka-Przybysz, M. *et al.* (2021) Where trees cannot grow–Particulate matter accumulation by urban meadows. *Science of the Total Environment* 785, 147310.

Révész, K., Gallé, R., Humbert, J.Y. and Batáry, P. (2024) Effects of uncut refuge management on grassland arthropods–A systematic review. *Global Ecology and Conservation* 57, e03381.

Robbins, P. and Birkenholtz, T. (2003) Turfgrass revolution: measuring the expansion of the American lawn. *Land Use Policy* 20, 181–194.

Rózová, Z., Pástorová, A. and Kuczman, G. (2023) Development of flower meadows in an urbanized environment. *Ekológia (Bratislava)* 42, 218–229.

Rust, W., Sotkewicz, M., Li, Z., Mercer, T. and Johnston, A.S. (2024) Soil–plant–pollinator relationships in urban grass and meadow habitats: Competing benefits and demands of tall flowering plants on soil and pollinator diversity. *Diversity* 16, 354.

Sehrt, M., Bossdorf, O., Freitag, M. and Bucharova, A. (2020) Less is more! Rapid increase in plant species richness after reduced mowing in urban grasslands. *Basic and Applied Ecology* 42, 47–53.

Smetana, S.M. and Crittenden J.C. (2014) Sustainable plants in urban parks: A life cycle analysis of traditional and alternative lawns in Georgia, USA. *Landscape and Urban Planning* 122, 140–151.

Smith, L.S. and Fellowes, M.D. (2015) The influence of plant species number on productivity, ground coverage and floral performance in grass-free lawns. *Landscape and Ecological Engineering* 11, 249–257.

Smith, C. and Hilaire, R.S. (1999) Xeriscaping in the urban environment. *New Mexico Journal of Science* 241–241.

Smith, L.S., Broyles, M.E.J., Larzleer, H.K. and Fellowes, M.D.E. (2015) Adding ecological value to the urban lawnscape. Insect abundance and diversity in grass-free lawns. *Biodiversity and Conservation* 24, 47–52.

Southon, G.E., Jorgensen, A., Dunnett, N., Hoyle, H. and Evans, K.L. (2017) Biodiverse perennial meadows have aesthetic value and increase residents' perceptions of site quality in urban green-space. *Landscape and Urban Planning* 158, 105–118.

Tilman, D., Reich, P.B. and Knops, J.M. (2006) Biodiversity and ecosystem stability in a decade-long grassland experiment. *Nature* 441, 629–632.

Townsend-Small, A. and Czimczik C.I. (2010) Correction to "carbon sequestration and greenhouse gas emissions in urban turf," *Geophysical Research Letters* 37, L06707.

Wang, R., Mattox, C.M., Phillips, C.L. and Kowalewski, A.R. (2022) Carbon sequestration in turfgrass – soil systems. *Plants* 11, 2478.

Wells T.C.E. (1980) Management options for lowland grassland. In: Rorison, I.H. and Hunt, R. (eds), *Amenity Grasslands: An Ecological Perspective*. Wiley, Chichester, UK.

Wolfin, J., Watkins, E., Lane, I., Portman, Z.M. and Spivak, M. (2023) Floral enhancement of turfgrass lawns benefits wild bees and honey bees (*Apis mellifera*). *Urban Ecosystems* 26, 361–375.

Yang, F., Ignatieva, M., Larsson, A., Zhang, S. and Ni, N. (2020) Public perceptions and preferences regarding lawns and their alternatives in China: A case study of Xi'an. *Urban Forestry and Urban Greening* 53, 126703.

9 Lawns and Sports Turf

Abstract

Lawns and sports turf are composed of short-stature, mown grass. Grass plants grow from their base which means that lawns can be cut frequently and kept short, without killing the plants. Thus, lawn grasses are conducive to be walked and played on. Sport turfgrass is the most frequent surface used in competitive sports and remains popular due to its cushioning action. Grass has extensive use too in non-formalised sport, recreational activities and landscape aesthetics. Indeed, it is considered that lawn/sports grass covers 50% of all urban green space. Turfgrasses are divided into 'cool' and 'warm' season grasses based on their ecological backgrounds and heat/drought tolerances. The purpose of the turfgrass determines the level of management. Elite sports turf is intensively managed, golf course greens being cut daily and having high nutritional demands, whereas grassed areas of parkland may only be cut monthly and never be fertilised or irrigated.

9.1 Introduction

Turfgrass covers a range of sward types, largely defined by the height of the grass and the species composition. These include the highly managed 'elite' sports turfs which epitomise golf, bowling and tennis as well as the close-mown 'fine' lawns of private residences and heritage gardens. But utilitarian public park swards and green space as well as 'rougher', less manicured turf used on roadside verges and banks are also 'managed' areas, albeit perhaps less intensively so. Once the sward is allowed to grow above a certain height, and broadleaf plant species become a significant component of the community, then these grass areas tend to be referred to as meadows (see Chapter 7). There is also commonality with pastures and 'rangeland' where the primary purpose of the sward is to provide grazing for livestock. It is perhaps the intensively managed, short grass swards, popular due to their aesthetics and consistency as sports surfaces, that represent the most challenged form of green space as far as sustainable environmental management is concerned.

The role of turfgrass within horticulture tends to be divided up into three categories, namely functional, recreational or for aesthetic reasons. Functional aspects include providing a walkable surface where there is an intermediate level or low frequency of pedestrian traffic (Fig. 9.1). Grassed areas are employed as temporary car parks, camping sites or for hosting fetes, fairs and other non-permanent activities. Aircraft runways that accommodate light aircraft can be composed of turf, especially in rural areas where landings may be infrequent. Grass verges provide an accessible and relatively safe zone at the side of roads or buildings. Such roadside verges may help trap dust and deactivate vehicle emissions and other pollutants. Grass perimeters around airports are thought to prolong aircraft engine life due to reducing the incidence of dust. Turf surfaces act as a catchment and filter out water-borne pollutants, including heavy metals, and both the activities of the plants themselves and the associated rhizosphere microflora can deactivate organic pollutants, including pesticides and light hydrocarbons. A mantle of turf aids soil to resist wind and water erosion and helps stabilise steep slopes or waterside embankments. Soil covered with turf (*Schedonorus arundinaceus,* tall fescue) was shown to reduce soil erosion (10–62 kg ha^{-1}) compared to bare soil alone (22 kg ha^{-1}) during a 30-minute intensive rainstorm (Beard and Green, 1994). Damaged, compacted or contaminated soils are improved though the action of grass, with root activity promoting fissures and pores that aid drainage and aeration. Organic matter released from root mucilage and leaf clippings encourages the formation of a favourable, 'crumb-like', soil structure.

DOI: 10.1079/9781800621763.0009

Fig. 9.1. A key aspect of grass swards is that they can be walked and played on. Village greens are functional as well as an important aesthetic complementing the surrounding built infrastructure.

A cover of turfgrass promotes the activity of earthworms and other invertebrates. Their burrowing habits, in turn, improve the drainage capacity of the sward, and by converting leaf litter into organic humus they enhance soil structure and fertility. Grass alters its surrounding microclimate, with lawns and meadows functioning as natural cooling/humidifying agents. Last but not least, functional benefits include direct economic impact, with the growing/retailing of turf and grass lawn seed being significant industries, augmenting the labour and employment centred around turf management.

Perhaps it is the recreational aspects associated with turf that are most in the public eye. Despite the advent of synthetic sports-playing surfaces, grass is the most frequent surface used in competitive sports (with sports turf management being a discrete professional discipline in its own right). Grass remains popular for sport, largely through its capacity to provide a cushioning, yet playable, sports surface. The cushioning properties help minimise injury to the athletes during tackling, sliding, falling, etc. It also facilitates an ideal (or unique!) playing surface for some sports. For example with tennis, the speed of the ball off the playing surface is accelerated when compared to alternatives such as clay (crushed brick or stone) or synthetic carpets. In golf, grass is the only feasible playing surface in temperate climates, due to the scale of the area involved.

Grass has extensive use in non-formalised sport, through its predominance in parks, children's play areas and domestic gardens. Here it also provides the forum for relaxation and social interaction through a wide spectrum of activities, spanning events from school fetes to family picnics.

9.2 Role of Turf in the Landscape

Within the landscape, lawns provide open space and opportunities for freedom of direction and movement. They act as an aesthetic foil to other key features including commonly buildings, trees, floral beds, pathways and water features. Indeed, humans seem to have a preference for landscapes where one-third of the area is composed of vertical structure and two-thirds is devoted to open space, and grass is invariably the preferred choice for the horizontal plane, open space component. Without the dominant role of grass many famous gardens and landscapes would have a claustrophobic or cluttered feel to them. They would lose their 'grand effect', namely the scale of the landscape that extensive lawns provide – as typified by Lancelot

'Capability' Brown through the 'English Landscape Style'. Similarly, many iconic landscape views and vistas would not exist if the open ground that the grass cover affords were to be lost. The capacity of grass to provide a sense of space and place is critical, e.g. the grass swards of the Champ de Mars act as a green 'royal carpet', progressing up to the feet of the Eiffel Tower in Paris, France.

Turf has been linked to a range of ecosystem services (Monteiro, 2017; Chawla *et al.,* 2018). Over and above its recreational, aesthetic and sport value, these include reducing rainwater runoff, increasing infiltration, purifying water from sediments, pollutants and immobilizing nutrients, controlling soil erosion and trapping dust, improving soil quality and reducing fire hazards. Close-mown turf, especially if moist, can act as a firebreak, stopping the propagation of fire across the landscape. For house owners, lawns enhance property aesthetics, provide increments in property value and offer a site for recreational activity.

Ornamental and sports turf is thought to cover approximately 50% of urban green space and in global terms would equate to an area equivalent to the land mass of Spain and England combined (1.4% of total grasslands) (Wiewióra and Żurek, 2023). In the USA, turfgrass covers a substantial amount of land, encompassing approximately 20 million ha and represents a real estate value of US$40 billion. This is largely due to the 80% (85 million) of households that participate in outdoor lawn or garden activities. To put this into context, amenity turfgrass is estimated to cover an area three times larger than any other irrigated crop type in the USA (Milesi *et al.*, 2005). Turf management varies in its intensity, and there are large areas of urban grass that are rarely, if ever, irrigated or fertilised. In contrast, some elite sports turfs need the almost daily attention of mowing, and frequent nutrient applications to meet 'conventional notions' of quality turf. Such notions may include the need for the sward to possess a deep green hue, i.e. arguably relatively superficial aspects, over and above their need to be the ideal playing surface. In terms of sports turf, golf courses are the heartbeat of the turfgrass industry worldwide and comprise 40% of total turfgrass, chemicals and services use (Fig. 9.2) (Chawla *et al.,* 2018).

It is partly due to the extensive use of turf and some lawns/sports surfaces being so intensively managed that has led to turf itself becoming a 'battlefield' between those wishing to promote its benefits and its severest critics (a real case of 'turf wars' – if ever there was one!). Those in favour cite its capacity to provide a strong aesthetic and to be integral to a quality recreational experience (formal sport or otherwise), while those against argue that the management of such 'green deserts' is non-sustainable and contributes to the global depletion of natural resources, including oil, minerals, aggregates and clean water. Indeed, for some, 'the lawn' has become a metaphor for society's lack of environmental awareness and our ability to exploit dwindling or increasingly costly resources, for largely aesthetic reasons. These conflicts are perhaps best illustrated by the numerous studies showing the advantages of turfgrass in terms of its carbon sequestration potential (e.g. Raciti *et al.*, 2011) and the disadvantages with respect to carbon dioxide and other greenhouse gas emissions associated with its traditional management (e.g. Townsend-Small and Czimczik, 2010a, 2010b) or its establishment

Fig. 9.2. The science of turf management is most advanced within the context of elite sports facilities. Golf has been instrumental in the development of new grass cultivars and optimising the management of turf. Note the differential sward height between the golf 'fairway' (bottom right) and the 'green' (centre and left).

in inappropriate locations (Selhorst and Lal, 2013). In a review of lifecycle analyses for different ground-cover types in Georgia, USA, Smetana and Crittenden (2014) suggested traditional lawn turf (composed of non-native species and intensively managed) performed poorly compared to planting with native prairie grasses. Other studies suggest the wider environmental, social and health aspects need to be considered too; for example, some of the arguments for and against lawns and intensively managed sports turf are summarised in Table 9.1.

9.3 Grass Genotypes – Physiology and Traits

So why is the grass sward such a predominate part of the horticultural landscape? The key is in the nature of grass itself, in that the growing points (intercalary meristems) occur in the stem nodes, many of which are located at the base of stem, allowing stems and associated leaves to regrow readily after being cut or grazed. This ability to regenerate means that a low-stature sward can be maintained perpetually allowing the lawn to be a walkable (or playable) surface. Turfgrasses spread either by developing daughter plants close to the base of the mother plant – in a process known as tillering – or through initiating lateral stems which can grow above ground (stolons) or below ground (rhizomes). The first 'lawns' would have developed out of natural grassland and meadows, where grazing had resulted in the low, easily traversed sward being maintained.

Abiotic factors such as moisture-deficient soil or exposure to wind result in the sward being naturally short in stature too. For example, early golf courses in Scotland, UK were reputably developed from coastal sand dune systems, where the combination of free-draining sandy soil, wind, salt exposure and grazing by sheep (and/or rabbits) resulted in a fine, short turf (Fig. 9.3), which was ideal to allow a small ball to roll and bounce. A short,

Table 9.1. The physical, social and environmental advantages and disadvantages associated with turf landscapes and lawns

Advantages	Disadvantages
Opportunities for sport – both formal and informal.	Extraction and transport of mineral deposits for construction: top-dressing and fertiliser.
Active sport seen by many policy makers as a means to combat sedentary lifestyles and reduce the risks of obesity-related diseases including stroke, late-onset diabetes and coronary heart disease.	Use of oil and natural gas in agrochemicals and fertiliser synthesis, and in producing energy for maintenance equipment.
Sequestration of atmospheric carbon.	Oxidation of carbon in soil disturbance.
Carbon storage is thought to be comparable to agro-ecosystems but not quite as good as forest systems. Carbon sequestration per year estimated as 0.52 t C ha^{-1} with permanent grassland over sandy loam soil.	
Rainfall capture and infiltration and attenuation of water flow into water courses.	Potable water for irrigation.
Can aid flood plain function acting as a capture point for flood waters.	Potential source of pollution to ground and surface water.
Contribution to urban cooling and heat island mitigation.	Release of nitrous oxide, N_2O, (a greenhouse gas) from volatilisation of nitrogen fertilisers.
Biodiversity – value depends on what it is being compared to as an alternative; biodiversity in a natural turf system is greater than in tarmacadam or synthetic turf alternatives.	Pesticide use, especially if not targeted to pest species, impacts on biodiversity and pollutant sources.
Grassland habitat of various forms. For example, certain golf courses have designated sites of special ecological interest.	Allergens including allergic rhinitis (hay fever) but mostly only when grasses flower.
Psychological benefits of green swards.	Source of noise pollution, due to routine mowing.
Location for children's play.	Location for dog fouling and potential to cause toxocariasis (infection of the roundworm parasite *Toxocara canis*), especially in areas where dog walking and children's play overlap.

Fig. 9.3. Many grass species are well adapted to coastal locations, where the combined factors of high wind, salt spray, thin sandy soils, periods of limited water availability (due to thin, rocky or sandy soils) and grazing pressure have naturally encouraged ecotypes that maintain short stature and fine-textured foliage – key characteristics for sports turf.

close-knit turf has greater resilience to wear and tear than turf composed of taller stems of grass – another factor that favours short turf being promoted for paths and playing surfaces. Grass species are the predominant constituents of formal lawns, although it should be noted that other 'green mantles' exist composed of species such as *Matricaria chamomilla* or *Chamaemelum nobile*, (chamomile), *Hypnum* or *Thuidium* spp. (moss) and *Trifolium* spp. (clover).

Turfgrasses are divided into 'cool-season' and 'warm-season' grasses depending on the climate. The cool-season grasses have evolved from alpine, acidic moorland or coastal strand conditions (Table 9.2). They tend to be fine-leaved and form close-knit swards when close-mown. Free-draining conditions, e.g. sandy subsoil, and the avoidance of organic matter, such as thatch (dead grass leaf-blades), help to keep pathogen pressure at bay. Many species are derived from European ecotomes and have been introduced to North America and elsewhere. Most 'cool season' grasses are C3 plants, i.e. they fix carbon into the three-carbon molecule 3-phosphoglycerate. In contrast, warm-season grasses predominate in tropical/semi-tropical regions and grow at temperatures >10°C, with optimum growth at temperatures between 25 and 35°C. They are active during spring and summer but often go dormant in cooler months, with a tendency to turn brown. Many warm-season grasses (Table 9.3) are quite drought tolerant and can survive temperatures >45°C. The warm-season grasses are C4 plants, i.e. carbon is fixed into the four-carbon molecule oxaloacetate. The biochemical pathways associated with the transport and fixation of carbon dioxide are more efficient than in C3 plants, so C4 plants have both a lower requirement for nitrogen and their stomata are not required to open for so long prior to meeting their carbon requirements. This latter aspect means they are likely to lose less water through the stomata and hence can be more efficient in terms of their water utilisation. The suitability of the main species for different functions are outlined in Table 9.4.

These tables outline the common species of grass used, but within each species there are numerous cultivars, specifically bred and selected for particular traits, e.g. improved pathogen tolerance, more compact habit or greater capacity to stay green when under stress. The development of new cultivars is a prolonged, costly and sophisticated business. Plant breeding (crossing, selecting and conducting trials) typically takes a decade or more, with further investment required in the production of seed, stock-holding, manufacturing and marketing before varieties can be released into the trade. Different breeders (seed houses) may have their own range of cultivars, covering the spectrum of requirements for different amenity functions, soil type, climate, pest/pathogen pressure and stress tolerance. Depending on where they are used, turfgrass species and cultivars may require tolerance to any one or more of the following: excessive heat, cold, drought, waterlogging, salinity, heavy metals, shade and physical wear (trampling) (Fan *et al.*, 2020).

The seed houses will promote mixtures of grass species/cultivars to provide a 'blend' of genotypes for a specific function. For example, a blend required for a fairway on a links golf course (i.e. one by the sea, where the sandy soils impose drought stress and the prevailing winds blow in salt) might be best

Table 9.2. Commonly used cool-season turfgrass species

Species	Common name	Habit and characteristics	Uses
Agrostis canina	velvet bent	Stoloniferous and forms a very dense turf. Relatively good drought tolerance, but more thatch production compared to other *Agrostis* spp.	Fine golf and bowling greens, where drought may be a problem.
A. capillaris	brown top bent	Rhizomatous and stoloniferous growth pattern, but with short sections between plants. Occurs on poor acid soils – under such conditions, can be an aggressive coloniser. Tolerates cool, damp conditions, but poor wear tolerance.	Fine-leaved, tufted habit makes it ideal for fine lawns. A cool temperate species, it is used in western Europe, New Zealand and north-west North America.
A. castellana	highland bent	Short rhizomes and very fine seed. Cooler climate areas with dense growth pattern. Poor wear tolerance.	Fine lawns and provides good winter colour
A. stolonifera	creeping bent	Leafy stolons and forms a tight-knit sward that tolerates close mowing.	Putting greens/bowling greens
Schedonorus arundinaceus	tall fescue	Tuft-forming although some rhizomatous types are now available. Coarse turfgrass, but generally not as good as *Lolium perenne* due to slower establishment and coarser leaf texture. Good in dry, low-fertility soils.	Useful in rough areas where there is requirement for taller, erect turf, such as airfields.
Festuca ovina	sheep's fescue	Tuft-forming. Very hardy – adapted to moorland conditions. Tolerates close mowing.	General lawn use plus finer lawns and greens.
F. ovina tenuifolia	fine-leaved sheep's fescue	Tuft-forming. Finer leaved than *F. ovina*. Relatively poor wear tolerance. Does not tolerate heavy, damp soils.	Good for ornamental lawns especially on well-drained soil. Due to poor wear tolerance, only for lawns with little traffic. Some cvs also suited for low-maintenance, environmentally sensitive areas, where parent species can be found.
F. rubra commutata	Chewings fescue	Tuft-forming. Quick to establish but poor competitor. Moderate tolerance to wear. Tolerance to shade, cold and drought. Cvs vary but some have greater tolerance to pathogens compared to *F. rubra littoralis*. Prone to forming thatch on acid soils. Colours up well in spring.	Does best on well-drained conditions.General lawn use plus finer lawns and greens.
F. rubra rubra	strong creeping red fescue	Rhizomatous with wide range of natural habitats and hence tolerant of salt and drought. Deep rooting but slow to recover from injury. Intolerant of wet or heavy, clay soils. Less tolerant of close mowing compared to finer-leaved fescues.	General lawns, without excessive requirement for close mowing

Continued

Table 9.2. Continued.

Species	Common name	Habit and characteristics	Uses
F. rubra litoralis	slender creeping red fescue	Rhizomatous with finer leaves than *F. rubra rubra*. Good binding capacity in the sward matrix. Not particularly disease tolerant.	Good on extreme environments. Close-mowing tolerance, so fine lawns and greens
F. longifolia	hard fescue	Tuft-forming. Drought, shade and heat tolerant – with some capacity to stay green over summer.	Fine turf, especially in infertile soils and hot, dry conditions.
Lolium perenne	perennial rye	Tufted perennial with broad leaves. Common, robust species. Quick to establish and withstands wear. Strong grower that requires nutrient-rich soil.	Sports turf due to hard-wearing character. Species does not tolerate close mowing, although modern cultivars more tolerant of this. Some cultivars have 'stay green' genes to maintain green hue when under stress.
Poa pratensis	Kentucky blue/ smooth stalked meadow	Rhizomatous growth. Can be slow to germinate and establish but once it does, good wear tolerance. Robust with some drought tolerance.	Used with *Lolium* to provide good wear resistance, e.g. soccer and rugby pitches. Also general grass areas – banks and golf fairways.

served by a mix of 35% *Festuca rubra litoralis*, slender creeping red fescue (e.g. 15% cv. Nigella to provide a dense sward and good visual quality and 20% cv. Seroa for high aesthetic value, but also salt and drought tolerance), 25% *Festuca rubra rubra*, strong creeping red fescue (with cv. Laverda due to its high shoot density), 20% *Festuca rubra commutata*, Chewings fescue (with 10% cv. Orionette to give a high-density sward to keep weeds in check and resist red-thread disease and 10% cv. Siskin with good disease resistance, salt tolerance and visual merit even under low-input management regimes); 10% *Festuca ovina*, sheep's fescue (cv. Quatro, again with salt and drought tolerance and good for year-round colour) and finally 10% *Festuca brevipila*, hard fescue (cv. Crystal due to its high wear and drought tolerance) (Anon., 2024). This blend allows each species/cultivar to exploit a distinct ecological niche within the golf turf matrix and ideally encompasses a robust community of genotypes that provide a resilient playing surface suited to the local conditions. As a matter of contrast, a football turf for a cool, temperate climate – where there is more trampling, physical wear and the grass needs to retain its integrity during the winter – might be composed of 100% *Lolium perenne* (perennial ryegrass) but still contain a mix of cultivars, e.g. 20% cv. Europitch, 20% cv. Eurodiamond, 25% cv. Gildara, 25% cv. Monroe and 10% Eurocordus.

Sowing a mix also ensures that if conditions are not conducive to some of the species/cultivars incorporated within the blend, then the complementary taxa will still ensure some form of sward develops. The blends help complement different characteristics, for example, in general, *Festuca* species can improve the aesthetics of the turf, promoting good colour and a fine texture to the sward, whereas the addition of *Lolium* improves wear, disease resistance and avoids bare patches developing in winter (Wolski *et al.*, 2021).

In reality, swards rarely grow in the exact proportions to the percentage of seed applied, and indeed over a period of time a significant component of the turf can be attributed to 'weed' grass species such as *Poa annua*, which naturally colonises (particularly damaged) turf. *P. annua*, however, is often considered undesirable in sports turf, due to its shallow rooting/poor drought tolerance, preponderance to turn yellow and its characteristic of forming unsightly flower/seed heads that can interfere with the playing capabilities of the turf, e.g. altering the natural roll of a golf ball.

Management has a role to play in these species interactions and the 'ecological' succession that takes place. In seed mixes comprising 90% (weight) *Schedonorus arundinaceus* (tall fescue) and 10% (weight) *Poa pratensis* (Kentucky bluegrass), Macolino *et al.* (2014) showed that the dominance of *P. pratensis*

Table 9.3. Commonly used warm-season turfgrass species

Species	Common name	Habit and characteristics	Uses
Cynodon dactylon	Bermuda	Leaves are borne on stems with long internodes giving a branched appearance. Close mowing is required to avoid this and promote a finer- textured sward. Species requires full sun and is drought tolerant. An aggressive species that can become invasive where it is not desired. Dies back if exposed to frost.	Sports turfs in tropics and for general robust lawns. Aggressive nature means will recolonise bare areas quickly.
Eremochloa ophiuroides	centipede	Originates from south-eastern Asia. Slow-growing species that grows prostrate to the ground and produces medium-textured, pale-coloured sward. Over-fertilisation can cause dieback. Turns brown when dormant. Poor salt, wear and cold tolerance. Prone to nematode damage.	Prefers acidic, infertile soils and has low maintenance requirements. General lawns and low-maintenance rough areas.
Paspalum notatum	Bahia	Native of South America – now common in south-east USA. High drought tolerance, doing well on sandy, infertile lawns. Invasive and may become a weed in other species lawns. Its own open habit means other species can invade. Good pest and pathogen tolerance.	Rather course- textured leaf. Used for erosion control alongside roads and banks. Robust habit means also used for low-maintenance lawn open areas and wildlife habitat.
Stenotaphrum secundatum	St Augustine	Water-use efficient with few pest problems. Broadleaf blades with creeping growth habit that nevertheless forms a dense, prostrate turf that is generally weed-free. Some drought tolerance, with good shade and salt tolerance but tends to produce thatch.	Lawns and general use but not wear tolerant enough for sports pitches. Shade tolerance means it can be used around buildings and near trees.
Zoysia japonica, *Z. matrella* and *Z. tenuifolia*	zoysia	Spreads by stolons and rhizomes, providing a very dense turf, with stiff leaf blades. Exceptional wear tolerance for a warm-season grass with both good drought tolerance and moderate shade tolerance. Better cold tolerance than other warm-season species, but turns brown after frost and does this earlier than other species. Generally low-water and nutritional requirements.	Lawn grasses and due to wear tolerance used on golf courses, parks and athletic fields. Adaptable to range of soil types and pH.

is favoured over time (at an agricultural research station in north-east Italy, at least). Its ascendency, however, is accelerated by closer mowing (lower height of cut) (Fig. 9.4) and by using cultivars of *S. arundinaceus* that are tuft-forming rather than rhizomatous – these being less competitive to the *P. pratensis* compared to the latter (Fig. 9.5).

9.4 Grass Genotypes for More Sustainable Management Practices

The drive for more sustainable approaches to turf management has activated interest in identifying new genotypes of grass that provide acceptable performance whilst reducing inputs and lowering

Table 9.4. Cool-season and warm-season grass species and their appropriate use in amenity and sports turf. Description: green, excellent suitability; blue, good suitability; yellow, possible to be used. Abbreviations: Fair. = fairway; Foot. = football pitches; Ten. = tennis courts (and cricket squares); Polo = horse polo fields and racecourses; Lawn = close-mown amenity lawns and Low Main. = low-maintenance grass areas. (Modified with permission from Wiewióra and Żurek, 2023.)

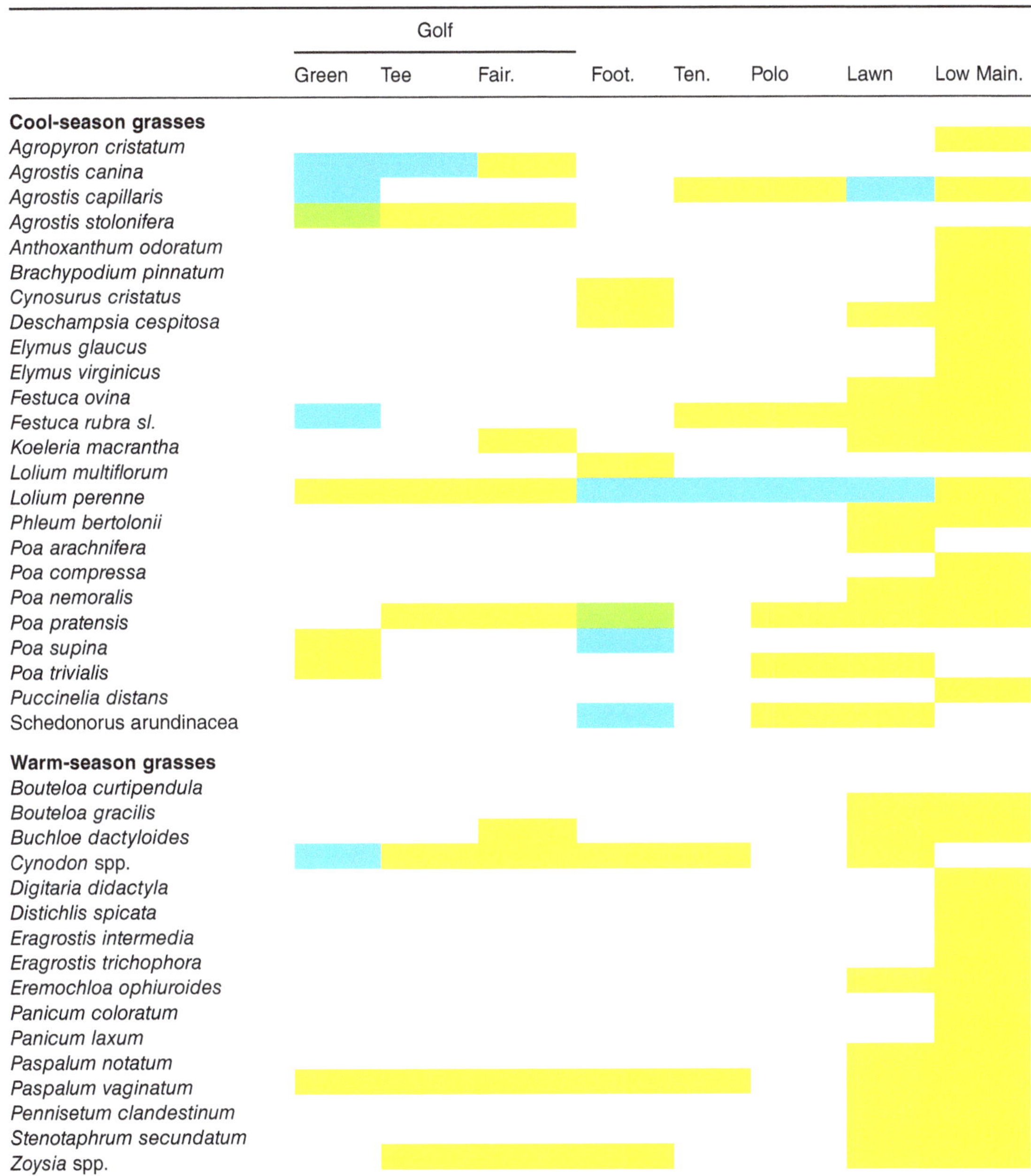

	Golf							
	Green	Tee	Fair.	Foot.	Ten.	Polo	Lawn	Low Main.
Cool-season grasses								
Agropyron cristatum								yellow
Agrostis canina	blue	blue	yellow					
Agrostis capillaris	blue				yellow	yellow	blue	yellow
Agrostis stolonifera	green	yellow	yellow					
Anthoxanthum odoratum								yellow
Brachypodium pinnatum								yellow
Cynosurus cristatus				yellow				yellow
Deschampsia cespitosa				yellow			yellow	yellow
Elymus glaucus								yellow
Elymus virginicus								yellow
Festuca ovina							yellow	yellow
Festuca rubra sl.	blue				yellow	yellow	yellow	yellow
Koeleria macrantha			yellow				yellow	yellow
Lolium multiflorum				yellow				
Lolium perenne	yellow	yellow	yellow	blue	blue	blue	blue	yellow
Phleum bertolonii							yellow	yellow
Poa arachnifera							yellow	
Poa compressa								yellow
Poa nemoralis							yellow	yellow
Poa pratensis		yellow	yellow	green		yellow	yellow	yellow
Poa supina	yellow			blue				
Poa trivialis	yellow					yellow	yellow	
Puccinelia distans								yellow
Schedonorus arundinacea				blue		yellow	yellow	
Warm-season grasses								
Bouteloa curtipendula								
Bouteloa gracilis							yellow	yellow
Buchloe dactyloides			yellow				yellow	yellow
Cynodon spp.	blue	yellow	yellow	yellow	yellow		yellow	
Digitaria didactyla								yellow
Distichlis spicata								yellow
Eragrostis intermedia								yellow
Eragrostis trichophora								yellow
Eremochloa ophiuroides							yellow	yellow
Panicum coloratum								yellow
Panicum laxum								yellow
Paspalum notatum							yellow	yellow
Paspalum vaginatum	yellow	yellow	yellow	yellow	yellow		yellow	yellow
Pennisetum clandestinum							yellow	yellow
Stenotaphrum secundatum							yellow	yellow
Zoysia spp.		yellow	yellow	yellow			yellow	yellow

the environmental impact of turf management. It has been long-recognised that 'low-input' grasses exist such as the warm-season bahiagrass (*Paspalum notatum*) and cool-season hard fescue (*Festuca longifolia*), but to date, these species have had only limited acceptance due to a lack of performance capability. Attention has now turned to selecting and breeding new cultivars that can provide a good

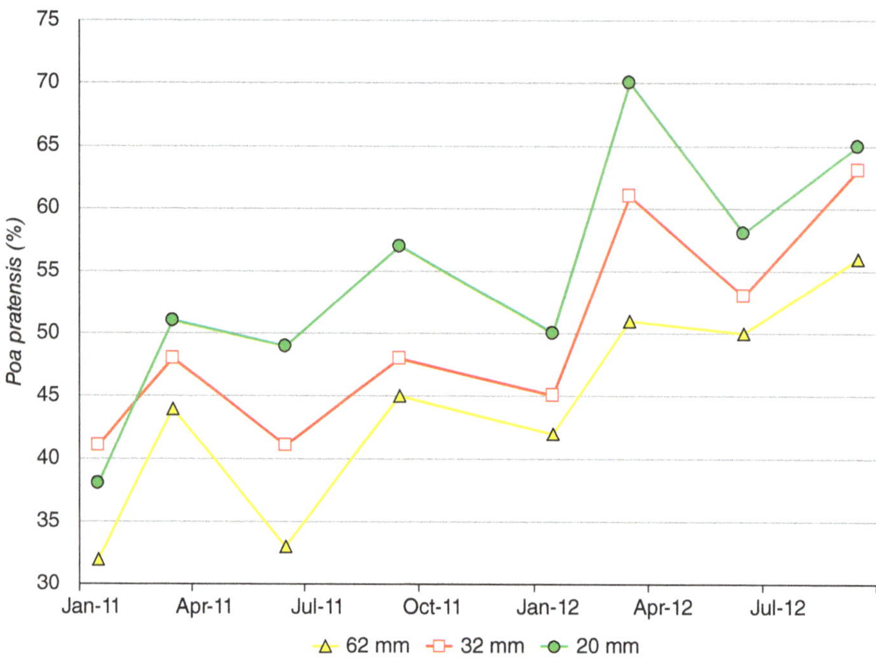

Fig. 9.4. Percentage of the turf sward area composed of *Poa pratensis* (Kentucky bluegrass) over time. The coloured lines reflect different heights of cut, with the longest being 62 mm (yellow), medium 32 mm (red) and the shortest, when the grass is cut at just 20 mm height (green). In a seed mix comprising 90% (weight) *Schedonorus arundinaceus* (tall fescue) and 10% (weight) *Poa pratensis* (Kentucky bluegrass), the *Poa* is favoured over time, and particularly under the close-mowing regime of 20 mm cut height. (Modified with permission from Macolino *et al.,* 2014.)

compromise between desired performance criteria and sustainable management regimes. Development of seashore paspalum (*Paspalum vaginatum)* during the 1990s resulted in cultivars (e.g. Salam, Sea Isle 1, Sea Isle 2000 and Sea Dwarf) with finer-textured leaves and greater tolerance to low mowing. These were considered to be improved sufficiently to meet the standard requirement for golf course turf. Moreover, they had traits including increased pest tolerance, an ability to tolerate poor-quality water (e.g. saline/brackish), lower nutrient requirement and better retention and reduced water requirements compared to traditional warm-season golf course turf (Duncan and Carrow, 2000). Native selections of *Lolium* with good stress tolerance characteristics have proved promising compared to commercial varieties in Iran. Research in Turkey has examined both cool- and warm-season genotypes for their use in sport turf within the dry Mediterranean climate. Cultivars were tested for their playing-quality traits (ball rebound, ball roll and shock absorption) as well as visual quality. Kir *et al.* (2019) concluded that *Cynodon dactylon* x *Cynodon transvaalensis* cv. Tifway-419 and one of the *Paspalum vaginatum* 'Sea' cultivars, namely cv. Sea Spray proved the most promising for the Mediterranean region. In the USA too, native genotypes are being examined for greater stress tolerance, in an attempt to select for, or breed in, traits that ultimately reduce energy, chemical or water inputs. These include *Buchloe dactyloides* (buffalograss), *Bouteloua gracilis* (Blue grama), *Bouteloua curtipendula* (Sideoats grama), *Distichlis spicata* (saltgrass), *Agropyron cristatum* (crested wheatgrass), *Koeleria cristata* (prairie junegrass) and *Deschampsia caepitosa* (tufted hairgrass) (Johnson, 2000). Dwarf selections of *Cynodon dactylon* have shown not only a compact growth habit but a reduction in mowing frequency and 25% less requirement for nitrogen fertiliser.

Certain grasses form symbiotic or mutualistic relationships with fungi. The fungus genus *Epichloë*, for example, commonly act as endophytes, living their whole life in the tissues of the grass without visible symptoms (Wiewióra and Żurek, 2023). *Epichloë* (*Neotyphodium*) species of endophytes are often present in *Lolium*, *Festuca*, *Poa* and *Agrostis*

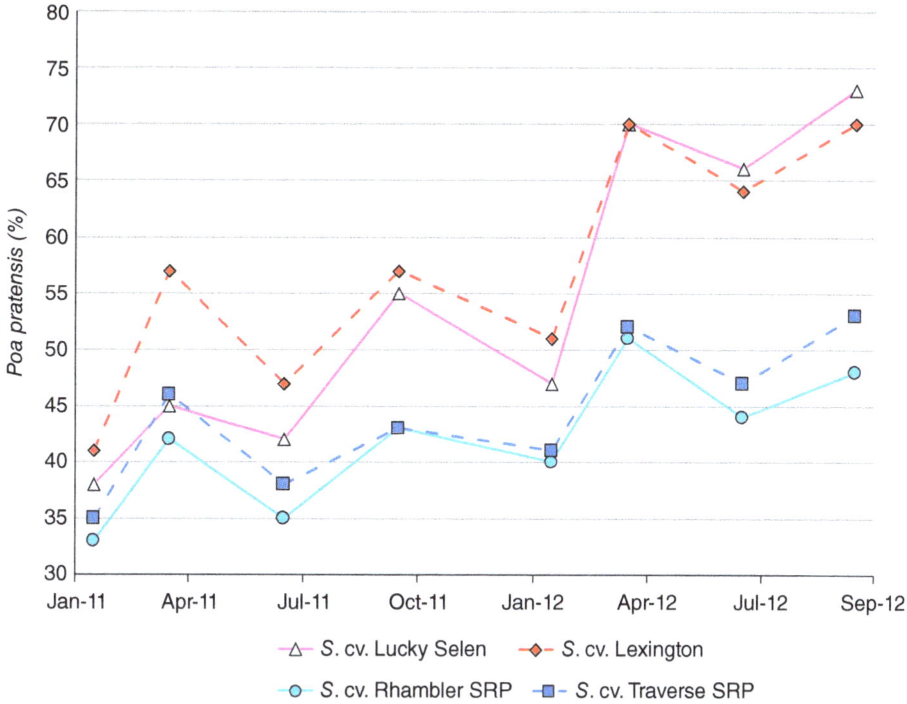

Fig. 9.5. Vegetational succession in turf showing percentage of *Poa pratensis* (Kentucky bluegrass) in the sward over time. The seed mixes comprise 10% (weight) *Poa pratensis* cv. Nublu Plus (Kentucky bluegrass) and 90% (weight) of one of the following *Schedonorus arundinaceus* (tall fescue) cultivars: tuft-forming *S.* cv. Lucky Selen or *S.* cv. Lexington or rhizomatous *S.* cv. Rhambler SRP or *S.* cv. Traverse SRP. The rhizomatous cultivars resist the dominance of the *Poa* more effectively than the tuft-forming types. (Modified with permission from Macolino *et al.*, 2014.)

grasses. The most often identified in lawn grasses are *Epichloë festucae*, *E. typhina* (*Sphaeria typhina*), *E. coenophiala*, or *E. uncinata*. The presence of the fungi may bring many benefits to the host grass, for example, inducing enhanced drought tolerance, better cell regeneration after damage and resisting pest/pathogen invasion/infection, as well as contributing to more efficient gathering and utilisation of nitrogen and phosphorus. The way the grasses are used/managed determines the nature of these interrelationships. Indeed, the benefits may not always be apparent under certain circumstances – for example, more intensive sward management regimes with heavy applications of water and fertiliser can undermine the role of the fungi. As conditions become more adverse for grass species due to climate change, and management regimes look for more sustainable use of resources, then the role of these endophytic fungi is likely to become more prominent.

It should be noted too that climate change will alter the geographical distribution and adaptive capacity of turfgrass taxa, with warm-season grasses being used in more northerly/southerly latitudes than is currently the case. Due to the nature of the C4 carbon capture mechanism, warm-season grasses use about 50% less water to fix carbon dioxide (CO_2) and have substantial lower evapotranspiration rates than their cool-season counterparts. Increasing the adaptive range of warm-season grasses (such as making *Cynodon dactylon* – Bermudagrass – varieties more cold tolerant) can therefore help with water conservation on golf courses and elsewhere (Taliaferro, 2000).

9.5 Cultural Procedures

Grass sward from seed

For extensive areas of lawn and sports fields, the generation of the sward is usually accomplished by sowing seed. This has the advantage over the other principal way of establishing a sward (i.e. by laying turf sods) through lower costs, more choice of grass species within the blend, and potentially allied to this, better adaptation of species to the site. Turf establishment on inaccessible locations, or where the landscape is convoluted by difficult angles or slopes, is easier too via seed. The amount of seed required is calculated by weight; the number of seed present per unit of weight can vary widely across taxa. For example, 1 g equates to 700 seeds of *Lolium perenne*; 1000 seeds of *Festuca rubra commutata*; 1500 seeds of *Poa pratensis*; or 14,000 seeds of *Agrostis castellana*. Rate of sowing depends on a balance between the need to cover the ground rapidly and the costs involved. Typical rates vary between 4 and 40 g m^{-2}. Species that subsequently self-propagate via stolons and rhizomes require lower rates of seeding compared to those that are tuft-forming.

Species, cultivar, germination rates and company/brand can all influence price with 1 kg of seed retailing typically at between £7.00 to £40.00 in the UK (US$9.60 to US$54.80). Time of sowing depends on local climatic conditions. In the UK, the seed of cool-season grass species has been traditionally sown in early autumn or mid–late spring, when temperatures are high enough, and there is also sufficient moisture held within the soil. In the northern hemisphere there are many favourable arguments for sowing in September. The ground retains warmth after the summer, whilst early autumnal rains increase soil moisture levels. High levels of germination in September will provide sufficient time for new roots to develop before the onset of frost, and competition from weeds will be minimal during winter. Strong establishment at this stage aids tolerance to drought in the following summer.

Where irrigation is an option, then sowing can take place at moderate temperatures throughout the year. Even within the cool-season grasses there is some variation in their temperature tolerances for germination between species, with *Festuca* tolerating cooler conditions (8–18°C) compared to *Agrostis* (12–21°C), *Lolium* (7–25°C) and *Poa pratensis* (15–25°C). Some specialist seed mixes are designed to germinate at temperatures as low as 3.5°C (e.g. certain *Lolium* cultivars). For species such as *Festuca* though, higher temperatures >22°C at night can inhibit germination. Successful establishment of the seedlings is reliant on avoiding drought or cold stress shortly after sowing. Again, depending on species, germination rates should be in the region of 70–95%.

Seed of warm-season genotypes is invariably sown in late spring/early summer. These species cope better with warm summer temperatures and the advantage of early establishment is that it helps to resist weed pressure from *Poa annua* during the subsequent cool autumn period.

Seed bed preparation

If grass is to be seeded on soil (in sports turf it is often sown on to a sand, peat or other free-draining medium) then residual weeds need to be removed. Theoretically repeated cultivation will exhaust the seed bank, but such organic cultivation principles are relatively rare in elite golf, for example. Here, the application of herbicides remains 'mainstream'. An application of herbicides is followed on by a fallow period and then cultivation. The non-selective glyphosate [N-(phosphonomethyl)glycine] is often the most commonly used herbicide for clearing up existing weeds and unwanted grass species. After weeds have died (usually about 10 days after application), these are raked off. The land is then left fallow and reapplications of herbicide employed as required. Following this, the soil is cultivated (tillage), though the extent of this will depend on the area and future function of the turf. The purpose of tillage is to create a fine seed bed that promotes seed germination and establishment, whilst also contouring the ground to provide a smooth, even surface. Successful grass seed germination is promoted by contact between the seed and the moist surface of the soil, whilst also ensuring adequate oxygen levels around the developing and respiring seed embryo. Hard clods of earth and stones should be removed at this stage. On larger-scale sites, the soil may be ploughed, disced and harrowed to improve the soil structure for seed germination. On smaller sites, the soil may be rotovated and then consolidated or even hand dug and settled with a rake. Either way the objectives are to allow re-contouring or aligning of the soil surface, the incorporation of any residual organic matter, breaking up of soil clods and preparation of the soil for seed germination.

Many lawns are still constructed with a 2% gradient to help encourage surface drainage. For sports pitches, this usually means the centre of the playing field is the highest point with a slight gradient to the edges. For domestic lawns, gradients are aligned to help avoid surface runoff being directed towards the dwelling property. A lawn surface should be true – i.e. no sharp changes of gradient or undulations as these encourage the turf to be ‘scalped’ (bare patches) due to the action of the passing blades of the lawn mower.

For elite sports grounds and golf course greens, an artificial soil surface is constructed to ensure optimum drainage, enhance play performance and reduce incidence of turf disease. A fundamental element is maximising drainage whilst maintaining sufficient moisture availability to support strong grass growth. The rationale behind these artificially constructed playing surfaces is to ensure play can continue during periods of heavy precipitation, not that such surfaces are necessarily easier to maintain than traditional turf. Golf greens represent the leading edge of these ‘manufactured’ soils for turf development. The United States Golf Association (USGA) recommend golf greens to be constructed by providing drainage channels and a number of layers of free-draining media above the parent soil (or if the topsoil has been removed – subsoil). The soil is contoured to fit the requirements of the pitch or green and then compacted to form a stable surface. A series of sub-surface drains are embedded within the parent soil ensuring that any main drains are placed in line with the greatest falls in height. This encourages water to flow to the lowest point on the site and exit from there. The main drains are connected to a series of lateral drains (held in trenches >150 mm wide × 200 mm deep) and usually spaced at regular intervals to maximise water capture. On regular, geometric sport facilities such as soccer or rugby pitches, these drains will be laid out in a formal lattice or herringbone fashion, whereas in irregular areas such as golf greens they may act as a series of ‘branches’ leading to the main drains. Lateral drains should be orientated to aid water movement towards natural, water-collecting depressions in the landscape. Although clay pipes were the traditional land drain in many countries, perforated plastic drains (>100 mm diameter) tend to predominate now. Drains should be angled to allow a gradient of at least 0.5%, placed with the perforations at the base and their trenches backfilled with gravel to a depth >25 mm.

Over the parent soil (sub-layer) a layer of gravel (blanket layer) is placed, again to a depth of at least 100 mm (Fig. 9.6). This is pre-washed crushed stone or pea gravel (6–12 mm diameter) that is likely to resist weathering. (This is in contrast to ‘softer’ sandstone and limestone aggregates that may degrade into smaller particle sizes over time). Much controversy exists as to the relationship between the gravel layer and the root zone layer

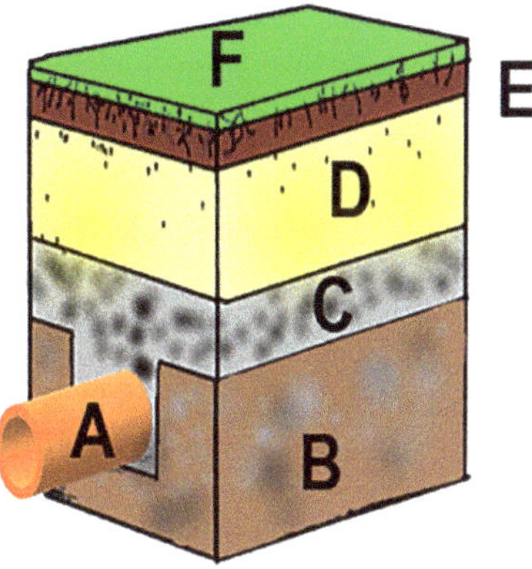

Fig. 9.6. Artificial soil structure as used on a golf or bowling green to ensure good drainage. Key: A = lateral drainpipe; B = parent soil; C = gravel (blanket layer); D = root zone; E = top of root zone where organic matter from root exudates/dead leaves accumulates; F = grass sward.

above it, in that inappropriate aggregate sizes in both allow fine sand particles to migrate downwards and infiltrate the gravel, thereby reducing the pore space for water to drain. Sharp contrasts in particle sizes between fine sand placed above coarse gravel is also thought to contribute to a ‘perched’ water table – where drainage from the top fine sand layers is impeded by a lack of conductivity and capillary action between the two media. To avoid this, gravel selected for the blanket layer should conform to a standard where the particle grades are 0% >12 mm; 10% maximum <2 mm; and 5% maximum <1 mm diameter. In addition, the smallest 15% of the gravel particles should overlap in size (‘bridge’) with 15% of the root zone’s largest particles. This helps to form pores that inhibit further migration of particles from the root zone into the gravel. Another factor that aids successful bridging is that the diameter that best represents the smallest 15% of gravel particles (D15) must be less than or equal to eight times the diameter that represents 85% of the smallest root zone particles (D85). For example, if 15% of the gravel particles are smaller than 4 mm, then 85% of the root zone particles need to be 0.5 mm or smaller. Similarly, to maintain good permeability of water across the root zone/gravel interface, the D15 value for gravel should be equal to or greater than 5 times the D15 value for the root zone. Where these criteria are not met, the USGA recommends an intermediate (blinding) layer to be placed between the gravel and root zone. This requires the gravel to have 65% of particles between 6 and 9 mm, with no more than 10% of particles >12 mm and no more than 10% <2 mm. This also requires the blinding layer to have 90% of particles between 1 and 4 mm diameter. Obviously, careful grading of the gravel and root zone particles is important to ensure compatibility.

The root zone invariably is composed entirely of sand, although older recommendations suggested 20% of either soil or peat in the mix. In reality, degradation of the dead grass (thatch) means that the organic component can increase in the top strata of the root zone mix, without intentional additions. As with the gravel, there are recommendations with respect to the root zone composition (Table 9.5).

The root zone mix needs to represent a media with a total pore space in the region of 35% to 55%, of which the air-filled porosity is 5% to 30%, and capillary (potential water-filled) porosity 15% to 25%. Hydraulic conductivity when saturated with water should equate to 150 mm h^{-1}. The root zone media is frequently blended by the company supplying the substrates or obtained direct from a media manufacturer whose product meets the desired specification. It is good practice, however, that the required specifications are validated by a soil science laboratory or similar professional body. Thoroughly mixed root zone material is placed over the gravel or blinding layer, then consolidated to provide a uniform depth of 300 mm. Care is required to avoid intermixing of the root zone with the basal gravel.

Seeding and base fertiliser application

Fertiliser requirements can be assessed from soil nutrient and pH samples taken from the site. For sand-based root zones, base fertiliser is normally a requirement. Nitrogen is mobile (i.e. highly soluble in water and easily leached from those soils that possess few binding sites for its anions) and may not be applied at this stage, but potassium and phosphorus are incorporated prior to sowing (to a depth of 120–180 mm). Lime may also be added at this early stage if the parent soil is excessively acidic. Nitrogen is usually applied at the time of sowing, or shortly after. Nitrogen forms a relatively small proportion of the fertiliser during early establishment phases e.g. 6% N; 9% P_2O_5; 10% K_2O at 35–70 g m^{-2}. The topsoil or root zone is consolidated by rolling and then raked to provide a fine tilth. Seed are applied and the area rolled again to ensure good contact between seed and soil, thus ensuring sufficient access to moisture for the developing seedling.

Ideally, seed should be placed approximately 6 mm below the soil surface; if they rest on the surface, they are prone to being dislodged by the runoff of excess surface water. Placed too deep and the seedling's carbohydrate and protein resources may be overtaxed as it grows through the soil to reach the light. Uniform distribution of seed is important to ensure an even cover to the new sward and avoid bare patches that will be prone to weed invasion. Mechanised seeders are available and cover a variety of scales of planting. These range from hand-held seeders through to tractor-mounted hopers and distributors. Cultipack seeders work on the principle of spreading seed evenly along a line from a hopper or series of hoppers but also at the correct planting depth using a ridged roller. This leaves characteristic parallel lines of grass seedlings along 'mini-furrows', but the grasses soon spread to colonise the inter-furrow spaces.

Where there is a requirement to sow grass on inaccessible sites, or where it is desirable to keep machinery off the soil to avoid compaction, then hydro-seeding can be employed. Hydro-seeding is

Table 9.5. Particle size distribution of USGA root zone mix. (Used with the permission of the USGA, Anon., 2004.)

Medium	Particle diameter (mm)	Recommendation (by weight)
Fine gravel	2.0–3.4	≤10% of the total particles in this range, including a maximum of 3% fine gravel (preferably none)
Very coarse sand	1.0–2.0	
Coarse sand	0.5–1.0	≥60% of the particles must fall in this range
Medium sand	0.25–0.50	
Fine sand	0.15–0.25	≤20% of the particles may fall within this range
Very fine sand	0.05–0.15	≤5%
Silt	0.002–0.05	≤5%
Clay	<0.002	≤3% (Total fines ≤10%)

the process in which grass seed is pumped onto the landscape in a stream of aqueous solution. The solution may contain fertiliser, mucilage gels to aid adhesion and even mulch materials (e.g. peat or vermiculite) as well as water.

Establishing a sward with turf (sod)

Lawns and sports pitches can be established by laying down sections of pre-grown turf (known as sods). These can be translocated from elsewhere on a given site, for example from one area of a golf course to another. Most frequently, however, they are grown on specialised turf farms. These turf farms are located in regions with easily cultivated sandy or peat-based soils and the turf grown on fields. Alternatively, some businesses now provide the sods using soil-less systems to cultivate the turf. The turf sections are lifted with the minimum amount of substrate (usually ≤10 mm) to keep weight down, but with a sufficient shoot/root matrix to provide integral strength to the turf sod (allowing for handling, transport and laying without sections breaking). Rolls of turf (typically 2.0 × 0.5 m) should be delivered from the farm, placed *in situ* and irrigated all within 24-48 h. The turf needs to be transported damp to minimise root death and stored during transit to avoid overheating. Heat stress is exacerbated by factors including high transport and storage temperatures, poor air movement through the roll or stack, high nitrogen ratios in the leaf tissues, the presence of pathogens and transporting the turf with an excessively long leaf-blade length (high height of cut) or where seed heads have formed prior to lifting.

Lawn establishment via turfgrass is more expensive compared to seed (turf sods costs vary between £2.95 to £7.50 m^{-2} [US$3.74–9.54]) but has the advantage of providing an instant visual effect. Additionally, if sods are well laid and managed carefully, especially through appropriate irrigation scheduling, they can offer a fully playable surface within a few weeks of laying. Under optimum growing conditions, new roots from the turf will start to integrate effectively with the parent soil on the site within a few days.

As with seed, successful establishment of the turf requires similar cultivation and preparation of the parent soil, with the added caveat that the parent soil should be uniformly moist (but not excessively wet) at the time of laying. Rolls of turf can be laid by hand or by unrolling from bars fixed to a tractor. Ideally, the sections of turf should be laid in a staggered manner to avoid four corners meeting at any one point, as these corners may shrink and 'turn up' if desiccated. Topsoil (or other substrates) is placed in any gaps between strips of turf, again to help avoid excessive drying. The turfs themselves need to be consolidated by light tamping to ensure effective contact between the root mat and the parent soil and to remove any air gaps or ruffles. On steep slopes, the turf may need to be staked in place to resist movement and slippage by gravity, with stakes being removed once roots have integrated with the soil beneath. Turf rolls placed across rather than down a slope will avoid rivulets of water forming channels between the rolls during heavy precipitation events. Newly laid turf needs to be irrigated (ideally within 30 min of laying) and then watered relatively frequently, e.g. daily in warm weather, over the subsequent 2–3 weeks. Once roots have penetrated the underlying soil, irrigation can be applied less frequently, but in larger, single volumes.

9.6 Lawn Maintenance Practices

Maintenance activities and time commitments are dictated by the use of the sward, size of area and the machinery available. Nevertheless, maintenance is dominated by mowing, irrigation nutrition management and controlling pests, weeds and pathogens.

Mowing

Mowing determines the playing quality of the turf and its ability to tolerate wear. For a typical, intensively managed sward, mowing will normally represent 60% of costs. A move to robotic mowers, however, is helping reduce costs in some situations (Fig. 9.7). Mowing is required to keep the sward neat, true, vigorous and in a suitable condition for its purpose. It is also a very effective means of eliminating weed species that are intolerant of frequent close cutting. Mowing is most frequent for those playing surfaces that require a very close-knit sward such as golf greens. Indeed, golf greens may be mown daily during the growing season to ensure the height of the sward remains between 4 and 5 mm. This being increased to 6 mm in winter. A very low cutting height of 3 mm is practiced for short periods, but if prolonged this reduces photosynthetic capacity, with corresponding reductions in shoot density, rhizome numbers, root number and length.

Such close-mown sward is more susceptible to drought stress and pathogens and has a lower wear tolerance. Similarly, excessively long grass swards (e.g. >50 mm) have very poor wear tolerance, and even a single passage of a pedestrian can result in lodging (bending) of the stems. The severity of cut influences the physiology of the grass plant, and there is a balance to be struck between the functionality and predisposition to stress; for example, there is a direct relationship between height of cut and the depth of the root system. So close-mown swards have shallower roots systems (Fig. 9.8) and hence require greater frequencies of irrigation and fertilisation to avoid water stress or nutrient imbalances.

Fig. 9.7. Self-automated, robotic mowers can take some of the labour (and associated costs) out of sports turf management. These lend themselves to 'green energy' power too, by being battery powered – and charged up from the grid via wind or solar power.

The height of cut is strongly influenced by the desired playing quality of different types of sports turf. The mowing regime on golf greens is designed to influence the speed of the ball across the grass – the more challenging championship courses aiming for very 'fast' greens. The speed of a ball across a green is measured via a stimpmeter – a rod with a groove that allows a demonstration ball to roll onto the green from a consistent height and angle. For tournaments, the roll distance determines the speed of the green, e.g. fast >2.75 m, moderate = 2.60–2.75 m, slow <2.60 m. Green 'speed' is affected by the uniformity, smoothness, firmness and the resilience of the sward. Resiliency (hardness or bounce) on the other hand affects the capacity of the turf to absorb the shock of a golf ball hitting it and to 'hold' the ball on the green.

In sports such as tennis and cricket, the bounce of the ball was traditionally aided by modifying the soil constituents too, for example by providing a harder surface through the use of clay, and then using a roller to compact this soil surface. The higher bulk density and poorer drainage characteristics of clay topsoils translates to more wear on the

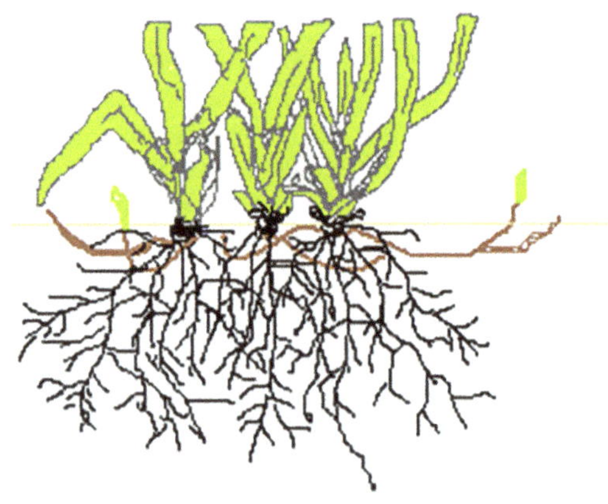

Fig. 9.8. Close mowing encourages a more tight-knit sward (left) but results in a shallower root system that is more prone to drought stress compared to sward that has a higher height of cut (right).

turf itself. To accommodate this, cricket squares are designed to incorporate a number of 'pitches': the area between the wickets in which the ball is bounced (and batted) and where there is the highest player footfall. Rather than renovate a pitch between consecutive matches, a new pitch is selected and the old pitch left unused for the grass to re-establish. In other sports, the roll or the bounce of the ball may be less important in comparison to the qualities the turf provides in terms of player safety and integral strength. Rugby pitches, for example, will have a longer sward length to provide greater cushioning when players tackle one another, and the longer leaf blades support a deeper root system that resists damage during scrumming.

Frequency and height of mowing relate to a compromise to maintain an effective playing surface, improve wear tolerance, avoid abiotic and biotic stress and ensure management, particularly labour, costs are not excessive. The height of cut associated with different sports reflects the requirements of each and the varying demands for wear tolerance (Table 9.6). Sports where there is a high footfall such as soccer, and where there are shearing and compression forces on the turf, have a high component of wear resistant *Lolium perenne* incorporated, maintained at medium cut height to aid resilience. Such pitches may also have artificial fibres (e.g. polypropylene) sown into the turf to improve physical robustness and help retain the playing quality should the grass die out in localised patches. For many intermediate lawns and sports turfs, mowing is implemented when the grass is 1.5 times the required height, so a lawn which ideally should be at a height of 4 mm would be cut when at 6 mm. Exceptions to this rule include the avoidance of cutting the sward when excessively wet, during frost episodes or indeed when unduly dry and plants have entered a quiescent state. Even in fine weather, removing moisture such as morning dew from the leaves (switching) prior to cutting has advantages. Not only is the effectiveness of the mowing improved, it limits problems due to clippings adhering to and clogging the machinery. At the onset of the growing season the first few cuts of the sward are kept high. This avoids excessive stress on the turf – it being wise to provide two or more successive cuts over a period of a few days, lowering the height of the blades on each occasion.

Table 9.6. Typical heights of cut for different sports

Turf	Height of cut (height after cutting, mm)
Golf green	4–6
Golf tee	7–8
Golf fairway	10–12
Bowling green/cricket square	4–6
Fine lawns	5–6
Soccer	12–25
Ornamental lawn	10–12
Rugby	25–50
Baseball	12–20

The intensely managed turf that is required for top-level playing performance receives the most management and hence has the highest environmental footprint. Using a lifecycle analysis approach, Tidåker *et al.* (2017) demonstrated that of the key activities on golf courses, mowing had the greatest CO_2e emissions profile (Fig. 9.9), with the greens and tees (potentially being mowed daily) correlating with the highest proportion of emissions per hectare. The overall greater area of the fairways and rough though, despite being cut less often, actually produces more CO_2e in total. Fertiliser production and emissions from applied fertilisers were also a relatively large contribution to CO_2e emissions. (Note CO_2e emissions is all the greenhouse gases that may be released, expressed as equivalents of CO_2, so for example methane CH_4 has an atmospheric warming potential 84 times greater than CO_2 itself).

Mowing of less intensively used swards

Where criteria for sport or other frequent use are less relevant, for example turfgrass around business parks, then considerably more sustainable approaches to mowing are feasible. Reductions in mowing frequencies are achievable with consequent savings in energy and labour inputs. Reducing the frequency of cut though means mowing at a higher sward height. For strong-growing, erect grass taxa such as *Lolium* (perennial ryegrass) and *Paspalum* (bahiagrass) this is acceptable; many varieties are still able to be cut readily at even 100 mm height. Naturally shorter-growing species, on the other hand, which may be outcompeted by weeds (*Zoysia*) or have unattractive features (*Cynodon*, Bermuda grass) are best cut at a lower height, e.g. 50 mm maximum.

Grass mowing machinery

The function of the sward not only influences the frequency of mowing, but also the type of mowing

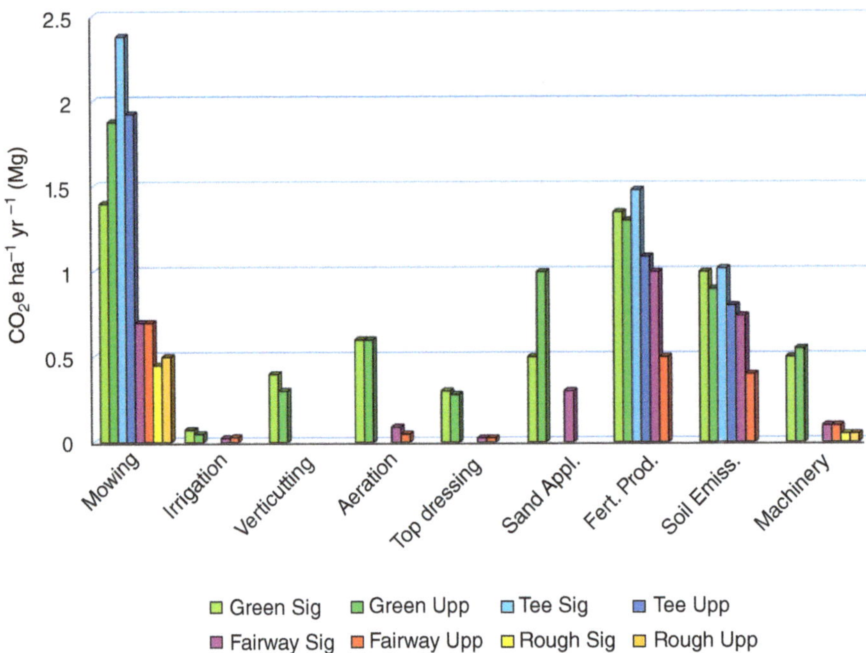

Fig. 9.9. The environmental footprint [Mega grams of carbon dioxide equivalents produced per hectare per year – CO_2e ha^{-1} y^{-1} (Mg)] through key activities on different parts of golf courses. Two courses were compared in Sweden (Sigtuna [Sig] and Uppsala [Upp]) with data for the golf greens, tee areas, main fairways and the rough grasslands that are adjacent to the fairways being presented (Sand Appl. = the transport and application of sand to aid drainage in the turf; Fert. Prod. = the manufacture of mineral fertilisers; Soil Emiss. = the emission of gases that contribute to atmospheric warming – mostly nitrous oxide (N_2O) in this case. 'Machinery' relates to the embodied energy in the golf course maintenance machinery). (Modified with permission from Tidåker *et al.* 2017.)

instrument used. There are five basic models of 'mower' to cut grass: the cylinder (reel) mower, rotory mower, flail mower, reciprocating mower and strimmer.

The cylinder mower comprises a stationary bedknife, a cylinder with rotating blades and one or more rollers. These can be 'driven' by one of the mower's side-wheels, which is attached to the cylinder via gears or chains. Cylinder mowers are ideal for small intricate areas and where there is a requirement for a low mowing height. They have a sheering action on the grass leaves with the rotating blades working in tandem with the fixed bedknife to give a clean cut. Mowing quality is a function of the sharpness of the cylinder blades and proper adjustment of the bedknife in line with these. Cleanness of cut is influenced not only by blade sharpness but also by the number of blades on the cylinder (more giving a finer, even cut), the rotational velocity of the cylinder and the speed of forward motion. The most effective mowing is accomplished when the mowing height (the height the blades are set above the ground) is equivalent to the distance between successive clips of the cylinder blades, i.e. the distance the mower travels between one cut and the next. Where this is not in proportion, the cut height is not uniform and a wave pattern can be left on the grass (marcelling). The precision 'clean-cut' of the cylinder mower results in this being the mower of choice for golf/bowling greens and the more aesthetically-demanding fine lawns. The rollers at the back of the mower too are used to extenuate the aesthetic aspects in that they leave the grass bending in the direction of cut and therefore can be used to form the characteristic stripe patterns frequently found on lawns and sports fields. The intricate cutting mechanism and the preponderance of the blades to notch when they hit a hard object means they are not so suited for longer grass and rougher areas.

Rotary mowers work on the principle of a horizontal blade rotating at high speed, and this blade fractures, rather than cuts, the grass. These are energy-efficient, robust machines, ideal for utility lawns, parkland and rough areas of grass along roadsides and other verges. Rotary mowers can either be placed on a frame with wheels or suspended on a cushion of air, in a similar fashion to a hovercraft. The rotating blade has some tolerance of impact with hard objects but can scalp the grass (cut into the substrate) on uneven topographies.

Flail mowers work on a similar principle to rotary ones but have the blades attached by pivots to the central horizontal arm, rather than the arm itself being the blade. These free-swinging blades are designed to deflect away from any hard foreign object they hit, thereby reducing risk of projectiles exiting from the skirting below the mower. The action of the blades and skirting means that the grass is finely chopped into a mulch, rather than being left as discrete stalks of grass. Maintenance focusses on ensuring the blades remain sharp and that there is free movement of the blade on the horizontal arm.

The reciprocating blade mower is less frequently encountered compared to other types but works on a similar principle to a hedge trimmer, i.e. there are two reciprocated saw-like blades that work in unison to clip the grass blades. These are ideal for meadow cutting, as they can deal with long grass and provide a uniform height of cut. The disadvantage is that foreign objects, sticks, stones, plastic debris. etc. get trapped between the blades.

The last common grass cutting instrument is the strimmer. This has a single shaft with either a high-speed rotating blade or nylon cord at the base. These fracture the grass stems. The advantage of using nylon cord is that the cord extends as it breaks off against hard objects, thus avoiding the need to stop the machine. Strimmers are primarily used for small areas of long or rough grass and are beneficial in sites that are uneven or where grass needs to be cut close to walls and other objects. Care is required, however, when operators are working close to trees as these can quickly strip a young tree of its bark.

Mowers are powered by a variety of means, including pedestrian-driven push mowers, four-stroke or two-stroke petrol engines, electrically powered machines and even solar-powered instruments. They can be mounted on, or be integral to, small tractors (sit-on mowers). Where large areas of turf need to be mown, 'gangs' or groups of mowers can be attached together and pulled by tractor, or as was previously done and is being promoted again by some landscape managers, by horse. Smaller areas such as domestic lawns are now often mown 'automatically' by radio- or computer-controlled systems. Increasingly, the form and power source of the mowers is not just determined by the function of the turf, labour available or direct cost, but also the running and maintenance costs, and indeed the 'environmental footprint' associated with the mower.

Alternative and low-energy mowers and other turf machinery

Lawn mowing and other turf maintenance activities are thought to contribute significantly to engine emissions, e.g. approximately 1% of total motor engine petrol consumption p.a. in the USA. Estimates vary between 2.7 and 5.5 billion litres of petrol and up to 500 million litres of diesel are used to mow and trim lawns each year. Two-stroke mowers also consume oil in their fuel, and most mowers use engine oil in their crankcases. Mowing of public land or sports facilities contributes 35% of this total fuel use, with maintenance of domestic lawns comprising the remaining 65% (USA Department of Energy, 2011). Professional lawn maintenance teams can mow for up to seven hours per day and use 4000 to 9000 litres of fuel each year, depending on land use, length of growing season and climate. Although lawn turf can sequester atmospheric carbon by as much as 0.14 tonnes C ha^{-1} $year^{-1}$, once fuel use, irrigation and emissions from fertilisers are take into consideration, emissions due to intensive lawn management can range between 2.2 and 6.3 tonnes C ha^{-1} $year^{-1}$ (Gillman *et al.*, 2023). Thus, there is a drive for alternative and more efficient forms of fuel. These include powering mowers with alternative hydrocarbon fuel sources, e.g. biodiesel, compressed natural gas, propane (liquefied petroleum gas), or electricity/battery-powered machines or solar-powered mowers. The alternative hydrocarbons still produce CO_2 but may burn more cleanly, reducing the proportion of other contaminants released to the atmosphere, or may be from more sustainable sources such as recycled fats and vegetable oils. Electricity is more sustainable too when the power stations that generate it rely on renewable sources such as wind, water or solar energy.

Grass clippings and thatch management

Clippings from mowing can either be collected by the mower catch-bin or left to be re-incorporated into the sward as organic matter. The pros and cons of this depend on the volume of clippings, the impact on the playability of the surface and the nutritional regimes employed. The removal of clippings to landfill sites, or even to centrally located composting facilities, results in additional transport costs and enhanced carbon footprint. It also results in a net loss of nitrogen from the lawn system, with a subsequent demand to replace this (usually through artificial fertilisers). Hence, mechanisms to deal with the clippings *in situ* are already popular and likely to grow as holistic, sustainable management approaches become more widely adopted.

Clippings are a source of plant nutrients, for example 3–5 % of dry matter is nitrogen, and so where incorporation is feasible then there are opportunities to reduce the volumes of additional fertiliser required. Incorporation of clippings is easiest when the volumes deposited are small but added on a frequent basis. In this way the dead material (thatch) is decomposed readily and incorporated into the soil organic matter. When thatch builds up though (>13 mm depth is considered undesirable), a thick layer of an organic matter 'mat' accumulates at the base and this can start to impede drainage and cause hydrophobic 'spots' in the turf. Similarly, the penetration of fungicides and liquid fertilisers are inhibited. Excessive thatch also alters the bounce of the ball, affecting playing quality, and may need to be removed mechanically.

The volume of thatch accumulation is influenced by a range of physical, chemical and biological factors. Excessive build-up is inhibited by the presence of a functional microbial population, the activity of macrofauna (primarily earthworms), good aeration and a medium to slightly alkaline pH (e.g. 6.7–8.5). Conversely, excessive thatch can result from compacted, poorly aerated and more acidic soil. Thatch accumulation is traditionally mitigated through cultural practice, such as topdressing (adding sand or other 'open' aggregates) and/or physical cultivation techniques, e.g. verticutting (also known as vertical mowing, scarification or dethatching) or core removal (aerification). As microbial degradation of the organic material is a key component, some products that promote microbial activation or even introduce new microorganism populations are commercially marketed for this purpose.

The effects of management on thatch development within a lawn composed of creeping bentgrass (*Agrostis stoloniferous* L. var. *palustris*) was studied by McCarty *et al.* (2007) over a 2-year period. In untreated swards, organic matter content increased by 32% but, in contrast, declined by 19% for turf where core removal (tining) was practiced in conjunction with verticutting procedures of various depths. Water infiltration rates increased with both verticutting (40–65%) and core removal (127–168%). Adding artificial populations of microorganisms in this study though was not found to enhance organic breakdown rates. The authors concluded that sand topdressing alone was insufficient for managing thatch-mat levels in an established creeping bentgrass lawn, and that coring plus verticutting provided significant additional advantages. Verticutting and similar activities are often carried out twice in the annual maintenance cycle – once at the start of the growing season and again at the end (i.e. in northern climates, March–April and September–October). Care should be taken, however, on those surfaces where the soil integrity is important – for example, verticutting and coring should be avoided on a cricket square where it is actually desirable to have a compacted soil (combined with a high clay content) to enhance the bounce of the ball.

Lignin is the principle residual component of thatch and adds to the hydrophobicity within the turf. Although the use of microorganisms to break down thatch has met with mixed results, the use of fungal-derived enzymes directly may be more promising. For example, laccase applications were used by Sidhu *et al.* (2022) and they found reductions in thatch thickness were similar to those achieved by physical cultivation, and indeed the optimum treatment was a combination of laccase and traditional cultural management. A single application of laccase was effective in improving water penetration too, but not as effective as the application of wetting agents.

Aeration and drainage

Heavy traffic and wear on turf causes soil compaction. Compacted soils reduce oxygen availability to the roots and increase bulk density – providing greater physical resistance to root development. They may also impede water infiltration and drainage. Puddling on the soil surface indicates a compacted soil. Different types of activity affect the degree of soil compaction and other stresses on the turf. Sheer volume of footfall traffic can cause compaction on sports turf, with certain locations on the pitch experiencing more compaction than others, e.g. the diamond pattern on soccer pitches (Fig. 9.10). Most compaction occurs in the top 30 mm of the soil profile, although movement of heavy machinery over a sward can induce compaction to a greater depth.

Over small areas compaction can be relieved by tining with a garden fork, although specialised machines cater for larger areas. As well as providing a macro-puncture hole in the soil surface, the tining action also helps create smaller fissures and micropores down the soil profile. Solid tines punch a hole into the soil surface, whereas hollow tines physically remove a plug (core) of soil. The advantage of the latter is that there is less likelihood of localised compaction adjacent to the hole. Tining machines can either comprise a roller with a series of solid, slit or hollow tines attached, or alternatively may utilise a vertical action, where soil plugs are removed at close spacing with minimal disruption to the turf surface. Normally tines are 75–100 mm long; however, longer, 150–200 mm tines are useful to deal with the afore-mentioned machinery-induced compaction. Cores removed can be anywhere between 6 and 18 mm in diameter.

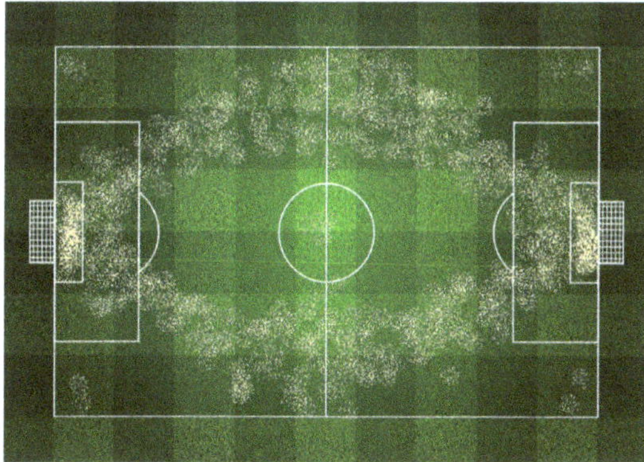

Fig. 9.10. The relative levels of soil compaction on a soccer pitch as determined by the typical footfall in different locations. This leads to a characteristic diamond pattern, with high wear and tear of the turf around active points such as the goalmouths.

At the top end of the machinery spectrum, some machines use vibrating tines pushed into soil that tilt slightly before removal – this tends to lift the turf surface as well as create airways beneath it. Hard, dry soils are not conducive to tining and most aeration/decompaction takes place in autumn, winter or early spring when the soil is moist but not waterlogged or frozen. Tining is also best avoided during dry periods as the activity increases the sward surface area by as much as 120%, hence increasing the evaporative demand on the grass and leaving it much more prone to desiccation. Topdressing (filling the holes with sand or similar substrate) and or irrigating immediately after tining can reduce this risk. Topdressing has the added advantage of filling the cores with sand, thus helping stop the soil particles themselves migrating back down into the voids once the soil plugs break down.

Turf reinforcement

This is used to improve wear resistance, particularly in sports turf, but systems also exist to facilitate vehicle movement over turf or for parking. Heavily walked grassed paths may also use polypropylene or steel wire reinforcement mesh to protect the turf and the soil structure from pedestrian traffic. In sports turf, sand root-zone reinforcement can be accomplished with thin polypropylene fibres or mesh to improve the structural strength of the substrates. Other systems, e.g. TSII (Motz Group) or Desso GrassMaster systems, reinforce the turf directly. These integrate living grasses with synthetic fibres either through the use of preformed mats that can be placed on the substrate, or as in the case of the Desso system, where synthetic turf fibres 200 mm long are injected (or 'sewn') by machine into the natural turf root zone. The advantages of these systems are that if the natural turf wears down then the synthetic materials can still allow some functionality and, as there are usually dyed green, they have some aesthetic attributes too.

9.7 Shade

Shade tolerance in amenity turf has always been a consideration when attempting to establish grass under heavy shade aligned with buildings or trees. It has become particularly relevant, however, over the last decade or so where sports stadia have become larger, and spectator stands or even enclosed roofs block out a proportion of the solar irradiance. Some turf species have greater shade tolerance than others, notably *Festuca arundinacea, Deschampsia cespitosa* and *Poa supina,* Species that are more shade tolerant can partially adapt better. Dong *et al.* (2022) showed that under increasing shade the more tolerant *Poa supina* cv. SupraNova increased its root length and root surface area, before these aspects declined in the deepest shade. In contrast, the intolerant *Lolium perenne* cv. Lark immediately reduced root development even under light shading. Overall though, the adaptations plants utilise to cope with shade tend to work against the characteristics of quality sports turf (e.g. stretched 'etiolated' leaf/shoot growth). Hence, most solutions tend to rely on technical approaches including the use of mobile lighting rigs, chemical growth inhibitors to encourage more compact plant habit, moveable pitches where turf is growing in full light then translocated in sections to the shaded area of the stadium, or indeed frequent re-turfing or using turf-stabilising materials to help maintain the playing quality. Many of the approaches adopted, of course, are not particularly sustainable in terms of resource management. Some stadia are designed to take account of these factors. For example, certain new stadia have translucent roofing panels on their stands to permit the transmission of photosynthetically active radiation to the pitch. Similarly, the Emirates Stadium in London, UK was designed to avoid complete enclosure of the stand sides, thus allowing irradiance to penetrate in above the seating areas. A number of strategies have involved growing 'alternative' pitches in full light in parallel to the one in the stadium, and these are moved in as pallet sections of pre-grown turf. In Phoenix, Arizona, USA, the whole pitch at the Arizona Cardinals American Football Stadium is grown and maintained outside the stadium and moved in under the stand via a complex roller system!

Stadia present other problems to the turf too. Other environmental problems include a lack of air movement, which can encourage fungal pathogens. There may be pressure to use the stadia for alternative sources of income. The Madejeski stadium in Reading, UK used to host soccer and rugby union on alternative weekends – sports that impose different and compounding pressures on the sward. Other stadia may be used for music concerts, boxing matches and even car racing. Often the turf is covered for protection under such circumstances, but nevertheless the exclusion of light, albeit only temporarily, can weaken the grass.

9.8 Nutrient Management

Fertilisation is one of the more controversial elements of turf management with respect to its environmental impact (see Chapter 3). As with other plant species, turfgrasses have a requirement for macronutrients, mainly nitrogen, phosphorus and potassium but also calcium, magnesium, and sulphur with essential micronutrients including iron, manganese, zinc, copper, boron, molybdenum and chlorine. Much of the plant growth is determined by the ratio of carbon (derived from photosynthesis) to nitrogen; indeed, this is a key factor that influences much of the dynamics of 'turf ecology' in general.

Optimum fertiliser management for turfgrass is a controversial topic. Many garden lawns and amenity areas are in effect self-sustaining; they rely on no artificial fertiliser at all, with the required nitrogen being supplemented by atmospheric or bacterial fixation or derived from 'recycled' organic matter degradation. Similarly, phosphate and potassium are sourced initially from the soil and again recovered from natural processes where grass clippings are retained and broken down *in situ*. In 'high-quality' sports turf, however, fertiliser may be applied regularly to ensure grass growth is strong and resilient to the wear imposed on it by the playing activities (or in ornamental lawns, regular close mowing). Even here, however, seemingly high-quality lawns may subsequently succumb to disease, often attributed to a high nitrogen regime.

Nutrition management then is essentially a response to the demands placed on turf when utilised as a sports surface or where great emphasis is placed on the aesthetics. This is particularly so where traffic levels are high, the play characteristics set specific criteria – ball roll speed, etc. – and/or where the environmental conditions are suboptimal for turf growth.

The level and frequency of nutrient inputs therefore reflect the function and management of the sward. Fertiliser application is required for intensively managed turf, as the constant wounding of the grass plant through mowing taxes its resources and puts the plant under stress (Hull *et al.,* 2020). Frequent mowing of a sward grown on a sand substrate and where clippings are removed continuously will require much greater fertiliser input than those situations where the turf is on an organic or clay soil (which can store nutrient ions through greater cation exchange capacity) and where the clippings are retained and re-incorporated within the turf. It should be noted that under certain circumstances fertiliser applications can exceed the requirements of the turf, and this not only poses potential for nutrient leaching and runoff, but also faster shoot growth rates occur than is actually desirable, hence more frequent mowing and higher labour costs.

Nitrogen, phosphate and potassium

Nitrogen is normally considered the instrumental element in determining turf nutrition regimes (it comprises 3–5% of plant dry weight). It is essential for healthy growth, being a component of chlorophyll, plant enzymes and DNA. Excessive application, however, is often blamed for inappropriate carbon-to nitrogen (C:N) ratios, excessive shoot growth, poor root development, increased susceptibility to certain pathogens and reduced tolerance to abiotic stress including cold, heat, drought and pedestrian or vehicle traffic. Nitrogen is absorbed as nitrate (NO_3^-), nitrite (NO_2^-) and/or ammonium (NH_4^+) ions. Nitrogen fertilisers can be either fast- or slow-releasing and are absorbed conventionally through the roots, but also as liquid feeds there is some uptake via the foliage. Fast-releasing forms which are readily available for absorption include ammonium nitrate (NH_4NO_3), ammonium sulphate ($NH_4)_2SO_4$), potassium nitrate (KNO_3) and urea ($CO(NH_2)_2$). These fast-releasing, highly available forms of nitrogen tend to be characterised by high solubility, rapid growth responses in the turf and being available at a range of temperatures, but they have potential to cause phytotoxic responses (leaf scorching) and be excessively leached through the soil or lost through volatilisation of gaseous ammonia. Slower-release forms include urea formaldehyde ($C_3H_8N_2O_3$), isobutylidine diurea ($C_6H_{14}N_4O_2$) and sulphur-coated urea (controlled-release granules), as well as organic materials such as activated sewage sludge (Milorganite), turkey litter, and soybean-based extracts. Although nitrogen can still be lost from the system through leaching, denitrification and volatilisation of slow-release forms, this is less of a problem compared to the faster-releasing compounds. The fate of nitrogen after fertiliser application is complex. Ammonium and nitrate ions may be immediately available to the plants or may require some further degradation of the source material. Plants may compete with microorganisms for the available nitrogen, and a proportion can be 'locked-up' by microbial dynamics. This may become available again at a later stage or alternatively be lost as nitrogen (N_2) and nitrous oxide (N_2O) gases and by denitrifying bacteria.

Phosphorus (P) has a fundamental role in the energy dynamics of cells through being a constituent of adenosine triphosphate (ATP) and so helps regulate cell division and growth. It tends to be recycled and better conserved within plant tissues compared to nitrogen. Similarly, it is relatively immobile in the soil and less readily leached. Although present in the soil/substrate, phosphate is not always readily available to the plant; it is often converted to insoluble forms such as aluminium phosphate ($AlPO_4$) or iron phosphate ($FePO_4$), where the solubility may be dependent on soil pH. Due to the dynamic equilibrium associated with these compounds in the soil matrix, however, as plant roots absorb phosphate ions and the concentration in the soil solution decreases, more are, in turn, released from the compounds, again becoming available to the plant in due time. To this extent, apart from new swards established on sand or other substrates with inherently low phosphate levels, phosphate is rarely a limiting factor to turfgrass development. Typical phosphorus-containing fertilisers include superphosphate and triple superphosphate (the active component in both being $Ca(H_2PO_4)_2$). These are formed from phosphate ores treated with sulphuric or phosphoric acid. The long-term sustainable use of these ores is now questioned, and alternative sources of phosphate may be required in future.

Potassium normally constitutes about 2.5% of turfgrass dry matter and is used to regulate cell osmotic relations and water uptake. Potassium is included as a fertiliser as it reputedly increases tolerance to wear and other stresses, although in reality this may be more to do with developing an appropriate nitrogen-to-potassium (N:K) ratio. Stress tolerance may relate more to avoiding an imbalance of nutrients, rather than individual ions specifically having a 'magic bullet' direct role in stress alleviation.

The ratio of the principal elements is varied depending on the time of fertiliser application and whether the aim is to address specific needs or deficiencies. In 'complete fertilisers' (ones that contain nitrogen, phosphate and potassium) the amount of nitrogen applied is often two to three times greater than potassium, with the phosphate ratio dependent on the soil type. So, a 10:1:5 $N:P_2O_5:K_2O$ may be applied in soils with high residual phosphate levels, whereas on a sand-based substrate, a ratio of 2:1:1 is more appropriate. Applications are best split over the year to even out availability and avoid excessive leaching, so a typical programme for a fescue lawn in the north-east USA (Anon., 2003) where the clippings are returned may follow:

Early June = (2:1:1) 50 kg ha^{-1} N, 25 kg ha^{-1} P_2O_5, 25 kg ha^{-1} K_2O
Early September = (2:1:1) 50 kg ha^{-1} N, 25 kg ha^{-1} P_2O_5, 25 kg ha^{-1} K_2O
Mid-November = (4:3:0) 50 kg ha^{-1} N, 37 .kg ha^{-1} P_2O_5

Late-autumn fertiliser application has been cited as improving the coverage and visual quality of turf, particularly in cold climates, although responses have not been uniform across species, e.g. *Festuca rubra* (red fescue) and *Agrostis stolonifera* (creeping bent grass) showed better responses compared to *Agrostis canina* (velvet bent grass) and *Poa annua* (annual meadow grass) (Kvalbein and Aamlid, 2012).

Leaching and runoff

As legislation increases in an attempt to tackle issues linked to leaching and runoff, alternative sources of fertiliser application or improved irrigation management are being investigated. Most sources of nitrogen leaching and runoff (although not exclusively so) relate to nitrogen application at higher-than-recommended rates, excess rainfall after fertilisation and fertilisation at a time when turf is not actively growing – it being argued that properly maintained lawns provide a relatively effective system for absorbing nutrients (Fig. 9.11).

Using more precisely controlled irrigation and fertigation systems (liquid fertiliser applied in the irrigation) was deemed promising by Snyder *et al.* (1984) in reduced nitrogen leaching. A tensiometer-controlled irrigation system was evaluated in the management of *Cynodon dactylon* × *C. transvaalensis* (Bermudagrass) grown on a free-draining sand substrate. Nitrogen was applied bimonthly at the

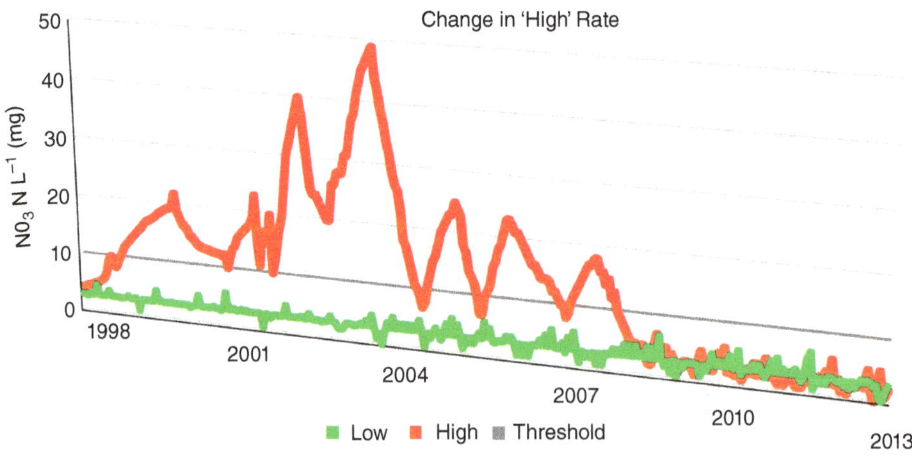

Fig. 9.11. Nitrate-nitrogen (NO_3 N) concentration (mg) in leachate from *Poa pratensis* treated at two nitrogen (N) rates. Low rate = 98 kg N ha^{-1}. High rate – originally 245 kg N ha^{-1}, then lowered to 196 kg N ha^{-1}. The threshold level of 10 mg L^{-1} relates to drinking water quality maximum contamination level. Lowering the higher rate brings the leachate into 'legal' levels after a period of time. (Modified with permission from Frank *et al.*, 2016.)

rate of 5 g m^{-2} $month^{-1}$ using either granular ammonium nitrate or sulphur-coated urea pellets, or as ammonium nitrate in solution within the irrigation. Irrigation itself was either applied daily or when the tensiometer 'sensed' a moisture deficit in the sand. A combination of tensiometer-controlled irrigation and nutrients being applied via fertigation or urea pellets resulted in the lowest levels of N leaching (<1% to 6% of total N) while maintaining acceptable turfgrass quality. In contrast, ammonium nitrate granules in combination with daily irrigation produced the greatest N losses (i.e. 22% to 56% of total N applied).

Even the grass species selected may influence nutrient leaching. Trenholm *et al.* (2012) found that *Stenotaphrum secundatum* cv. Floratam (St Augustine grass) was more successful than *Zoysia japonica* cv. Empire (zoysia) in mitigating nitrogen losses. Increasing frequency of irrigation of *S.* cv. Floratam across a range of different nitrogen fertiliser rates (49–490 kg ha^{-1}) had little effect on the amount of nitrate leached from this sward, whereas similar regimes with *Z.* cv. Empire corresponded with more leaching as the frequency of irrigation increased (Fig. 9.12). Subsequently, lower rates of nitrogen fertiliser were recommended for *Zoysia* spp. not only to reduce leaching but also to help avoid disease and improve the coverage of the turf. For both genotypes, lawn quality was improved by applications of nitrogen in early summer (Fig. 9.13).

Soil conditions and differences in root morphology and/or architecture play an important role in nutrient uptake capacity. Evaluations on *Poa pratensis* (Kentucky bluegrass/smooth-stalked meadowgrass) cultivars showed that nitrate losses averaged 4.7% of applied nitrogen in non-compacted soil across a series of trials, whereas subsurface soil compaction increased losses to 8.9%. Tining and verticutting have also been cited as improving infiltration rates and reducing runoff potential.

Phosphorus, despite being less mobile than nitrogen in the soil column, contributes to runoff pollution

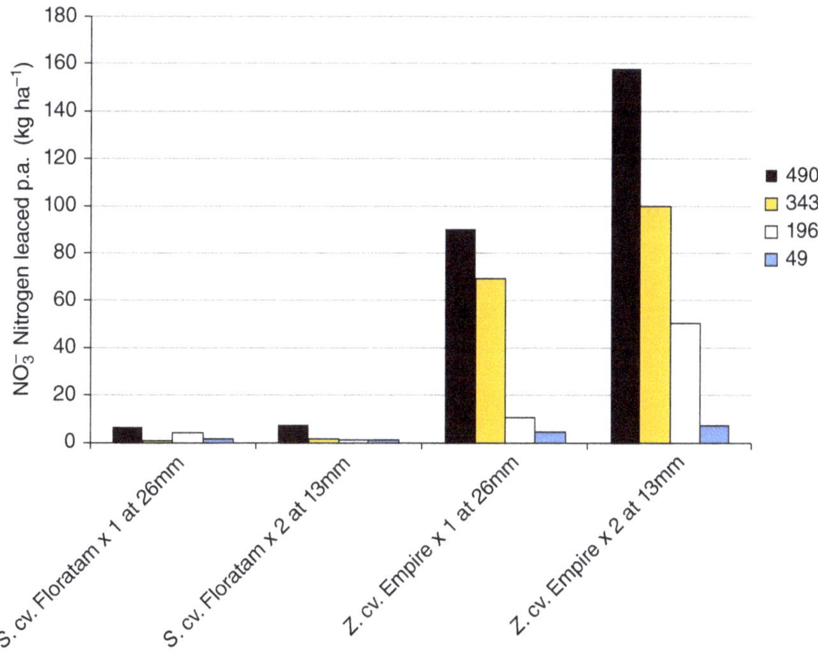

Fig. 9.12. The effect of fertiliser application rate (490, 343, 196 or 49 kg ha^{-1} nitrogen) on nitrogen leaching on lawns composed of *Stenotaphrum secundatum* cv. Floratam (St Augustine grass) or *Zoysia japonica* cv. Empire (zoysia grass), when irrigation applied was once per week at 26 mm depth or twice per week at 13 mm depth. (Modified with permission from Trenholm *et al.*, 2012.)

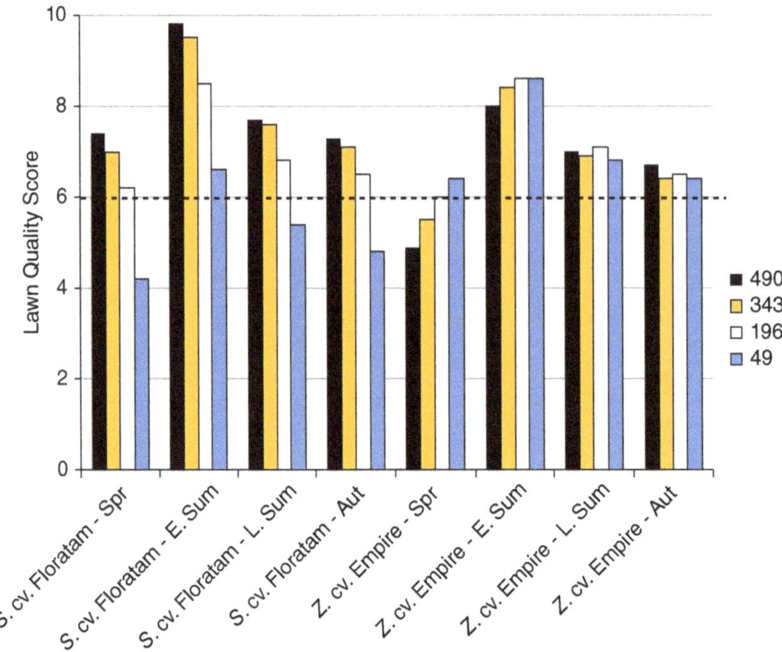

Fig. 9.13. The effect of fertiliser application rate (490, 343, 196 or 49 kg ha^{-1} nitrogen) on lawn quality (6 being deemed acceptable lawn quality) of *Stenotaphrum secundatum* cv. Floratam (St Augustine grass) or *Zoysia japonica* cv. Empire (zoysia grass), when fertiliser was applied either in Spring (Spr), early summer (E. Sum), late summer (L. Sum) or Autumn (Aut). (Modified with permission from Trenholm *et al.,* 2012.)

and subsequent eutrophication of water courses. This is especially so where the extractable P levels are high, such as in a recently applied liquid fertiliser, and where there is limited opportunity for the ions to bind to soil particles, e.g. pure sand zones as the sand particles tend to 'fix' P less effectively than clay particles. Unlike nitrogen, where the pollutant source can dissipate relatively easily with time, phosphate pollution is more problematic due to the persistent nature of this element. Mismanagement of fertiliser application on turfgrass is an important component of this. Column leachate studies have shown that P levels from intensively managed turf can be as high as 1 to 13 mg l^{-1}, well in excess of designated standards for surface water (0.3 mg l^{-1}). Tissue phosphate levels have been cited as being optimal at 3.4–3.9 g kg^{-1} in grasses (e.g. *Agrostis stolonifera*) and fertiliser applications often exceed that which is needed. Controlled-release forms of phosphate tend to provide less leachate than soluble fertilisers, especially when applied at low rates (e.g. 5 kg ha^{-1}) and irrigated at low frequency. King *et al.* (2012) evaluated conventional phosphate fertiliser regimes for golf courses, with a lower input regime (reduced rate, low dose applications, and organic formulations). Management practices that involved lower application rates and the use of organic phosphorus formulations resulted in substantial reduction in the dissolved, reactive and total soil phosphorus concentrations. Due to the perceived problems of high P leaching in sand-based substrates, it has been recommended that a two-pronged approach is adopted; turf managers need to explore the options available through organic formulations and lower dose applications, and fertiliser manufacturers should develop and make available commercial fertiliser blends with reduced or zero amounts of phosphorus incorporated.

9.9 Irrigation

Increasing demand for fresh water from domestic and industrial sources has decreased the allocation of water for irrigation of non-food crops, especially in warmer/drier climatic regions. A lack of irrigation water is a limiting factor to the development of new amenity turf areas in such locations, and existing swards now experience water rationing and leaner irrigation regimes, as well as alternative sources of water (saline water or recycled waste water) being examined. In parts of the USA, 75% of natural water supplies may be considered unfit for human consumption due to moderate or severe degrees of salinity and a proportion of this could be used for irrigation. Likewise, recycled water accounts for irrigation in 37% of all golf courses in the southwestern USA (Throssell *et al.*, 2009). In addition to alternative water sources, developing strategies to maintain acceptable turf quality with considerably less water use have been a primary goal for turfgrass researchers and managers.

Irrigation is not just required on turfgrass when there is inadequate precipitation. It may also be called upon when plant evapotranspiration exceeds the available supply in the substrate (i.e. substrate moisture-holding capacity is not great), or where moisture movement upwards through capillary action is insufficient to meet the demands of the transpiring leaf. Hence, the characteristics of the root zone, particularly depth of rooting, influences how much moisture is available to the leaf canopy. Shallow-rooted turf requires more frequent, but less intensive, irrigation than deep-rooted turf. In a somewhat circular relationship, root growth is very much determined by soil moisture and conversely irrigation practices profoundly affect root growth and survival. There is a form of irrigation known as 'syringing'; this is used to offset temporary wilting and is largely targeted at keeping the foliage hydrated or cool, rather than penetrating the substrate. It is employed to help the turf avoid the excessive heat of midday or is used on a newly established piece of turf where root penetration into the substrate is limited. As such, these irrigation schedules tend to apply only limited volumes of water at any one time. Not all irrigation practices just relate to moisture requirements and leaf cooling either. Irrigation is used to help better distribute fertilisers down the soil profile and avoid localised high concentrations of nutrients sitting on the foliage, potentially caused shoot and leaf damage.

Irrigation control and scheduling

The principle behind effective irrigation control is that irrigation should be applied whilst the substrate matric potential is high enough that the soil can still supply water to meet plant demand, in essence avoiding excessive moisture deficits and the potential of incurring hydrophobic zones in the sward. The aim is to avoid true drought stress to the grass whilst still also allowing moisture levels to decrease between consecutive irrigation events. Irrigation traditionally was managed by time clocks or the grounds personnel's perceptions of irrigation needs. However, feedback systems that monitor soil moisture status or algorithms based on climatic data (evapotranspiration and rain-based sensors) can evaluate moisture status accurately. Soil moisture sensors, evapotranspiration controllers and/or rain sensors improve irrigation effectiveness and are known as 'smart controllers'. These are designed to irrigate based on measured or estimated water depletion so that water is added to meet plant water needs while minimising losses. Soil moisture sensors can be based on electrical capacity or impedance signals initiated by a probe (frequency domain) and where the signal alters based on the soil moisture content. Alternatively, neutron probe and tensiometer-based systems can be utilised.

The advantage of sensor-based systems is that they help control irrigation supply more effectively, reduce the volume of water wasted as well as maintain acceptable turf quality. Compared to standard timer-based control clocks programmed on the basis of historical water use, irrigation controlled by soil moisture sensors has shown savings of anywhere between 0% and 70% based on location, sward composition and other management regimes. Rain-based sensor systems reduce water use over conventional controls by 7–45%, again depending on local circumstances. The latest generation of controllers are using machine learning (ML) software (artificial intelligence) to improve water use efficiency yet further, whilst minimising turf injury. Dahl *et al.* (2024) showed that using a trained ML algorithm in an autonomous smart irrigation system (ASIS) controller proved to be robust and effective in regulating irrigation cycles and soil moisture content over time and reduced runoff by 74% compared to the latest conventional controllers.

Estimates and actual evapotranspiration (ET) rates are used to control irrigation schedules. Treatments based on ET controllers (Grabow *et al.*, 2012) resulted in good quality turf but applied 11% more water, on average, than timer-based systems. This may have been due to an overestimation of the reference evapotranspiration value. In reality, different turf species transpire water at different rates. For similar heights of turf, Braun *et al.* (2022) estimated rates were greatest for *Lolium* spp., 7.79 mm day^{-1}; then *Agrostis* spp., 6.12 mm day^{-1}; *Schedonorus arundinaceus*, 5.90 mm day^{-1}; *Festuca* spp., 5.52 mm day^{-1}; and *Poa* spp., 5.35 mm day^{-1}. They argue that crop quality can be maintained even when the deficit irrigations imposed relate to as little as 59% to 74% of normal ET values – these being the irrigation regimes that allow species to adapt and regulate their water use, without causing visual injury. Reduced requirements have been noted in warm-season grasses too, without loss of quality, e.g. 75% to 83% of normal ET for *Cynodon dactylon* × *Cynodon transvaalensis* (Bermudagrass) and *Paspalum vaginatum* (seashore paspalum). The *Cynodon* cv. Midiron was found to be particularly tolerant of the lower irrigation regimes, but one downside was that deficit irrigation delayed the time of growth activation in the spring ('greening-up') for example by 6 weeks with 60% of the standard ET rate (Bañuelos *et al.*, 2011). In a review of irrigation management covering 83 studies, Gómez-Armayones *et al.* (2018) showed that in most cases regulating deficit irrigation to 50–60% of potential ET resulted in turf quality being maintained at acceptable levels but with lower water consumption. Conversely, high irrigation levels (above field capacity) increased the risk of nutrient leaching. They conclude turf irrigation should be scheduled to apply water at moderate levels of deficit irrigation, sufficient to maintain turfgrass quality but limited enough to promote a deep and extensive root system.

Frequency as well as volume of irrigation alter turf dynamics too. Less frequent irrigation has been recommended to improve turf quality through enhanced shoot density and encouraging deeper rooting, thus mitigating drought impacts (Jordan *et al.*, 2003). Where irrigation is controlled automatically, applications tend to take place during the night or early morning, thus reducing losses from direct evaporation but also avoiding the periods where the sward is in use or other maintenance activities are taking place. To avoid excessive surface runoff, the irrigation may be delivered in a series of cycles, thus allowing time for the water to infiltrate fully into the turf and substrate.

The over-application of irrigation in practice can be a consequence of poor distribution of water across/within the turf. Hydrophobic patches (dry spots) encourage grounds personnel to prolong irrigation in an attempt to ensure effective irrigation coverage. This means that non-problematic areas receive considerably more water than required and that water is wasted. The use of surfactants or wetting agents can overcome this problem, by reducing the surface tension of the dry spots and aiding the penetration of water. Park *et al.* (2005), for example, claim that surfactants improve *Cynodon* (Bermudagrass) performance whilst reducing the requirement for irrigation overall. During the drier seasons, surfactants reduced irrigation volumes by 50% and were associated with increased photosynthetic rates, stronger shoot growth, higher soil moisture levels and reduced the symptoms of hydrophobicity.

Water quality and salinity stress

Turfgrass experiences salinity in four major ways. These are growing in soils that are naturally saline, locations that experience sea breezes and salt deposition, exposure to salts in recycled water used as irrigation and regular fertilisation processes. Grass species vary in their tolerance to saline soils and water sources. Those adapted to coastal locations or saline deserts having greatest tolerance. Both the chloride (Cl^-) and the sodium (Na^+) ions can be toxic in high concentrations and interfere with other important ion uptake, e.g. potassium (K^+) ions. Tolerance is reportedly high in grasses such as *Paspalum, Puccinella, Sporobollus* and *Zoysia* (Liu *et al.*, 2023; Fig 9. 14).

Of the common turfgrasses, *Agrostis stolonifera* (creeping bentgrass), *Cynodon dactylon* (Bermudagrass) and *Stenotaphrum secundatum* (St Augustine grass) tend to have greatest tolerance, with *Schedonorus arundinaceus* (tall fescue) and *Lolium perenne* (perennial ryegrass) intermediate, and *Poa pratensis* (Kentucky bluegrass), *Agrostis capillaris* (browntop bentgrass) and certain forms of *Festuca rubra* (red fescue) least tolerant. While using saline water has become an increasingly accepted strategy for conserving potable water, the detrimental effects to plants and the underlying soil ecology and physical structure resulting from salt accumulation has become apparent. The level of salinity determines potential use for turfgrass situations. Electrical conductivity (EC) values of 0.75–2.25 dS m^{-1} should be restricted to non-sensitive species and used in situations where the soil drainage characteristics encourage ion leaching; 0.25–0.75 dS m^{-1} is used where there is some

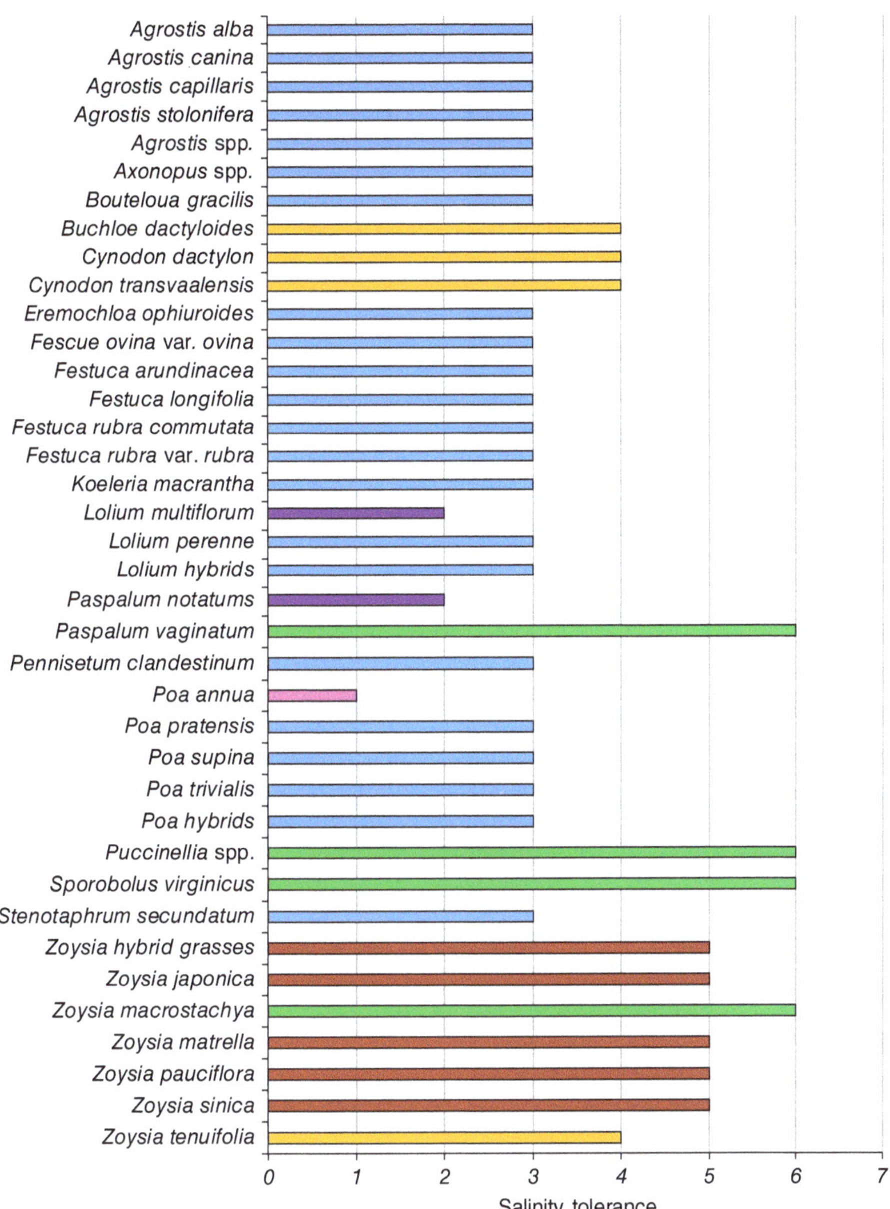

Fig. 9.14. Grass species and salinity tolerance ranges: 1 = 0 to 4 dS m^{-1}; 2 = 2 to 8 dS m^{-1}; 3 = 2 to 30 dS m^{-1}; 4 = 4 to 30 dS m^{-1}; 5 = 8 to 30 dS m^{-1}; and 6 = 8 to >30 dS m^{-1}. (Modified with permission from Liu *et al.*, 2023.)

leaching potential and less than 0.25 dS m^{-1} presents water supplies that are deemed low risk. In addition to problems induced by high sodium and chloride ion concentrations, poor-quality water can also be defined by high boron and bicarbonate ion levels that likewise cause phytotoxicity effects.

Salinity stress is particularly problematic at seed germination and early establishment stages. Studies have evaluated the type of irrigation application (overhead sprinkler vs subsurface capillary) with water of either saline (EC = 2.8 dS m^{-1}) or potable (EC = 0.6 dS m^{-1}) quality (Schiavon *et al.*, 2013).

Results with *Poa pratensis* (Kentucky bluegrass) and *Schedonorus arundinaceus* (tall fescue) showed that sprinkler-irrigated plots were more favourable; highest EC and sodium concentrations, 26.3 $mmol_c$ l^{-1} (millimoles of charge per litre), correlated with sub-irrigation with saline water. Despite this, sub-irrigation was shown to be a feasible way of establishing cool-season turfgrass, but subsequent development was delayed compared to sprinkler irrigation. In similar approaches Sevostianova *et al.* (2011a) used established plots of seven different cool-season grass species; however, in this case both sprinkler and sub-surface irrigation with saline water in semi-arid conditions was deemed to reduce turf quality below an acceptable level, with the exception of *Festuca arundinacea*. In contrast, evaluations of warm-season grasses generally showed greater tolerances to saline irrigation, although poor quality was evident with seasonal effects, i.e. during winter periods when cultivars were exposed to both saline stress and suboptimal temperatures (Sevostianova *et al.* 2011b).

In addition to naturally saline water sources, reclaimed water from other industrial processes or from agricultural land or practices (e.g. cleaning fruit and vegetables) can be used as an irrigation source. Reclaimed water that receives treatment and a high level of disinfection is used for turf irrigation and not deemed a threat to public health. Such sources of water will come with additional chemical elements, some of which will be N, P and K, so fertigation regimes for the turf need to take account of this. Depending on the turf type and volume, applications of these waste waters can, of course, help improve the quality of water entering ground aquifers or local rivers due to the filtering effect of the grass. Snyder and Cisar (2000) indicated that golf courses could tolerate reclaimed water with up to 10 mg/l nitrogen without the sources of nitrogen leaching into aquifers.

Lawns and sustainable water use

If irrigation management of recreation areas is a political topic in terms of environmental management, almost more so is the domestic lawn. Many homeowners link a neat, well-tended lawn to aspects of affluence and even social respectability or see the lawn as a cultural link back to a colonial past – where grand houses were offset by verdant lawns. Hurd (2006) suggested that for the western USA, water cost, education and regional culture are significant determinants of landscape choice and allied irrigation requirements. A number of studies report that volumes of water used in the landscape relate to the proportion of public to private land and income level within different residential groups – more affluent areas using more water. It is perhaps no surprise then that even in relatively dry regions such as the central USA, surveys reported that 45–60% of homeowners thought it was moderately to very important their lawns looked green all the time. Of those surveyed 65–77% ranked water conservation at the same level of importance as maintaining a green lawn (Bremer *et al.*, 2013). Water demand for landscape purposes places a strain on water resources. Landscape irrigation in the USA had been cited as between 9% and 60% of total municipal water use, whereas in parts of Spain, it contributes 38% in spring to 69% in summer (46% of annual total water use). In the more arid parts of the USA, for example, Los Angeles, California and Albuquerque, New Mexico, residential water use constitutes >50% of each city's total water use, and outdoor water use is often a large component of overall demand (Wang and Chermak, 2021). In southern Nevada, which includes Las Vegas, residential water use accounts for 60% of the entire region's water consumption and, of that, 70% is used for landscaping. The Environmental Protection Agency in the USA reported that a typical lawn consumes 37,854 litres of irrigation water beyond rainwater per year (Dobbs *et al.*, 2013).

The lack of understanding about turfgrass requirements for water and how much is actually being applied are significant issues relating to excessive water use. Despite potable water being a precious resource, 61–63% of residents did not know how much water their lawns required and 71–77% did not know how much water they applied to their lawns when they irrigated. Over-irrigation is a common problem in domestic landscapes, especially in the transition periods between summer and autumn when ET demand decreases quickly, but irrigation systems schedules are not re-set accordingly (Salvador *et al.*, 2011). Irrigation management can be poor too, when irrigation control systems are too complex, and homeowners mis-set them through a lack of comprehension or not bothering to read the instructions too closely!

There are opportunities to save water through better design too. In city centre streetscapes, lawns are often incorporated in almost a tokenistic manner, with relatively small areas of turf being used alongside roads and pavements, as short strips or squares in formal civic gardens, or as centres to traffic roundabouts. Such limited areas of turf are 'high maintenance', both due to the demands placed on them from an aesthetic perspective and from the extreme environmental factors they may be exposed to (high evaporative demands due to urban heat island effects, air pollution, traffic wear and compaction, road-salt stress, etc.). Irrigating such small and irregular-shaped swards via sprinklers can be very inefficient, due to the volume of water that misses the target area. In these circumstances, larger areas of grass or other forms of green infrastructure may represent a 'better return' for the resources, time and effort required to maintain them to a sufficiently high standard.

Attitudes to turf and water conservation

Due to its relative rarity in natural landscapes, paradoxically, turf is valued highly in countries with a Mediterranean climate, despite its high water-demanding characteristics. In Spain, lawn use was linked to young and wealthy homeowners (Padullés Cubino *et al.,* 2016). The price of water though does seem to be a moderating factor on the domestic use of water for irrigation. Poor irrigation performance in Spain correlated with situations where high homeowner income was combined with low water charges. There is some evidence that increasing charges per unit of water can result in savings (Baker, 2021). In parts of Spain though, attitudes have been hard to change, with some residents considering that the aesthetics of landscapes and irrigation were more important than saving water per se, this despite Spain being considered as a drought-prone country (Salvador *et al.,* 2011). Nevertheless, strategies are being developed to reduce water use in lawn turf management in such locations; these include drought awareness campaigns, lawn irrigation restrictions and subsidies for water-saving capital investments (Price *et al.*, 2014). Research by Hurd *et al.* (2006) in the drier regions of the USA indicated that education programmes and 'moral persuasion' can have a positive influence toward implementing water-conserving landscapes. Financial incentives are now also used in parts of the USA for homeowners to either reduce their area devoted to lawns or to replace them completely with xeric landscape designs and flora – 'cash-for-grass conversion'. (Zhang and Khachatryan, 2020).

Studies in New Mexico showed homeowners were conscious of the water resource challenges faced by communities and were prepared to help regulate use of the state's water resources. There was support to limit traditional turfgrasses (e.g. *Poa pratensis* [Kentucky bluegrass]) and to increase the areas planted to native, natural and water-conserving landscapes; for example, 92% favoured limiting turfgrass to less than 25% of the area around public buildings. Financial incentives or penalties can be important. Baker (2021) found that converting gardens dominated by lawns to desert xeric ones in Las Vegas (USA) reduced monthly water use by 19–21%. Moreover, these savings remain stable over time, suggesting that homeowners do not increase other water-intensive activities after conversion but simply reduce outdoor irrigation. This study also showed that the homeowners with the highest pre-conversion water demand achieve the greatest unit savings. In parts of California (USA), there is the incentive of US$150 rebate for every 50 m^2 of natural turf replaced with synthetic turf as local governments attempt to reduce water consumption. Other schemes encourage the removal of turf but replace it with other forms of land cover. Domestic landscapes covered by woodchips, bare soil and gravel as well as artificial turf are becoming common (Pincetl *et al.,* 2019); such approaches are not necessarily particularly 'environmentally friendly' either though, i.e. covering surfaces with artificial fibres or aggregates does little for biodiversity or microclimate cooling! Overall, lower-income households tended to show simpler changes, including simply letting the turf die. Higher-income homeowners responded by initially replacing the grass with the alternative landscape surface materials, then over time adding in pathways and xeric gardens later. These subsequent changes resulting in more plant diversity with 64% of properties contained six or more species. Shrubs became more common, including the use of Californian native species.

Water use for turf is increasingly likely to be determined by both the cost of the resource and legislation. In San Diego, California, USA, local authorities already restrict irrigation to less than 12 minutes per week. Water can be saved by restricting the use of the facility too. In Victoria and New South Wales, Australia, authorities restrict the quantity and timing of irrigation on all sports surfaces, resulting in the temporary closure of some facilities to focus irrigation onto a limited number of surfaces (James, 2011). In Spain, new golf course developments must be self-sufficient for irrigation water, and these are encouraged to link up with housing schemes in their locality where waste water from the housing can be collected and recycled for irrigation purposes.

In many parts of the world, climate change will increase the severity of drought that turfgrass experiences, and there will be even more emphasis on management that preserves adequate root zone soil water. Most research to date on drought tolerance has focused on *Poa pratensis* (Kentucky bluegrass – considered medium ranking for drought tolerance – Fig. 9.15) or *Schedonorus arundinaceus* (tall fescue – considered to have good tolerance – Fig. 9.15) and more research is required on the other cool-season grasses (Braun *et al.,* 2022). To help ensure turfgrass plantings survive in the warmest regions, further drives on cultivar selection/breeding for greater drought and heat tolerance will be necessary (Hatfield, 2017). Changes in temperature and precipitation variation will increase the potential for additional abiotic and biotic stresses on turfgrasses, with even more emphasis on appropriate management.

One aspect that also needs further research is responses to CO_2. As most grasses respond well to higher CO_2 levels in controlled environments, the impact of higher atmospheric CO_2 may be beneficial to the turfgrasses *in situ,* but whether this additional 'fertiliser affect' can offset the drawbacks due to more variable and extreme weather events is less clear. Increases in CO_2 promote turfgrass growth and by doing so, positively affect water use efficiency (more biomass for the same unit of water used). This may decrease the potential negative effects of more variable precipitation regimes (a consequence of climate change) because of the revised relationship between grass plants and soil moisture availability. Indeed, genotypic variation in response to soil water deficits provides a foundation for screening turfgrass species to adapt to climatic stresses.

9.10 Pesticide Use and Integrated Pest Management

Pesticides are still regularly used in turf management, but the intensity is again strongly dependent on the function of the sward, being higher on golf courses and other intensively managed turf. Attitudes to pesticide use and actual practice vary strongly based

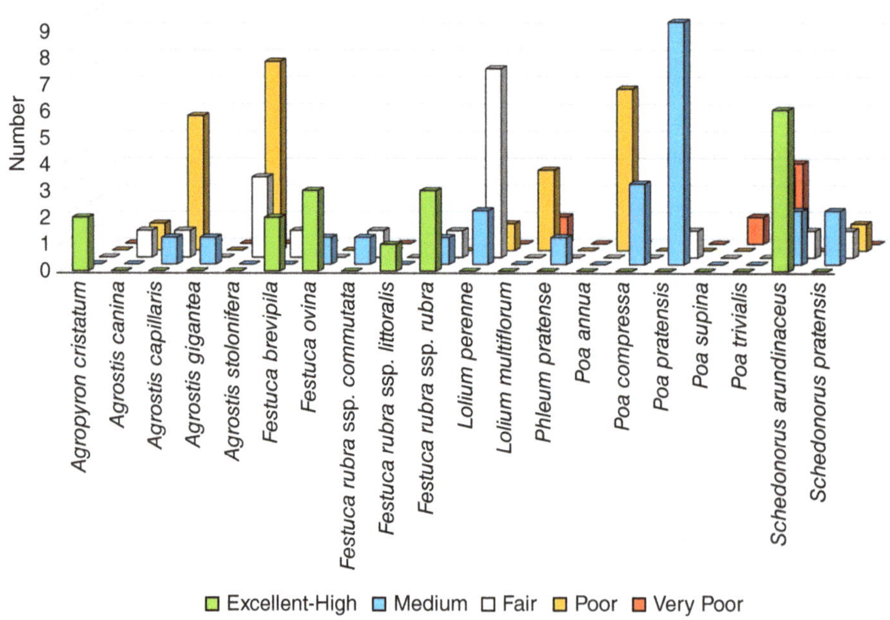

Fig. 9.15. The number of citations that have ranked different 'cool-season' turfgrass species on their drought tolerance (green – high to excellent tolerance compared to red – very poor tolerance). Not all taxa have experienced the same number of trials. (Modified with permission from Braun *et al.,* 2022).

on regional policies. As early as 2003, the government environmental agency in Quebec, Canada enacted a 'Pesticide Management Plan' (Cisar, 2010) out of an urgency to reduce non-target impacts of agrichemicals. This resulted in bans on pesticide use on government-owned land, withdrawal of pesticide application within 3 m of any water body and prohibition of pesticide use on non-golf playing fields where these were used by children under the age of 14. Golf courses in the province were required to submit a detailed pesticide use report and provide a 3-year plan for their reduction of pesticides. In another Canadian province, Ontario, the use of pesticides on domestic lawns has been banned on a community-by-community basis (Cisar, 2004). Attitudes, policies and application rates do vary from country to country. Pesticide use on golf courses particularly raises concerns, both because of the real or perceived toxicity of chemicals used and the potential exposure to those playing/working on and living around golf courses. Several European countries have either entirely banned (e.g. Spain and the Wallonia region of Belgium) or severely restricted (e.g. Italy and Norway) pesticides available to golf course managers. In Norway, Denmark, and the UK, there are fewer than 20 active ingredients available to golf green-keepers, whereas in the USA, there are still 200–250 pesticide active ingredients registered for use on golf courses (Bekken *et al.*, 2023). However, the picture is complex. Doherty *et al.* (2024) suggested that more modern pesticides were reducing exposure/toxicological risk to golfers in the USA. In contrast, Lee *et al.* (2022) felt South Korea needed to adopt more integrated pest management techniques to help reduce toxicological risks to golfers and damage to the wider environment. In terms of risk to the environment, different parts of the golf course pose different threats. Bekken *et al.* (2021) identified fairways as being particularly problematic. They suggested that variation in pesticide risk on golf courses relates to economic factors, such as maintenance budgets across different courses, and that overall the risks can be effectively lowered by (i) reducing pesticide use on fairways and (ii) selecting more 'benign' alternative products. In the southern USA, golf greens (i.e. the most intensively managed turf) were also seen as locations where pesticide risk notably increased (Bekken *et al.*, 2023).

Increasingly though, the use of chemical pesticides, especially those compounds that possess higher toxicity or may have some persistence in the ecosystem, will be hard to justify for non-food crops and amenity horticultural situations.

Strategies being adopted by professional land managers to reduce pesticide use include:

- Complying with local/national regulations on handling, storage and use of pesticides.
- Implementing cultural control programmes that minimise the likelihood of pathogen/pest problems arising. This includes understanding 'cause and effect' between management and potential problems, i.e. avoiding creating those environmental conditions that favour pathogens, invertebrate pests or weeds. Such control programmes have included integrated pest management (IPM), biocontrol, and best management practices (BMP) that have evolved to help reduce pesticide application and improve efficacy with tangible impacts. IPM, for example, takes a 'holistic' approach to pest management by understanding how predation rates and lifecycles are influenced by environmental influences and wider management strategies, and not just the efficacy of pesticides per se.
- Setting tolerance levels and implementing a regular monitoring programme to highlight problems earlier in their development and deal with them in a more timely manner.
- Seek expert advice where appropriate to properly identify problems so that the most integrated approach to managing them can be achieved.

Biocontrol approaches include the development of soil microbial populations that are either aggressive or competitive to the pathogenic organism. An alternative is to exploit mycorrhizae fungal species that confer better protection to the turfgrass under attack. Biocontrol agents are being developed to control nematodes that damage grass roots. These bionematicides include bacteria such as *Pasteuria* spp., *Purpureocillium lilacinus, Bacillus firmus* and *Myrothecium verrucaria* (Wilson and Jackson, 2013). Success with biological control agents in a laboratory situation, however, does not always translate into effective application in the field and further evaluations are necessary for some biocontrol strategies. For example, McGraw *et al.* (2010) investigated the use of nematodes as biocontrol for the annual bluegrass weevil, *Listronotus maculicollis.* Combinations of three nematode species (*Steinernema carpocapsae, S. feltiae* and *Heterorhabditis bacteriophora*) were evaluated across different application rates and timings. Results showed strong numerical, yet few statistically significant, reductions. *H. bacteriophora* reduced first-generation late instars between 69% and 94% in at least one field trial. At low instar densities, *S. feltiae* provided a high level of control (94%) but gave inadequate control for higher densities (24% to 50% suppression). Time of application was also important, with applications timed to coincide with the peak of larvae entering the soil (fourth instars) being optimal (McGraw *et al.*, 2010).

In practice, the use of IPM and biocontrol agents seems 'patchy' in elite turf grass management. Results are not always guaranteed, and many green-keepers still rely on more routine pesticide application. In countries where pesticides have been restricted, a degree of turf damage may need to be tolerated. In southern Sweden, Norway and in Denmark, chafer grubs and leatherjackets can cause damage to golf courses. Damages from chafer grubs are occasional, but damage from leatherjackets tends to be increasing. Restrictions on insecticides have necessitated the use of alternative control methods. Many experiments with microbiological agents like entomopathogenic nematodes and strains of *Bacillus thuringiensis* have been conducted, but Hesselsøe *et al.* (2022) argue that monitoring, warning systems and methods for biocontrol application, spraying equipment and technique, formulation of and effective species of microbiological agents, must all be improved. Good communication with golfers is essential too, as more damage from insect pests will occur in the future, and IPM methods are often more expensive and less effective than relying on synthetic insecticides alone.

Where biocontrol and IPM options work, there are tangible reductions in pesticide use. It is thought that golf greens managed via IPM often receive 30–50% less pesticide applications than the unrestricted pest management greens. Due to the nature of the processes involved, such as complex biological and environmental interactions, results vary with location, predominant grass species and year. When *Agrostis canina* (velvet bentgrass) was managed by IPM it outperformed *Agrostis palustris* (annual bluegrass/creeping bentgrass). It has been observed that control via IPM is more challenging during wet years, especially during warmer summer months.

Indeed, IPM recommendations can vary with location and context. For example, in the warmer states of the USA, mole crickets, e.g. *Neoscapteriscus borellii* (southern mole cricket) can be a problem. Here, IPM recommendations (Kerr *et al.*, 2021) include:

1. Use more tolerant cultivars or species, e.g. centipedegrass or zoysia.
2. Maintain healthy grass with proper irrigation and cutting.
3. Perform routine soil testing and add fertiliser or lime as needed.
4. Reduce watering during winter months; mole crickets require moist soil.
5. Plant a nectar source, such as *Spermacoce verticillata* (larraflower) or *Chamaechrista fasciculata* (partridge pea), to attract and support *Larra bicolor* (Larra wasp) populations.
6. Eliminate lights from sunset to well past dark during months of peak mole cricket flight.
7. Sample regularly for mole crickets; densities of 18–24 per m^{-2} may require management.
8. Apply insecticides at infested sites only if plant damage thresholds are exceeded; evaluate their effectiveness.
9. Target and map areas that become infested. Rotate insecticide active ingredients to delay pesticide resistance.

Three biocontrol predators of mole crickets are currently being evaluated in Florida (USA). Widespread applications have been made of the entomopathogenic nematode *Steinernema scapterisci* in addition to releases of two parasitoids – *Larra bicolor* (Larra wasp) and *Ormia deplete* (Brazilian red-eyed fly). These non-native 'natural' enemies were imported, tested for safety and released through a special Mole Cricket Biological Control Program. All these species are now present in Florida, but none are available commercially. The complexities of IPM are highlighted here; *Larra bicolor* adults require nectar for their survival and to establish a viable population, which include good numbers of the females that then lay their eggs in the mole cricket larvae. So, the IPM involves planting species that support nectar production for this wasp species. These include *Spermacoce verticillata* (shrubby false buttonweed or larraflower), *Pentas lanceolata* (white-flowered pentas) and *Chamaechrista fasciculata* (partridge pea). If these plants or other nectar sources are available, larra wasps will appear and forage at least 200 yards from them to locate mole crickets (Kerr *et al.*, 2021).

Despite these approaches there may be incidences where pesticide application is considered necessary. Here too, mitigating actions include reducing the toxicity of the chemicals used through decreasing concentrations or the range of active ingredients in each agrichemical. Similarly, developing compounds with more specific modes of action that reduce the impact on non-target species is more commonplace. Despite the advances in IPM, there is still some way to go for the turf sector to be truly sustainable in some parts of the globe. A survey in the USA in 2023 (Shaddox *et al.*, 2023) indicated that reliance on conventional chemical pesticides actually increased from 2015 to 2021. The reliance on biological control products declined to 14% in this period.

9.11 Less Intensive Management

Although turf culture is one of the more challenging areas for environmental and sustainable management practices to be adopted, especially in relation to sports turf, advances are being made in some countries at least. New concepts and approaches are challenging the need for heavy chemical use, many local authorities and other landowners are questioning their requirement to mow extensive areas of infrequently used grass on a regular basis and some homeowners are now more tolerant of non-grass species becoming present in their garden lawns. Not all solutions, however, are deemed to be held within the true philosophy of sustainability, for example the use of artificial turf to avoid mowing and pesticide applications (but it still has strong negative connotations for biodiversity), or even the application of growth-retarding chemical compounds (plant growth regulators) to help reduce the carbon footprint of mowing. On a more positive note, however, attention is being directed to changes in attitude and philosophy, for example the adoption of genotypes better adapted to local conditions and changing attitudes to water, pesticides, fertiliser use and indeed more fundamentally what a lawn should look like and be used for. Considerations over cost and available labour to look after high-quality lawns, e.g. in parks, also means that lower input, less intensive systems are being considered by local authorities, with some evidence that this is preferred by the public, over alternatives such as plastic artificial turf (Barnes and Watkins, 2022).

Genotype selection for 'low-input' systems

It is argued that as restrictions on water use, fertilisation, and pesticide applications continue to increase, then the turf managers will need to re-evaluate the choice of species used for top-quality lawns. Using golf fairways in the northern USA as the context for their research, Watkins *et al.* (2011) examined which grass species performed best when managed under carefully controlled, low-input regimes. Seventeen grass genotypes were evaluated for their performance in relation to differential traffic levels (zero, three or six passes per week using a drum-type traffic simulator) and two mowing heights (19 and 25 mm). *Festuca ovina* spp. *ovina* (sheep's fescue), *Festuca rubra* ssp. *cummutata* (Chewings fescue), *Agrostis capillaris syn. tenuis* (browntop bentgrass) and *A. canina* (velvet bentgrass) proved to be the most promising. Turf managers in amenity grass situations (parks, residential areas and gardens) especially have shown a positive attitude to using lower-maintenance species such as *Festuca* spp. more extensively (Barnes *et al.*, 2018).

Genetics can also help aesthetics and attitudes. The University of Aberystwyth, Wales, UK discovered a naturally occurring 'stay-green' gene in *Schedonorus pratensis* (meadow fescue), which through conventional cross-breeding has been introduced to *Lolium* sports turf cultivars. This allows the turf to remain greener for longer when exposed to drought stress or low irradiance conditions during mid-winter, thus making these cultivars more acceptable for premier sports stadia.

Even the concept of grass-only swards is now being challenged. Utilising nitrogen-fixing clover, especially the small-leaved white clover, *Trifolium repens* form *microphyllum*, has received attention for its contribution to 'low-input' turf communities. Not only does this species provide nitrogen to the soil matrix when there is root turnover but the small leaves also improve sward colour in the winter (Monteiro, 2017). How these incorporated species affect play quality is not yet known.

Transgenic lines of turfgrass have been evaluated in the USA. These have been designed to improve the stature of the grass (less vertical growth and more compact) and potentially reduce mowing frequency. For example, new transgenic *Poa pratensis* (Kentucky bluegrass) cultivars with a dwarf growth habit have been developed by overexpressing the gibberellin 2-oxidase pathway (Fidanza *et al.*, 2022). Similarly, in *Stenotaphrum secundatum* (St Augustine grass) genetically modified version of cv. Floratam are characterised by reduced levels of endogenous gibberellic acid by the insertion of the gibberellic acid 2-oxidase gene (Buhlman *et al.*, 2022). In the case of the *Poa*, mowing frequency was reduced by approximately 50% compared to conventional cultivars (≤21 mowing events per season compared to ≥41 for other turfgrasses on average). Of eight new transgenic lines of S. secundatum cv. Floratam, two required less frequent mowing than conventional Floratam across all seasons and at both locations tested. One of these also showed turf quality comparable to Floratam.

Artificial turf

An alternative to irrigated lawns that it is now being seriously considered on its environmental benefits (rather than just playing quality) is artificial turf. Artificial turf can be the preferred surface for certain sports, such as hockey where there is a high requirement to ensure a consistent role of the ball and avoid it 'embedding' in the turf. Artificial turf stands more wear and tear in contact sports such as rugby, especially if first-tier competitive matches are played on a regular basis (Fig. 9.16). Certain households in California, USA have adopted artificial turf to help save water. Some golf courses have used artificial turf for the greens, but there are now golf courses that are entirely composed of synthetic grass (e.g. The Echo Basin Golf Course, Colorado, USA). Although such courses are likely to reduce their environmental impact via lower energy, water and chemical use, they are likely to have negative ecological impact at a local level due to loss of animal/plant habitat, especially if 'the rough' (the peripheral areas that flank the fairways) are put over to artificial turf too!

Whilst artificial turf may have a role in sports facilities, Cameron (2023) was negative about its use in domestic properties. Not only does it kill soil fauna and reduce habitat for birds and other animals (Sanchez-Sotomayor *et al.*, 2023), but the prevalent use of artificial turf undermines opportunities for local microclimate cooling, rainwater penetration and evaporation from the soil (i.e. flood mitigation strategies). The use of synthetic surfaces in the garden can increase plastic pollution (especially mesoplastics) as well as interfere with people's exposure to beneficial soil/microbial populations (Robinson *et al.*, 2020).

Fig. 9.16. A number of rugby stadia have moved over to artificial turf, due to the need to hold top-tier matches on a very frequent basis. Rugby scrums and lineouts particularly put a lot of physical stress on conventional grass swards.

Areas where less intensive management is warranted

Despite all the emphasis on high-quality turfgrass and the requirements for intricate management regimes, not all areas of turf need to be managed intensively (see also Chapter 8). Many local governments are realising the costs of frequent mowing regimes and are considering land use change, or even just a relaxation in the intensity of current management activities. In countries such as the UK, Australia and the USA, roadside verges are frequently managed as 'rough grassland', with perhaps only a single annual cut to stop succession to woody plant species taking place. Similarly, there may be no requirement for pesticides, fertiliser or irrigation, and the mixed plant communities that evolve can be resilient to invasive weed invasion (Nemec *et al.*, 2021). Management of such sites is sometimes referred to as integrated vegetation management (IVM) where the aims are to reduce financial costs, reduce resource use and encourage wildlife. This holistic approach requires an understanding of the life cycle, seasonal cycle, species composition and population dynamics of grassland vegetation (Navie *et al.*, 2010). When managed well, this approach can reap tangible benefits to local wildlife, with the development of more biodiverse plant communities and a more heterogeneous physical structure to the sward. IVM has been utilised recently to restore prairie plant communities as well as encourage populations of specific fauna taxa including ants, bees, butterflies, diurnal moths, beetles and some birds and small mammal species.

This change of attitude has also been reflected in the sphere of the domestic lawn too, in some countries at least. Whereas traditionally the inclusion of broadleaved dicots in fine lawns was seen as an anathema, and such 'weeds' were an indication of poor management (or gross negligence!), greater tolerance is now forthcoming for the inclusion of dicot species. To survive, many of these dicot species either possess some tolerance to low mowing akin the grasses or have a prostrate plant habit that allows them to avoid the mower blades (Fig. 9.17). Whether the presence of these species within the sward is intentional or not, surprising numbers of species can be found in domestic lawns. Surveys have shown 159 species in UK lawns (Thompson *et al.*, 2004), 83 in western Germany (Muller, 1990) and 139 in the lawns of Christchurch, New Zealand alone (Horne *et al.*, 2005). Although the inclusion of certain dicots can disrupt the uniformity of the lawn, others that are more effective at retaining their colour (e.g. *Trifolium* spp., clovers) during stress may contribute to retaining the 'greeness' of the sward. Similarly, these and related 'nitrogen-fixing' species contribute to the nitrogen balance of the turf, thereby reducing the reliance on artificial forms of nitrogen and reducing the chance of nitrate leaching (Monteiro, 2017). As the term 'weeds' used above implies, some species of dicots are readily integrated into the grass community, but the likelihood of any specific species becoming prominent will be dependent on the key ecological factors associated with the lawn, namely location, soil type and nutrient status, moisture availability, irradiance intensity, length of grass, cutting and other disturbance regimes. Typical dicots found or easily incorporated into non-intensively managed European cool-season lawns are shown in Table 9.7. By using the appropriate types of dicots, lawns can stay green for longer during drought periods. Some 'sustainable' mixed species lawns may only require 50% of the conventional volume of irrigation applied, with mowing frequencies also being reduced to once every two to three weeks (VanDerZanden and Cook, 2010).

Fig. 9.17. *Viola lutea* (mountain pansy) growing in a garden lawn. The prostrate nature of the foliage allows some tolerance to infrequent mowing.

Grasslands for conservation and biodiversity are covered in more detail in Chapter 8.

Future directions

Developing truly sustainable systems of construction and management for turfgrass remains challenging.

Table 9.7. Herbaceous perennial dicots (forbs) that can be effectively integrated with lawns of 30–60 mm height.

Species	Common name	Characteristics
Achillea millefolium	yarrow	Drought tolerant; pale cream flowers.
Ajuga reptans	bugle	Moist, well-drained soils. Can be invasive in North America. Blue flowers.
Anthyllis vulneraria	kidney vetch	Dry soils of high pH; yellow – sometimes red/orange – flowers.
Bellis perennis	common daisy	Moist soils, although has some drought tolerance. White flowers sometimes flushed pink.
Campanula rotundifolia	harebell	Adaptable to wide pH, acid and chalk-land grass communities. Distinctive pale blue bell-shaped flower.
Cardamine pratense	cuckoo flower	Moist/wet lawns, pale mauve flowers in spring.
Galium verum	lady's bedstraw	Very drought tolerant. Compatible with lawn grasses – acid yellow flowers.
Chamaemelum nobile	chamomile	Pale green leaves, white flowers – some tolerance to drought. Also used as a mono-culture sward – chamomile lawn.
Glechoma hederacea	ground ivy	Moist soils; tolerates shade. Broadleaved, but blends well with lawn grasses.
Helianthemum nummularium	common rock-rose	Dry grassland, will not tolerate excessive low mowing – sulphur yellow flowers.
Hieracium brunneocroceum	orange hawkbit or fox and cubs	Medium/moist soils. Basal rosette of green, oval, hairy leaves and tall stems topped with orange, daisy-like flowers.
Hieracium pilosella	mouse-ear hawkweed	Light soils, basal oval leaves, with yellow flowers. Can be invasive in non-native countries.
Hypochaeris radicata	cat's ear	Tolerant of a wide range of soil conditions. Daisy-like yellow flowers.
Lotus corniculatus	bird's foot trefoil	Sandy free-draining soils. Tolerates a degree of trampling and mowing, yellow and red pea-like flowers.
Lotus pedunculatus	greater bird's foot trefoil	Moist soils and wet pastures. Similar, but slightly larger than *L. corniculatus*.
Leontodon hispidus	rough hawkbit	Dry soils with neutral or high pH. Rosette-forming perennial with ability to regenerate after close mowing. Yellow flowers.
Medicago lupulina	black medick	Short-lived perennial that dies out if sward gets too long. Does best when there is a degree of soil disturbance. Small, yellow 'ball-like' floret head.
Oxalis corniculata	yellow woodsorrel	A number of prostrate, spreading *Oxalis* species can survive in lawn turf. The yellow flowered *O. corniculata* is often associated with drier or disturbed soil.
Plantago lanceolata	ribwort plantain	Rosette-forming perennial with leafless, silky, hairy flower stems. Commonly found in pastureland and rough lawns, espescially on high pH soil. Brown/grey flower panicles appear on stalk above the grass.
Plantago media	hoary plantain	A more refined plant than *P. lanceolata, P. media* grows in moist lawns. Produces a slender stalk with pink-white flowers.
Potentilla erecta	tormentil	Neutral to low pH soils. Wet and moist soils; often associated with grasses of acidic heathlands. Small, open cup-shaped yellow flowers.
Primula veris	cowslip	Neutral to high pH soils. Medium to dry lawns, with yellow clusters of flowers appearing on a stalk held above the sward.
Primula vulgaris	primrose	Moist and heavier soils than *P. veris*. Pale yellow flowers amongst the foliage.
Prunella vulgaris	self-heal	Fairly drought tolerant; neutral to high pH soils. Purple conical flowers at mowing height.
Ranunculus repens	creeping buttercup	Moist to wet lawns. Creeping habit, with bright yellow, buttercup flowers displayed just above the foliage.
Taraxacum officinale	dandelion	Moist and heavy clay soils. Persistent, deep, fleshy tap-root, allows for quick regeneration. Robust leaves and yellow daisy flower, with conspicuous globe seed head.
Trifolium repens	white clover	Low growing and tolerant of mowing – prefers clay soil, but tolerant of others. Can 'fix' atmospheric nitrogen. Leaves stay green under stress and white flowers are nectar source for insects.

Continued

Table 9.7. Continued.

Species	Common name	Characteristics
Trifolium fragiferum	strawberry clover	Tolerant of wet and saline soils. Pink fruiting head resembles a strawberry.
Veronica filiformis	slender speedwell	Grows from prostrate rhizomes, below the height of cut. Prefers moist soils. Attractive blue or mauve flowers.
Veronica chamaedrys	germander speedwell	More refined than *V. filiformis*, with a deeper blue flower.
Viola odorata	sweet violet	Shady or moist lawns. Found in woodland edges, banks and hedgerows but can invade turf swards. Blue-purple or white flowers.
Viola riviniana	dog violet	Wide range of soils, except very low pH or very wet. Foliage can be conspicuous within the lawn. Blue- or purple-coloured pansy-like flower. No scent to the flower.

Priority needs to be given to the notion that not all grassed areas require the same approach. Much of the public green space can be managed in a more benign, less energy- and resource-intensive manner, with reductions in frequency of mowing and adopting systems that allow for better 'cycling' of carbon and nitrogen, such as the re-incorporation of mower clippings. Similarly, mindsets need to change with respect to the private lawn, where nutrient inputs, fossil fuel energy consumption and water are often considerably more excessive than they need be. Greater consideration also needs to be given to the siting of lawns in the first instance, the component species and the composition of the physical attributes, as well as the type and intensity of the maintenance to improve the environmental performance of turfgrass. Elite sports turf will always be the most demanding aspect of turf management, but even here utilising slower-release fertilisers, recycled or low-grade irrigation water, sustainable and locally sourced aggregates, integrated pest management programmes and developing grass genotypes bred for their stress tolerance and low-input requirements (rather than largely playing quality and aesthetics attributes) should go some way to improving the environmental credentials of the sector.

Conclusions

- Turfgrass provides a range of functional uses – a playing surface for formal and informal sport, aesthetic value and psychological benefits, land stabilisation and avoidance of soil erosion, a foil to buildings and other infrastructure, a safe haven adjoining roadways, low-use pathways and a walkable surface, habitat, microclimate modification and the ability to trap aerial and water-borne pollutants.
- The ability of grass plants to grow from active meristems at the base of the plant and essentially to continually 're-new' themselves allows them to play a unique role in the landscape, not least being resilient against considerable wear and tear. This also means though that maintenance and management issues can be significant under certain circumstances and the criteria imposed on the grass sward.
- Cool-season grasses associated with turf in mild, temperate climates tend to be C3 plants, whereas their tropical/sub-tropical counterparts, warm-season grasses, are usually C4.
- Sport places the most demand on turf, and the interaction between the soil (substrate) and the grass genotypes used is critical in ensuring the playing quality of the turf is maximised for any given sport. Often this requires the turf species to grow in exceedingly sharp-draining substrates and be dependent on artificial sources of nutrition (fertilisers), additional water (irrigation) chemical additives (e.g. pesticides and growth regulators) and environmental manipulation (mowing, tining, etc.) to establish and thrive. The sustainability of some of the approaches is open to question and new management regimes are being developed to minimise inputs, reduce waste and avoid negative environmental impacts.
- New grass species and cultivars continue to be selected to meet the demands of sports turf performance and also additional pressures brought on by a changing climate and the need for more sustainable land management.
- Sport turf surfaces and lawns can be established either from seed or laying down sections of turf (sods) pre-grown elsewhere.
- Mowing accounts for most of the labour and fiscal costs associated with maintaining elite sports turf. Handheld and human-driven machinery is

being replaced in some situations with robotic systems.
- Domestic lawn management is seeing alterations in traditional management methods, with emphasis on a changed rationale for the lawn in the first place, reducing mowing frequency, less reliance on irrigation and tolerance of non-grass plant species as part of the lawn community.

References

Anon. (2003) Turfgrass Fertilization: A Basic Guide for Professional Turfgrass Managers. Available at: http://plantscience.psu.edu/research/centers/turf/extension/factsheets/turfgrass-fertilization-professional (accessed 28 November 2024).

Anon. (2004) USGA Recommendations for a Method of Putting Green Construction (2004). Available at: chrome-extension: //hbgjioklmpbdmemlmbkfckopochbgjpl/https://www.usga.org/content/dam/usga/images/course-care/2004%20USGA%20Recommendations%20For%20a%20Method%20of%20Putting%20Green%20Cons.pdf (accessed 28 November 2024).

Anon. (2024) J Premier Fairway Seed Mix. For High Performance, Sustainable Golf Fairways and Tees (non rye). Johnsons Sports Seed. Available at: https://www.johnsonssportsseed.co.uk/mixtures/johnsons-sports-seed/mixtures/tees-fairways-and-roughs-mixtures/j-premier-fairway-prod618 (accessed 19 June 2024).

Baker, J.E. (2021) Subsidies for succulents: Evaluating the Las Vegas Cash-for-Grass rebate program. *Journal of the Association of Environmental and Resource Economists* 8, 475–508.

Bañuelos, J.B., Walworth, J.L., Brown, P.W. and Kopec, D.M. (2011) Deficit irrigation of Seashore paspalum and bermudagrass. *Agronomy Journal* 103, 1567–1577.

Barnes, M.R. and Watkins, E. (2022) Differences in likelihood of use between artificial and natural turfgrass lawns. *Journal of Outdoor Recreation and Tourism* 37, 100480.

Barnes, M.R., Nelson, K.C., Meyer, A.J., Watkins, E., Bonos, S.A. *et al.* (2018) Public land managers and sustainable urban vegetation: The case of low-input turfgrasses. *Urban Forestry and Urban Greening* 29, 284–292.

Beard, J.B. and Green, R.L. (1994) The role of turfgrasses in environmental protection and their benefits to humans. *Journal of Environmental Quality* 23, 452–460.

Bekken, M.A., Schimenti, C.S., Soldat, D.J. and Rossi, F.S. (2021) A novel framework for estimating and analyzing pesticide risk on golf courses. *Science of the Total Environment* 783, 146840.

Bekken, M.A., Soldat, D.J., Koch, P.L., Schimenti, C.S., Rossi, F.S. *et al.* (2023) Analyzing golf course pesticide risk across the US and Europe—The importance of regulatory environment. *Science of The Total Environment* 874, 162498.

Braun, R.C., Bremer, D.J., Ebdon, J.S., Fry, J.D. and Patton, A.J. (2022) Review of cool-season turfgrass water use and requirements: II. Responses to drought stress. *Crop Science* 62, 1685–1701.

Bremer, D.J., Keeley, S.J., Jager, A.L. and Fry, J.D. (2013) Lawn-watering perceptions and behaviors of residential homeowners in three Kansas (USA) cities: Implications for water quantity and quality. *International Turfgrass Society Research Journal* 12, 23–29.

Buhlman, J.L., Freshour, S.L., Harriman, R.W., Koch, M.J., Unruh, J.B. *et al.* (2022) Mowing frequency and performance of genetically modified St. Augustinegrass and standard commercial cultivars. *International Turfgrass Society Research Journal* 14, 652–662.

Cameron, R. (2023) “Do we need to see gardens in a new light?” Recommendations for policy and practice to improve the ecosystem services derived from domestic gardens. *Urban Forestry and Urban Greening* 80, 127820.

Chawla, S.L., Agnihotri, R., Patel, M.A., Patil, S. and Shah, H.P. (2018) Turfgrass: A billion dollar industry. In: *Proceedings of the National Conference on Floriculture for Rural and Urban Prosperity in the Scenario of Climate Change*, Gangtok, University of Sikkim, India, pp. 16–18.

Cisar, J.L. (2004) Managing turf sustainably. In: *Proceedings of the 4th International Crop Science Congress* 26. Brisbane, Australia, September 2004. Australian Society of Agronomy.

Cisar, J.L., (2010) Reducing environmental impacts of fertilizers and pesticides through sustainable high performance turfgrass systems. *Acta Horticulturae* 938, 105–111.

Dobbs, N.A., Migliaccio, K.W., Dukes, M.D., Morgan, K.T. and Li, Y.C. (2013) Interactive irrigation tool for simulating smart irrigation technologies in lawn turf. *Journal of Irrigation and Drainage Engineering* 139, 747–754.

Doherty, J.J., Putnam, R.A., DeFlorio, B.A. and Clark, J.M. (2024) Golfer exposure to traditional pesticides following application to turfgrass. ACS *Agricultural Science and Technology* 4, 414–423.

Dong, L., Xiong, L., Sun, X., Shah, S., Guo, Z. *et al.* (2022) Morphophysiological responses of two cool-season turfgrasses with different shade tolerances. *Agronomy* 12, 959.

Duncan, R.R. and Carrow, R.N. (2000) *Seashore Paspalum: The Environmental Turfgrass*. John Wiley & Sons.

Fan, J., Zhang, W., Amombo, E., Hu, L., Kjorven, J.O. and Chen, L. (2020) Mechanisms of environmental stress tolerance in turfgrass. *Agronomy* 10, 522.

Fidanza, M., Baldwin, C., Koch, M., Freshour, L. and Harriman, B. (2022) Required mowing events of transgenic Kentucky bluegrass cultivars compared with conventional turfgrass cultivar blends and species mixtures. *Crop, Forage and Turfgrass Management* 8, e20146.

Frank, K.W., Crum, J.R., Bryan, J.M. and Hathaway, A.D. (2016) Fifteen years of nitrogen leaching from a Kentucky bluegrass turf. *Crop Science* 56, 3338–3344.

Gillman, L.N., Bollard, B. and Leuzinger, S. (2023) Calling time on the imperial lawn and the imperative for greenhouse gas mitigation. *Global Sustainability* 6, e3.

Gómez-Armayones, C., Kvalbein, A., Aamlid, T.S. and Knox, J.W. (2018) Assessing evidence on the agronomic and environmental impacts of turfgrass irrigation management. *Journal of Agronomy and Crop Science* 204, 333–346.

Grabow, G.L., Ghali, I.E., Huffman, R.L., Miller, G.L., Bowman, D. and Vasanth, A. (2012) Water application efficiency and adequacy of ET-based and soil moisture–based irrigation controllers for turfgrass irrigation. *Journal of Irrigation and Drainage Engineering* 139, 113–123.

Hatfield, J. (2017) Turfgrass and climate change. *Agronomy Journal* 109, 1708–1718.

Hesselsøe, K.J., Menzler-Hokkanen, I., Petersen, T.K. and Edman, P. (2022) *Chafer Grubs and Leatherjackets on Golf Courses–Integrated Management in Scandinavia*. Norwegian Institute of Bioeconomy Research (NIBIO) Report. Oslo, Norway

Horne, B., Stewart, G.H., Meurk, C.D., Ignatieva, M. and Braddick, T. (2005) The Origin and Weed Status of Plants in Christchurch Lawns.

Hull, R.J., Alm, S.R. and Jackson, N. (2020) Toward sustainable lawn turf. In: *Handbook of Integrated Pest Management for Turf and Ornamentals*. CRC Press, Boca Raton, Florida, USA, pp. 3–14.

Hurd, B.H., Hilaire, R.S. and White, J.M. (2006) Residential landscapes, homeowner attitudes, and water-wise choices in New Mexico. *HortTechnology* 16, 241–246.

James, I.T. (2011) Advancing natural turf to meet tomorrow's challenges. Proceedings of the Institution of Mechanical Engineers, Part P: *Journal of Sports Engineering and Technology* 225, 115–129.

Johnson, P.G. (2000) An overview of North American native grasses adapted to meet the demand for low-maintenance turf. *Diversity* 16, 40–41.

Jordan, J. E., White, R.H., Vietor, D.M., Hale, T.C., Thomas, J.C. and Engelke, M.C. (2003) Effect of irrigation frequency on turf quality, shoot density, and root length density of five bentgrass cultivars. *Crop Science* 43, 282–287.

Kerr, C.R., Leppla, N.C., Buss, E.A. and Frank, J.H. (2021) Mole Cricket IPM Guide for Florida: IPM-206/IN1021, 8/2021. *EDIS* 2021–4.

King, K.W., Balogh J.C., Agrawal S.G, Tritabaugh, C.J. and Ryan J.A. (2012) Phosphorus concentration and loading reductions following changes in fertilizer application and formulation on managed turf. *Journal of Environmental Monitoring* 14, 2929–2938.

Kır, B., Avcıoglu, R., Salman, A. and Ozkan, S.S. (2019) Turf and playing quality traits of some new turfgrass alternatives in a Mediterranean environment. *Turkish Journal of Field Crops* 24, 7–11.

Kvalbein, A. and Aamlid, T.S. (2012) Impact of mowing height and late autumn fertilization on winter survival and spring performance of golf greens in the Nordic countries. *Acta Agriculturae Scandinavica, Section B-Soil and Plant Science* 62, 122–129.

Lee, H.G., Kim, M.S., Kim, J.Y., Kim, H.K., Jo, H.W., Moon, J.K. and Kim, J.I. (2022) Status of pesticide usage on golf courses in Korea and optimal pesticide usage plan. *Sustainability* 14, 5489.

Liu, H., Todd, J.L. and Luo, H. (2023) Turfgrass salinity stress and tolerance—A review. *Plants* 12, 925.

Macolino, S., Pignata, G., Giolo, M. and Richardson, M.D. (2014) Species succession and turf quality of tall fescue and Kentucky bluegrass mixtures as affected by mowing height. *Crop Science* 54, 1220–1226.

McCarty, L.B., Gregg, M.F. and Toler, J.E. (2007) Thatch and mat management in an established creeping bentgrass golf green. *Agronomy Journal* 99, 1530–1537.

McGraw, B.A., Vittum, P.J., Cowles, R.S. and Koppenhöfer, A.M. (2010) Field evaluation of entomopathogenic nematodes for the biological control of the annual bluegrass weevil, *Listronotus maculicollis* (Coleoptera: Curculionidae), in golf course turfgrass. *Biocontrol Science and Technology* 20, 149–163.

Milesi, C., Running, S.W., Elvidge, C.D., Dietz, J.B., Tuttle, B.T. and Nemani, R.R. (2005) Mapping and modeling the biogeochemical cycling of turfgrasses in the United States. *Environmental Management* 36, 426–438.

Monteiro, J.A. (2017) Ecosystem services from turfgrass landscapes. *Urban Forestry and Urban Greening* 26, 151–157.

Müller, N. (1990) Lawns in German cities. A phytosociological comparison. *Urban Ecology* 209, 222.

Navie, S.C., Hampton, S.J. and Bloor, N. (2010) Integrated management of mown vegetation in eastern Australia. In: *17th Australasian weeds conference proceedings: new frontiers in New Zealand, together we can beat the weeds* pp. 26–30.

Nemec, K., Stephenson, A., Gonzalez, E.A. and Losch, M. (2021) Local decision-makers' perspectives on roadside revegetation and management in Iowa, USA. *Environmental Management* 67, 1060–1074.

Padullés Cubino, J., Kirkpatrick, J.B. and Vila Subirós, J. (2016) Do water requirements of Mediterranean gardens relate to socio-economic and demographic factors? *Urban Water Journal* 14, 401–408.

Park, D.M., Cisar, J.L., McDermitt, D.K., Williams, K.E., Haydu, J.J. and Miller, W.P. (2005) Using red and infrared reflectance and visual observation to monitor turf quality and water stress in surfactant-treated Bermudagrass under reduced irrigation. *International Turfgrass Society Research Journal* 10, 115–120.

Pincetl, S., Gillespie, T.W., Pataki, D.E., Porse, E., Jia, S. *et al.* (2019) Evaluating the effects of turf-replacement programs in Los Angeles. *Landscape and Urban Planning* 185, 210–221.

Price, J.I., Chermak, J.M. and Felardo, J. (2014) Low-flow appliances and household water demand: An evaluation of demand-side management policy in Albuquerque, New Mexico. *Journal of Environmental Management* 133, 37–44.

Raciti, S.M., Groffman, P.M., Jenkins, J.C., Pouyat, R.V., Fahey, T.J., Pickett, S.T. and Cadenasso, M.L. (2011) Accumulation of carbon and nitrogen in residential soils with different land-use histories. *Ecosystems* 14, 287–297.

Robinson, J.M., Cando-Dumancela, C., Liddicoat, C., Weinstein, P., Cameron, R. and Breed, M.F. (2020) Vertical stratification in urban green space aerobiomes. *Environmental Health Perspectives* 128, 117008.

Salvador, R., Bautista-Capetillo, C. and Playán, E. (2011) Irrigation performance in private urban landscapes: A study case in Zaragoza (Spain). *Landscape and Urban Planning* 100, 302–311.

Sanchez-Sotomayor, D., Martin-Higuera, A., Gil-Delgado, J.A., Galvez, A. and Bernat-Ponce, E. (2023) Artificial grass in parks as a potential new threat for urban bird communities. *Bird Conservation International* 33, e16.

Schiavon, M., Leinauer, B., Serena, M., Sallenave, R. and Maier, B. (2013) Establishing tall fescue and Kentucky bluegrass using subsurface irrigation and saline water. *Agronomy Journal* 105, 183–190.

Selhorst, A. and Lal, R. (2013) Net carbon sequestration potential and emissions in home lawn turfgrasses of the United States. *Environmental Management* 51, 198–208.

Sevostianova, E., Leinauer, B., Sallenave, R., Karcher, D. and Maier, B. (2011a) Soil salinity and quality of sprinkler and drip irrigated cool-season turfgrasses. *Agronomy Journal* 103, 1503–1513.

Sevostianova, E., Leinauer, B., Sallenave, R., Karcher, D. and Maier, B. (2011b) Soil salinity and quality of sprinkler and drip irrigated warm-season turfgrasses. *Agronomy Journal* 103, 1773–1784.

Shaddox, T.W., Unruh, J.B., Johnson, M.E., Brown, C.D. and Stacey, G. (2023) Survey of pest management practices on US golf courses. *HortTechnology* 33,152–156.

Sidhu, S., Huang, Q.J., Carrow, R.N., Jesperson, D., Liu, J. and Raymer, P.L. (2022) A review of a novel enzyme system for the management of thatch and soil water repellency in turfgrass. *International Turfgrass Society Research Journal* 14, 450–461.

Smetana, S.M. and Crittenden, J.C. (2014) Sustainable plants in urban parks: A life cycle analysis of traditional and alternative lawns in Georgia, USA. *Landscape and Urban Planning* 122, 140–151.

Snyder, G.H. and Cisar, J.L. (2000) Monitoring vadose-zone soil water for reducing nitrogen leaching on golf courses. *Fate and Management of Turfgrass Chemicals* 743, 243–254.

Snyder, G.H., Augustin, B.J. and Davidson, J.M. (1984) Moisture sensor-controlled irrigation for reducing N leaching in Bermudagrass turf. *Agronomy Journal* 76, 964–969.

Taliaferro, C. (2000) Bermudagrass has made great strides-and its diversity has barely been tapped. *Diversity* 16, 23–24.

Thompson, K., Hodgson, J.G., Smith, R.M., Warren, P.H. and Gaston, K.J. (2004) Urban domestic gardens (III): Composition and diversity of lawn floras. *Journal of Vegetation Science* 15, 373–378.

Throssell, C.S., Lyman, G.T., Johnson, M.E., Stacey, G.A. and Brown, C.D. (2009) Golf course environmental profile measures water use, source, cost, quality, management and conservation strategies. *Applied Turfgrass Science* 6 0–0.

Tidåker, P., Wesström, T. and Kätterer, T. (2017) Energy use and greenhouse gas emissions from turf management of two Swedish golf courses. *Urban Forestry and Urban Greening* 21, 80–87.

Townsend-Small, A. and Czimczik, C.I. (2010a) Carbon sequestration and greenhouse gas emissions in urban turf. *Geophysical Research Letters* 37(2), L02707.

Townsend-Small, A. and Czimczik, C.I. (2010b) Correction to "Carbon sequestration and greenhouse gas emissions in urban turf". *Geophysical Research Letters* 37(2), L06707.

Trenholm, L.E., Unruh, J.B. and Sartain, J.B. (2012) Nitrate leaching and turf quality in established 'Floratam' St. Augustinegrass and 'Empire' zoysiagrass. *Journal of Environmental Quality* 41, 793–799.

U.S. Department of Energy (2011) Clean Cities Guide to Alternative Fuel Commercial Lawn Equipment, https://afdc.energy.gov/files/pdfs/52423.pdf#:~:text=Mowers (accessed 26 August 2025).

VanDerZanden, A.M. and Cook, T.W. (2010) *Sustainable Landscape Management: Design, Construction, and Maintenance*. John Wiley & Sons, New York, USA.

Wang, J. and Chermak, J.M. (2021) Is less always more? Conservation, efficiency and water education programs. *Ecological Economics* 184, 106994.

Watkins, E., Fei, S., Gardner, D., Stier, J., Bughrara, S. *et al.* (2011) Low-input turfgrass species for the north central United States. *Applied Turfgrass Science* 8, 1–11.

Wiewióra, B. and Żurek, G. (2023) Amenity grasses – A short insight into species, their applications and functions. *Agronomy* 13, 1164.

Wilson, M.J. and Jackson, T.A. (2013) Progress in the commercialisation of bionematicides. *BioControl* 58, 715–722.

Wolski, K., Markowska, J. Radkowski, A., Brennensthul, M., Sobol, K. *et al.* (2021) The influence of the grass mixture composition on the quality and suitability for football pitches. *Scientific Reports* 11, 20592.

Zhang, X. and Khachatryan, H. (2020) Investigating monetary incentives for environmentally friendly residential landscapes. *Water* 12, 3023.

10 Small Urban Green Spaces

Abstract

As urbanisation increases, there is a demand for, and a desire to access, natural or green spaces. Large space is at a premium, however, and many ambitions for green infrastructure have had to 'downsize' to fit with the space available. There are a wide variety of small-scale urban green typologies; some are designed to meet the demands for aesthetics and recreation, but others justify their inclusion by providing vital services to the city – such as affording space for flood water. Some of these typologies have long historical roots (e.g. public and private gardens) while others are more contemporary (e.g. modular living walls). This chapter illustrates and defines in more detail some of the smaller, external green spaces now present in many cities. Spaces that may require more unique management approaches to optimise their aesthetics and functions.

10.1 Introduction

Green infrastructure includes the large city parks, tree-lined avenues, cemeteries and peri-urban nature reserves but also smaller-scale green spaces. Often, these spaces have been designed to complement the nearby buildings (Fig. 10.1) or provide some specific function – such as holding runoff rainwater or providing food for city residents. Over the last 30 years, interest has risen in green roofs and roof gardens, green walls, city allotments and urban farms, and small urban nature reserves as well as green spaces used as part of an integrated approach to sustainable water management: so-called sustainable drainage systems – SuDS. The latter include features such as swales, rain gardens and reed beds. Some of these features are not necessarily new phenomena (e.g. the Hanging Gardens of Babylon – now thought to have possibly been located at Nineveh in 700 BCE) but their prominence is rising as humans attempt to grapple with a range of environmental and social issues aligned with population growth, urbanisation, non-sustainable consumerism and climate change. These and other factors are drivers for change too with respect to people's attitudes to green space and how they value this space. The last 20 years have seen the concept of the sky garden be introduced: green oases placed on the sides or tops of multi-level tower blocks. At the other extreme, guerrilla gardening has taken hold in a number of cities as citizens 're-utilise' unused urban open space and attempt to make their immediate neighbourhood more vibrant and engaging through the 'spontaneous' planting of flowers, fruit and vegetables. This chapter briefly explores a range of small, specialised urban landscapes in more detail.

10.2 Public Gardens

Today's public city gardens may have evolved from a number of routes, In the Medieval period, early town gardens were primarily vegetable plots outside city walls, used for food production. In time, these are likely to have become places for citizens to meet, relax and socialise. Public squares in warm Mediterranean climates were likely to be planted to improve their microclimate and provide an amiable meeting or trading place, with in time the amount and variety of planning increasing to create small garden spaces. In the Victorian era, areas within public parks were more intensively 'gardened' to create colour and interest. More intricate designs, features and the utilisation of more unusual plants often resulted in these locations being fenced or hedged off from the rest of the park, becoming a specific destination in themselves. Parkland was also devoted to specific plant collections becoming botanic gardens or heritage landscapes. Some were linked to collections of exotic animals, i.e. zoological gardens (zoos) – where the plants provided both a more pleasant environment for the human visitors, and, in time, some seclusion for the animals. Maintenance of 'high horticulture' in public gardens

DOI: 10.1079/9781800621763.0010

Fig. 10.1. There is a long tradition of growing plants in close proximity to buildings.

was more often than not reliant on public money (town councils and local authorities), although in recent years many have come to depend more on charities and, as professional workforces have been reduced due to cost-cutting, come to rely on volunteer labour. Public gardens array themselves in a variety of styles, but usually a key theme is around education and providing access to green space for recreational reasons.

10.3 Private Gardens

A private garden, also known as a domestic garden (or yard in North America), is an area of land near a private residence that is cultivated mainly for ornamental purposes and/or non-commercial food production. Residents have autonomy over the garden. These gardens serve as personal sanctuaries, offering space for relaxation, contemplation and connection with nature. They can also be very pragmatic – a place to dry the laundry or allow children to play securely. The private garden varies in prominence across different cultures and counties. Design options are diverse and are tailored to suit individual preferences, available space and environmental conditions as well as reflect different cultural priorities. Private gardens can be formal, featuring symmetrical designs and geometric shapes, or informal, embracing a more naturalistic approach with meandering pathways and diverse plantings. For keen gardeners in the UK, key styles may be adopted. These include the following:

- Formal garden: geometric and structured, descended from European gardens like Versailles, featuring straight pathways, vistas, topiary, mazes, and sculptures. This style can work well even in small spaces and in close proximity to buildings.
- Contemporary/modern garden: features innovative use of materials like concrete, stone, steel and geometric shapes. Planting is carefully planned with a limited palette of plants – often a case of 'less is more'. Some plants may have bold architectural forms.
- Urban garden/outdoor room: focuses on hard landscaping and garden furniture, suitable for small gardens in towns and cities. Raised beds can double as seating and storage benches maximise space. Features are functional and relate to socialising, e.g. patio seating areas, firepits, clay ovens and sheltered alcoves.
- Cottage garden: an easy, relaxed, ad hoc style with lots of flowers, often mixed with vegetables and fruit trees. Garden furniture does not need to match. Often conspicuous by the absence of a lawn.
- English country garden: combines elements of traditional and formal styles with wide paths, deep herbaceous borders, topiary, lawns, traditional garden furniture, trellised areas, roses, urns and perhaps a sundial.
- Woodland garden: includes trees, shrubs and ground cover, with paths and seating areas. More likely to have a sunny glade with patches of *Digitialis purpurea* (foxgloves), ferns and the odd woodpile rather than a lawn.
- Wildlife garden: often designed to mimic a woodland edge with a few large trees and shrub layer present, with a more open mini-meadow nearby. Simple, single-flower species and cultivars will be used as these allow insects to access pollen and nectar. Ponds, wetland areas and weed patches maximise the biodiversity, and birds may be encouraged through the use of feeders and nesting boxes.
- Exotic/jungle garden: features may include *Dicksonia antarctica* (tree ferns), *Musa basjoo*

(banana palms), *Canna speciosa* (canna), *Hedychium* spp. (ginger lilies) and other brilliantly coloured flowers. Large leaves planted densely, with some plants needing special winter care.

- Naturalistic/prairie garden: involves choosing plants that grow well in the area and planting them in large blocks with lots of repetition and informal style. Prairie gardens often associated with movement of the stems in the wind. More likely to include ornamental grasses and paths instead of lawns.
- Mediterranean/dry/xeric garden: ideal for areas with low rainfall, using drought-tolerant plants like succulents, cacti and ornamental grasses.
- Rustic garden: a country-themed garden with elements like chestnut hurdle fences and recycled agricultural equipment:
- Seaside garden: open and sparsely planted. If actually by the sea, may rely on salt-tolerant halophyte plants such as *Armeria maritima* (thrift), *Glaucium flavum* (yellow horned poppy) and *Crambe maritima* (sea kale), as well as plants that look wind tolerant such as *Eryngium x zabellii* (sea holly) and *Erigeron glaucus*. Almost obligatory in such gardens to have old pieces of fishing gear and the odd defunct rowing boat, with the centre planted up with appropriate wind-sculpted plants.

10.4 Allotments and Community Gardens

An allotment is a plot of land rented to individuals for non-commercial gardening, primarily to grow food crops. It serves as a productive garden away from the user's residence. Allotments in the UK are usually run by local councils, parish councils, community groups or charities, with the gardeners often organised in an allotment association. Each member has their own allocated plot. Short-term vegetable crops tend to dominate these landscapes. There is often a waiting list until someone is allocated an allotment.

Community gardens may also have individual plot areas, but in general, plant growing is more of a shared responsibility, with groups of gardeners cultivating fruit, vegetables and sometimes flowering plants. These gardens are generally managed and controlled by unpaid individuals or volunteers from the local community rather than a single family or individual. Collegiate activity and social bonding are thus important parts of the ethos of community gardens. Community gardens provide opportunities to learn about sustainable growing practices, build relationships and develop a sense of belonging. They offer fresh products and plants and contribute to a sense of community and connection to the environment. They can be set up specifically to provide access to fresh food, enhance community engagement and enhance positive mental health.

10.5 Mini and Pop-up Parks

These are small-scale green spaces generally no more than 0.5 ha in size, often situated in locations like road junctions, derelict land between houses, car parks or where an individual building has been removed. They can be permanent parks but are often temporary in nature. The flexible nature of these parks means they can test ideas and explore new approaches as to how people might use the locations in the longer term or reactivate interest in a given area. They often include artwork, seating, planted beds and containers with trees and flowers, and programs of events can be planned around them to help engage with the local community.

At a more informal level, any small urban 'uncared' for space can be procured for planting by guerrilla gardeners, essentially 'illegal' gardeners, who take it upon themselves to beautify a spot or use the location to grow crops and encourage spontaneous community interest. Despite being illegal technically, many local authorities turn a blind eye to such activity as they appreciate the area is being cared for, even if not by the legal civic authority.

10.6 Green Roofs and Podiums

Green roofs are any building roof that hosts some form of vegetation. The original green roofs would have been composed of grass turf or heather draped over a cottage, byre or similar building, and where the vegetation had managed to stay alive, largely due to a wet or humid climate. Such sod or turf roofs provided insulation and fire resistance for homes. The modern green roof movement began in Germany during the 1960s and 1970s. This era saw the development of 'extensive' green roofs on modern buildings using lightweight materials and shallow soil layers, making them suitable for urban environments. There are various forms of these now, including some evolved to hold larger plants (Table 10.1). Advances in waterproof membranes

Table 10.1. Characteristics of different green roof types and the typical plant/animal communities they host

Green roof type	Characteristics	Typical plant/animal communities
Extensive lightweight	No added substrate per se, just pre-grown ≤25 mm medium attached. Limited biodiversity potential. Low water-holding capacity, some with irrigation installed.	Bryophyta (moss), *Sedum* spp. and other xerophytes.
Super lightweight	Consists of thin (12 mm) drainage board, a filter, fleece/ water retention mat and pre-grown vegetated mat about 25 mm in thickness. Minimum building loadings required and can be used on some retrofitted buildings. Can be drought prone and irrigation sometimes supplied.	*Sedum* and *Sepervivum* spp. grown on the vegetated mats.
Extensive	Substrate ≤100 mm. Not normally irrigated. Planted with plug plants into substrate or pre-grown vegetation mats. Limited water-holding capacity. Low maintenance. Can be undulated topography with deeper and shallower areas of substrate. Flat roofs and those with a gentle slope. Typical weight: 60–150 kg m^{-2}.	As above, but also flowering annuals, e.g. *Centaurea, Coreopsis,* and *Linaria*; low-growing perennials, e.g. *Aubretia, Campanula, Delasperma, Dianthus, Sysyrinchium* and *Thymus*; geophytes, e.g. *Allium schoenoprasum, Iris pumila, Muscari azureum* and *Tulipa turkestanica.*
Semi-extensive	Substrate depth 100–200 mm. Many of the advantages of extensive roofs whilst allowing a wider range of species. Sometimes irrigated. Relatively good rainwater retention and detention capacity. Often used when there is access to the roof or the roof is overlooked. Slightly higher maintenance, e.g. during plant establishment or if weed species become problematic. Weight: 120–200 kg m^{-2}.	
Brown roof	Substrate depths vary e.g. 100–200 mm and sites are usually sculptured to provide a heterogeneous landform. Designed to replicate the ecological niche of terrestrial ex-industrial brownfield sites and utilise aggregates such as crushed brick and concrete rubble, sands and gravels. Important for certain invertebrate species and bird species that colonise such sites. Relatively good rainwater retention and detention capacity.	For example, in north-west Europe useful habitat for *Phoenicurus ochruros* (black redstart) and *Motacilla alba* (pied wagtail).
Biodiverse/ wildlife (extensive)	Where depth and substrates allow, a range of natural/ semi-natural vegetation communities can be replicated, e.g. chalk grassland, heathland and mesic meadow. Natural colonisation often encouraged. Depending on habitat and species present, human access may be restricted or limited at times. Frequently include a pond and wetland areas. Relatively good rainwater retention and detention capacity.	Due to the ability to manipulate the substrates effectively, green roofs provide a good opportunity to alter conditions to suit particular species or communities. Flower-rich meadows have been created, including populations of terrestrial orchids. Even amphibian/reptile species have been found on these roofs, despite being many storeys up in some cases.
Intensive	Substrate depth ≥200 mm. Intensive maintenance required. Sometimes irrigated. Composed of woody or herbaceous ornamental plantings. Good rainwater retention/detention capacity. Usually good access for people provided.	Wide range of herbaceous perennials, smaller woody perennials, geophytes and flowering annuals.
Lawn roof	Laid down to turf and used for amenity functions or aesthetics. High maintenance. Good rainwater retention/detention capacity.	Species vary with climate and depth of substrate.

Continued

Table 10.1. Continued.

Green roof type	Characteristics	Typical plant/animal communities
Blue-green roof	Combines green roof technology with water management systems to maximise storm water retention. Can have permanent ponds or temporary puddles	The plant community depends on the degree of wetness, which in turn is affected by the climate and depth of substrate or ponds present. Permanently wet areas can hold bog and wetland plant species. Can attract in flying aquatic insects.
Biosolar roof	Integrates solar panels for sustainable energy capture with vegetated green roofs. The cooling aspect of the plants enhances energy efficiency while supporting biodiversity.	Plants need to be small to avoid shading the panels. Conversely, some plants need to be shade adapted.
Roof food garden	Plants can be grown in soil (e.g. ≥200 mm) or in planters/containers. Soils need to be nutrient-rich organic media. Access to irrigation required. Usually useful social space.	Usually annual vegetables and vines such as *Pisum* (pea)/*Lycopersicon* (tomato), although full fruit trees have been used, e.g. the Reading International Solidarity Centre (RISC) garden, UK.
Roof gardens	Replicating a ground-level garden at height. Buildings need to be designed to cope with the weight of deep substrates and large plants (including small trees) when soils and canopies hold their maximum amount of water. Some depth may exceed 500 mm. Where depths are not sufficient, plants are grown in raised beds and planters too. Various styles, but often reflecting the most sought-after penthouse locations, i.e. highly stylised. Usually very good rainwater retention/detention capacity.	Depending on site/climate, plants may need to have some tolerance to wind and high light exposure. Typical plants might include *Betula* spp. (birch, including coloured-bark asiatic types), dwarf *Pinus* spp. (pine) and grasses such as *Stipa* and *Festuca* spp.
Podium garden	A green space constructed over the top of an underground car park or at the base of a block of apartments. Often there is an assumption the beds and plantings are separated in some way from the ground-level, natural soil.	Often associated with business buildings or modern residential living, so 'corporate'-style planting with strong architectural features is common. Some, such as the Barbican in London, UK, may have more semi-natural plantscapes

and drainage systems made green roofs more practical and resilient. A podium garden is a landscaped green space constructed on top of a podium structure, such as the roof of a car park or the base of a residential or commercial building. In a manner of speaking, they are the equivalent of the traditional terrace garden, i.e. a planted area above ground level, but associated with the lower tiers of a building. These gardens are often designed as communal outdoor areas for relaxation and recreation, offering a unique way to integrate greenery into urban environments. Green roofs provide ecological, aesthetic and energy-efficient benefits by improving insulation, reducing urban heat islands, managing storm water and enhancing biodiversity.

Green roofs vary in complexity based largely on the depth and characteristics of the substrate placed on the roof, as this determines the amount of water and nutrients available to the plant communities that develop. Their raison d'être too varies significantly from city to city and building to building. Most green roofs are formed from a series of layers placed horizontally above a building's roof, and these layers comprise a waterproof sheet to protect the roof, a root barrier to stop roots penetrating the roof zone itself, a mat or insulating layer to further protect the roof, a drainage layer and a filter sheet to stop particle migration to the lower zones (Anon., 2014). Above this is the growing substrate and vegetation. Some systems will contain modular structures that are clipped together and may contain wells to hold water and provide structural integrity to the green roof. The angle of the roof will determine the construction principles as well as physical matters, such as volume of moisture held within the system.

The rationale for green roofs is multifaceted. They are promoted because of their environmental benefits (see Chapters 3 and 5). This includes providing wildlife habitat in space that would otherwise be fairly inhospitable to most species. They improve the energy efficiency (Bevilacqua, 2021) and noise mitigation (Van Renterghem, 2018) of buildings, reduce the thermal flux of a building, contribute to urban cooling, have potential to absorb certain aerial pollutants and extend the life of the physical roof material, as well as help detain rainfall and reduce subsequent runoff (Fig.10.2). They also prove popular due to being architectural features in their own right and adding to the distinctiveness of the building. Depending on the green roof design, they may also be social space for the residents or workers of the building as well as a much-needed space to grow food in a congested urban world. In light of the fact that roofs can represent up to 32% of the horizontal surface area of urban conurbations, there is tremendous potential to mitigate many of the environmental problems associated with urbanism, through a wider adoption of green roof technology (Oberndorfer *et al.*, 2007).

Fig. 10.2. Cameron and others argue that the benefits of green roofs increase as the substrate gets deeper and the plant communities more complex. Deeper substrates provide greater resilience to the planting, but come at a cost, as they require more structural support from the building. Some of the trade-offs between the ecosystem services and the increased carbon footprint in constructing the building still need to be discerned. The notion of viewing a city from above and seeing mostly vegetation is compelling though! (Artist impression using Adobe Firefly – Author's own.)

Green roof typology

The extent to which vegetation grows on a green roof depends on substrate, type, depth and climate (Nagase and Dunnett, 2010; Monteiro *et al.*, 2017; Seyedabadi *et al.*, 2021). These points frequently define the function and type of roof employed (Fig. 10.3). Costs tend to increase with the more complex systems and the depth of substrate required (largely due to weight considerations – see below), but in theory, large areas of a city's horizontal surface area could be converted to green using extensive and semi-extensive vegetated roof systems aligned with contemporary building design. This could be the case for cities within temperate climates and adequate rainfall but will remain challenging for those in drier Mediterranean or semi-tropical climates, where rainfall may be more intermittent (Esfahani *et al.*, 2022).

Fig. 10.3. Due to the thin substrate layers, and limited water-holding capacity on many green roofs, xerophytic plants are often a wise choice.

Weights and load bearings

Deeper substrates correlate with greater physical mass, more water-holding capacity and larger plants – factors that all add to the weight on a roof. It is absolutely paramount that a build structure can tolerate the weight placed on it by the addition of a green roof system. This is where the environmental horticulturist needs to bring in a structural engineer to seek the right guidance on approaches to individual buildings. For new buildings, the potential weight load can be calculated and the building engineered sufficiently to meet these demands, although additional costs may be significant. For retrofitting an existing building, an independent assessment will need to be carried out to determine whether the installation will meet the existing structural capacity inherent in the building, or whether this will need to be modified. Load bearing characteristics are divided into three categories. The 'deadload', namely the weight of all the built elements and all components associated with the roof assembly, including plants, substrate and water at full capacity in the soil and also on the plants. This needs to take account of the fact that trees and shrubs may put on considerable biomass over the lifespan of the building. The 'liveload' takes account of any human activities on the roof; this may be weight of people, e.g. during social gatherings, but also maintenance or other mobile equipment taken onto the roof from time to time. The final aspect is 'transient load', which is the weight conferred when the building may move in the wind or even during seismic activity.

The State of Victoria in Australia calculates that typical weight loadings for ornamental plants are turf = 5.1 kg m^{-2}, succulents and low grasses = 10.2 kg m^{-2}, herbaceous perennials and subshrubs = 10.2–20.4 kg m^{-2}, larger shrubs = 30.6 kg m^{-2}, trees – small (≤6 m) = 40.8 kg m^{-2}, medium (≤10 m) = 61.2 kg m^{-2} and large (≤15 m) = 150 kg m^{-2} (Anon., 2014).

Many green roofs are placed on flat roofs, but pitched or sloped roofs are feasible too (Fig. 10.4). Shallow roof gradients, up to a 10° pitch, tend to use the same materials for a flat roof. At steeper pitches than this, however, additional support and infrastructure is usually required to ensure sheeting layers, substrates and plants all stay in position. More complex infrastructure increases direct costs, and the steeper angles have implications for access, with knock-on effects for maintenance costs. Some

Fig. 10.4. Flat roofs are often preferable for green roofs, but gentle roof gradients can be vegetated too.

roofs have sloping angles greater than 45°; this does not preclude the use of green roofs, but the technology begins to resemble that of green walls. As with flat roofs, the best way of improving the structural integrity of the system is to have an effective integration of plant roots and substrate, and good foliage cover of the roof surface, plant roots and stem structure helping to bind the system together.

Roofs, whether flat or otherwise, can be dangerous places to carry out maintenance activities. Most roofs that are accessible require a balustrade or retaining fence structure to ensure personnel are safe. Some parts of roofs are stronger than others and restricted zones may need to be identified and cordoned off in some manner. There may also be instrumentation and building infrastructure 'plants' such as air conditioning units or fume vents, again where it is inappropriate for maintenance staff to be close to. Equipment and even water/plant material may be difficult to get onto the roof, and these factors need to be considered when designing and managing the green roof.

Substrate technology

Growing substrates for green roofs are typically more inorganic in nature that those used in most other scenarios covered by environmental horticulture. This partially reflects the history of green roofs where drought-adapted species (e.g. houseleeks – *Sempervivum* spp.) were encouraged to colonise shallow (or even no) substrates, or where the ambitions were to replicate the stony aggregate nature of many brownfield sites. There is also the assertion that coarse inorganic materials such as gravels and stones allow rapid drainage, thus not holding excessive weights of water on the roof, before filtering into the physical drains. Many green roof substrates do have an organic component, but frequently it is as low as 15–35%, at least for extensive systems. Organic materials used include coir, composted bark and green waste compost. Inorganics may use a local quarried gravel, sand or lava rock (scoria) or 'recycled' materials such as crushed brick, crushed tiles, powdered fuel ash, pumice, chemically inert foams or furnace slag. Although, if some of these materials are technically deemed as a 'waste', there use may not comply with local or national regulations. The organic component is in the minor part of the ratio for a variety of practical reasons, including limited longevity before degradation, physical slumping (loss of oxygen and water-holding capacity) and it may become hydrophobic once it dries out (highly likely in a non-irrigated, shallow depth, extensive roof system). Once dry the organic component may be difficult to re-wet. Fear of fire on a roof also disinclines the use of flammable organics such as bark.

Specialised lightweight materials may also be utilised where the weight on the roof is a concern. These include leca (expanded clay granules), rockwool and expanded shale as well as pumice and lava rock. Some of these materials, however, do not come from sustainable sources or involve high energy inputs to achieve their desired properties.

Specifications for substrates for extensive/semi-extensive roofs include the ability to drain freely, whilst holding enough moisture to sustain plant growth outwith heavy rainfall events. The ability to detain and retain precipitation water during storms is one of the advantages of green roofs, but the responsibility for this falls largely on the substrate, hence the need for these attributes to be 'fit for purpose'. Substrates need to retain their structural integrity and be stable and consistent in their performance over time. Not least, they need to support plant growth with adequate amounts of nutrition, moisture and air to encourage a strong root system (Fig. 10.5.). This is especially important in shallow substrates, where the plant may need to regenerate from its root system or dormant crown after drought.

Specifying the substrate and its parameters are only part of the problem. Moving tonnes of substrate onto a roof is extremely challenging and is more than likely to overtax the domestic elevators

Fig. 10.5. Free-draining substrates composed of aggregates and sand favour Mediterranean-adapted species such as *Delosperma dyeri* cv. Red Mountain.

in a building. The logistics of moving substrate and other materials to the roof needs to be considered early in the planning process. High-rise roofs will require the hire and siting of large construction cranes; those lower to the ground may have the substrates pumped up from street level via a powered blower. Either way, space is required on the streets surrounding the building for the machinery and the storage of bulk bags holding the substrate. Substrates, especially the lighter types, can be blown up onto the roof using a compressor pump attached to a hose and the hose hoisted onto the roof. The blowing process, however, can redistribute the particle distribution, lighter particles being drawn up first, and some re-mixing may be required once the substrate is on the roof. Ensuring the substrate is moist beforehand will reduce this risk and also stop dust blowing around the worksite. It is also important to ensure the entire volume of substrate is not piled in the one location putting pressure on the roof at that point. Wherever possible the substrate should be evenly distributed as it is transported onto the roof.

Irrigation

Not all green roof systems need irrigation infrastructure, although some sort of watering facility would be useful during the establishment phase of the vegetation, even for the most highly adapted xerophytic species. The requirement for irrigation is based on climatic factors, the site's own microclimate (degree of shade or exposure to prevailing winds), substrate depth and constitution, plant choice and anticipated final size of plants and wider functional requirements. Green roofs that are designed to keep their building's occupants cooler during summer months, for example, may benefit significantly if water is applied to the plants and the substrates. In most cases, irrigation design is influenced by the nature of the water supply resource; for example, is it a potable water supply or can rainwater be harvested and stored? Important factors to consider will include quantities of water available and the pressure of the supply. If some form of grey water is used, the quality of that source becomes a consideration, even if there is some dilution with potable or rainwater. Plant selection should be strongly determined by the likely availability of the water supply. Having an inexpensive and environmentally sound, regular supply of water is key to widening the plant species selection available, optimising growth and improving establishment rates as well as improving the long-term success of the roof.

Irrigation, in general, follows the principles for other landscape typologies (e.g. see Chapter 6 for woody plants and Chapter 9 for turfgrass and lawns), although there are a few specific factors to consider. Preferred irrigation distribution systems will depend on the area to irrigate, volume and frequency of irrigation required, site characteristics (wind strength and direction), plant type (e.g. overhead spray lines and sprinklers may not be conducive to Mediterranean and other drought-adapted species with pubescent leaves, or plant species prone to leaf disease when the leaves are frequently saturated), cost and ease of maintenance. Sprinkler systems which would normally have acceptable distribution ranges at ground level may end up watering only half the roof area (and perhaps those pedestrians on the leeward side of the building at street level below) due to the high wind speed encountered on roof tops. Micro-sprays and pop-up sprinklers though, located close to the roof level, may be acceptable on roofs if there is a protective parapet to shield them from the wind. Perforated drippers and seep hose is useful, notably if utilised under low water pressure scenarios (as may be the case on a roof). Surface drip/seep systems are readily visible and accessible, sub-surface may be more difficult to maintain and can be accidently dug up, but this needs to be counterbalanced with the advantages of less water lost due to evaporation.

Unlike most other landscape situations, the use of water retention cells below the drainage layer can be capitalised on. The excessive water that is stored here during high rainfall events can be re-utilised to supply the vegetation with water. In some systems/substrates this is directly from capillary action; however, this is usually poor in very stony aggregates and an additional capillary wick system can be exploited to raise the water to the levels of the roots.

Most extensive systems use *Sedum* or other xerophytic plant species; drainage layers/substrates need to ensure these plants do not 'sit in water' after heavy rainfall events, and that they are designed to let excess water flow into the building's drainage pipes. Although irrigation can help keep the building cool, e.g. by maintaining plant evapotranspiration during warm periods of the day, it also needs to be recognised that water flowing through the substrate can itself transfer heat onto the building's roof surface.

How 'green' are green roofs?

This has been asked a number of times: do the benefits of green roofs actually outweigh the carbon equivalents used in their construction? Green roofs add to the infrastructure and construction costs of a building. This may involve the modification of columns and beams within the building and the roof tiles/slabs. In an attempt to reduce weight loadings lightweight plastic polymers are increasingly used in the construction of green roofs (Vijayaraghavan, 2016). These themselves, however, may have an environmental cost. Bianchini and Hewage (2012) reviewed the environmental benefits of green roofs by making a comparison between pollutants generated during their construction and their ability to absorb pollutants from the air, once fully functional. Pollutants from the polymers used included emissions of NO_2, SO_2, O_3 and particulate matter (PM_{10}). It was calculated that a green roof needs to be *in situ* for between 13 to 32 years to offset the emissions associated with the manufacture of these polymers. The variation in the years calculated depended on the details of the roof construction and whether some or all the polymers used were re-cycled. For newly manufactured polymers, the air pollutant rates were 1820 and 3960 kg ha^{-1} for extensive and intensive green roofs, respectively. This dropped to 680 and 1750 kg ha^{-1} when recycled materials were used. Intensive green roofs have higher values due to the more extensive use of polymers in their construction. Green roofs, in comparison, may remove 70–85 kg ha^{-1} of aerial pollutants per year, although this value is likely to be greater with intensive roofs where the plant biomass increases considerably over time.

Shahmohammad *et al.* (2022) outlined how they thought green roofs could be developed to maximise their ecosystem services and reduce their negative environmental impacts. The main points included using recycled, reused or locally available materials to reduce the environmental footprint, while being aware that substandard or life-expired materials may result in a reduction in the green roof performance. Substrate and drainage materials significantly affect storm water retention potential, leachate quality and plant survival and also influence the environmental footprint of the roof (e.g. the distance materials are transported). Plant choice is important too. Different plants have varying potential for CO_2 sequestration, air pollution absorption, temperature reduction, storm water retention, local habitat provision and improving water quality and consumption. Replacing plants regularly is not ideal and having a high initial plant diversity helps with sustainability and natural vegetation dynamics. Perennial herbs, with good drought and heat tolerance, often seem optimum.

10.7 Green Walls

Green walls or vertical green systems (VGS) tend to be divided into six main types:

1. Green façades: These are vertical greenery systems where climbing plants grow on building walls, supported by structures such as trellises, wire cables or meshes. Normally the plant roots are in the natural soil at the base of the building or in some form of large container adjacent to the wall base. Green façades usually comprise perennial climbing plants (vines), annual climbing species or wall-shrubs. Climbing species can either fix themselves to the supporting trelliswork through morphological features such as leaf tendrils, adhesion pads or aerial roots.
2. Living walls: These support plants by having modules or cells of substrate embedded in/on the wall, and the plant rootball grows in these modules. These compartments are often supplied with water and nutrients through artificial irrigation/fertigation systems.
3. Bio-walls: These are similar to living walls but tend to be designed to improve indoor air quality and humidity; they can be composed of higher plants, microorganisms or populations of primitive plants (e.g. Bryophyta – mosses).
4. Natural wall plantings: These are essentially conventional walls that are naturally colonised by plants. Damp locations may encourage the colonisation of mosses and lichens on the surface of the wall, or cracks within the brickwork and loose mortar allow seedlings to gain a foothold.
5. Retaining living walls: These are systems used for specific problematic sites and are part of an engineered solution to steep slopes, where vegetation is used to stabilise the soil and prevent erosion. Retaining living walls are designed to provide structural strength which resists the lateral forces on the slope that would normally result in soil particles moving. Some systems can perform on slopes up to 88° and many have capacity for variable slope angles as low as 45°. Most systems are modular in nature and combine inert physical elements with plant material, e.g. geotextile bags in conjunction with interlocking units, metal, concrete, plastic cellular confinement mats or woven plant mats. The eventual aim, however, is to provide a green mantle when mature, where the underlying structural elements are no longer visible.
6. Curtain plantings: Some green façade systems have plants grown in troughs either at the top of the wall or at intermediate levels, and the plants trail down the wall.

Green walls are a component of modern urban green infrastructure, although façade types have a long history (Fig. 10.6). They contribute to a range of ecosystem services including habitat provision for urban biodiversity (Collins *et al.*, 2017) providing thermal insulation in winter (Cameron *et al.*, 2015), screening out aerial particulate matter and improving air quality (Tomson *et al.*, 2021), attenuating noise (Pérez *et al.*, 2016), aiding psychological well-being (Yeom *et al.*, 2021) and improving the aesthetics of the cityscape A further role is their potential to reduce urban air temperatures, helping to mitigate urban heat island effects, and lower surface temperatures of buildings thereby reducing the reliance on mechanised air conditioning (Cameron *et al.*, 2014; Kunasingam *et al.*, 2024). (See Chapters 3 and 4 for more extensive reviews of these subjects.) The taxa used and the scale of planting depends on the green wall system employed.

10.8 Green Façades

Although the term has not always been used historically, green walls have a long pedigree when the use of climbing plants or vines against a building

Fig. 10.6. Green walls come as traditional façade types (top) or more contemporary living walls (bottom).

wall are taken into consideration. In Europe, *Hedera helix* (English ivy) is a common woodland plant that will happily adhere itself to stone and brickwork as well as tree bark. As it self-seeds itself into cracks and apertures, this species would have had to be actively removed from walls if undesired, rather than needing any active encouragement to grow. Old, ruined buildings today are often covered with *Hedera*, and it is a point of contention as to whether these ruins are being further damaged by the plant, or indeed, whether the stems of the plant are actually keeping the ruin intact! Despite the arguments that rage around *Hedera* and whether it damages masonry and increases dampness around walls, it is still a common enough veneer to houses today. A point reinforced in that there are numerous cultivars of various leaf forms and colours available from any garden centre or nursery. These include cultivars of *Hedera helix*, but also of other *Hedera* species. Due to its evergreen nature, close, dense foliage and relatively rapid growth rate, *Hedera* is a common species used by the professional green wall sector. To offset concerns, it is often grown on a trellis or framework that fixes to the wall, rather than encouraged to grow up the wall itself. *Hedera* is not the only species with tendril or rootlets that may explore weaknesses in the masonry. Other species include *Ipomoea* (morning glory), *Parthenocissus quinquefolia* (Virginia creeper), *P. tricuspidata* (Boston ivy), *Calystegia sepium* (hedge bindweed) and *Hydrangea anomala petiolaris* (climbing hydrangea).

Other 'traditional' wall-climbing plants used on housing, or other domestic structures (e.g. to screen unsightly garages, sheds and outhouses) have been selected either due to their interesting form and colour of foliage or because they have attractive flowers (Table 10.2). Over and above the aesthetic criteria, fruit trees, fruiting vines and vegetables have been well utilised against walls, due to the favourable microclimate associated with these locations. In the northern hemisphere, walls with southern aspects are advantageous as heat traps and hence have been used in more temperate climates to help early-season shoot development, to protect blossom from frost and to allow fruit to ripen successfully. As such, the growing of fruit such as *Prunus* (apricots, peaches, nectarines, and almonds), *Ficus* (figs), *Vitis* (grapes) and *Actinidia* (kiwi) are feasible at latitudes much further north than would be the case if they were grown as free-standing bushes/trees. Although, it should be noted that these genotypes are on occasions still protected by additional lean-to glasshouse structures covering the wall to further increase temperatures.

10.9 Living Walls

Living walls take their inspiration from those plants in nature that can survive on stone or brick walls, by seeding into cracks in the wall and deriving enough moisture and nutrients from the apertures within the wall. Such plants are normally adapted for rock screes or arid stony soils (lithosols) and include species such as *Cymbalaria muralis* (ivy-leaved toadflax), *Phlox subulata*, *Aubretia* x *cultorum* and *Centranthus ruber* (valerian). But that is where the analogy ends, as most living walls have water supplied to the plants through drip irrigation, seep hose or hydroponic systems using rockwool or some other material that allows effective capillary water movement. This enables a much wider range of species to be grown, including species such as *Carex* spp. (sedges) that would normally be associated with quite wet locations.

Table 10.2. Examples of taxa commonly grown as green façades. (C = climbing habit; S = free-standing; T = often trained against wall to maximise benefits.)

Species	Common name	Attributes
Foliage		
Celastrus scandens	American bittersweet	C – Distinctive leaves that turn golden yellow in autumn.
Garrya elliptica cv. James Roof	silk tassel bush	S – Grey-green leaves, and long male flower, 'tassels'.
Hedera colchica	Persian ivy	C – Evergreen climber with some forms variegated gold and green, e.g. *H.* cv. Sulphur Heart.
Hedera helix	English ivy	C – A wide variety of cultivars including variations based on leaf size and shape.
Hedera hibernica	Irish Ivy	C – Similar to *H. helix* and adapted to mild, maritime climates.
Parthenocissus henryana	Chinese Virginia creeper	C – Vigorous, large, deciduous climber, with dark green leaves turning red in autumn. Produces dark blue berries.
Parthenocissus quinquefolia	Virginia creeper	C – Famous for its leaves that turn flame red and orange in autumn.
Parthenocissus tricuspidata	Boston ivy	C – Ovate or three-lobed leaves that colour up purple and crimson in autumn.
Flowers		
Apios americana	American groundnut	C – Reddish brown flowers, but also edible tubers.
Akebia quinata	chocolate vine	C – Semi-evergreen, scented brownish purple flowers in early spring, followed by fleshy purple fruits.
Bignonia capreolata	cross vine	C – Related to the *Catalpa* spp. (Indian bean trees). Long, tubular, red and yellow flowers which can have a coffee-like fragrance.
Campsis radicans	Trumpet vine	C – Red or yellow tubular, trumpet-like flowers.
Ceanothus spp.	Californian lilac	S – Used as a wall shrub in temperate areas, although often derived from arid zones in the south-western USA. Flowers usually various shades of blue or purple.
Clematis spp.	clematis	C – Wide range of cultivars from 10 m high *C. montana* types to new dwarf 'patio' cultivars. Flower colours across red, pink, blue, purple, white and even yellow ranges.
Coronilla valentina		C – Blue green leaves with yellow pea-like flowers.
Hydrangea anomala subsp. *petiolaris*	climbing hydrangea	C – A self-clinging, deciduous species. It possesses 'heads' of white lace-cap flowers in summer as well as attractive purplish bark and heart-shaped leaves.
Jasminum spp.	jasmine	C – Various species and cultivars including the yellow-flowered winter jasmine - *Jasminum nudiflorum*. *Jasminum officinale* and cultivars have strong scent. Some species have a semi-evergreen habit depending on climatic conditions.
Lonicera spp.	honeysuckle	C – Includes *L. periclymenum, L. japonica, L. henryi, L. sempervirens* and *L. x tellmanniana.* Various colours of yellow, white, red and orange, often with combinations of more than one. Also famous for their sweet fragrance in the evening, designed to attract moths and other nocturnal pollinators.
Passiflora spp.	passion flower	C – Includes *P. incarnate* and *P. caerulea.* Usually blue or purple flowers, but white and red colours (*P. alata*) exist too.
Rosa	rose	T – Includes the rambling (very vigorous, single flowering period) roses and climbing roses, of less stature, but often repeat flowering habit. Wide range of colours.
Solanum crispum	potato vine	C – Cultivars tend to have purple, blue or white flowers with an orange centre.
Trachelospermum spp.		C – *Jasminum*-like plants with white flowers.
Wisteria		C – Cultivars derived from *W. sinensis* and *W. floribunda.* Panicles of purple, white or pale pink flowers depending on cultivar.

Continued

Table 10.2. Continued.

Species	Common name	Attributes
Fruit		
Citrus spp.		T – A wide range of *Citrus* genotypes can be trained against walls, including lemon, lime, orange and grapefruit in Mediterranean-type climates (or under glass in temperate regions).
Ficus carica	fig	S – Large, green, lobed leaves and green. purple 'fruit'. Fruiting is enhancing by restricting the root growth, so can be grown in a pot beside the wall. In *Ficus* the 'fig fruit' is actually the stem of an inflorescence, with the flowers inside, these being pollinated in nature by the wasps within the family Agaonidae.
Malus	apple	T – Apples can be trained against a wall through techniques such as espalier and cordon. Depending on the vigour of the rootstock and the degree of pruning, trees can ultimately reach 5–6 m in height
Pyrus	pear	T – Culinary and dessert pears. Can be trained in similar way to *Malus*.
Prunus spp.	plum, cherry, apricot, peach	T – Many 'stone' fruit such as cherry benefit in terms of ripening from the warmer microclimate around a wall
Vitis	grape	T – Grape vines can rapidly clothe a wall in foliage and with careful pruning also yield acceptable quality fruit.

Living wall systems vary in their construction and design, with commercial companies frequently designing and patenting their own system. The market also varies depending on context and situation. At one end of the spectrum, there are 'do-it-yourself' kits for the domestic market that can be fixed together and put on a house or garden wall and are often no more elaborate than a range of planting troughs and modules linked together to form a single unit. At the other extreme, there are very elaborate bespoke systems that are custom-fitted to a wall and may cost the equivalent of £2–3 million. This is particularly so when the wall is an integral part of a prestigious new iconic building or forms a city centre landmark piece. Broadly speaking, however, two approaches dominate the designs utilised. These are hydroponic systems and substrate-based systems (Anon., 2014).

Hydroponic-based living walls can be constructed from either modular containers or large panels. To avoid risk of dampness to the building (and to provide an extra layer of thermal insulation) hydroponic systems are usually discrete from the wall, with an air gap being left between the building and the irrigated wall. The wall is attached to the building by way of brackets or a stand-alone framework. The back of the living wall is composed of a waterproof sheet or lining, and immediately in front of this is the inert medium that conducts the water and anchors the plants. This comprises materials such as rock or stone wool (basaltic rock or industrial slag that is heated and spun into fibres), horticultural foams or felt fabrics. These materials conduct and supply water to the plants but also drain freely, allowing oxygen to pass through to the plant roots. As most hydroponic media are inert, they allow effective control of nutrient concentrations and any structural decay is a prolonged process. Where slabs of rockwool, or similar material are used, it is the nature of such material that the base holds more water or is wetter for longer, due to the action of gravity, than areas near the top. Designers take advantage of this, and often more drought-adapted species are placed at the top (e.g. *Helianthemum*, *Sedum* and *Silene*) and those more tolerant of prolonged wetting or which require more consistent and higher volumes of water are located nearer the base (e.g. *Asplenium*. *Hosta* and *Primula vulgaris*). Nutrition in hydroponic systems needs to be carefully managed to avoid excessive or inadequate ions in the solution, and this is controlled via fertigation systems that dose the irrigation at intermediate intervals with nutrients, when ionic concentrations drop below certain thresholds.

Living walls with substrate-based systems use discrete cells, modules or troughs. These are orientated on an independent, structurally secure metal rack or framework, anchored directly to the wall or placed on a grid structure adjacent to the building wall. The advantage of these systems is that individual cells or modules can be removed, if plants fail or become overcrowded. Irrigation in module systems may be via pipes linking the modules or from drip lines run along the tops of the modules and an individual dripper supplies water to each unit. Substrates vary in composition and are still being researched for optimum performance, but as with any successful medium the notion is to maximise water-holding capacity, whilst minimising anaerobic conditions, i.e. keeping them well-drained and aerated. Substrate systems by nature have a greater buffering capacity than hydroponic systems, for example if a pipe should fail, there usually is some moisture reserve within the module to keep plants alive for a few days. The downsides, however, include that nutrients will become exhausted after a time, and conversely, there can be a concentration of excessive salts if the nutrition is not carefully managed or if there are additional ions in the water supply (calcium, sodium, etc.).

Excessive water dripping off a building, due to over-irrigation or poorly positioned drippers, can be a problem, particularly so during cold conditions or where periods of frost occur, as the resultant 'ice-slide' on the pavement below poses something of a hazard to pedestrians. Excessive water shedding off the building may also prove problematic to any plantings at the base of the wall or indeed to the building fabric itself. As such, drip trays are used frequently to capture excess irrigation water from the growing medium as well as water droplets that drip off foliage. Often this water is recycled and pumped back to the top of the building for re-use. The capacity of the drip trays and holding tanks should be sufficient to hold an entire irrigation cycle's water volume.

The relatively restricted root volume available within a living wall, plus exposure to high irradiance (if in full sun) and wind, has resulted in a predominance of low-growing plants forms being favoured for living walls. Although there are different rationales for living walls, aesthetics plays a significant role in many walls systems, and as most walls have a high public profile the 'need to look good' is often at the forefront of a client's mind. This is somewhat in contrast to the underlying philosophes that have driven the green roof movement – where early principles were based around ecological agendas, for example replicating rare and unusual habitats associated with urban brownfield sites.

The fact that living walls are viewed from the streetscape, however, has dictated that plantings should be interesting, attractive, uniform in cover and easily maintained. Although many plants die back to a rootstock or become leafless in winter, large brown patches on the sides of walls are not always appreciated by the clients who have commissioned the wall. As such, an extra dimension to the plant selection is the inclusion of evergreen species, or species where the dying foliage remains attractive, e.g. grass and *Carex* spp. Other plants may be included because they have a prostrate growth habit and cover the wall fabrics and structural parts, provide a few weeks of colour and contrasting form through flowering, have attractive evergreen foliage or foliage of an unusual colour or help provide a specific ecosystem service, for example supplying nectar or pollen for invertebrates or even birds such as hummingbirds. These factors determine species choice and, as a consequence, alpine plants (Table 10.3), small woody plants including evergreen species (Table 10.4), small herbaceous perennials (Table 10.5) and grasses and ferns (Table 10.6) dominate living walls in temperate climates. Despite the wide range of plants available it needs to be remembered that walls, by and large, are fairly inhospitable environments, with wind and excessive temperatures in summer placing a strain on the plant's ability to regulate water distribution and hence leaf temperature. Similarly, unlike natural soil, substrates in containers and hydroponic mats have a relatively small volume, meaning that thermal buffering capacity is low; hence, roots can be exposed to frost which would not normally be a problem with plants in natural soil at ground level. Compared to stems and leaves, roots are much more susceptible to injury from sub-zero temperatures. As outlined above, for many urban wall systems aesthetics is important as they are in the public eye. Complementing functionality with an attractive colour scheme or range of leaf forms is therefore a consideration (Fig. 10.7).

In the tropics (and indoors in buildings of temperate climates) epiphytes such as bromeliads and orchids become more important in walls systems, and augment evergreen vines and shrubs.

Due to the inaccessible nature of the plantings on wall systems, maintenance can be one of the more costly activities. Physically locating personnel close to the wall may require the provision of cherry picker, genie lift or even scaffolding. This will only be feasible if there is space on the ground for the equipment to be moved in to. Some larger green wall structures are maintained through the use of special gantries that are held by cables from the roof. Safety is also a paramount consideration, and personnel may need to be kitted out with a harness, descent and safety rope, rope protector, rope-grabbing tool,

Table 10.3. Examples of alpine plant species used in green walls within temperate climates and their attributes

General attributes/traits. Dwarf growing with usually either prostrate or hummock-forming growth habits. Some are 'silvered leafed' or have numerous leaf hairs (hirsute) to avoid moisture loss and protect from UV light. Often flowering in spring. Alpines are useful for green walls due to their tolerance to drought and exposed conditions, low temperatures but also high wind speeds.

Groups/species (cultivars)	Attributes/prominent colour (F = flower, L = leaf)
Armeria maritima (thrift)	F – pink, clump-forming.
Aubretia x *cultorum*	F – purple, blue or magenta.
Dianthus spp.	F – pink, white or red. L – deep green or grey.
*Herniaria glabra (*green carpet)	L – small leaves, deep green. This plant is known simply as 'green carpet' due to its habit of forming a uniform sward of mid-green.
Phlox subulata, P. douglasii (alpine phlox)	F – blue, purple, pink, pink/white striped. L – fine-leaved texture, mat-forming.
Pulsatilla vulgaris (pasque flower)	F – blue, white or wine red.
Saxifraga spp.	F – white, pink and yellow. Some hummock/clump-forming, others prostrate 'mats' of foliage.
Sedum acre (mossy stonecrop)	F – sulphur yellow. Very low growing/drought tolerant.
Sedum album	F – white. Foliage persists throughout winter but turns red to bronze as temperatures become cooler. *S. album* cv. Coral Carpet is a selected form that colours up cherry red in summer.
Sedum ewersii	F – pink. L – blue-green.
Sedum hispanicum var. *minus*	F – pink or white. *S. hispanicum* var. *minus* cv. Blue Carpet has grey/blue foliage which turns deep blue to purple in winter.
Sedum hybridum cv. Immergrunchen	F – yellow.
Sedum kamtschaticum	F – yellow. *S. kamtschaticum* var. *floriferum* cv. Weihenstephaner Gold has orange tips to the flowers giving an overall gold appearance.
Sedum reflexum	F – yellow. Branches look like miniature conifer trees. Forms with deep ice-blue foliage contrasting well with the yellow flowers.
Sedum sexangulare	F – yellow. L – mid-green in summer turning orange to red in autumn and eventually brown. It derives its name from the fact that the leaves are arranged in rows of six.
Sedum spurium	F – red, white and pink. L – rosettes of green or bronze-green leaves turning red and orange in some cultivars. *S. spurium* cv. Summer Glory has deep pink/red flowers in midsummer contrasting well with deep green foliage.
Sempervivum montanum	F – purple/red. Rosette leaves with flowering stems held well above the canopy. A very variable species with numerous forms.

descent mechanism, lanyard and suction cups, depending on the height and size of the wall, as well as the ubiquitous 'high-viz' jacket and hard hat. High buildings are not only prone to exposure to the wind and high wind speeds, but their geometry often encourages 'unexpected' gusts of wind, and workers need to be attached to the building in some way at all times. Hazards due to dropped equipment or loose modules also need to be considered, and road/pavement areas below the wall may need to be fenced off to protect the public.

Perhaps a more significant concern for the future viability of green walls, at least on tall buildings, is a greater fire risk to buildings associated with a mantle of vegetation. This is a consequence of the Grenfell Tower tragedy in London, UK in 2017, where tower block safety was compromised by partially flammable solid external cladding. Although plant material is >90% water, tissue moisture is quickly evaporated off under the intense heat of building fires, rendering the remaining organic matter highly flammable and a source of fire propagation along the outside of a building. In living wall systems, plastic modules and other infrastructures could also contravene fire risk standards. Some of these issues may be solvable, such as limiting plants to only

Table 10.4. Examples of shrubs and dwarf sub-shrubs used in green walls within temperate climates and their attributes

General attributes/traits. Woody plants used on walls tend to be those that are prostrate in habit or the dwarf forms of larger genotypes. Many are from Mediterranean, semi-arid climates or from heathlands, with sclerophylus or hirsute leaves. Other types used though are shrubby climbers or ground cover plants, often adapted to the shade of a forest floor environment. Flowering, foliage and even fruiting characteristics can be important for selection on wall systems.

Species/cultivars	Attributes/prominent colour (F = flower, L = leaf)
Ceanothus repens	F – blue. L – deep green. Evergreen or semi-evergreen in exposed locations.
Centranthus ruber (valerian)	F – red, pink or white flowers. Drought adapted – can grow in stone and rubble walls without irrigation.
Cerastium tomentosum (snow in summer)	F – white. L – grey. Vigorous growing ground cover, for high-light locations.
Convolvulus cnoerum	F – white. L – grey.
Cotoneaster damerii	F – white. L – green. One of the more prostrate Cotoneasters spreading to 2 m, flowers in early summer, followed on by bright red berries.
Cistus spp. (sunrose)	F – white, pink. Various hybrids including *Cistus* × *hybridus* (*C. populifolius* x *C. salviifolius*) and *C. purpureus* (*C. ladanifer* x *C. creticus*) types – even hybrids formed with *Halimium* spp., i.e. x *Halimiocistus* which can bring in yellow flowers. Not all genotypes are cold tolerant.
Erica spp. (heath)	F – purple, pink, white. L – green, bronze. Includes *E. carnea* hybrids such as cv. Myretoun Ruby and cv. Springwood White which have a degree of lime tolerance. Careful selection of cultivars can allow for year-round flowering.
Euonymus fortunei	L – green or variegated with silver or gold.
Hedera spp. (ivy)	L – green or variegated with silver and gold. Trailing or climbing habit over the wall.
Hebe spp.	F – white, pink, purple, blue. L – various forms and colours in blue-green but also variegated forms, some with pink edges to leaves e.g. *H.* cv. Magicolors. Wind tolerant where temperature not excessively low.
Helianthemum spp. (rockrose)	F – white, red, pink, orange, yellow. L – mid-grey.
Hyssopus officinalis (hyssop)	F – spikes of whorled, tubular blue flowers. L – aromatic, linear leaves. Popular with pollinators.
Iberis sempervirens (perennial candytuft)	F – white. L – mid-green. Short-stature bush.
Lavandula spp. (lavender)	F – purple, mauve, white. L – green-grey. Both *L. angustifolia* (English) and *L. stoechas* (French) lavenders do well on walls. Good for attracting pollinating insects when in bloom.
Lithodora diffusa	F – electric blue. L – mat-like mid-green. *L.* cv. Heavenly Blue is common and reliable.
Origanum spp. (marjoram)	F – pink. L – aromatic, grey/green.
Pachysandra terminalis (Japanese spurge)	F – small white flowers. L – pale green, serrated leaves. *P.* cv. Green Carpet particularly low-growing variety; shade tolerant.
Potentilla fruiticosa	F – orange, yellow, white, pink, red. L – green.
Rosmarinus officinalis Prostratus Group	F – blue. L – mid-green. Prostrate forms of common rosemary.
Salvia officinalis (sage)	F – purple. L – blue-green or silver or purple hues.
Thymus spp. (thyme)	F – purple. L – green, bronze, gold.
Vinca minor (periwinkle)	F – blue-purple. L – green. Trailing habit and shade tolerant.

certain parts of the building or even examining plant taxa for relative fire loading, e.g. excluding only those species with high levels of volatile oils or resins. It is still not clear how practical implementation of green walls might be affected by future policy in this context.

10.10 Bio-walls

Bio-walls is the term used to describe indoor living walls, where the original objectives centred around the use of plants and the microbial populations

Table 10.5. Examples of herbaceous perennials and geophytes used in green walls within temperate climates and their attributes

General attributes/traits. Herbaceous or semi-herbaceous plants that tend to keep a low growing habit, again selected for their tolerance to exposed conditions and ability to spread across the wall. Aesthetic characteristics such as interesting leaf colour, form or highly impact or prolonged flowering capabilities also important.

Groups/species (cultivars)	Attributes/prominent colour (F = flower, L = leaf)
Ajuga reptans	F – blue. L – green, bronze. Shade tolerant. *A.* cv. Black Scallop has dark red/purple leaves.
Allium schoenoprasum (chives)	F – purple, blue. L – Green, strap/needle-like.
Artemisia spp. (tarragon)	F – small clusters of yellow flowers. L – green or blue-grey. Aromatic.
Bergenia spp.	F – white, red, pink. L – green with large, bold, rounded form.
Chamomile nobile (chamomile)	F – white. L – small, fern-like, scented. *C.* cv. Treneague is a very low growing, non-flowering selection that can be used to represent a uniform green plane.
Erigeron spp.	F – white, daisy-like flowers. *E. karvinskianus* gives a pleasing multi-tone affect with pink and yellow in flowers too.
Euphorbia amygdaloides cv. Purpurea (wood spurge)	F – acid yellow/green. L – deep purple.
Fragaria vesca (alpine strawberry)	F – white. L – mid-green. Also enhanced by small red fruit in summer.
Helleborus spp.	F – white blooms in winter. Shade-adapted species that dies back in summer.
Hemerocallis spp. (day lily)	F – wide range of colours, with individual blooms only lasting one day. L – strap-like leaves. Dwarf forms most suitable for walls.
Heuchera spp.	F – loose panicles of red, pink or white flowers. L – very wide range of colour tones including lime green, deep red, almost black, yellow, copper and tawny.
Hosta spp.	F – purple, white. L – green, variegated, gold. Large, bold leaves.
Liriope muscari	F – blue spikes. L – green strap-like leaves. *L.* cv. Monroe White has white flowers.
Mentha spp. (mint)	F – white or mauve. L – mid-green through to olive grey/green. A wide range of species, most having strong aromatic scent. Responds to trimming back in summer to re-shoot from base.
Omphalodes verna (blue – eyed Mary)	F – bright blue with white centre similar to forget-me-nots. L – green.
Petroselinum crispum (parsley)	F – small umbels of white flowers. L – finely divided leaves of bright green. When used en masse leaves resemble miniature forest.
Pratia pedunculata	F – blue. L – fine-textured green. Can be very vigorous when conditions suit.
Primula vulgaris (primrose)	L – green or variegated with silver and gold. Trailing or climbing habit over the wall.
Silene schafta (autumn catchfly)	F – purplish pink. L – mid-green, lance-shaped leaves. Mat-forming semi-evergreen perennial.
Sisyrinchium striatum	F – pale yellow in slender spires. L – grey-green strap-like.
Stachys byzantina (lamb's ears)	F – purple. L – soft-textured and light grey.
Tellima grandiflora	F – spires of pale yellow/green flowers. L – emerald-green, scalloped. Shade tolerant.
Teucrium chamaedrys	F – purple. L – medium green, similar appearance to *Nepeta* and *Salvia*.
Tiarella spp.	F – panicles of small star-shaped flowers in white or pale pink. L – green or purple-centred green. Shade tolerant and even flowers well in shade.
Viola spp.	F – blue, purple. L – mid-green. Perennial forms used rather than the bedding plant types, including cultivated forms such as *V.* cv. Martin, *V.* cv. Irish Molly, *V.* cv. Columbine and *V. odorata* cv. Queen Charlotte.

associated with their roots to remove aerial pollutants from the interior space. Over time this definition has been blurred as more bio-walls have been constructed for both aesthetic and well-being objectives. Walls implemented to 'clean' air are designed to have air passing over the plant/microbial communities and so act as natural filters. This is usually accomplished by a circulation fan, although passive air systems have been used too.

Table 10.6. Examples of ferns, grasses and grass-like plants used in green walls within temperate climates and their attributes

General attributes/traits. Ferns often do well in shaded areas, or locations with high humidity. Grasses can vary in their tolerances, but many prefer free draining situations. These plants are often used in green walls for their ability to 'lift' or 'lighten' the planting composition, due to their fine texture and movement in the wind.

Groups/species (cultivars)	Plant type/attributes (F = fern, G = grass, GL = grass-like)
Acorus gramineus cv. Variegatus (sweetflag)	GL – Variegated leaves of green and cream; prefers the wetter locations in walls.
Asplenium trichomanes (maidenhair spleenwort)	F – Hardy, evergreen fern native to north-west Europe. Well-suited to planting in a dry or shaded wall. Forms a rosette of dark-stemmed, pinnate fronds with small, rounded or oblong segments.
Blechnum spicant (deer fern)	F – Hardy, shade loving and adapted to cool moist locations. Large, pinnate fronds give a bold image. Needs to be in a medium where the pH remains low, i.e. requires acidic irrigation water.
Carex albula syn. *comans* cv. Frosted Curls	GL – Pale silver/green foliage. Does best in regions with mild winters.
Carex oshimensis cv. Evergold	GL – Characterised by dark green leaves with a central strip of creamy yellow. Forms low evergreen hummocks.
Cyrtomium fortunei (Japanaese holly fern)	F – Requires rich, moist but well-drained medium. Semi-evergreen with arching fronds of a mid-green colour.
Deschampsia flexuosa cv. Goldtau	G – Forms a compact tuft of foliage, with spires of flowering shoots with pale-cream flower heads, followed by brown seeds.
Dryopteris affinis (golden shield fern)	F – Bright yellow-green when unfolding, followed by fronds coloured rich green. Semi-evergreen habit.
Festuca glauca	G – Prefers full sun and free-draining conditions. Improved versions have strong consistent blue-colour foliage, e.g. *F. glauca* cv. Intense Blue.
Luzula nivea, L. sylvatica (woodrush)	GL – Prefers moist conditions and tolerates shade. Gold forms such as *L. sylvatica* 'Aurea' help lighten dark locations.
Phyllitis scolopendrium (hart's tongue fern)	F – Evergreen, non-pinnate leaf type of mid-green hue. Moist, cool, shady conditions.
Polypodium vulgare	F – Requires shade but otherwise copes well with range of moisture conditions. Adapted to the cool, moist climates of north-west Europe. Long, leathery, dark-green fronds with a slight sheen, and illustrates an attractive, lacy texture when viewed from a distance.
Polysticitum polyblepharum (Japanese tassel fern)	F – Provides interest all year round. As the fronds unfold, they turn a glossy deep green and are presented in a slightly recurved rosette. The fronds have a slight reflective sheen.
Uncinia rubra (red hook sedge)	GL – Mahogany-red sedge grass, moist free-draining media in sun or part shade to exploit the foliage hue.

Bio-walls are used to reduce levels of indoor volatile organic compounds (VOCs) that are derived from materials and activities commonly found in offices (Table 10.7, Anon., 1989 – see also Chapter 11). As the rhizosphere bacteria require carbon sources to develop, the small-chain carbon atoms in the VOCs are utilised and catabolised to release energy, thus degrading their activity as pollutants. Although, the rhizosphere has a high diversity of microbial species, it is believed that some of these microorganisms developed their ability to degrade VOCs based on environmental pressures or the lack of availability of other carbon sources. Although much is made about the plants in a bio-wall, in many ways their primary function is to act as hosts for the microorganisms and increase the surface area that these microbial communities can grow on. Increasing the microorganism's access and time of exposed to the VOCs is important. The VOCs are readily soluble and a film of moisture is important to improve contact. Slow rates of air passage as well as an operating temperature of approximately 20°C are optimal for maximising microbial activity and hence removing the VOCs from the air.

Fig. 10.7. Most green walls are green, but colour can be introduced through flowers and foliage to give variation in interest. Artistic impression of how alpines, geophytes, low-growing herbaceous and woody sub-shrubs can be added to a *Sedum* spp. wall to add colour. A = *Aurinia saxatilis*, B = *Heuchera* cv. Wildberry, C = *Iberis sempervirens*, D = *Allium aflatunense* cv. Purple Sensation, E = *Aubrieta* cv. Purple Cascade, F = *Genista lydia*, G = *Santolina* cv. Yellow Buttons, H = *Potentilla fruticosa* cv. Lundy, I = *Lavandula stoechas* cv. Devonshire Compact, J = *Phlox subulata* cv. Red Wings, K = *Armeria maritima*, L = *Helianthemum* cv. Cerise Queen, M = *Dianthus deltoides* cv. Flashing Light, N = *Phlox condensate*, O = *Sedum ternatum* and P = *Lithodora diffusa* cv. Heavenly Blue. (Designed using Adobe Firefly – Author's own.)

Table 10.7. Canadian guidelines for common indoor contaminants: contaminant maximum exposure limits (ppm), as absolute values or concentrations over certain exposure periods (Anon., 1989)

Contaminant	Maximum exposure limit (ppm) or other as stated	Duration
Carbon dioxide	3500	
Carbon monoxide	11	8 h
	25	1 h
Formaldehyde	0.1	
Acetaldehyde	5.0	
Nitrogen dioxide	0.05	1 h
	0.25	
Ozone	0.12	1 h
Sulphur dioxide	0.019	5 min
	0.38	
Benzene	10	
Toluene	200	
Trichloroethylene	100	
Naphthalene	9.5	
Particulate matter ($PM_{2.5}$)	40 $\mu g\ m^{-3}$	

Although bio-walls are also sometimes quoted as being useful for providing oxygen and removing carbon dioxide, it is unlikely such factors are significant, not least because irradiance levels in interior environments are often too low to optimise photosynthesis in the plants contained within the wall.

Bio-walls are a novel way of introducing greenery to the interior environment and have the advantage of taking up less room than a conventional interior planted landscape. This, however, needs to be balanced with the costs of specific systems and ease of access to areas behind the wall (e.g. access to irrigation pipes and drippers). Nevertheless, bio-walls are becoming a more common feature in lobbies, reception spaces and other communal areas within buildings, with the psychological health benefits of viewing plants being one of the main motivators for the inclusion of such features (see Chapter 4).

10.11 Sustainable Drainage Systems (SuDS)

A number of urban green (or green-blue) spaces are designed with the intent to manage urban water flows and stop/decrease the chance of flooding. These can work at various scales from a storm water planter being attached to a building drainpipe up to the scale of whole parks being used as sacrificial land to hold excess water temporarily during storm events.

10.12 Storm Water Planters

A storm water planter is a contained, vegetated area designed to collect and treat storm water runoff coming off a building roof or other similar hard surface. Water from a roof is collected in the guttering and led down to ground level by one or more drainpipes. Planters are commonly installed by the side of buildings or on pavements. They are filled with gravel or stone and topped with soil and vegetation. Runoff water enters and fills the planter, with any excess water being directed to the sewer system via an overflow pipe. A planter's main purpose is to act as a 'brake' on water entering the main drain/sewage system, thus reducing the flow volume and rate at a critical period during a storm. Therefore, numerous planters help the sewage system from being overloaded. The role of the plants is to dry out the substrate through evapotranspiration and make the feature more visually attractive. The planter can be sunk into the soil with the top rim level with the pavement, thereby also trapping any surface water flow or be a raised feature with the down drainpipe terminating at some distance above ground level. The latter can be relatively easy to maintain – with easy access to replace plants and re-fresh the substrate.

10.13 Swales

A swale is a shallow depression or channel in the landscape and can be natural or designed. They are places to store excess surface runoff water. They typically have gently sloping sides and are normally covered in grass or other ground-cover vegetation. In urban situations, swales may now be designed and constructed as part of a housing development or other built infrastructure. Here the base may be composed of gravel to improve filtration. The surrounding land may be contoured or pipes added to ensure water runs off the landscape and accumulates in the swale. Swales facilitate rainwater infiltration into the soil, replenishing groundwater. Swales can hold relatively large volumes of water, help relieve the pressure on surrounding rivers by reducing the flow rate and also help trap any pollutants that are suspended or dissolved in the runoff water. Swales are generally cheaper to construct and maintain compared to underground drainage systems.

10.14 Rain Gardens

Rain gardens are a component of sustainable drainage systems which help regulate water flows across the urban environment and intercept sources of water-borne pollution. They are essentially a shallow depression in the landscape, with absorbent, yet free-draining, soil. They are important for detaining water and slowing the rate of water flowing into conventional drainage pipes and sewers and also for retaining water (reducing the volume of water entering drains, by losing a proportion to groundwater reserves and back to the atmosphere as evapotranspiration). Unlike swales, bioswales and rain gardens imply both some form of engineered soil to help water infiltration and some functional planting scheme. Rain gardens have been promoted most in the USA and Australia, where urban design has been closely engaged with managing storm water events and also ensuring enough water is available to maintain urban green infrastructure.

Rain gardens are seen as an intervention (Fig. 10.8) – a place to temporarily trap and store excess water runoff – and many are still integrated with existing drainage systems, so once full, any excess water is still carried into the conventional sewage system and does not overflow into adjacent roadways and houses. Although water and diffuse pollution management are the key raison d'etre for rain gardens, they also provide a range of other ecosystem services including improved aesthetic value to an area, localised cooling and recreational opportunities, including, where pollution is not a problem, opportunities for food growing. They can also make great habitat for wildlife, especially if the planting is informal and the plant species selected or naturalised contributing to faunal feeding opportunities.

Rain gardens are usually situated in a low depression where water would normally accumulate naturally or on a gentle slope in the landscape. The greater the gradient of the slope, the more difficult it will be to site the rain garden, as remodelling the land to create a level perimeter becomes more challenging; slopes in excess of 1:8 are best avoided as these would require retaining structures and may look incongruous within the wider landscape. Unlike storm water planters, rain gardens are usually situated some distance from a building (>3 m) to avoid any problems associated with overflow. Water coming off building roofs via downpipes can be led to the rain garden by additional pipes or channels cut out in the soil or even large sloping swales. These channels themselves are frequently brick-, pebble- or clay-lined to stop soil eroding as water rivulets form.

Fig. 10.8. Rain gardens can be linked to conventional drainage systems, added as a place to detain runoff water and allowing it to infiltrate the ground or dissipate via evapotranspiration.

Water capture and infiltration

As it is desirable for a proportion of the water to percolate down through the soil, naturally free-draining soils, such as sands and chalks, are more appropriate for rain gardens than areas where the water table is naturally high or where heavy clay soils impede the movement of water downwards. Ideally if the parent soil is permeable enough, then water should drain away at rates >50 mm h^{-1}. The media that is used above the parent soil is also important. Poorly draining media reduce infiltration and drainage rates even in those rain gardens that have a high surface area relative to their associated drainage area. Prolonged ponding of the area may not only impact on the drainage characteristics but also erode public attitudes relating to unsightly algal growth and concerns about mosquito infestation. Often engineered media with a high sand content or gravel with only a limited proportion of smaller fine particles is advocated. Good *et al.* (2012) indicated that media composed of sand or sand–topsoil mixes demonstrated adequate hydraulic conductivity (sand: 800–805 mm h^{-1}, sand–topsoil: 290–302 mm h^{-1}). Slate gravel has been shown to have superior infiltration and conductivity rates than sand, but sometimes this can be too fast, and amendments with layers of organic matter such as green compost or pine bark achieve a good balance between conductivity and moisture retention (Riley *et al.*, 2014).

For soils that have poorer drainage capacity, then a larger rain garden may be required than in locations where water percolates through quickly. Irrespective of this, the width of rain gardens should be at least 3–5 m to provide enough vegetation to retain pollutants and help deactivate them (phytoremediation), as well as stopping water overflowing too quickly. Areas for rain gardens are dictated by the space available, but Dussaillant *et al.* (2004) suggested that the space required to maximise the re-charge of groundwater was 10–20% of the area of the contributing impervious surfaces.

Planting

The plants used in rain gardens need to be able to tolerate inundation with water but also periods where rainfall may be absent and the free-draining soil/substrate retain little moisture. The depth of the medium/storage zone affects the volume of moisture retained but seems to have little impact on maximum duration of saturated conditions in the root zone, in practice. Perhaps this is due to the fact most subsoils are permeable and free-draining. Detailed evaluations of plant species for rain gardens are still in their infancy, although riverside species that are associated with temporary river systems (i.e. those that dry out in the dry season) may be a useful habitat for environmental horticulturists to explore. Species such as *Eucalyptus camaldulensis* (river red gum) are thought to be able to tolerate both high levels of drought as well as periods of temporary inundation as they experience occasional and often drastic flooding following extreme rainfall events. Taxa recommended, e.g. within the USA, have included *Betula nigra* (river birch), *Betula* cv. Duraheat, *Magnolia virginiana*, *Magnolia* cv. Sweet Thing, *Itea virginica*, *Cornus sericea* (dogwood), *Juncus effuses* cv. Frenzy, *Panicum virgatum* cv. Shenandoah, *Helianthus angustifolius* (swamp sunflower), *Helianthus* cv. First Light and *Lobelia cardinalis* (cardinal flower); *Eupatorium purpureum* subsp. *maculatum* (Joe Pye weed), *Eutrochium purpureum* (gravel root) and *Rudbeckia fulgida* var. *sullivantii* cv. Goldsturm (black-eyed Susan) were useful for rain garden situations (Turk *et al.*, 2014). Where climates are more predictable, for example consistently wet, then specialised 'bog' plants can be exploited (Fig. 10.9).

Maintenance activities centre around removing weeds until the desirable vegetation becomes established. As rain gardens have a higher probability of being moist compared to conventional gardens, predation pressure from slugs and snails can impair the establishment of young plants and careful plant choice may be required. Annual cutting back of dead material from herbaceous plants will encourage space for new shoot growth in the following spring. Although the build-up of plant biomass over time will reduce water flow strength and rate, the development of a hydrophobic organic layer on the substrate surface may impede water infiltration. Removing organic matter on an annual basis and light cultivation of the soil surface can reduce the likelihood of this.

Pollutant control

The scale, substrate type and depth and plant composition within rain gardens influences their effectiveness as a pollution control mechanism. This is also determined by rate and volume of runoff and its sources. In a review of pollution capture,

Fig. 10.9. Where the base of the rain garden is consistently wet, then specialised species can thrive. Here *Primula bulleyana*, *P. japonica* and *P. pulverulenta* have gained a foothold.

Sharma and Malaviya (2021) showed that rain garden systems captured 52–89% of total nitrogen, 87–91% of nitrates and 60–80% of ammonium ions. Phosphate levels retained varied between 85% and 100%. Organic pollutants such as naphthalene and toluene could be trapped with as much as 90% being retained. Rain garden systems were also effective at trapping agrochemicals, with retention rates for 2,4D, 90%; dicamba, 92%; atrazine, 84–100% and glyphosate, 99%. Heavy metals were extracted from the runoff with accumulation rates in the topsoil being cadmium 0.2–2.5 mg, copper 3–290 mg, lead 1–310 mg and zinc 11–3900 mg per kg of topsoil.

Conclusions

- As demand for land increases within cities, the divisions between built and green infrastructure become blurred. Small-scale interventions such as gardens, pocket parks, green walls, roofs and SuDS systems are used to provide contact with nature and deliver some degree of ecosystem service.
- Small-scale landscape typologies vary in their degree of sophistication and reliance on technology. Those based on natural soil may be easiest to manage, whist those remote such as green walls and roofs may be more reliant on appropriate functional substrates and controlled irrigation systems.
- Such factors dictate the type of vegetation that dominates in any given system.
- Access to different typologies vary – private gardens are for the exclusive use of the residents and green roofs and walls may only be visited by a professional environmental horticulturist, while, in contrast, pocket parks or community gardens may be at the very heart of the social life of any given neighbourhood.
- There are sub-categories for each broad typology, so for example green walls divide into green façades, living walls, bio-walls, naturalised wall plant communities, retaining living walls and curtain plantings.

- Some green typologies are becoming more common due precisely to their functionality and are key nature-based solutions – for example, the embedding of rain gardens along urban roadways to help draw off surface rainwater.

References

Anon. (1989) Exposure Guidelines for Residential Indoor Air Quality. A Report of the Federal-Provincial Advisory Committee on Environmental and Occupational Health, Health Canada, Minister of Supply and Services Canada 1995.

Anon. (2014) A Guide to Green Roofs, Walls and Facades in Melbourne and Victoria, Australia. Available at: http://www.growinggreenguide.org/wp-content/uploads/2014/02/growing_green_guide_ebook_130214.pdf (accessed 12 February, 2025).

Bevilacqua, P. (2021) The effectiveness of green roofs in reducing building energy consumptions across different climates. A summary of literature results. *Renewable and Sustainable Energy Reviews* 151, 111523.

Bianchini, F. and Hewage, K. (2012) How "green" are the green roofs? Lifecycle analysis of green roof materials. *Building and Environment* 48, 57–65.

Cameron, R.W., Taylor, J.E. and Emmett, M.R. (2014) What's 'cool' in the world of green façades? How plant choice influences the cooling properties of green walls. *Building and Environment* 73, 198–207.

Cameron, R.W., Taylor, J. and Emmett, M. (2015) A *Hedera* green façade – Energy performance and saving under different maritime-temperate, winter weather conditions. *Building and Environment* 92, 111–121.

Collins, R., Schaafsma, M. and Hudson, M.D. (2017) The value of green walls to urban biodiversity. *Land Use Policy* 64, 114–123.

Dussaillant, A.R., Wu, C.H. and Potter, K.W. (2004) Richards equation model of a rain garden. *Journal of Hydrologic Engineering* 9, 219–225.

Esfahani, R.E., Paco, T.A., Martins, D. and Arsenio, P. (2022) Increasing the resistance of Mediterranean extensive green roofs by using native plants from old roofs and walls. *Ecological Engineering* 178, 106576.

Good, J.F., O'Sullivan, A.D. Wicke, D. and Cochrane T.A. (2012) Contaminant removal and hydraulic conductivity of laboratory rain garden systems for stormwater treatment. *Water Science & Technology* 65, 2154–2161.

Kunasingam, P., Clayden, A. and Cameron, R. (2024) How does plant taxonomic choice affect building wall panel cooling? *Building and Environment* 256, 111493.

Monteiro, M.V., Blanuša, T., Verhoef, A., Richardson, M., Hadley, P. and Cameron, R.W.F. (2017) Functional green roofs: Importance of plant choice in maximising summertime environmental cooling and substrate insulation potential. *Energy and Buildings* 141, 56–68.

Nagase, A. and Dunnett, N. (2010) Drought tolerance in different vegetation types for extensive green roofs: Effects of watering and diversity. *Landscape and Urban Planning* 97, 318–327.

Oberndorfer, E., Lundholm, J., Bass, B., Coffman, R.R., Doshi, H. *et al.* (2007) Green roofs as urban ecosystems: Ecological structures, functions, and services. *BioScience* 57, 823–833.

Pérez, G., Coma, J., Barreneche, C., de Gracia, A., Urrestarazu, M., Burés, S. and Cabeza, L.F. (2016) Acoustic insulation capacity of Vertical Greenery Systems for buildings. *Applied Acoustics* 110, 218–226.

Riley, E.D., Kraus, H.T. and Bilderback, T.E. (2014) Physical properties of varying rain garden filter bed substrates affect saturated hydraulic conductivity. *Proceedings of the International Plant Propagator's Society 2013* 1055, 485–489.

Seyedabadi, M.R., Eicker, U. and Karimi, S. (2021) Plant selection for green roofs and their impact on carbon sequestration and the building carbon footprint. *Environmental Challenges* 4, 100119.

Shahmohammad, M., Hosseinzadeh, M., Dvorak, B., Bordbar, F., Shahmohammadmirab, H. and Aghamohammadi, N. (2022) Sustainable green roofs: A comprehensive review of influential factors. *Environmental Science and Pollution Research* 29, 78228–78254.

Sharma, R. and Malaviya, P. (2021) Management of stormwater pollution using green infrastructure: The role of rain gardens. *Wiley Interdisciplinary Reviews: Water* 8, p. e1507.

Tomson, M., Kumar, P., Barwise, Y., Perez, P. and Forehead, H. (2021) Green infrastructure for air quality improvement in street canyons. *Environment International* 146, 106288.

Turk, R.L., Kraus, H.T., Bilderback, T.E., Hunt, W.E. and Fonteno, W.C. (2014) Rain garden filter bed substrates affect stormwater nutrient remediation. *HortScience* 49, 645–652.

Van Renterghem, T. (2018) *Green roofs for acoustic insulation and noise reduction*. In: *Nature Based Strategies for Urban and Building Sustainability*. Butterworth-Heinemann, Oxford, UK, pp. 167–179.

Vijayaraghavan, K. (2016) Green roofs: A critical review on the role of components, benefits, limitations and trends. *Renewable and Sustainable Energy Reviews* 57, 740–752.

Yeom, S., Kim, H. and Hong, T. (2021) Psychological and physiological effects of a green wall on occupants: A cross-over study in virtual reality. *Building and Environment* 204, 108134.

11 Interior Landscapes

Abstract

Plants are grown indoors as well as outdoors. Indoor (interior) plantscapes are used to enhance the aesthetics and functionality of interior spaces. The advantage of growing plants indoors is that they are not vulnerable to the 'vagaries' of the climate or prevalent weather, with plants from tropical wet and dry biomes being particularly popular in buildings within temperate and cold regions. More challenging is supplying indoor plants with appropriate light (irradiance) and matching their environmental needs to that of the other building occupants, i.e. people. Interior plants are utilised across a spectrum of situations – often being companion planting (to humans) within domestic housing, as well as being used to improve office, shop, hotel, restaurant, foyer, etc. environs. Plants are a key component of biophilic design where their use within buildings contributes to people connecting to nature and where they help promote a healthier environment.

11.1 Introduction

Interior landscapes (plantscapes) constitute the use of plants indoors or in semi-protected environments. These range from a single houseplant placed on a windowsill within a domestic dwelling to extensive iconic biomes and atria that are significant tourist venues. The latter include the Gardens by the Bay, Singapore or the Eden Project, UK. In Singapore, the protected structures are required to either keep plants cool as in the 'Cloud Forest Biome' or grow them at lower humidities than would occur naturally outdoors, i.e. in the 'Flower Dome'. At the Eden Project, almost the converse is true as the 'Tropical Biome' is designed to maintain plants within warm, high-humidity conditions (Fig. 11.1) and the 'Mediterranean Biome' where the regimes are also warm but relatively dry compared to outdoors. Such venues have been specifically designed to accommodate plant collections, but many other public plant displays are incorporated within existing buildings and structures, either exploiting light from nearby windows or being supplied with irradiance through artificial illumination.

Interior plants cover a range of contextual situations and are used in offices, schools, leisure centres, shopping malls, hotels, cafes and restaurants. Frequently, they are utilised as part of a company's marketing strategy or for reinforcing a 'brand image' or philosophy of an organisation. Plantscapes are regularly utilised to help create a unique 'sense of place' and add to the visitor experience of a given destination. Small trees and large shrubs with bold forms and strong textures are exploited to create an impression of grandeur, wealth, power or influence, for example, in the foyer of a large, commercial company or bank or within an iconic public space. Robust and strongly textured species such as *Monstera deliciosa*, *Fatsia japonica* and *Ficus elastica* promote feelings of strength, reliance and power.

In contrast, more intimate, relaxed or flowing planting styles as typified by taxa such as *Schefflera schizophylla* or *Ficus benjamina* might be used to convey a sense of tranquillity or restfulness – such a design style being appropriate for a hospital entrance or in the offices of a charitable organisation. Restaurants and hotels may deploy plants to reinforce their style of interior design, for example recreating an Edwardian ambience with palms and ferns that perhaps complements the wicker furnishings, e.g. *Howea forsteriana*. Alternatively, strongly architectural forms of plants are used to augment impressions of minimalist design and architecture, making good use of simple elegant shapes such as fastigiate *Sansevieria trifasciata* or bold 'spikes' of *Dracaena draco* and *Aechmea fasciata*. Somewhat in contrast to the designed landscapes and gardens found outdoors, greater emphasis is placed on foliage indoors to provide form and texture (Fig. 11.2), with colour being deemed less important. However, there are exceptions to every rule, and the icon

DOI: 10.1079/9781800621763.0011

Fig. 11.1. The 'Tropical Biome' at the Eden Project, Cornwall, UK. Interior plants here are being used to replicate tropical environments, with strong educational narratives around the environment, biodiversity conservation and sustainable and equitable development.

Fig. 11.2. Plant form is important in interior displays either to complement or contrast with the geometric shapes of buildings and rooms. Strong forms help to create their own sense of place within a building.

'indoor' plant – the orchid – comes in a spectacular range of colours. The orchid display at Changi Airport, Singapore is world famous.

11.2 Purpose and Function

There is evidence that plants have been grown indoors, or at least at the transition between outdoor spaces and building interiors, as earlier as 1000 BCE in China as well as perhaps in parts of the Middle East. It was the development of the modern glasshouse and improved technologies associated with protected growing of plants that largely resulted in the boom in popularity of 'houseplants' and interior plant displays during the Victorian era, 1837–1901. This is a trend that still exists today, with two distinct markets being associated with the use of plants in interior spaces. On the one hand, there is the domestic houseplant market where plants are purchased by individuals and used within the home setting (Fig. 11.3) on windowsills and within private conservatories; on the other hand, business offices, local authorities, retail arcades, leisure venues and other organisations will use large plant displays in specifically designed

Fig. 11.3. Plants are a key component of interior design in many houses. (Used with permission from R. Cameron Nicol.)

areas such as lobbies and atria to improve their interior decor. The role of interior plants in education and science should not be underestimated either. Northern Europeans, including the Victorians in the UK, in the 18th and 19th centuries went to elaborate lengths to design glasshouses within the new botanic gardens that would house and display the 'exotic' tropical and semi-tropical species being discovered in other parts of the world – a trend that has been repeated in recent years with new protected structures constructed in Wales, Cornwall and Surrey, UK for conservation and educational purposes.

Most plants in a domestic household are used in small-scale displays, with plants held within individual pots and grown on a windowsill or well-illuminated table. A number of these plant types are treated as disposable in that they are grown for a few months and then thrown away, for example, *Euphorbia pulcherrima* (poinsettia) or *Rhododendron simsii* (pot azalea). Such plants are commonly 'forced' into flowering for instant appeal and used as presents on special occasions such as Christmas, Easter or Mothering Sunday. Skilled gardeners can encourage re-flowering (or in the case of *Euphorbia*, the recolouring of the young bracts), but most people find it easier to purchase another plant the following year.

In the case of plants used in large office complexes, shopping malls, etc., the plants may not actually be owned per se but rented from an interior landscaping company. Traditionally the main driving force for interior plantscapes has been aesthetics. They complement the other criteria underpinning interior design helping to contribute to a 'unique location' or 'atmosphere' and reflecting the tastes, attitudes and ethics of the building owner or occupier. With careful design, plantscapes can influence environmental factors and improve the interior dwelling or working space, by for example, filtering out direct sunlight, reducing glare from reflective surfaces or acting as a baffle against noise sources. In offices they are designed to enhance privacy and break up large, open spaces into smaller 'rooms'. They provide focal points of interest or are carefully sited to visually block unsightly objects. They are also frequently exploited to help direct pedestrian traffic through a public space, channelling pedestrian flows through airports, railways stations, shopping malls, etc. to avoid queues and congestion. Moreover, interior plantscapes facilitate an opportunity to engage with living objects and promote a sense of nature. Architects employ them as part of their biophilic designs – an increasingly important role as the average urban dweller now spends 90% of their time indoors. This latter aspect underpins the now widely cited benefits of including plants within interiors, most notably, working environments.

Health and well-being aspects of interior plant displays

Interior plantscapes are increasingly marketed, not only on their contribution to interior design, but also as a health-promoting factor within a building. This includes their capacity to modify the aerial environment (see below) and also provide psychological and physical health benefits (see also Chapter 4). As people, on average, spend 80% of their time in interior spaces (Deng and Deng, 2018; Moya *et al.*, 2019), it is important these environments promote good health, not undermine it. Interior plants deliver a more attractive working (Smith *et al.*, 2017) or school environment (Liu *et al.*, 2022), provide visual relief to hard surfaces and geometries and are associated with providing a more relaxing ambience. They are linked with improved mood, better attention span, improved productivity at work (Aydogan and Cerone, 2021), greater creativity in the workplace and fewer days taken as sick leave from work (Larsen, *et al.*,1998; Khan *et al.*, 2005; Smith *et al.*, 2017; Chatakul and Janpathompong, 2022). Plants added to windowless office environments correlated with a 12% increase in response time for office workers working on their computers (Lohr *et al.*, 1996). Kim *et al.* (2018) showed that the presence of plants increased positive perceptions and also reduced task response times. Shibata and Suzuki (2004) noted a gender difference; when conducting a series of experiments with Japanese college students, they found that female participants performed better on a primary response task, as well as on an association task, when they were exposed to indoor plants but found no difference in task performance with male participants. It is not just positive uplift or better attention span; interior plants provide a calming environment, reduce anxiety (Han, 2009) and allow for more rapid recovery from stress (Chang and Chen, 2005). Plants have been linked to reduced health complaints in employees, e.g. 23% less symptomatic physical complaints in office employees (Fjeld *et al.*, 1998) and 25% reduction in those working in a hospital X-ray department (Fjeld, 2000).

Interior plants or having a view of exterior plants seems important for overall employee satisfaction and perceptions about their jobs (Dravigne *et al.*, 2008: de Vries *et al.*, 2023).

Context may be important in some of these studies Thatcher *et al.* (2020) found in their laboratory studies that the presence of indoor plants statistically improved performance, but the positive outcomes could not be replicated in two field studies based at telephone call centres in South Africa. When using various proxy measures for performance and well-being (namely, perceived productivity, perceived physical and psychological health, work engagement, job satisfaction and evaluations of the work environment), there was no evident advantage based on plants being in the centres. So, some office environments may be conducive to the benefits but others less so.

Physiological responses to the presence of interior plants include lower systolic blood pressure (Lohr *et al.*, 1996; Han *et al.*, 2022), lower electrodermal activity (Park *et al.*, 2004), lower levels of fatigue and fewer symptoms of physical discomfort (dry throat, coughing and dry skin) (Park and Mattson, 2008; Jiang *et al.*, 2024). Offices and other buildings with air conditioning and central heating systems create low humidities, and the addition of plants counteracts this.

In hospitals and other medical settings, the presence of plants is thought to reduce stress and anxiety levels in patients. Park and Mattson (2008, 2009) found that both flowering and foliage plants increase the pain tolerance of patients following surgery, indicating that plants could be an inexpensive and effective aid to convalescence. Similarly, Dijkstra *et al.* (2008) conducted an experiment in a hospital room and found that the indoor plants reduced stress compared to paintings of the urban environment. For patients with psychological health problems having rooms with plants present was correlated with more social interactions and higher levels of engagement with others (Talbott *et al.*, 1976).

In urban societies, there has been a recent shift in more people working from home over the last 5 years. Irrespective if the home is being used as an office, a living space or for relaxation, houseplants again improve the ambience. Houseplants in the home are linked to improved mood (Lohr and Pearson-Mims, 2000; Dzhambov *et al.*, 2021), happiness (Bermejo and Sparke, 2018), less aggression (Taghipour and Rahimi, 2023) and overall better mental health (Mousavi Samimi and Shahhosseini, 2021). Increasing the proportion of plants within an individual's viewpoint is also thought to enhance the positive effects, with greater reduction in anxiety and more positive responses reported (Han, 2009). These results allude to the fact that interior plants appear to have a significant role to play in people's homes as well as other locations such as hospitals, hospices, educational facilities and the working office.

Many interior landscapes are dominated by green foliage (rather than golden, variegated or bi-coloured plants), and this may be no surprise if the research carried out by Elsadek and Fujii (2014) holds true generically. These authors studied people's response to interior plant colour variation with *Spathiphyllum wallisii* (green), *Cordyline terminalis* (green and red) and *Aglaonema pictum* (green and white) being utilised to assess visual stimulation. The plant genotypes were selected to represent similar forms and habits but different colours. Overall, most respondents preferred green plants over green and red or green and white plants (Fig. 11.4). Using eye-tracking technology and monitoring brain stimulation, it was evident that people exhibited different responses to the plant colours. The researchers concluded that there are practical implications to this research. Entirely green plants should be used predominately where stress relief and reducing anxiety are the overriding criteria, e.g. in doctors' waiting rooms or hospitals where the plantscape can help provide a relaxing environment. In contrast, the more striking green and red plants, which stimulated the motor area in the brain that controls muscles, may be better used in office environments to improve employee productivity or in children's play areas to help stimulate creativity.

Plant taxa, form and degree of health also appears to have an effect on people's perceptions of subjective well-being. Berger *et al.* (2022) displayed images of indoor plants of different form and asked participants their preferences and how they thought the plant might affect their subjective well-being. Comparisons were of different taxa, but also different shapes within one taxon (*Ficus benjamina* cv. Danielle – weeping fig) and included a healthy and unhealthy specimen of *Dypsis lutescens* (Areca palm). Plant type affected preference (Fig. 11.5A), but most participants perceived that all plants tested, except the unhealthy *Dypsis,* would benefit their subjective well-being as the mean scores were

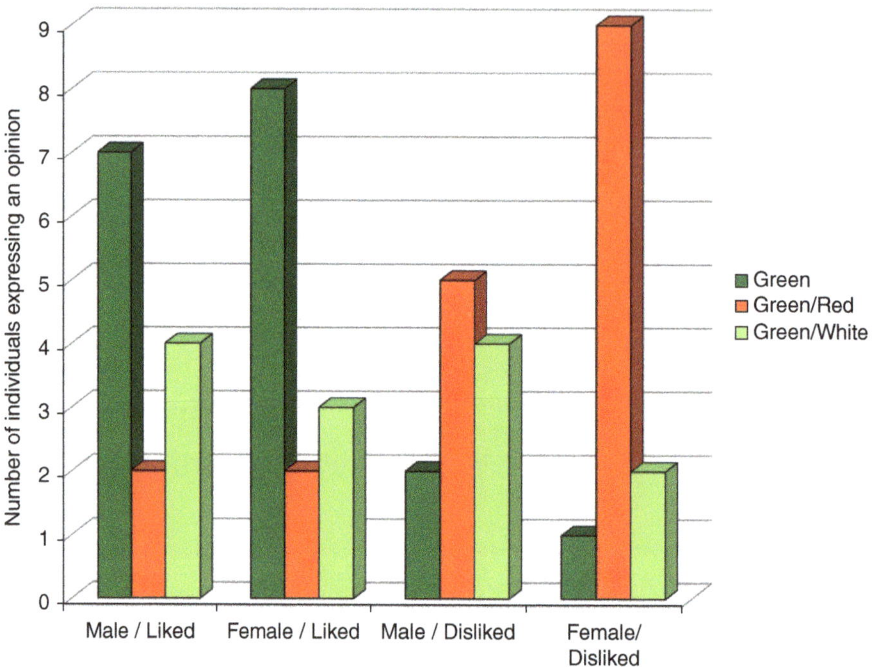

Fig. 11.4. Number of people expressing an opinion about plant colours when presented with an image of plants with either uniform green colour or variegated with red or white patterns. (Modified with permission from Elsadek and Fujii, 2014.)

all higher than the mid-point of the scale. *Ficus benjamina* (globe, columnar and pyramidal forms), *Dypsis lutescens* (healthy specimen) and *Epipremnum aureum*, however, achieved significantly higher scores than all other plants ($P < 0.05$) for perceptions about improving well-being (Fig. 11.6B). Interestingly, this was the case for *Ficus benjamina* cv. Danielle despite quite wide ranges in preference for the different shapes in this plant (Fig. 11.5A). The neglected palm scored significantly lower than all other plants, suggesting that neglected plants may have a low or negative impact on participants' perceived well-being.

What plants and how they are perceived may affect other aspects of human comfort. Plants have been used to determine if a classroom is perceived as warmer or cooler than it actually is (Hui, 2025, personal communication) (Fig. 11.6).

Sales of houseplants have been rising for the last decade or so. This includes increased popularity with young adults. In countries such as the UK, this may be due to an inability to buy a property (and garden?) due to housing shortages and the increased costs of everyday living. People renting, however, have capacity to look after a houseplant, even if they cannot buy a house of their own. There is also an increased awareness that contact with nature can be good for mental health; the nurturing requirements of plant care help in this respect. Houseplants were steadily becoming more popular before the SARS COVID-19 pandemic, but then sales jumped in 2020 (e.g. by 500% in the UK) as lockdowns were enforced in an attempt to help stop the spread of the virus.

Modifying the interior aerial environment

Implications on physical health also relate to the way that interior plantscapes influence the indoor aerial environment. Plants contribute to improving air quality within buildings through modifications of humidity and an ability to remove or detoxify certain aerial pollutants. Atmospheric gases such as carbon monoxide (CO) and carbon dioxide (CO_2), sulphur oxides (SO_x), nitrous oxides (NO_x), ammonia (NH_3) and organic volatiles can diffuse indoors from the exterior environment. These are augmented by a cocktail of other aerial pollutants generated indoors. These pollutants include particulate matter (dust) and also volatile organic compounds (VOCs), of which 300 have been identified and include

Fig. 11.5. Preference (A) and perceptions about a taxon's capacity to provide subjective well-being; (B) across a range of taxa and plant forms. Different forms of *Ficus benjamina* cv. Danielle included, i.e. columnar (Col.), globular (Glob.), pyramidal (Pyr.) or spreading (Spread) types. (Modified with permission from Berger *et al.*, 2022.)

compounds such as formaldehyde, toluene, benzene and xylene. So-called 'sick building syndrome' has been attributed to VOCs and is a phenomenon where occupants of the building suffer symptoms including headaches, nausea, dizziness, respiratory problems, dry throat and eyes, and complain of loss

Fig. 11.6. Plants alter interior environments but also people's perception of that environment: a study taking place at the University of Sheffield, UK to determine if the presence of plants in the classroom affect occupants' perceptions of thermal comfort.

of concentration. These aspects have increased as buildings have become more air-tight, with greater emphasis on reducing draughts and regulating temperature with air conditioning/central heating. VOCs are generated from office features such as furniture, printers and laminated surfaces over time or are released from products commonly used within the building such as detergents, paints, varnishes and polishes. Houses with adjoining garages may be predisposed to increased hydrocarbons, emanating from the fuel tanks of parked cars, diffusing through the connecting door to the house. In contrast to outdoor spaces, poor ventilation in buildings contributes to enhanced levels of common gases such as CO_2. Even this relatively benign gas has been associated with reduced workplace productivity (Seppänen and Fisk, 2006) and mental ability (Shaughnessy *et al.*, 2006).

Plants remove particulate matter from the air by adsorbing it onto their leaf surfaces (Lohr and Pearson-Mims, 2000). They also metabolise a range of aerial pollutant gases, either by direct action, or by hosting microbial organisms that themselves metabolise and break down the organic molecules. These may be located on the leaf surfaces (phylloplane bacteria and fungi) or in the growing media and around the roots, namely rhizosphere microorganisms. It is clear that plants absorb some of the gases and metabolism them directly – CO_2 of course being essential for photosynthesis. Tarran *et al.* (2007) found that by placing just three plants in an office environment CO_2 levels were reduced by 25% (and 10% in an air-conditioned building), and these plants were also effective at removing any CO present (86–92% reduction).

The capacity of plants alone to remove VOCs is perhaps more limited. Although pot plants can remove significant levels of VOCs and other pollutants in experiments using enclosed, sealed chambers, their ability to do the same in real buildings effectively has come under question. In their review, Cummings and Waring (2020) indicated that single plants have clean air delivery rates ($m^3\ h^{-1}$) that span orders of magnitude (with a median of $0.023\,m^3\ h^{-1}$), but in reality, somewhere in the region of 10–1000 individual plants per m^2 of floor space are required for the VOC-removing ability to achieve the equivalent of opening a window (exchanging rates ~$1\,m^3\ h^{-1}$).

The ability to deal with small organic molecules, however, appears enhanced by the presence of bacteria and fungi (Wolverton and Wolverton, 1993; Tarran *et al.*, 2007), By artificially enhancing the population of *Pseudomonas putida* bacteria on leaves of *Rhododendron indica*, De Kempeneer *et al.* (2004) showed that airborne toluene could be removed much more rapidly compared to leaves with only a background level of microorganisms. The role of the plant was essential, however, as the same effect was not apparent when the *Pseudomonas* was placed on an artificial surface. Indeed, removal rates tend to be optimised when plants are growing well, i.e. with sufficient irradiance, at appropriate temperatures and humidities and when not under water or nutrient stress.

Rhizosphere (root-associated) microorganisms are likely to be even more effective at VOC removal/degradation, justifying further work on plant-based systems that allow room air to be filtered over the root systems of plants. Thus, green wall filtration systems that pull contaminated air over an active (i.e. micro-biotically active) root system may be much more effective at cleaning air, than simple pot plants (Mata *et al.*, 2022). Several plant-based systems have emerged to support the improved indoor environment, such as the biofiltration system, where air is drawn through organic material (such as moss, soil and plants), resulting in the removal of organic gases and contaminants involving a mechanical system (Moya *et al.*, 2019). In other situations, such as 'the nature-based' air filtering system employed at a students' residence, a mini-greenhouse containing more than 30 plants was used to replace an existing window with the glasshouse connected to an air circuit to treat the indoor air (Gattringer *et al.*, 2021).

Plants working with their allied microbiota have been shown to aid the removal of SO_2 (Lee and Sim, 1999), NO_2 (Coward *et al.*, 1996) and mercury vapour (Bastos *et al.*, 2004) in addition to CO_2 and VOCs from interior environments (Wolverton and Wolverton 1993, Dingle *et al.*, 2000; Orwell *et al.*, 2004; Tarran *et al.*, 2007). Various combinations of indoor environments (ventilation, low light) and plant types (C3 and Crassulacean acid metabolism [CAM] plants) acting as botanical biofilters in tandem with conventional ventilation systems reduced levels of CO_2 by 76%, total VOCs by 87%, formaldehyde by 75%, particulate matter $PM_{2.5}$ by 52% and PM_{10} by 51% (Sharma *et al.*, 2022).

Which plants do this best is open to some scrutiny. Most studies have been with potted plants (in experimental chambers) rather than with biofiltration living walls (in real buildings). Liu *et al.* (2007) reported that for VOC removal *Hemigraphis alternata, Tradescantia pallida, Hedera helix, Asparagus densiflorous, Hoya camosa,* and *Crassula portulacea* were best, whilst Aydogan and Montoya (2011) recorded a fourfold difference between the most (*Chrysanthemum morifolium*) and least (*Hedera helix*) efficient species that these authors tested for formaldehyde removal. When comparing relative removal rates by different plant taxa, Wolverton and Wolverton (1993) ranked the order for formaldehyde ($\mu g\ h^{-1}$) as *Nephrolepis exaltata* cv. Bostoniensis, *Chrysanthemum morifolium, Dracaena deremensis cv.* Janet Craigs, *Hedera helix*, *Chamaedorea elegans* and *Cyclamen persicum* with *Sansevieria trifasciata* least effective. For xylene, however, the order changed to *Chamaedorea elegans*, *Nephrolepis exaltata* cv. Bostoniensis, *Chrysanthemum morifolium*, *Cyclamen persicum*, *Sansevieria trifasciata*, *Dracaena deremensis* cv. Janet Craigs and finally *Hedera helix*. In similar approaches, Sriprapat *et al.* (2014) evaluated 12 common houseplants and demonstrated that *Sansevieria trifasciata, Kalanchoe blossfeldiana* and *Dracaena deremensis* were most effective at removing toluene, whereas *Chlorophytum comosum*, *Sansevieria ehrenbergii*, *Aglaonema commutatum* and *Sansevieria hyacinthoides* would be the taxa of choice for ethylbenzene removal. Some of the reactions in degrading the VOCs are concentration dependent, however, with metabolism decreasing when background levels fall, thus plants will not necessarily eliminate the VOCs completely from a room. Removal rates are also under the influence of other factors including the size/volume of plant material present, air movement, irradiance, temperature, moisture status of the growing media and whether the plants or pots themselves release their own VOCs. Indeed, when Yang *et al.* (2009) assessed uptake rates and took account of different leaf areas, certain trends contradicted previous studies, although this approach was useful in identifying 'universally' effective taxa, e.g. *Hemigraphis* and *Hedera* (Fig. 11. 7).

As with exterior plants, there are counterclaims that certain interior plants may be detrimental from a human health perspective. As well as some species being toxic – *Dieffenbachia amoena* (dumbcane) *Solanum pseudocapsicum* (winter cherry) *Nerium oleander* (oleander) (Der Marderosian *et al.*, 1976) – others are linked to contributing to respiratory disorders in humans (e.g. asthma, rhinitis and laryngitis). This includes complaints associated with common species such *as Ficus benjamina* (Axelsson *et al.*, 1985) and *Spathiphyllum wallisii* (Kanerva *et al.*, 1995), yet exposure levels may need to be high before symptoms become apparent. Indeed, most cases of illness or discomfort are associated with plant maintenance workers, rather than office employees, suggesting people need to have regular direct contact with the plants or be working in close proximity to them for prolonged periods.

The capacity to sequester carbon through interior planting has been raised, although the benefits may have more to do with reducing CO_2 levels from offices and other rooms, rather than contributing significantly to global atmospheric CO_2 targets. Controlled environment studies suggesting that larger woody plant species or individual specimens were more effective at absorbing carbon than either smaller species/specimens or herbaceous subjects.

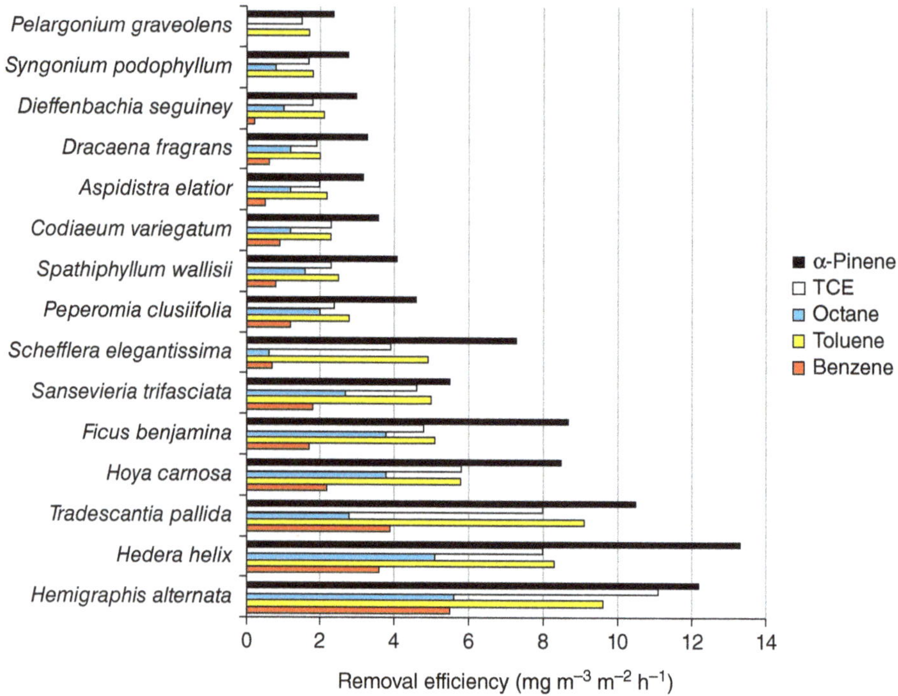

Fig. 11.7. Removal efficiency (mg m^{-3} m^{-2} h^{-1}) of five representative volatile organic compounds (VOCs) (benzene, toluene, octane and trichloroethylene [TCE]) when assessed on leaf area per species. (Modified with permission from Yang *et al.*, 2009 and only representing a sample of species tested.)

11.3 Interior Plant Requirements

The success of growing plants within buildings and other enclosed areas depends on a matrix of factors including the levels of irradiance, temperature regimes, air quality and movement and humidity. In addition, root systems need to be provided with appropriate levels of moisture, nutrients and oxygen, as well as a medium that provides sufficient anchorage. Tropical/semi-tropical plants housed in buildings located in temperate or cool climates will require heating; conversely, other plants may require air conditioning or misting systems to keep them cool, this being a common problem for botanical glasshouses and atria in more tropical regions, due to the high amounts of solar energy these capture and retain.

Temperature

The choice of interior plant species will be determined by the predominate temperature in the building. For most buildings this will be set for human thermal comfort, e.g. in temperate regions, this is considered to be 21°C during the day and 18°C at night. In locations specifically designed to cater for the plant collections, however, the temperature can be dictated by plant requirements, not necessarily that of the humans. Different plant collections will have different temperature requirements. It is not always safe to assume either that plants require temperatures comparable to or higher than the human thermal comfort zone; some may prefer slightly cooler temperatures.

Interior plants can be broadly divided into four groups based on their natural distribution and biome:

- Cool temperate: Day temperatures that vary with season from approximately 12–18°C with night temperatures of approximately 4–6°C. This includes some of the flowering house/interior plants such as *Rhododendron simsii* (pot azalea) and *Cyclamen persicum*.
- Warm temperate/cooler tropical (e.g. tropical montane regions): Day temperatures that vary with season from approximately 21–24°C, with night temperatures of approximately 8–12°C.

This includes taxa such as *Solanum capsicastrum* (winter cherry) and some of the ferns (who also like high humidity), e.g. *Adiantum raddianum* (maidenhair fern) and *Nephrolepis exaltata* cv. Bostoniensis (Boston fern).

- Moist tropical and sub-tropical: Relatively stable temperatures across the season with day values 28–32°C and night 14–16°C. This includes the broadleaved 'jungle plants', e.g. *Aglaonema commutatum* (Chinese evergreen), *Aechmea fasciata* (bromeliad) and *Caladium bicolor.*
- Arid tropical (desert/semi-arid regions): High day temperatures up to 45°C but tolerating 2–5°C at night. This includes the desert cacti (e.g. *Kroenleinia grusonii* [barrel cactus] and *Lobivia aurea* [golden lily cactus] and succulents (e.g. *Aloe variegata* [partridge-breasted aloe] and *Echeveria glauca* [blue echeveria]).

Diurnal variations are feasible in most office blocks and similarly used buildings, with heating systems being turned down at night. Most plants appreciate a cooler night temperature than day temperature. Recreating season fluxes in temperature though may be harder to achieve, especially for cooler-adapted species, at least without inducing the wrath of the human workforce or occupants. Not all rooms in a building, however, are managed in the same way and entrance lobbies, hallways and some corridors may experience greater influence from exterior temperatures, hence making them more suitable for species that require some cool-season effect. Knowing both the thermal and irradiance properties of different parts of a building are useful in siting plants in the most appropriate locations.

Temperature and irradiance interact, and excessively high temperatures are more detrimental when irradiance is suboptimal, as this induces relatively high rates of respiration but not photosynthesis, stressing the plant through a negative carbon balance. In such situations supplementary light will be required. Plants require high irradiance and consequently are placed beside windows; however, this is where central heating radiators are also frequently placed. Exposure to direct sources of heat such as this can scorch leaves, as well as cause localised drying of the air. Room temperature will also influence overall humidity levels and higher temperatures are likely to correspond to greater transpiration and the need for more frequent irrigation. Although most of the focus is on air temperature, root temperatures should not be forgotten. Planters that are located against exterior walls or sunk into the floor of the building below ground level may experience lower temperatures than the rest of the building, and plants may run the risk of root chilling injury.

Irradiance

Irradiance (light) is needed by interior plants for photosynthesis, photomorphogenesis (regulating plant development), synthesising chlorophyll and other pigments and regulating stomatal behaviour, as well as irradiance influencing factors such as leaf temperature, rate of transpiration and mineral uptake. In practical terms, irradiance is measured as the human eye perceives it, i.e. as lux (lumens per square metre), but plants have a preference for certain specific wavelengths of irradiance (for example, photosynthesis is highly dependent on red and blue spectral wavelengths). Irradiance as required by plants is therefore defined by the photosynthetically active region (PAR, i.e. 400–700 nm) and this is measured as μmol photons m^{-2} s^{-1}. Lux can be converted to μmol m^{-2} s^{-1}, but the conversion factors can vary depending on the light source, i.e. it will be different from daylight (multiply lux by 0.018) compared to say 'cool white' fluorescent-tube lighting (multiply lux by 0.013).

It is thought that most plants will not survive below 400 lux, with shade-tolerant species coping with 400–750 lux (e.g. *Philodendron scandens*), medium-light species requiring 750–1500 lux (e.g. *Dracaena marginata*) and high-light species needing 1500–2500 lux (e.g. *Hibiscus rosa-sinensis*). To put these values into context, on a clear, sunny day outdoors irradiance can exceed 50,000 lux, whereas a typical office environment may only be 500 lux. It is often considered that house plants placed more than 1.5–2m from a north-facing window or perhaps 3 m from a southern aspect one will begin to experience suboptimal irradiance. In atria, protected courtyards and conservatories in the domestic home environment, energy for plant development and display is exploited through natural light. Protection from cold, wind and even excessively low humidity is provided by glass roofs/walls (or by other translucent materials) whilst still allowing the relevant irradiance wavelengths to pass through and be utilised by plants for photosynthesis and photomorphogenesis. In the design of such structures, however, care is required to ensure that features such as positional aspect, degree of shading (e.g. from adjacent walls or buildings) and depth and width of glass (transmission) area are fully taken into consideration. Too much solar irradiance may lead to excessive temperature rise, and too much shade may lead to suboptimal irradiance for some species. A deep, narrow atrium, for example, with a small ground area but surrounded by many storeys of built structure, may still not provide enough light to facilitate plant requirements. The planting design may itself also influence irradiance dynamics, with tall plants with wide canopies adding to shade problems for the lower strata of vegetation. Species of plants included in the design for their flower displays or fruit (e.g. *Citrus*) may particularly require higher irradiance levels. Similarly, greater irradiance intensity may be a prerequisite to ensure coloured or variegated foliage plants retain their characteristic hues.

Although emphasis in plantscape design is often placed on maximising irradiance (especially to ensure adequate irradiance levels in winter), in some regions of the world excessive irradiance (and heat) can be equally problematic. For tropical 'forest floor' species adapted to mid- or lower irradiances, direct summer sunlight and high temperatures (e.g. when combined with low humidity) is damaging, and provision is required for venting of the glasshouse structures or providing additional shading. Even when these tropical woodland or understorey plants are grown in temperate regions (where irradiance intensities are generally less), care is required to not exceed irradiance thresholds. For example, some orchid species/hybrids grow better when not exposed to direct strong sunlight during the middle of the day so are best placed on a west-, north- or east-facing window of a house or office (in the northern hemisphere – Fig. 11.8). (NB not all orchids have the same requirements. Indeed, growers categorise them based on their irradiance requirements; *Phalaenopsis* spp. [moth orchids] and *Paphiopedilum* spp. [slipper orchids] are considered low-light requiring. *Cattleya* spp. [corsage orchid] needs medium light, whereas *Vanda* and *Brassavola* groups require high levels of light.) Elsewhere, such as with production glasshouses in the tropics, shading can be accommodated by whitewashing glass, automated blinds or the use of electrically controlled glass ('electrochromatic glass' or 'smart-glass') that changes its colour or transparency.

The Gardens by the Bay project in Singapore (a tropical, humid climate) wished to create publicly accessible conservatories that were cooler than ambient conditions, with both cool-dry and cool-moist biomes being created. This involved innovative technologies that included utilising glass which allowed high irradiance penetration but reduced the normal heat load associated with solar radiation (Er *et al.*, 2008). The roofs were also fitted with sensor-operated retractable sails that provide shade to the plants when it gets too hot. In tandem, cool-water pipes were laid at the base of the biomes and excess heat vented from the top of the glass structures, thus helping to keep the main body of air below outdoor temperatures. One of the biomes – the Mediterranean zone – additionally needs to be drier than the outdoor conditions in Singapore. Here, the air is dehumidified by a liquid desiccant (drying agent) before cooling – both to make the cooling process more efficient and to reduce relative humidity to what the Mediterranean flora require. The desiccant used is re-dried using waste heat from an on-site electricity-generating station – itself powered by

Fig. 11.8. Despite being tropical plants, some orchids (top: *Phalaenopsis* hybrid; bottom: *Miltonia* Goodale Moir hybrid) do not tolerate direct sunlight and are best positioned in the shade of other plants or on a west-, east- or even north-facing window (in the northern hemisphere).

the burning of city biomass. The on-site-generated electricity powers the refrigerant plant that keeps the circulation water cool. In essence, the project keeps two glasshouses permanently cool, through combined green and sustainable technologies 'feeding off' each other.

In many conventional buildings though, natural solar irradiance is absent or suboptimal. In these situations, artificial sources of light are required to provide either the plants' entire needs or to supplement the natural irradiance. Selecting the proper lighting system for indoor plant growth is a demanding process that requires an accurate prior study. It should ensure certain characteristics in terms of intensity (the amount of light received by the vegetation) and quality (the spectral composition of the light source). Incandescent, fluorescent, halogen and metal halide (high-intensity discharge) lamps have all been traditionally used to illuminate indoor spaces and provide plants with photosynthetic energy. Low-energy light-emitting diodes (LEDs), however, have tended to dominate the sector in recent years. These can be orientated to provide the appropriate spectra and quantity of irradiance (Olle and Viršile, 2013). LEDs show several advantages over other lamp types such as a much longer lifespan and producing a high luminous flux with a low radiant heat output. This makes them more competitive in energy efficiency and economic terms (Singh *et al.*, 2015).

The fact that LEDs emit radiation in a controlled spectral composition is an advantage when growing plants. Given that LEDs emit in over a very narrow window (20–40 nm) of wavelength, this point can be exploited to match the wavelengths emitted to those that are preferentially absorbed by chlorophyll (absorbance by chlorophyll peaking with red and blue wavelengths). This improves the use of energy as most emitted light can be aligned to maximise photosynthesis. Nevertheless, such narrow-wavelength LEDs give interior plants an unnatural appearance due to the dominance of red and/or blue light (and most people feel rather unnerved sitting in a red light!), so they tend not to be used to promote the aesthetic qualities of the plant. In addition, better plant growth may be achieved when using a broader spectrum with additional wavelengths (Kim *et al.*, 2005), which help photomorphogenic responses (controlling stem orientation, flower initiation, etc.). This makes white-light LEDs more appropriate. In order to obtain white LEDs, blue LEDs are usually coated with phosphor, which makes them more inefficient lights but improves the visuals of the plant considerably (Massa *et al.*, 2008).

Illumination from the lamps themselves may contribute to the interior ambience, but it should be noted that the spectrum of irradiance that is required for successful plant development may be somewhat different to that perceived by the human eye and appreciated by the human brain! Conversely, there will be many instances where humans will also be experiencing (requiring) the irradiance produced and this should be taken into account when specifying the spectrum of light used in the interior space. Providing irradiance intensity is optimal, illumination durations in line with office working hours are usually sufficient to provide daily requirements of PAR (8–14 h being ideal).

Air quality

The presence of plants may improve interior air quality, but contaminants within buildings can also be phytotoxic. Aerial pollutants and gaseous fumes can result in chlorosis and necrosis of foliage, epinastic growth, early flower bud abscission and even complete defoliation in severe cases (Ingels, 2009). Phytotoxic chemicals commonly used in offices include cleaning products, paints, varnishes and preservatives which may release ammonia, chlorine and volatile hydrocarbons. Chlorine used to disinfect bathing water can be a problem too in swimming pools and leisure centres, although this is offset by ensuing the plants are sufficiently far away from the pool to avoid direct splashing, and that there is forced air movement/ventilation to help reduce the background levels of chlorine in the air. In other circumstances, ethylene may induce epinasty and chlorosis in interior plants, due to concentrations building up in confined spaces with limited air circulation or replacement. Ethylene is synthesised from a variety of organic sources (including decaying plant material) but is also a by-product of hydrocarbon combustion. Even common household dust can be problematic – covering leaves, disrupting irradiance capture and blocking stomatal pores, thereby interfering with photosynthesis and transpiration. Dust is removed from the foliage by washing down the leaves periodically, or where feasible, cleaning them with a fine cloth.

Growing media

In contrast to most exterior plants in the wider landscape, many of which ultimately will grow in soil, interior plants tend to be reliant for their entire lifespan on artificial, constructed substrates held within containers, planters and planting beds. To avoid excessive weight on building floors, growing media need to be lightweight whilst retaining the capacity to retain moisture. Coir or similarly lightweight, organic materials are often used, but so too are inert materials, such as clay leca. Leca is an expanded clay pellet with a honeycomb microstructure that contributes to its light weight but also enables it to hold sufficient volumes of air and water. Leca is also used in hydroculture; this involves the leca being placed in a planter which is then part-filled with water or a weak, nutrient solution. Other inert material can be used as alternatives (e.g. perlite, rockwool or glass beads) and plants access the water through capillary action associated with these media. Although the media used in plantscapes is often lightweight, a sufficient bulk density (e.g. 0.15–0.75 g cm^{-3}) is still required to anchor the plants and avoid instability problems as they grow. Unlike natural soils, the substrates in interior plantscapes do not experience natural flushing through of nutrients with rainfall, and nutrient levels in the rhizosphere can build up over time. As such, care is required not to over-fertilise plants and avoid root injury through excessive accumulation of ions. Liquid fertilisers are frequently favoured over 'base dressings' of N, P_2O_5 and K_2O to help control nutrient application, with opportunities to flush through the media with excess water, or change old media for new, being sought. An additional requirement of the growing media is that it needs to be aesthetically pleasing, contributing to the overall design package. As such, ornamental mulches of cocoa shells, gravel or glass beads are utilised to cover the substrate surface.

Irrigation

The old adage that most houseplants are killed by overwatering or underwatering is probably not far from the truth. Interior landscapes experience fierce heat and drying if located in too close a proximity to a radiator or adjacent to a south-facing (in the northern hemisphere) window (or north aspect one in the southern hemisphere). Often, as they are grown in pots or containers, there are limited reserves of water – meaning that irrigation needs to be frequent. Conversely, plants located some distance from the windows experience little air movement and have limited transpiration resulting in the growing media staying damp and overly wet. So, applying water based on a plant's needs is doubly important in interior landscapes. Irrigation systems vary depending on the medium used and the size of the planting area. Large plantscapes may possess their own automated irrigation system with drip lines being strategically placed to supply water to individual specimens. In large displays, these are commonly linked to sophisticated control systems, automatically irrigating after set thresholds for light integrals, evapotranspiration demand or media moisture availability. Such systems still require occasional checking to ensure roots, detritus or precipitating calcium salts do not accumulate and block drainage areas or the application drippers. One disadvantage with discrete dripper-type systems is that a moist 'cone' within the substrate develops, with limited lateral movement of moisture – this tending to encourage root growth around this moist zone, without encouraging more extensive longitudinal root development. In contrast, hydroculture systems provide a more uniform supply of water and employ a sub-surface reservoir, which irrigates the medium via capillary wicks or other mechanisms. The reservoir may have gauges to allow notification of when it requires refilling manually or may be self-regulating with a feeder water pipe controlled by a ballcock to maintain water levels at a set height. With such hydroculture systems, a certain depth of container is required to ensure the capillary action of water rise is effective. For smaller-scale plant displays and individual pot plants, then hand-watering by watering can or hose may be employed, but this is labour-intensive if the plantscape area is large, or many offices need to be attended to. Such irrigation approaches also run the risk of water being spilled on inappropriate places including employees' computers!

11.4 Acclimatisation to Interior Environments

Plants used in interior landscapes are conventionally cultivated in intensive production systems, with the usual stipulation to maximise crop growth and quality, before being sold on and moved indoors. As many of the plant species used are derived from tropical/sub-tropical biomes, they tend to either be produced in outdoor nurseries in such climates or within heated, well-lit glasshouse production systems in more temperate regions. Either way, plants regularly experience higher irradiance and humidity levels during production than they will subsequently experience in their final planting location. With the advent of large corporate plantscapes in the 1970s, it became evident that plants moved rapidly from nurseries with high irradiance to interior locations with lower irradiances often experienced leaf chlorosis and abscission. This was due to a lack of acclimatisation (acclimation) to the lower light irradiance within individual leaves of the plant. Changes in relatively humidity too can exacerbate the stress on leaves. During the nursery stage if plants are grown in glasshouses, they may experience relative humidity (RH) of 85–90%, whereas office environments are closer to 45–50% RH, with some domestic homes as low as <25% RH. Today, more care is taken in adapting plants from nursery conditions to those of their ultimate destination. During the last 6–8 weeks of nursery production, plants are now exposed to 'step-down' irradiance levels and raised ventilation to reduce relative humidity. Irradiance intensity is reduced by as much as 50% on two or three separate occasions over this period. The plant adapts physiologically to these changes; for example at the cellular level, the chlorophyll is rearranged within the chloroplast granum to capture more of the limited (irradiance) and nitrogen metabolism alters to deal with less photosynthates being produced; photosynthesis itself is downregulated but is more efficient in capturing carbon. At the whole plant level, leaves tend to be thinner and the plant develops smaller and more widely spaced stomata on the underside of the leaves. Root-to-shoot ratios also frequently alter (Rodríguez-Calcerrada *et al.*, 2008). Acclimation can be aided by less frequent irrigation and lower nutrient supply, in addition to reducing irradiance intensity and background humidity (Chen *et al.*, 2005).

Once in the interior environment, inappropriate humidity or rapid, constant air movement can still be problematic to the foliage. Draughts and direct placement close to open windows, heaters or air-conditioning units should be avoided. Not only will these aspects affect temperature, but they may decrease humidity or cause excessive movement of the foliage, resulting in necrotic lesions on leaves. Some species are more susceptible to low RH and air movement than others (Fig 11.9). Whereas, most broadleaved species tolerate the humidity typical of offices, i.e. 45–50%, bamboos, ferns and other fine-leaved plants may require 70% RH once *in situ*. Species from arid climates on the other hand, e.g. cacti, require lower values than those from moist tropics. In large or high-quality plantscapes, mist or fog nozzles can be specified in the design and automated to counteract low humidities. Systems tend to rely on mains water and compressed air is used to create the mist or fog. Spray nozzles should be orientated in the direction of the plant canopies to maximise the humidity effect but also avoid accidental wetting of seating areas, paths or electrical equipment. As the system works to create a vapour of moisture (which is easily inhaled by humans), prolonged storage of water in tanks needs be avoided to minimise risks due to *Legionella pneumophilia* (legionnaires' disease).

11.5 Pests and Pathogens

Interior plantscapes have led the way in the use of biocontrol agents to help control pests and pathogens. This has largely been out of necessity as the use of chemical pesticides in close proximity to humans (especially the general public, who may be unaware of the chemical application) and within enclosed spaces has required that alternative solutions should be sought. Where chemicals are used, they tend to be relatively benign products such as detergents or spot treatment with alcohols or oils.

Minimising stress through careful regulation of irradiance, humidity and irrigation is key to reducing the chances of infection and pathogen spread (Lockwood, 2000). For example, root pathogens such as *Pythium, Phythophthora* and *Rhizoctonia* can be more prevalent after plants have been overwatered or roots have been damaged by an excessive accumulation of fertiliser salts. Common plant pests are now regularly controlled by biocontrol means, for example, *Tetranychus urticae* (glasshouse red spider mite) by *Phytoseiulus persimilis*, *Trialeurodes vaporariorum* (glasshouse whitefly) by *Encarsia formosa*, *Planococcus citri* (citrus mealybug) by the *Leptomastix dactylopii* wasp and *Coccus hesperidium* (soft brown scale insect) by *Metaphycus alberti* wasps (Table 11.1).

Where plants are heavily infested with pests or badly damaged by pathogens, they can be dug up and removed from the interior landscape and placed back in a glasshouse or quarantine area to deal with the problem. In such locations the use of more virulent pesticides may be appropriate. Plants can then be restored to full health before being placed back in the plantscape (assuming

Fig. 11.9. Bathrooms can have higher humidities than other parts of the domestic house, and these are often conducive to the growth of tropical understorey plants. In both these cases, the owners' preoccupation with the plants seems to have undermined the functionality of the bath. (Author's own, left. Used with permission from R. Cameron-Nicol, right).

the original cause of the disease/pathogen in the interior landscape has been remedied in the meantime).

11.6 Managing the Interior Landscape

Most of the larger plantscapes are designed, installed and maintained by specialised landscape design or horticultural companies. These may provide a 'complete service', employing a wide range of personnel to cover the design and maintenance aspects. Some may have their own production nursery, whereas others buy in plant specimens from specialist nurseries across the globe.

In terms of planting and maintenance, a range of service levels are offered. Plants may simply be sold to the client. Alternatively, the client may rent them from the interior landscape company, thus helping the client replace plants relatively easily if quality begins to deteriorate. Some companies offer design solutions too, including developing a plantscape that fits with the existing or proposed interior, taking account of such infrastructure factors as site access, weight loading, colour schemes, design styles and suitability of locations for different plants – irradiance and photoperiods, draughts, seating arrangements, etc. Once a plan is agreed the company will install and maintain the landscape, usually offering a service whereby landscape technicians will visit the planting on a regular basis to water, clean, prune, fertilise and inspect plants for pathogens, disease or stress effects. During such arrangements the plantscape companies may wish to insure themselves against any damage or other liabilities potentially caused by its personnel when working within a client's building. The personnel need training, not only in the horticultural skills required, but also in ensuring 'high customer service' and being able to respond in a rapid and professional manner to the client's needs.

11.7 Environmental Sustainability

Interior plants are seen as a component of biophilic design. Biophilically designed buildings are designed

Table 11.1. Pests commonly found in interior landscapes and their biocontrol agents

Pest	Symptoms	Biocontrol agents
Glasshouse whitefly (*Trialeurodes vaporariorum*)	Sap-sucking white 'ghost-like' fly. Reduces plant vigour and multiplies rapidly. It transmits plant viruses and produces honeydew with corresponding sooty moulds.	Parasitic wasps *Encarsia formosa* – one of the pioneer biocontrol agents – known to be able to parasitize 16 different species of whitefly around the globe. Each adult accounts for about 100 whitefly nymphs in its lifetime.
Mealybug (*Planococcus* and *Pseudococcus* spp.)	Mealybugs cause leaf chlorosis through sap-sucking activity, but plants are also damaged by honeydew excretions and the sooty moulds that form on them.	Ladybird beetles *Cryptolaemus montrouzieri* – Australian orange and black ladybird. Parasitic wasps *Leptomastix dactylopii* – yellow parasitoid from Brazil. *Leptomastidea abnormis* – from southern Europe; copes with lower temperatures than other species. *Anagyrus pseudococci* – Middle East origin.
Peach-potato aphid (*Myzus persicae*)		Ladybird beetles. *Adalia bipunctata* – two-spotted ladybird, usually red with two black spots, although it can vary in pattern and colouring Parasitic wasps *Aphidius colemani* – will also control *Aphis gossypii*, the melon or cotton Aphid. *Aphidius ervi* – this parasite is known to track aphid colonies by the volatile chemicals released from plants when under attack from aphids. At shorter distances the aroma of the honeydew allows the wasp to 'home' in on its host. *Aphelinus abdominalis* – a species that often walks rapidly, rather than flies, to its host. Once the female finds an aphid, she touches it with her antennae, turns around, raises her wing tips and injects the ovipositor in the aphid. Parasitoid gall midges *Aphidoletes aphidimyza* – tend to be active at night. Eggs are deposited in the aphid colonies, and once hatched, the larvae seek out an aphid, paralyse it and suck out the contents. Lacewings *Chrysoperla carnea* – larvae prey on a wide range of species and are a good general pest control agent known to attack several species of aphids, red spider mites, long-tailed mealybugs, thrips, whiteflies, leafminers, small caterpillars and beetle larvae, as well as the eggs of leafhoppers and moths.
Red or two-spotted spider mite (*Tetranychus urticae*)	Yellow spots followed by faint white webbing across the leaf	Predatory mites *Phytoseiulus persimilis* – fast-moving voracious predatory mite; confusingly, a more obvious red-orange colour than *Tetranychus.* *Neoseiulus californicus* – works better than *P. persimilis* at low humidity.

Continued

Table 11.1. Continued.

Pest	Symptoms	Biocontrol agents
Soft scale insects Brown soft scale (*Coccus hesperidium*) Hemispherical scale (*Saissetia coffeae*) Tessellated scale (*Eucalymnatus tessellatus*)	Brown soft scale is probably the most frequently encountered scale on plants indoors. Heavy infestations can encrust the stems and petioles of their host plant. They also settle on leaves, usually along midribs. Honeydew serves as a medium for the growth of sooty molds which in turn inhibit photosynthesis and make plants unsightly.	Parasitic wasps *Metaphycus helvolus* – small black and yellow wasp; attracted to light sources, but will go back to plants at night to feed. *Metaphycus stanleyi* – for tessellated scale as well as some control of brown soft and hemispherical when it can access the scales before hardening. *Encyrtus infelix* – small, black wasp with characteristic large head. Ladybird beetles *Harmonia axyridis* – harlequin ladybird that has become a concern in the natural environment of the UK due to its competitive nature against native ladybird spp. *Chilocorus nigritus* – a small, round, entirely black ladybird spp. from India. Good for controlled armoured scale insects as well as soft scale. *Lindorus lophanthae* – lustrous black ladybird, with grey 'alligator shaped' larvae. Will tackle mealybugs as well as scales.
Thrips Western flower thrip (*Frankliniella occidentalis*) Onion or tobacco thrip (*Thrips tabaci*)	White or transparent mottling on leaves, leaf abscission	Predatory mites *Amblyseius cucumeris* – need a temperature of 15°C plus. They hibernate in winter unless artificial irradiance is supplied. True bugs *Orius insidiosus* – minute pirate bug, within the order Hemiptera (the true bugs): generalist predators consuming mites, aphids and small caterpillars in addition to thrips.
Vine weevil (*Otiorhynchus sulcatus*)	Characteristic semi-circular cut sections from leaf edges. Larvae damage roots of young plants	Nematodes *Heterorhabditis megidis* – requires temperatures above 12°C, so generally good for interior landscapes

to break down the barriers between human living and the experience of nature. It follows the philosophy that incorporating natural elements into built environments can improve health, happiness and overall quality of life. The main goal is to create spaces that foster positive interactions between people and nature. Plants are key to this but so are physical natural objects – wood, stone and water, as well as human exposure to natural light. Colours within the buildings may be the more natural hues found in nature and the shapes of objects drawn from nature (fractals). Effective biophilic design should also take account of where the energy is being drawn to heat, cool and ventilate the building. Plant support systems are themselves dependent on energy. In many areas, heat energy is already supplied by the requirements imposed on satisfying human thermal comfort (i.e. the central heating of occupied buildings), but ideally this should be from sustainable 'green' sources such as heat pumps, photovoltaic solar panels, wind turbines or sustainable biofuels. Low-power LED lamps help reduce energy use and passive cooling can be used to cool building interiors and ventilate the living/working spaces, but energy will still be required to some extent.

Consideration also needs to be given to energy use in the production of house/interior plants. Often this is in tropical environments, and the 'footprint' of production may relate more to the transport of bulky, heavy plant material across the globe; plants tend to be flown too to their markets as not much would be left alive after a 2-month voyage in the hold of a bulk container ship! Where production systems themselves require energy then alternative sources of energy to carbon-based fuels are appropriate for glasshouse complexes. Even if these do not always completely replace conventional forms of energy, they can at least augment them. Heat pump systems, for example, can reduce reliance on oil-based fuels by 30–40% in glasshouse systems used for commercial crop production. Using solar energy within the complex can also be utilised to heat solid interior objects such as walls and paths, or indeed water baths and pipes during daylight hours. These can be exploited as heat stores, allowing energy to be re-radiated out again and hence maintain high air temperatures during the night, without the requirement for additional heating. Finding alternative sources of energy is particularly important in some atria, such as the Eden Project in Cornwall, UK, where the practice needs to align with the philosophy being promoted through their educational activities, namely around developing sustainable approaches to modern living and food production.

The concept of growing plants indoors has in itself helped develop technologies that have wider application (LED lights being an example). Growing plants effectively within protected structures is a key component of space travel and exploration for example. Indeed, some of the advances in building/plantscape technology in recent decades have evolved out of research conducted by the National Aeronautics and Space Administration (NASA) as well as other space-related organisations and allied technological industries. This has included the development of interior bio-wall filtration systems to improve air quality and raise humidity within buildings. The use of plantscapes in atria, biomes and conservatories therefore allow us to better understand plant requirements when grown under highly artificial conditions (and in turn, this contributes to our knowledge base on aspects such as high-tech crop growing and future plant conservation strategies). Allied to this, plantscapes have been very much at the forefront in the development of genuinely effective biocontrol strategies and enabling complex plant communities to be sustained without reliance on pesticides. Equally importantly, such environments provide ready access to green space and help inform us about our relationship and reliance on nature. Most interior landscapes have a raison d'être based on aesthetics, but their value to society is much deeper than this.

Conclusions

- Interior plants are used in a wide variety of settings, including offices, domestic properties, leisure centres, shopping malls, hotels, cafes and restaurants. They are part of the interior decor and are utilised not only to improve the aesthetics of a room but also to reinforce an image or style. Their popularity as houseplants has grown in recent years.
- Plants in interior landscapes provide relaxing surroundings, with evidence of their ability to reduce stress, and provide a purposeful connection to nature. They modify the humidity within a room making it more amenable to occupants. They also help improve air quality directly and also through their capacity to support rhizosphere microorganisms, with ventilated green wall systems being used in some buildings to actively remove volatile organic compounds and other pollutants.
- Growing plants indoors can be challenging and careful thought is required as to how appropriate irradiance, temperature and humidity are going to be provided. Similarly, infrastructure within the interior environment means that irrigation, nutrition and pest/pathogen control need be astutely managed so as to avoid problems such as water leaks or the application of toxic chemicals.
- Biocontrol of pest and pathogens is commonplace in interior landscapes, as the use of high-toxicity pesticides has been largely restricted due to these locations being public space. As such, this sector of horticulture has been at the forefront of implementing integrated pest management and pioneering the development and use of biocontrol agents.
- Most corporate companies are not experts in plant husbandry, and office plantscapes tend to be managed by specialist horticultural and landscape businesses. As such, the plants may be either rented or purchased outright. Often in the latter case, a maintenance contract is put in place to ensure the plants are maintained to a professional standard.

References

Axelsson, G., Skedinger, M. and Zetterströ, O. (1985) Allergy to weeping fig -A new occupational disease. *Allergy* 40, 461–464.

Aydogan, A. and Cerone, R. (2021) Review of the effects of plants on indoor environments. *Indoor and Built Environment* 30, 442–460.

Aydogan, A and Montoya, L.D. (2011) Formaldehyde removal by common indoor plant species and various growing media. *Atmospheric Environment* 45, 2675–2682.

Bastos, W.R., de Freitas Fonseca, M., Pinto, F.N., de Freitas Rebelo, M., dos Santos, S.S., da Silveira, E.G. and Pfeiffer, W.C. (2004) Mercury persistence in indoor environments in the Amazon Region, Brazil. *Environmental Research* 96, 235–238.

Berger, J., Essah, E., Blanusa, T. and Beaman, C.P. (2022) The appearance of indoor plants and their effect on people's perceptions of indoor air quality and subjective well-being. *Building and Environment* 219, 109151.

Bermejo, G. and Sparke, K. (2018) Happier in a home with plants? A study on the relationship between human wellbeing and ornamental plants in private households. In: XXX International Horticultural Congress IHC2018: XIX Symposium on Horticultural Economics and Management, VII Symposium on 1258, 47–54.

Chang, C.-Y. and Chen, P.-K. (2005) Human responses to window views and indoor plants in the workplace. *HortScience* 40, 1354–1359.

Chatakul, P. and Janpathompong, S. (2022) Interior plants: Trends, species, and their benefits. *Building and Environment* 222, 109325.

Chen J., McConnell D.B., Henny R.J. and Norman, D.J. (2005) The foliage plant industry. In: Janick, J. (ed.) *Horticultural Reviews*, John Wiley and Sons, Oxford, p. 31.

Coward, M., Ross, D., Coward, S., Cayless, S. and Raw, G. (1997) Pilot study to assess the impact of green plants on NO_2 levels in homes. Healthy Buildings IAQ '97: Global issues and regional solutions pp 129–134, Fachinformationsdienst BAUdigital, Darmstadt, Germany.

Cummings, B.E. and Waring, M.S. (2020) Potted plants do not improve indoor air quality: A review and analysis of reported VOC removal efficiencies. *Journal of Exposure Science and Environmental Epidemiology* 30, 253–261.

De Kempeneer, L., Sercu, B., Vanbrabant, W., van Langenhove, H. and Verstraete, W. (2004) Bioaugmentation of the phyllosphere for the removal of toluene from indoor air. *Applied Microbiology and Biotechnology* 64, 284–288.

de Vries, S., Hermans, T. and Langers, F. (2023) Effects of indoor plants on office workers: A field study in multiple Dutch organizations. *Frontiers in Psychology* 14, 1196106.

Deng, L. and Deng, Q. (2018) The basic roles of indoor plants in human health and comfort. *Environmental Science and Pollution Research* 25, 36087–36101.

Der Marderosian, A.H., Giller, F.B. and Roia, F.C. (1976) Phytochemical and toxicological screening of household ornamental plants potentially toxic to humans I. *Journal of Toxicology and Environmental Health* 1, 939–953.

Dijkstra, K., Pieterse, M.E. and Pruyn, A. (2008) Stress-reducing effects of indoor plants in the built healthcare environment: The mediating role of perceived attractiveness. *Preventive Medicine* 47, 279–283.

Dingle, P., Tapsell, P. and Hu, S. (2000) Reducing formaldehyde exposure in office environments using plants. *Bulletin of Environmental Contamination and Toxicology* 64, 302–308.

Dravigne, A., Waliczek, T.M., Lineberger, R.D. and Zajicek, J.M. (2008) The effect of live plants and window views of green spaces on employee perceptions of job satisfaction. *HortScience* 43,183–187.

Dzhambov, A.M., Lercher, P., Browning, M.H., Stoyanov, D., Petrova, N., Novakov, S. and Dimitrova, D.D. (2021) Does greenery experienced indoors and outdoors provide an escape and support mental health during the COVID-19 quarantine? *Environmental Research* 196, 110420.

Elsadek, M. and Fujii, E. (2014) People's psycho-physiological responses to plantscape colors stimuli: A pilot study. *International Journal of Psychology and Behavioral Sciences* 4, 70–78.

Er, K.B.H., Kishnani, N., Kessling, W. and Koo V.Y.B. (2008) An empirical approach to the delineation of growing conditions within a cooled conservatory in Singapore. *Acta Horticulturae* 801, 156.

Fjeld, T. (2000) The effect of interior planting on health and discomfort among workers and school children. *HortTechnology* 10, 46–52.

Fjeld, T., Veiersted, B., Sandvik, L., Riise, G. and Levy, F. (1998) The effect of indoor foliage plants on health and discomfort symptoms among office workers. *Indoor and Built Environment* 7, 204–209.

Gattringer, H., Efthymiou-Charalampopoulou, N., Lines, E. and Kolokotroni, M. (2021) Nature based solution for indoor air quality treatment. *Journal of Physics, Conference Series* 2042, 012133.

Han, K.T. (2009) Influence of limitedly visible leafy indoor plants on the psychology, behaviour, and health of students at a junior high school in Taiwan. *Environmental Behaviour* 41, 658–692.

Han, K.T., Ruan, L.W. and Liao, L.S. (2022) Effects of indoor plants on human functions: A systematic review with meta-analyses. *International Journal of Environmental Research and Public Health* 19, 7454.

Ingels, J.E. (2009) *Landscaping Principles and Practices*, 7th edn, Delmar, New York.

Jiang, J., Irga, P., Coe, R. and Gibbons, P. (2024) Effects of indoor plants on CO_2 concentration, indoor air temperature and relative humidity in office buildings. *Plos One* 19, e.0305956.

Kanerva, L., Mäkinen-Kiljunen, S., Kiistala, R. and Granlund, H. (1995) Occupational allergy caused by spathe flower (*Spathiphyllum wallisii*). *Allergy* 50,174–178.

Khan, A.R., Younis, A., Riaz, A. and Abbas, M.M. (2005) Effect of interior plantscaping on indoor academic environment. *Journal of Agricultural Research* 43, 235–242.

Kim, H.H., Wheeler, R.M., Sager, J.C., Gains, G.D. and Naikane, J.H. (2005) Evaluation of lettuce growth using supplemental green light with red and blue light-emitting diodes in a controlled environment-a review of research at Kennedy Space Center. *Acta Horticulturae* 711, 111–120.

Kim, J., Cha, S.H., Koo, C. and Tang, S.K. (2018) The effects of indoor plants and artificial windows in an underground environment. *Building and Environment* 138, 53–62.

Larsen, L., Adams, J., Deal, B., Kweon, B.-S. and Tyler, E. (1998) Plants in the workplace: The effects of plant density on productivity, attitudes, and perceptions. *Environmental Behaviour* 30, 261–281.

Lee, J.-H. and Sim, W.-K. (1999) Biological absorption of SO_2 by Korean native indoor species. In: Burchett, M.D. (ed.) *Towards a New Millennium in People-Plant Relationships*. Contributions from International People-Plant Symposium, Sydney, pp. 101–108.

Liu Y.-J., Mu Y.-J., Zhu Y.-G., Ding H. and Crystal Arens, N. (2007) Which ornamental plant species effectively remove benzene from indoor air? *Atmospheric Environment* 41, 650–654.

Liu, F., Yan, L., Meng, X. and Zhang, C. (2022) A review on indoor green plants employed to improve indoor environment. *Journal of Building Engineering* 53, 104542.

Lockwood, S.L. (2000) *Interior Planting: A Guide to Plantscapes in Work and Leisure Spaces*. Gower Publishing Ltd., Aldershot, UK.

Lohr, V.I. and Pearson-Mims, C.H. (2000) Physical discomfort may be reduced in the presence of interior plants. *HortTechnology* 10, 53–58.

Lohr, V.I., Pearson-Mims, C.H. and Goodwin, G.K. (1996) Interior plants may improve worker productivity and reduce stress in a windowless environment. *Journal of Environmental Horticulture* 14, 97–100.

Massa, G.D., Kim, H.H., Wheeler, R.M. and Mitchell, C.A. (2008) Plant productivity in response to LED lighting. *HortScience* 43, 1951–1956.

Mata, T.M., Martins, A.A., Calheiros, C.S., Villanueva, F., Alonso-Cuevilla, N.P., Gabriel, M.F. and Silva, G.V. (2022) Indoor air quality: A review of cleaning technologies. *Environments* 9,118.

Mousavi Samimi, P. and Shahhosseini, H. (2021) Evaluation of resident's indoor green space preferences in residential complexes based on plants' characteristics. *Indoor and Built Environment* 30, 859–868.

Moya, T.A., van den Dobbelsteen, A., Ottele, M. and Bluyssen, P.M. (2019) A review of green systems within the indoor environment. *Indoor and Built Environment* 28, 298–309.

Olle, M. and Viršile, A. (2013) The effects of light-emitting diode lighting on greenhouse plant growth and quality. *Agricultural and Food Science* 22, 223–234.

Orwell, R.L., Wood, R.L., Tarran, J., Torpy, F. and Burchett, M.D. (2004) Removal of benzene by the indoor plant / substrate microcosm and implications for air quality. *Water, Air, Soil Pollution* 157, 193–207.

Park, S.-H. and Mattson, R.H. (2008) Effects of flowering and foliage plants in hospital rooms on patients recovering from abdominal surgery. *HortTechnology* 18, 563–568.

Park, S.-H. and Mattson, R.H. (2009) Therapeutic influences of plants in hospital rooms on surgical recovery. *HortScience* 44, 1–4.

Park, S.-H., Mattson, R.H. and Kim, E. (2004) Pain tolerance effects of ornamental plants in a simulated hospital patient room. *Acta Horticulturae* 639, 241–247.

Rodríguez-Calcerrada, J., Reich, P.B., Rosenqvist, E., Pardos, J.A., Cano, F.J. and Aranda, I. (2008) Leaf physiological versus morphological acclimation to high-light exposure at different stages of foliar development in oak. *Tree Physiology* 28, 761–771.

Seppänen, O.A. and Fisk, W. (2006) Some quantitative relations between indoor environmental quality and work performance or health. *Heat, Ventilation, Air-Conditioning and Refrigeration Research* 12, 957–973.

Sharma, S., Bakht, A., Jahanzaib, M., Lee, H. and Park, D. (2022) Evaluation of the effectiveness of common indoor plants in improving the indoor air quality of studio apartments. *Atmosphere* 13, 1863.

Shaughnessy, R.J., Haverinen-Shaughnessy, U., Nevalainen, A. and Moschandreas D. (2006) A preliminary study on the association between ventilation rates in classrooms and student performance. *Indoor Air* 16, 465–468.

Shibata, S. and Suzuki, N. (2004) Effects of an indoor plant on creative task performance and mood. *Scandinavian Journal of Psychology* 45, 373–381.

Singh, D., Basu, C., Meinhardt-Wollweber, M. and Roth, B. (2015) LEDs for energy efficient greenhouse lighting. *Renewable and Sustainable Energy Reviews* 49, 139–147.

Smith, A.J., Fsadni, A. and Holt, G. (2017) Indoor living plants' effects on an office environment. *Facilities* 35, 525–542.

Sriprapat, W., Suksabye, P., Areephak, S., Klantup, P., Waraha, A., Sawattan, A. and Thiravetyan, P. (2014) Uptake of toluene and ethylbenzene by plants: Removal of volatile indoor air contaminants. *Ecotoxicology and Environmental Safety* 102, 147–151.

Taghipour, M. and Rahimi, F. (2023) The role of apartment live plants to control anger and aggression. *Asian Journal of Multidisciplinary Research and Review* 4, 61–75.

Talbott, J.A., Stern, D., Ross, J. and Gillen, C. (1976) Flowering plants as a therapeutic/environmental agent in a psychiatric hospital. *HortScience* 11, 365–366.

Tarran, J., Torpy, F. and Burchett, M. (2007) Use of living pot-plants to cleanse indoor air – Research review. In: *Proceedings of 6th International Conference on Indoor Air Quality, Ventilation and Energy Conservation – Sustainable Built Environment*, Sendai, Japan, 3, 249–256.

Thatcher, A., Adamson, K., Bloch, L. and Kalantzis, A. (2020) Do indoor plants improve performance and well-being in offices? Divergent results from laboratory and field studies. *Journal of Environmental Psychology* 71, 101487.

Wolverton, B.C. and Wolverton, J.D. (1993) Plants and soil microorganisms: Removal of formaldehyde, xylene, and ammonia from the indoor environment. *Journal of the Mississippi Academy of Science* 38, 11–15.

Yang, D.S., Pennisi, S.V., Son, K.C. and Kays, S.J. (2009) Screening indoor plants for volatile organic pollutant removal efficiency. *HortScience* 44, 1377–1381.

Index

Note: Page numbers in **bold** type refer to **figures**, page numbers in *italic* type refer to *tables*.

www.ingramcontent.com/pod-product-compliance
Lightning Source LLC
LaVergne TN
LVHW060628110826
845147LV00014B/873

9781800621749